MORPHOLOGICAL ORGANIZATION IN EPITAXIAL GROWTH AND REMOVAL

SERIES ON DIRECTIONS IN CONDENSED MATTER PHYSICS

Published

Vol. 1: Directions in Condensed Matter Physics — Memorial Volume in Honor of Shang-keng Ma
ed. G. Grinstein and G. Mazenko

Vol. 2: Ionic Solids at High Temperatures
ed. A. Stoneham

Vol. 3: Directions in Chaos (Vol. 1)
ed. Hao Bailin

Vol. 4: Directions in Chaos (Vol. 2)
ed. Hao Bailin

Vol. 5: Defect Processes Induced by Electronic Excitation in Insulators
ed. N. Itoh

Vol. 6: Spin Glasses and Biology
ed. D. Stein

Vol. 7: Interaction of Electromagnetic Field with Condensed Matter
ed. N. N. Bogolubov, Jr., A. S. Shumovsky and V. I. Yukalov

Vol. 8: Scattering and Localization of Classical Waves in Random Media
ed. Ping Sheng

Vol. 9: Geometry in Condensed Matter Physics
ed. J. F. Sadoc

Vol. 11: Quasicrystals — The State of the Art
ed. P. Steinhardt and D. DiVincenzo

Vol. 12: Spin Glasses and Random Fields
ed. A. P. Young

Series on Directions in Condensed Matter Physics – Vol. 14

MORPHOLOGICAL ORGANIZATION IN EPITAXIAL GROWTH AND REMOVAL

Editors

Zhenyu Zhang
Oak Ridge National Laboratory

Max G Lagally
University of Wisconsin-Madison

World Scientific
Singapore • New Jersey • London • Hong Kong

Published by

World Scientific Publishing Co. Pte. Ltd.

P O Box 128, Farrer Road, Singapore 912805

USA office: Suite 1B, 1060 Main Street, River Edge, NJ 07661

UK office: 57 Shelton Street, Covent Garden, London WC2H 9HE

British Library Cataloguing-in-Publication Data
A catalogue record for this book is available from the British Library.

MORPHOLOGICAL ORGANIZATION IN EPITAXIAL GROWTH AND REMOVAL

ISBN 981-02-3471-6

This book is printed on acid-free paper.

Printed in Singapore by Uto-Print

PREFACE

Crystal structure is the foundation of all other properties of materials in the condensed state. Before this century, Nature was the primary grower of crystalline materials. As we have learned more about the processes involved in crystal growth, we have been able to devise ways to grow ever better materials, characterize their structure, and relate structure to chemical, mechanical, electronic, optical, electrical, and magnetic properties. But only in the last 15 years have we begun to develop a true understanding of crystal growth at the atomistic level. This development has come about through a confluence of superb techniques for observing growth at the level of individual atoms and of a technological driver, the increased need for thin films with precisely known structure and morphology.

Developments at an early stage (up to 1989) of this booming field were summarized in a widely circulated volume, the product of a NATO Advanced Study Institute. No summary exists, however, of the much higher level of understanding we have achieved since then. This volume intends to fill this gap. It provides a critical assessment of the current status of and likely future directions in thin-film growth. The volume emphasizes the important conceptual developments made in the 1990s, focusing on descriptions of the atomic-scale mechanisms that control the dynamics and thermodynamics of the morphological evolution of the growth front during growth of thin films in diverse systems of fundamental and technological significance.

The articles are arranged in five major categories. Section I provides the theoretical basis, covering approaches at a variety of length scales. Sections II and III focus on experimental aspects of semiconductor-on-semiconductor and metal-on-metal growth, respectively. Section IV combines these materials in considering metal-on-semiconductor growth. Finally, Section V considers the reverse of growth, the controlled removal of atoms from the surface.

Of the many scholars whose achievements have aided in the development of this field and who make possible a volume such as this, one name stands out far above the rest, that of Professor George Comsa. Professor Comsa has been a singular source of physical insights and outstanding experimentation, a person dedicated to good science for sciences sake, and, for those who have had the opportunity, an absolute delight to work with. George has recently retired from active research. We dedicate this volume to him.

The publication of a book such as this depends on the efforts of many people. We thank the authors for their timely contributions and for their patience with the editors and the editorial process. We are obligated to Dr. Christian Teichert, who obtained the image used as the cover photo. We thank Ms. Lakshmi Narayan, the in-house Editor for this volume at World Scientific, for her expert and enthusiastic help; Ms. Sharron King for technical help; and Ms. Betty Waddell for secretarial assistance. We also wish to express our deepest appreciation to our family members for permitting us to spend the extra hours working on this project. Finally, we would like to thank the U.S. Department of Energy, the U.S. National Science Foundation, the U.S. Office of Naval Research, and the U.S. Air Force Office of Scientific Research for support of our editorial activities.

Zhenyu Zhang
Oak Ridge National Laboratory

Max G. Lagally
University of Wisconsin–Madison

CONTENTS

THEORETICAL BASIS

Study of Strain and Temperature Dependence of Metal Epitaxy

C. Ratsch, P. Ruggerone, and M. Scheffler
Fritz-Haber-Institut der Max-Planck-Gesellschaft, Faradayweg 4-6, D-14195 Berlin-Dahlem, Germany

1 Introduction

Metallic films are important in catalysis, magneto-optic storage media, and interconnects in microelectronics, and it is crucial to predict and control their morphologies. The evolution of a growing crystal is determined by the behavior of each individual atom, but technologically relevant structures have to be described on a time scale of the order of (at least) tenths of a second and on a length scale of nanometers. An adequate theory of growth should describe the atomistic level on very short time scales (femtoseconds), the formation of small islands (microseconds), as well as the evolution of mesoscopic and macroscopic structures (tenths of seconds). Different theoretical approaches to the description of growth phenomena address different time and length scales, and an overview is given in Table 1. These different techniques should be regarded as complementary to each other rather than as alternatives, and we will show in this article how some of them may successfully be combined.

The development of efficient algorithms combined with the availability of cheaper and faster computers has turned density functional theory (DFT) into a reliable and feasible tool to study the microscopic aspects of growth phenomena (and many other complex processes in materials science, condensed matter physics, and chemistry). These calculations provide the so-called potential energy surface (PES, also often called total energy or Born-Oppenheimer surface) that is the potential energy experienced by a particle during chemical reactions and diffusion. The basic concepts behind DFT are sketched in Section 2. A single microscopic process on such a PES can be characterized by either doing a molecular dynamics (MD) simulation or by using transition state theory. Both approaches typically will give the same result. Problems may occur when the energy barrier is comparable to the thermal energy and correlated effects cannot be neglected. An overview over the different processes and how they can be described within transition state theory is given in Section 3.

Technologically relevant applications of growth often involve disparate materials with different lattice constants. However, it is not understood so far how strain in a system influences the most fundamental process that occurs during epitaxial growth, the diffusion of a single adatom on a substrate. It is the aim of this article to discuss the current understanding of this aspect. In Section 4

Table 1: The time and length scales handled by different theoretical approaches to study growth.

	Type of information	Time scale	Length scale
Density functional theory calculations	Microscopic	-	$\leq 10^3$ atoms
Ab initio molecular dynamics	Microscopic	$t \leq 10$ ps	$\leq 10^2$ atoms
Semi-empirical molecular dynamics	Microscopic	$t \leq 1$ ns	$\leq 10^3$ atoms
Kinetic Monte Carlo	Microscopic to Mesoscopic	1 ps $\leq t \leq$ 1 hour	$\leq 1\ \mu$m
Rate equations	Averaged	0.1 s $\leq t \leq \infty$	All
Continuum equations	Macroscopic	1 s $\leq$ t $\leq \infty$	≥ 10 nm

some recent results for the strain dependence of surface diffusion that is determined by an energy barrier and a prefactor will be presented. While the DFT results were obtained for only a few metallic systems, we assert that the physical trends discussed are rather general.

A DFT total energy calculation by its nature is primarily a *static* calculation. An accurate way to describe the *dynamical* evolution of a growing crystal is given by MD simulations. It is not possible to carry out such a simulation on mesoscopic and/or macroscopic time and length scales. Ab initio MD runs usually can cover at most times of picoseconds, whereas semi-empirical MD runs may extend up to some nanoseconds (at the cost of the accuracy). This is because an MD time step cannot be longer than the inverse of a typical phonon frequency. For example, a MD run simulates the entire sequence of unsuccessful attempts occurring between two successful diffusion events that may be separated by interval of the order of nanoseconds. Thus, MD can model only very few events, and a proper statistics cannot easily be obtained for growth processes. Moreover, since growth patterns usually develop on a time scale of seconds, the inadequacy of MD is evident. Finally, the growth structures involve large numbers of particles ($\sim 10^2$ to 10^4), far beyond the reach of MD (simulations with $\sim 10^2$ atoms are hardly feasible, and only for very short times). Because an adequate treatment of the statistical nature of the problem is typically not achieved, we believe that the value of MD is often overstated, and the method of choice for studying the spatial and temporal development of growth is kinetic Monte Carlo (KMC).

The growing crystal defines a lattice system onto which the non-crystalline growth units (atoms or molecules) are deposited. In the case of growth from solution or from vapor one may assume with sufficient generality that the growth units impinge on the surface (or, more exactly, on the solid-vapor) interface at random with an average frequency. The same random character belongs to the diffusion of growth units on the surface. It is therefore natural to describe deposition, desorption, and diffusion mechanisms as stochastic processes, and this matches very well with the stochastic nature of the Monte Carlo method. A brief introduction to this method and its combination with microscopic parameters obtained from *ab initio* parameters is given in Section 5. It is shown that realistic *ab initio* kinetic Monte Carlo simulations are able to predict an evolving mesoscopic structure on the basis of microscopic details.

In order to compare quantitatively results from a Monte Carlo simulations with experimentally measured quantities, such as for example the density of islands on the surface, one needs to take a statistical average over this quantity. In a KMC simulation this is obtained by doing a simulation on a sufficiently large lattice. An alternative and more elegant approach is the construction of mean field equations that by definition only contain averaged quantities. An example are rate equations discussed briefly in Section 2. Numerical solution of these equations can give the density of islands, but without microscopic details. This ansatz currently is not sufficiently advanced to make reliable predictions for experimentally relevant quantities and needs to be developed further. As the final goal a proper mean field approach should yield the same results as the statistical average over the results of a more detailed microscopic approach. The inclusion of more details (that in principle is possible) unfortunately makes a mean field or continuum approach less tractable, so that new techniques have to be developed.

2 Basic Concepts of Density Functional Theory

The total energy of an N-electron, poly-atomic system is given by the expectation value of the many-particle Hamiltonian using the many-body wave-function of the electronic ground state. For a solid or a surface the calculation of such an expectation value is impossible within the framework of a wave-function approach. However, Hohenberg and Kohn [1] have shown that the ground-state total energy can also be obtained without explicit knowledge of the many-electron wave-function by minimizing an energy functional $E[n]$. This is the essence of density-functional theory (DFT), which is primarily (though in principle not exclusively) a theory of the electronic ground state, couched in terms of the electron density $n(\mathbf{r})$ instead of the many-electron wave-function $\Psi(\{\mathbf{r}_i\})$.

The important theorem of Hohenberg and Kohn [1] (see also Levy [2]) states: The specification of a ground state density $n(\mathbf{r})$ determines the corresponding external potential $v^{\rm ext}(\mathbf{r})$ *uniquely* (to within an additive constant),

$$n(\mathbf{r}) \rightarrow v^{\rm ext}(\mathbf{r}) \quad . \tag{1}$$

The external potential $v^{\rm ext}(\mathbf{r})$ is typically (and definitely for our purpose here) the Coulomb potential due to the nuclei, assumed at fixed positions as they are much heavier than the electrons (Born-Oppenheimer approximation). While the other direction $[v^{\rm ext}(\mathbf{r}) \rightarrow n(\mathbf{r})]$ is well known to exist (because $v^{\rm ext}(\mathbf{r})$ determines the many-particle Hamiltonian H) Eq. (1) is less obvious. The theorem enables to transform the variational problem of the many-particle Schrödinger equation into a variational problem of an energy functional:

$$E_0 \leq \langle\Psi|H|\Psi\rangle = E_v[\Psi[n]] = E_v[n] \quad , \tag{2}$$

with E_0 the energy of the ground state, and $E_v[n] = \int d\mathbf{r}\, v^{\rm ext}(\mathbf{r}) n(\mathbf{r}) + G[n]$. The variable is $n(\mathbf{r})$ (the electron density of any N-electron system), and $v^{\rm ext}(\mathbf{r})$ is kept fixed. $G[n]$ is a *universal* functional independent of the system, i.e., independent of $v^{\rm ext}(\mathbf{r})$. The main advantage of this approach is that $n(\mathbf{r})$ only depends on three variables, while $\Psi(\{\mathbf{r}_i\})$ depends on all the coordinates of the N electrons [3]. Thus, it is plausible that the variational problem of $E_v[n]$ is easier to solve than that of $\langle\Psi|H|\Psi\rangle$, yet the result for the ground-state energy and the ground state electron density will be the same. The total energy is [4]

$$E^{\rm tot}(\{\mathbf{R}_J\}) = E_0(\{\mathbf{R}_J\}) + \frac{1}{2} \sum_{J,J',J\neq J'} \frac{Z_J Z_{J'}}{|\mathbf{R}_J - \mathbf{R}_{J'}|} \quad , \tag{3}$$

where $\{\mathbf{R}_J\}$ includes all atoms, and Z_J is the nuclear charge.

An important problem remains, namely that the exact form of the functional $G[n]$ is unknown. Earlier work (in particular the Thomas-Fermi approach) had shown that the treatment of the kinetic energy $\langle\Psi| -\frac{1}{2}\nabla^2|\Psi\rangle$ is of particular importance and Kohn and Sham [5] therefore wrote the energy functional in the form

$$E_v[n] = T_s[n] + \int d\mathbf{r}\, v^{\rm ext}(\mathbf{r}) n(\mathbf{r}) + \frac{1}{2}\int d\mathbf{r}\, v^{\rm H}(\mathbf{r}) n(\mathbf{r}) + E^{\rm xc}[n] \quad , \tag{4}$$

where $T_s[n]$ is the functional of the kinetic energy of a system of non-interacting electrons with density $n(\mathbf{r})$, and $v^{\rm H}(\mathbf{r}) = \int d\mathbf{r}' \frac{n(\mathbf{r}')}{|\mathbf{r}-\mathbf{r}'|}$ is the Hartree potential that describes the electrostatic interaction between electrons. $E^{\rm xc}[n]$, the so-called exchange-correlation functional, accounts for the Pauli principle, dynamical correlations due to the Coulomb repulsion, and the correction of the self-interaction included for convenience in the Hartree term. With Eq. (4) the problem of the unknown functional $G[n]$ is mapped onto one that involves $T_s[n]$ and $E^{\rm xc}[n]$. Although the functional $T_s[n]$ is not known explicitly in a mathematically closed form, it can be evaluated exactly by using the following "detour" proposed by Kohn and Sham. The variational principle applied to Eq. (4) leads to

$$\frac{\delta E_v[n]}{\delta n(\mathbf{r})} = \frac{\delta T_s[n]}{\delta n(\mathbf{r})} + v^{\rm eff}(\mathbf{r}) = \mu \quad , \tag{5}$$

where μ is the Lagrange multiplier associated with the requirement of a constant particle number and thus equals the electron chemical potential. The effective potential is defined as

$$v^{\rm eff}(\mathbf{r}) = v^{\rm ext}(\mathbf{r}) + v^{\rm H}(\mathbf{r}) + v^{\rm xc}(\mathbf{r}) \quad , \tag{6}$$

with $v^{\rm xc}(\mathbf{r}) = \delta E^{\rm xc}[n]/\delta n(\mathbf{r})$, and $n(\mathbf{r})$ is a ground-state density of any non-interacting electron system, i.e.,

$$n(\mathbf{r}) = 2\sum_{i=1} f_i|\phi_i(\mathbf{r})|^2 \quad , \tag{7}$$

where we introduced the occupation numbers f_i. The factor 2 accounts for the spin degeneracy. Since $T_s[n]$ is the kinetic energy functional of non-interacting electrons, Eq. (5) [together with Eq. (7)] is solved by

$$\left[-\frac{1}{2}\nabla^2 + v^{\rm eff}(\mathbf{r})\right]\phi_i(\mathbf{r}) = \epsilon_i\phi_i(\mathbf{r}) \quad . \tag{8}$$

These are the Kohn-Sham equations, that are to be solved self-consistently together with Eqs. (6) and (7). In principle, this gives the exact ground-state electron density and total energy of a system of interacting electrons. However, the functional $E^{\rm xc}[n]$ is still unknown. Some general properties of this functional and values for some special cases are known. Detailed and very accurate understanding exists for systems of constant electron density. These results for $\epsilon^{\rm xc} := \epsilon^{\rm xc}_{\rm LDA}(n)$ are then used in the functional

$$E^{\rm xc}_{\rm LDA}[n] = \int d\mathbf{r}\, n(\mathbf{r})\epsilon^{\rm xc}_{\rm LDA}(n(\mathbf{r})) \quad , \tag{9}$$

which is the local-density approximation (LDA) [5]. Thus, in the LDA the many-body effects are included such that for a homogeneous electron gas the treatment is exact and for an inhomogeneous system exchange and correlation are treated by assuming that the system can be composed from many small systems with a locally constant density.

The LDA can be improved by including the dependence on the density gradient which leads to the generalized gradient approximation (GGA). Several different GGA's were proposed in the literature [6,7,8,9,10] and have been used successfully for DFT calculations for atoms, molecules, bulk solids, liquids, and surfaces (an overview can be found in Refs. [11,12]), but also limitations have been pointed out [13,14]. It is by now clear that the lattice constants calculated with a GGA are typically larger than those obtained with the LDA, with the experimental values usually being in between. Binding energies (or cohesive energies) of molecules and solids are clearly improved by the GGA as well as energy barriers of chemical reactions. (see Ref. [15] and references therein). A general rule of thumb for the use of LDA or GGA is still missing and tests should be carried out for each specific problems. The total energies are changed when going from the LDA to the GGA, thus for surface diffusion the changes in energy barriers, i.e., in total energy *differences*, may be pronounced as well although the rank of the processes is not altered (see Refs. [16,17]). The GGA data are usually more reliable, but still careful calculations with various forms of GGA must be performed.

The general method described above can be transformed into an operative tool following diverse schemes that essentially differ in the basis employed to expand the electronic wave function. A widely used basis set (also chosen for the calculations presented below) consists of plane waves (see Refs. [18,19]) but other choices are possible [20,21]. A deeper discussion of the implementations of the DFT plane-wave basis set method for the study of growth phenomena can be found in Ref. [22].

3 Atomistic Processes

The different atomistic processes encountered by adatoms that are deposited onto a surface are illustrated in Fig. 1. After deposition (*a*) atoms can diffuse across the surface (*b*) and will eventually meet another adatom to form a nucleus (*c*) or get captured by an already existing island or a step edge (*d*). Once an adatom has been captured by an island, it may either break away from the island (*reversible aggregation*) (*e*) or remain bonded to the island (*irreversible aggregation*). An atom that is bonded to an island may diffuse along its edge (*f*) until it finds a favorable site. As long as the coverage of adsorbed material is low (say $\Theta \leq 10$ %), deposition on top of islands is insignificant and nucleation of islands on top of existing islands practically does not occur. However, if the step down motion (*g*) is hindered by an additional energy barrier, nucleation of island on top of islands (*h*) becomes likely at some critical coverage. In this article we will mainly focus on the migration

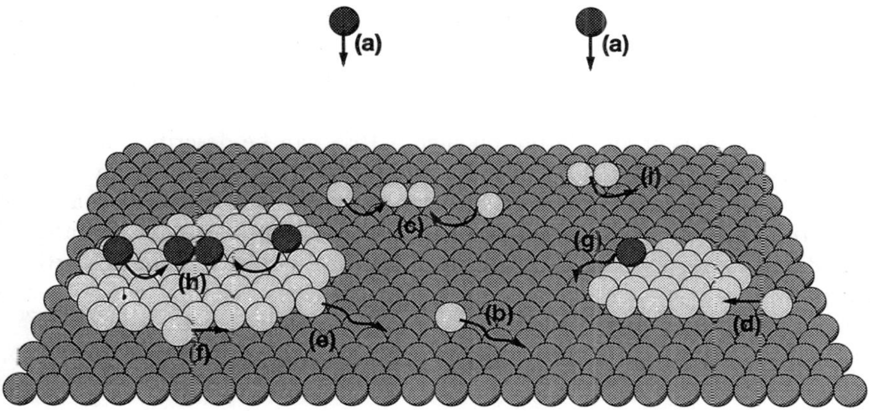

Figure 1: The different atomistic processes for adatoms on a surface. See the text for a discussion.

of adatoms on the flat terrace or along step edges and will not discuss effects that are due to mass transport between different layers. Moreover, we will say nothing about the possibility of long jumps even though it most likely plays a role for certain systems as it has been observed with the field ion microscope for Pd on W (211) by Senft and Ehrlich [23], and very recently in a scanning tunneling microscopy study for Pt on Pt (011) by Linderoth *et al.* [24].

3.1 Transition State Theory

To understand the diffusion of an adatom on a surface we need to calculate its potential-energy surface (PES):

$$E^{\mathrm{PES}}(X_{\mathrm{ad}}, Y_{\mathrm{ad}}) = \min_{Z_{\mathrm{ad}}, \{\mathbf{R}_I\}} E^{\mathrm{tot}}(X_{\mathrm{ad}}, Y_{\mathrm{ad}}, Z_{\mathrm{ad}}, \{\mathbf{R}_I\}) \quad , \tag{10}$$

where $E^{\mathrm{tot}}(X_{\mathrm{ad}}, Y_{\mathrm{ad}}, Z_{\mathrm{ad}}, \{\mathbf{R}_I\})$ is the ground-state or the total energy of the many-electron system at the atomic configuration $(X_{\mathrm{ad}}, Y_{\mathrm{ad}}, Z_{\mathrm{ad}}, \{\mathbf{R}_I\})$. According to Eq. (10), the PES is the minimum of the total energy with respect to the z-coordinate of the adatom Z_{ad} and all coordinates of the substrate atoms $\{\mathbf{R}_I\}$. Assuming that vibrational effects can be neglected, the minima of the PES represent stable and metastable sites of the adatom. Note that this PES refers to slow motion of nuclei and assumes that for any atomic configuration the electrons are in their respective ground state. Thus, it is assumed that the dynamics of the electrons and of the nuclei are decoupled (Born-Oppenheimer approximation).

The dynamics of an adatom on such a PES can be described by different approaches. In a molecular dynamics simulation the forces on each atom are computed and the atoms are moved accordingly. Such a simulation is very computer intensive, and we will discuss the advantages and disadvantages in more detail in Section 5. An alternative idea that will be used in this article is that processes such as diffusion are described by rates. Within transition state theory (TST) the rate of a microscopic process j usually has the form[25,26,27]

$$\Gamma^{(j)} = \frac{k_B T}{h} \exp(-\Delta F^{(j)}/k_B T) \quad , \tag{11}$$

where $\Delta F^{(j)}$ is the difference in the Helmholtz free energy between the maximum (saddle point) and the minimum (equilibrium site) of the potential curve along the reaction path of the process j. T is the temperature, k_B the Boltzmann constant, and h the Planck constant. The free energy of activation $\Delta F^{(j)}$ needed by the system to move from the initial position to the saddle point is given by

$$\Delta F^{(j)} = E_d^{(j)} - T\Delta S_{vib}^{(j)} \quad . \tag{12}$$

Here $E_d^{(j)}$ is the sum of the differences in the static total and vibrational energy of the system with the particle at the minimum and at the saddle point, and $\Delta S_{vib}^{(j)}$ is the analogous difference in the vibrational entropy. The rate of the process j can be cast as follows:

$$\Gamma^{(j)} = \Gamma_0^{(j)} \exp(-E_d^{(j)}/k_B T) \quad , \tag{13}$$

where $\Gamma_0^{(j)} = (k_B T/h)\exp(\Delta S_{vib}^{(j)}/k_B)$ is the effective attempt frequency. In the case of isotropic motion of an adatom on the surface it follows from Eq. (13) that the diffusion constant is $D = D_0 \exp(-E_d^{(j)}/k_B T)$[28]. The prefactor $D_0 = 1/(2\alpha)\Gamma_0^{(j)} l^2$, where l is the jump length and α the dimensionality of the motion ($\alpha = 2$ for the surface).

The two basic quantities in Eq. (13) are the attempt frequency $\Gamma_0^{(j)}$ and the activation energy $E_d^{(j)}$. TST[26,27] allows an evaluation of $\Gamma_0^{(j)}$ within the harmonic approximation:

$$\Gamma_0^{(j)} = \frac{\prod_{i=1}^{3N} \nu_i}{\prod_{i=1}^{3N-1} \nu_i^*} \quad , \tag{14}$$

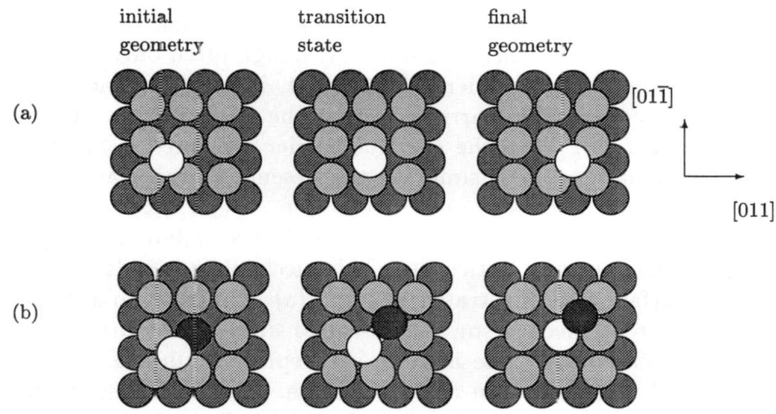

Figure 2: Diffusion via hopping (a) and exchange (b) on a fcc (100) surface.

where ν_i and ν_i^* are the normal mode frequencies of the system with the adatom at the equilibrium site and at the saddle point, respectively, and $3N$ is the number of degrees of freedom. The denominator in Eq. (14) contains the product of only $3N - 1$ normal frequencies, because for the adatom at the saddle point one of the modes describes the motion of the particle toward the final site and has an imaginary frequency. TST is only valid when $E_{\mathrm{d}}^{(j)}$ is larger than $k_{\mathrm{B}}T$.

The attempt frequency $\Gamma_0^{(j)}$ shows a much weaker temperature dependence than the exponential and for typical growth temperatures it is of the order $10^{12} - 10^{13}s^{-1}$. When the barriers for two diffusion events are different a *compensation effect* may occur, i.e., $\Gamma_0^{(j)}$ is larger for processes with a higher energy barrier. Indeed, a higher energy barrier usually implies a larger curvature of the potential well around the equilibrium site of the adatom. The corresponding vibrational frequencies of the adatom in such a potential are larger as well, which implies [see Eq. (14)] that the attempt frequency increases. Boisvert *et al.*[29] found that for metal diffusion the prefactor might decrease by up to one order of magnitude for very low energy barriers. Now consider all possible paths l that describe process j to get from one stable or metastable adsorption site, $\mathbf{R}_{\mathrm{ad}}^{(j)}$, to an adjacent one, $\mathbf{R}_{\mathrm{ad}}^{(j)\prime}$. The energy difference $E_{\mathrm{d}l}^{(j)}$ between the energy at the saddle point along l and the energy of the initial geometry is the barrier for this particular path. If the vibrational energy is negligible compared to $E_{\mathrm{d}l}^{(j)}$, the diffusion barrier $E_{\mathrm{d}}^{(j)}$ then is the minimum value of all

$E_{dl}^{(j)}$ of the possible paths that connect $\mathbf{R}_{ad}^{(j)}$ and $\mathbf{R}_{ad}^{(j)\prime}$, and the lowest energy saddle point is called the *transition state*. Although often only the path with the most favorable energy barrier is important, it may happen that several paths exist with comparable barriers or that the PES consists of more than one sheet (e.g. Ref. [30]). Then the *effective* barrier measured in an experiment or a molecular dynamics (MD) simulation represents a proper average over all possible pathways.

The common view for surface diffusion is that an adatom hops from one lattice site to a neighboring lattice site. This mechanism is called *hopping* and for the (100) surface it is illustrated in Fig. 2(a). Diffusion might also occur with a very different mechanism, the so-called *diffusion by atomic exchange* (or *exchange mechanism*). The adatom can replace a surface atom and the replaced atom then assumes an adsorption site. This was first discussed by Bassett and Webber [31] and Wrigley and Ehrlich [32]. Even for the crystal bulk exchange diffusion has been discussed [33]. This mechanism is actuated by the desire of the system to keep the number of cut bonds low along the diffusion pathway. On fcc (100) surfaces diffusion by atomic exchange was observed experimentally for Pt and Ir [34], and theoretically predicted for Al [35]. The geometry for exchange diffusion at a fcc (100) surface is shown in Fig. 2(b).

If one looks at a diffusion event in an experiment one can often not distinguish the different mechanisms that were operative, so that an experimentally deduced quantity is only an averaged quantity over the different mechanisms. Within TST, one can define an individual rate for each process, for example for surface diffusion via hopping and exchange. In a temperature range where several of these mechanisms are important, one has to take the correct average of them to compare the theoretical values with the experimental ones [30].

3.2 Analysis of Experimental Results

Before discussing results that were obtained for growth parameters from ab initio calculations and comparing them with data obtained from experimental measurements, we would like to briefly outline how experimental data often are analyzed. For simplicity, we will only focus on the surface diffusion constant.

Processes (a), (b), (c), (d), and (e) of Fig. 1 can be described by phenomenological rate equations of the form

$$\frac{dN_1}{dt} = \Phi - 2\kappa_1 N_1^2 - N_1 \sum_{j>1} \kappa_j N_j + 2\gamma_2 N_2 + \sum_{j>2} \gamma_j N_j \tag{15}$$

$$\frac{dN_j}{dt} = N_1(\kappa_{j-1} N_{j-1} - \kappa_j N_j) - \gamma_j N_j + \gamma_{j+1} N_{j+1} \quad . \tag{16}$$

Such a set of equations constitutes the basis of nucleation theory and describes the time evolution of the adatom density, N_1, and the density of islands of size j, N_j, for growth on a flat surface in the submonolayer regime. Adatoms are deposited onto the substrate at a rate $\Phi = F\mathcal{N}$ adatom/s where F is the flux in ML/s and $\mathcal{N}$ the number of atoms pro ML. The second and third term in Eq. (15) account for isolated adatoms being "lost" because two adatoms can meet at a rate κ_1 to form a new nucleus, or adatoms get captured at a rate κ_j by an island of size j. The last two terms in Eq. (15) describe further supply sources of adatoms and are gain terms because dimers may dissociate and adatoms detach from an island of size j at a rate γ_j. Equation (16) reflects the fact that the number of islands of size j increases because islands of size $j-1$ grow and islands of size $j+1$ shrink. The number of islands of size j decreases when islands of size j either shrink or grow. Note that no evaporation into the gas phase is included in Eqs. (15) and (16) (that means that the description in terms of Eqs. (15) and (16) is appropriate only at not too high temperatures).

With the assumption that agglomerates of $i^* + 1$ and more adatoms are stable against break-up ($\gamma_j = 0$ for $j > i^*$) one derives the scaling relation[36]

$$N^{\text{is}} = C_0 \left(\frac{D}{F}\right)^{-i^*/(i^*+2)} \tag{17}$$

where $N^{\text{is}} = \sum_{j>i^*} N_j$ is the island density and D the diffusion coefficient of an adatom on the flat surface. The number i^* is called the size of the critical nucleus. The main steps to derive Eq. (17) are as follows: A steady state for the adatom concentration is assumed, i.e., the left-hand side of Eq. (15) is neglected. This means essentially that the number of adatoms deposited equals the number of atoms captured when all islands on the surface are larger than i^*. Eq. (15) can then be inserted into Eq. (16), and the latter is integrated to obtain Eq. (17). The total island density thus depends on the ratio of the adatom diffusion constant and the deposition flux. When the diffusion constant increases, a deposited adatom can sample a larger area before the next adatom is deposited. Therefore, the density of island decreases. The same results from a smaller deposition rate, because the time between two successive deposition events increases, and thus again the area sampled by a diffusing adatom is enlarged.

It is widely believed that scaling laws that result from nucleation theory can be applied straightforward, so that measurements of the island density as a function of the growth temperature can yield the barrier for surface diffusion and the prefactor. The results presented in Section 4 of this article show that this bears quite some risks and may lead to incorrect conclusions. Furthermore, it has been shown[37] that relation (17) is problematic for $i^* > 1$ because not always a well defined integer i^* exists. As long as the temperature is small enough (or the deposition rate is large enough), a dimer is essentially stable and $i^* = 1$. When D/F increases, i^* eventually becomes larger than 1, because atoms can break away from any island. In particular, it is plausible that an atom that is singly bonded at the edge of a large cluster is more likely to detach than an atom that is part of a small cluster with all atoms in the cluster being highly coordinated. Thus, the quantity i^* should be interpreted as an effective, averaged quantity, and it is clear that it does not have to take on an integer value. When D/F is small enough $i^* = 1$, and we restrict ourselves to this regime in the present discussion.

In an STM experiment the island density can be measured as a function of the temperature. In a log-log plot of N^{is} as a function as D/F one can obtain E_{d} from the slope of the data, and D_0 from the intercept with the y-axis using Eq. (17) (as long as T is low enough so that $i^* = 1$). This is precisely what is often done in experimental studies. Thus, we would like to stress that a quantity that is often called an *experimental value* is in fact only the result of an *analysis of experimental data*, but not a quantity that is measured directly. An experimentally deduced value for a diffusion barrier and prefactor can not distinguish between different mechanisms such as hopping or exchange, and therefore only represents an effective value that is an average over the different mechanisms.

It is important to note that the analysis is based on certain assumptions. For example, it is possible that not only adatoms, but also clusters diffuse on the surface. In that case, one can still measure the island density, but since cluster diffusion is not included in Eqs. (15) and (16), the scaling law described by Eq. (17) is not applicable anymore, and numbers extracted with it are meaningless. An analysis by Villain *et al.* [38] predicts that the exponent in Eq. (17) can increase in a certain range to 0.5 (for $i^* = 1$) when the diffusion of dimers is allowed. A recent Monte Carlo simulation by Kuipers and Palmer [39] showed that the island density N^{is} decreases significantly when clusters are allowed to diffuse. A decrease of the island density by one order of magnitude leads to an error for the prefactor of $(i^* + 2)/i^*$ orders of magnitude.

The quantity C_0 in Eq. (17) is also not known. It essentially depends on the capture numbers κ_j, and it can be estimated from Monte Carlo simulations to be between 0.1 and 0.5. Using a self consistent approach for the κ_j [40] one gets $C_0 \simeq 0.25$. However, an uncertainty of ~ 5 for C_0 translates also into an uncertainty for the prefactor of two orders of magnitude (for $i^* = 1$). We therefore believe that particularly results obtained for the prefactor from measured island densities using Eq. (17) can be significantly flawed. An improvement of scaling laws that are derived under less restrictive assumptions remains a challenge for the future.

4 Strain Dependence of Surface Diffusion

In this Section we will show how the concepts of DFT outlined above can be applied to study surface diffusion. In particular, we will concentrate on the effects of strain on energy barriers and attempt frequencies for metallic systems. We will point out that the analysis of experimental data based on simple scaling laws is more complicated than usually believed.

4.1 Ag on Ag (111) and Pt (111)

Growth of Ag on Pt (111) and Ag on a thin Ag film on Pt (111) has been the focus of recent studies [41,42]. This system has a lattice mismatch of 4.2 % and appears to be a well suited system that can provide important information about the effects of strain during growth. Brune *et al.* [41] measured the island densities for Ag on Pt (111), Ag on one monolayer (1 ML) of Ag on Pt (111), and Ag on Ag (111), and employed Eq. (17) to extract the diffusion barriers and prefactors. They obtained $E_{\mathrm{d}}^{\text{Ag-Pt}} = 157$ meV with $\Gamma_0 = 1 \times 10^{13}\ \mathrm{s}^{-1}$ for Ag on Pt (111), $E_{\mathrm{d}}^{\text{Ag-Ag/Pt}} = 60$ meV with $\Gamma_0 = 1 \times 10^{9}\ \mathrm{s}^{-1}$ for Ag on 1 ML Ag on Pt (111), and $E_{\mathrm{d}}^{\text{Ag-Ag}} = 97$ meV with $\Gamma_0 = 2 \times 10^{11}\ \mathrm{s}^{-1}$ for Ag on Ag (111). In particular for the lowest diffusion barrier the prefactor was found to decrease by 4 orders of magnitude, which the authors of Ref. [41] claim is an unusually high compensation effect.

Not many studies have been devoted to investigate the influence of strain on surface diffusion. The first thorough study we are aware of has been a molecular dynamics simulation for Si on Si (100) that employed a Stillinger-Weber potential [43]. In this study it was found that the barrier for diffusion along the fast channel parallel to the dimer rows is lowered by approximately 10% for 3% compressive as well as for 2% tensile strain, and that atop the dimer rows (in the same direction) it is increased by about 10%. A general trend can not be seen, which might also be due to the fact that the reconstructions for semiconductor systems make the behavior more complicated. For metallic systems there have been a few theoretical studies in the last few years that employed different methods with different degrees of sophistication.

In a recent first principle study [17,44] the influence of strain on the diffusion constant for Ag on Ag (111) has been discussed. It has been found that in the range of ±5 % strain the DFT results for the diffusion barrier exhibit a linear dependence with a slope of 0.7 eV as it is displayed in Fig. 3. The calculated diffusion barrier for the unstrained system, $E_{\mathrm{d}}^{\text{Ag-Ag}} = 81$ meV, is in

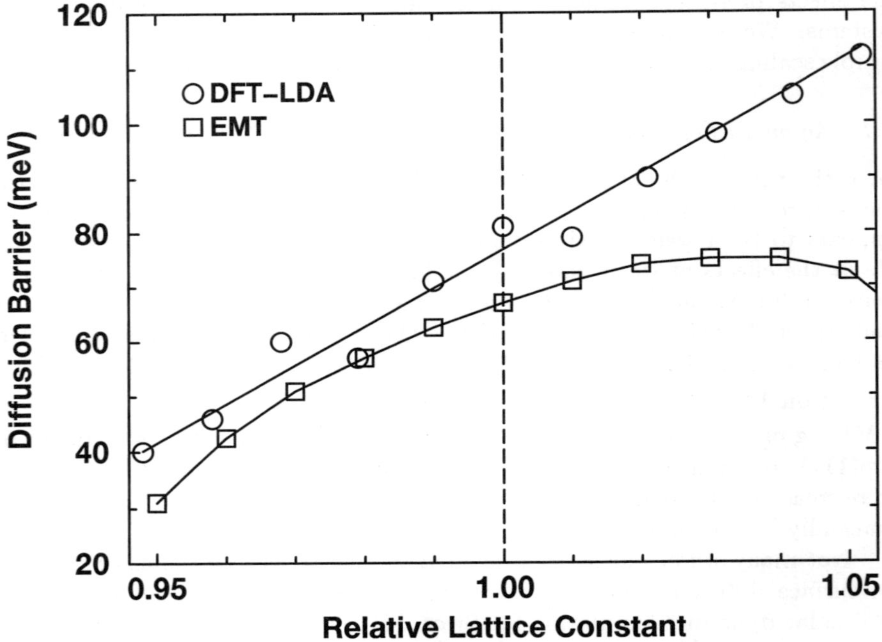

Figure 3: Diffusion barrier (in meV) for Ag on Ag (111) as a function of strain. The circles are DFT-LDA results from Ref. [17] and the squares are EMT results from Ref. [41].

Table 2: Diffusion barriers (in meV) for Ag on Pt (111), Ag on one monolayer (ML) Ag on Pt (111), and Ag on Ag (111).

System: Ag on	Experiment [41]	EMT [41]	DFT [17]
Pt (111)	157	81	150
1ML Ag on Pt (111)	60	50	65
Ag (111)	97	67	81

good agreement (within the error margins of the experiment and the calculations) with the results obtained from the analysis of the STM measurements of $E_{\mathrm{d}}^{\mathrm{Ag\text{-}Ag}} = 97$ meV. The accordance between experiment and theory extends to the system Ag/Pt (111) and Ag/1ML Ag/Pt (111). These results are summarized in Table 2. In Fig. 3 the DFT-LDA results are compared to those of an EMT study [41]. The EMT results exhibit a linear dependence only for smaller values of strain (± 2 %) and the diffusion barrier starts to decrease for values of misfit larger than 3 %. Indeed, it is plausible that a decrease of the diffusion barrier occurs when the atoms are separated far enough that eventually bonds are broken. However, as the DFT-LDA results show, for Ag/Ag (111) this happens at values for the misfit that are larger than 5 %. Additionally, when comparison with experiment is possible [i.e., Ag on Ag (111), and Ag on a monolayer Ag on Pt (111)] the EMT results are off by a factor that varies from 1.2 to 2.

The general trend of an increasing energy barrier for hopping diffusion with increasing lattice constant is quite plausible (for exchange diffusion see the next Section). Smaller lattice constants correspond to a reduced corrugation of the surface, and as result the atom is not bonded much stronger at the adsorption sites than at the bridge site. In contrast, when the surface is stretched the corrugation increases and the adsorption energy at the three-fold coordinated hollow sites increases. In an effective medium theory study a linear dependence of the surface diffusion constant as a function of the lattice constant has been found as well for the (111) surfaces of Ni, Pd, Pt, Cu, Ag, and Au [45]. The picture will change when the strain is so large that bonds are broken and then it is expected that the hopping diffusion barrier will start to decrease again at very large tensile strain.

It is worth noting that the diffusion barrier for Ag on top of a pseudomorphic layer of Ag on Pt (111) is substantially lower than it is for Ag on Ag (111). A question that arises is whether this reduced diffusion barrier is a result of the compressive strain or should be ascribed to electronic rearrangements induced by the Pt substrate. The diffusion barrier for Ag on Ag (111) with a

Table 3: Prefactor Γ_0 (in THz) and energy barrier E_d (in meV) for strained Ag on Ag (111), and for Ag on one monolayer (ML) Ag on Pt (111). The prefactors are calculated within the harmonic approximation of transition state theory. All results shown were obtained by considering 15 degrees of freedom.

Substrate	Ag (111) $a = 3.92$ Å	Ag (111) $a = 4.05$ Å	Ag (111) $a = 4.22$ Å	1 ML Ag on Pt (111) $a = 3.92$ Å
Γ_0	1.3	0.8	0.3	7.1
E_d	60	81	105	63

lattice constant that is compressed to the value of the lattice constant for Pt is $E_d^{\text{Ag-Ag}} = 60$ meV while that for Ag on Pt (111) (also with the Pt lattice constant of 3.92 Å obtained from DFT) is $E_d^{\text{Ag-Ag/Pt}} = 65$ meV. The agreement of these two values suggests that the reduction of the diffusion barrier for Ag on a layer of Ag on Pt (111) is mainly a strain effect and that the diffusion barrier on top of a layer of Ag is essentially independent of the substrate underneath.

The prefactor for strained Ag on Ag (111) using transition state theory within the harmonic approximation [cf. Eq. (14)] has also been calculated [44]. In this calculation up to 99 degrees of freedom were considered, and it is interesting to note that within a factor of 2 the result agreed with that obtained by taking into account only the degrees of freedom of the moving adatom. It was found that the prefactor changes only very little when the lattice constant (and thus the diffusion barrier, cf. Fig. 3) changes, so that there is no significant compensation effect. The results are summarized in Table 3. All the values calculated are of the order of 1 THz, as expected, since this is a typical surface phonon frequency. Thus, the calculations do not confirm the claim by the authors of Ref. [41] that there is a very large compensation effect for this system. An additional test has been done to investigate whether the prefactor (in contrast to the diffusion barrier) might be affected by the substrate underneath, but calculations for the prefactor for Ag on 1 monolayer Ag on Pt (111) also gave a prefactor of approximately 1 THz (cf. Table 3).

The reason for the large discrepancy between the theoretical and the experimentally deduced value for the prefactor for the system under compressive strain is not completely understood yet. We believe that the calculations are correct, and that the interpretation of the experimental data may be flawed. Possible reasons for this have been given in the previous Section when we introduced Eq. (17). A low diffusion barrier implies that the PES has a very small corrugation so that it is likely that small clusters diffuse. Also, long range effects (such as reconstructions or strain fields) may become important. The quantity C_0 in Eq. (17) depends on the capture numbers κ_j, but it is not clear how the κ_j behave in a strained system. We thus conclude that one has to be rather careful if one wants to use simple scaling laws to determine microscopic quantities from experimental data.

4.2 Hopping and Exchange Diffusion on (100) Surfaces

As it has been pointed out already in Section 3, diffusion may occur by hopping or atomic exchange on the (100) surface. Fig. 2 shows that the transition state of the exchange mechanism can be described by a dimer, and each atom of this dimer can form three chemical bonds, so that a simple bond counting argument suggests that the exchange mechanism is always the energetically preferred mechanism. It has been calculated by Feibelman [35] that the covalent nature of aluminum which allows the formation of directional bonds at certain geometries lowers the total energy of the system in the transition state for atomic exchange. Thus, the desire of the system to maximize the coordination of all atoms involved is the reason why diffusion proceeds via atomic exchange.

This is certainly a plausible argument for Al (100), but in general chemical bonding is a lot more complicated than counting nearest neighbor bonds and this explanation can not account for atomic exchange as the energetically preferred mechanism for Pt and Ir. Moreover, it is puzzling that for the transition metals Ir and Pt diffusion proceeds via exchange, but according to calculations by Yu and Scheffler [16] for Ag the preferred diffusion mechanism is hopping. The authors of Ref. [16] find that the barrier for hopping is 0.52 eV, while that for exchange is 0.93 eV when the exchange-correlation energy is approximated within the LDA (these numbers change to 0.45 eV and 0.73 eV within the GGA).

In a subsequent paper [46], Yu and Scheffler provided the following explanation for the different behavior of different transition metals. They realized that for $5d$ metals (Ir and Pt, with preferred exchange mechanism) the surface stress is significantly higher than it is for $3d$ and $4d$ metals (Ag, where hopping is energetically favored) [47]. This is caused by relativistic effects that play a more important role for the heavier $5d$ metals. Higher tensile surface stress has a tendency to pull the dimer in the transition state for atomic exchange closer toward the surface and the layer underneath to lower the total energy of the system. Thus, under tensile surface stress the barrier for atomic exchange decreases, in contrast to the barrier for hopping, that in analogy to the behavior on the (111) surface is expected to increase under tensile surface stress.

To test this claim, Yu and Scheffler calculated the energy barrier for hopping and exchange as a function of the lattice constant for Ag on Ag (100), and the results are shown in Fig. 4. Indeed, the two curves cross at approx-

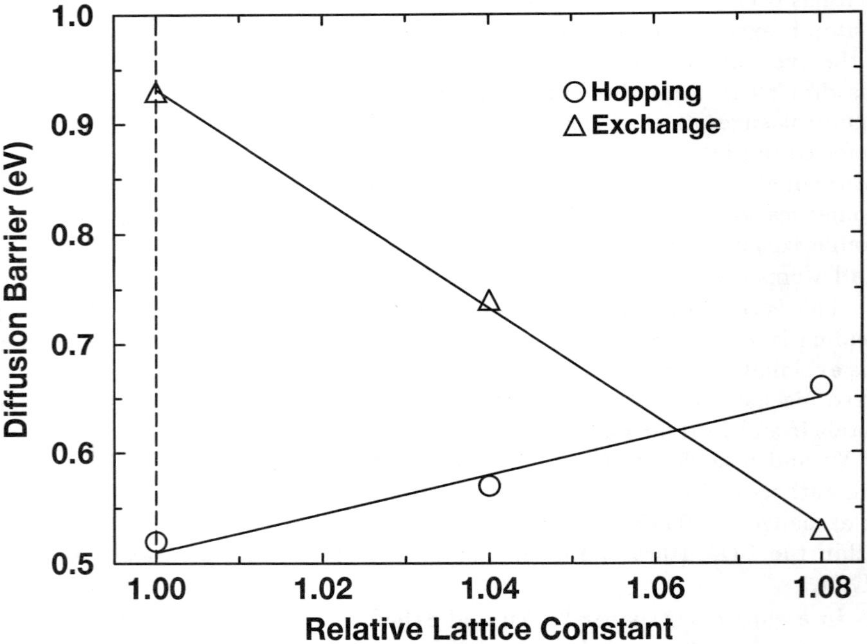

Figure 4: The energy barrier for the hopping and the exchange mechanism for Ag on Ag (100) as a function of the relative lattice constant a/a_0.

imately 6.5% lattice mismatch, and it is predicted that under high tensile surface stress the energetically preferred mechanism is exchange. Yu and Scheffler also claimed that for unreconstructed Au (100) the energetically preferred mechanism should be exchange, with a barrier of 0.65 eV (compared to 0.83 eV for hopping). This again is due to the fact that the dimer in the transition state of the the exchange mechanism is attracted to the surface; it is approximately 37% closer to the surface than the interlayer spacing in the bulk would suggest. This is similar to Al (100), where the dimer is 55% closer to the surface, but in contrast to the (unstrained) Ag (100), where the dimer is only 25% closer to the surface. When Ag (111) is strained by 8% the dimer is 57% closer to the surface.

It seems that there is clear evidence that atomic exchange is stabilized by the correlation of the local coordination of all atoms involved and the strength of the bonds (as for Al), and by the tensile surface stress. Following this argument, the opposite should happen under compressive surface stress: When the surface is compressed, the dimer is "squeezed" away from the surface, and as a result the energy barrier for exchange increases. This is again in contrast to the behavior of the barrier for hopping, that is expected to decrease. Recent results for Al on Al (100) [48] show that indeed the energetically preferred diffusion mechanism changes from exchange to hopping under compressive strain.

Before closing this Section, we would like to point out that based on knowledge for the barrier for surface diffusion it is not appropriate to conclude that one mechanism is the preferred mechanism. The diffusion constant also depends on the prefactor, and there are indications in the literature [49] that the prefactor for diffusion by exchange might be 1 or 2 orders of magnitude larger than the prefactor for hopping. In fact, the consequences of this are discussed in more detail in the example in the next Section. From the results for diffusion of Ag on Ag (111) it can be expected that there is not a significant strain dependence of the prefactor, but more research should be directed toward a better knowledge of the different prefactors for different mechanisms and different surfaces.

5 Ab initio Kinetic Monte Carlo Simulations

The individual microscopic processes can be used to get a deeper understanding of epitaxial growth, provided that their interplay as a function of temperature and deposition rate can be modeled properly. The kinetics during growth is a stochastic process that is well described by a master equation and the Monte Carlo technique yields a numerical solution of this equation [50,51]. The master equation governs the dynamic evolution of $P(\sigma, t)$, the probability that the system is in state σ at time t, and it has the following general form:

$$\frac{\partial P(\sigma,t)}{\partial t} = -\sum_{\sigma'} W(\sigma \to \sigma')P(\sigma,t) + \sum_{\sigma} W(\sigma' \to \sigma)P(\sigma',t) \quad . \tag{18}$$

Here $W(\sigma' \to \sigma)$ has the meaning of a transition probability from state σ' to state σ per unit time. On the right-hand side the first sum describes the rate of all processes where the system jumps out of the considered state (and hence decreases its probability), while the second sum describes the rate of all processes where the system jumps into the considered state (and hence increases its probability). The solution of the master equation is achieved computationally by choosing randomly among several possible transitions and accepting or refusing the particular events according to an appropriate probability. In thermal equilibrium the detailed balance condition

$$W(\sigma \to \sigma')P_{\text{eq}}(\sigma) = W(\sigma' \to \sigma)P_{\text{eq}}(\sigma') \tag{19}$$

ensures that the two sums on the right-hand side of Eq. (18) cancel and $P_{\text{eq}}(\sigma)$ is the steady-state distribution of the master equation.

The obvious question concerns the choice of a pertinent transition probability $W(\sigma \to \sigma')$. The detailed balance criterion, however, only determines the ratio $W(\sigma \to \sigma')/W(\sigma' \to \sigma)$. When static properties of the system are sought, the Metropolis algorithm [52] is appropriate. According to this scheme the probability that a new configuration is accepted is proportional to $\exp(-\Delta E/k_{\text{B}}\text{T})$, where ΔE is the difference between the total energies of the system in the new and old configuration. Such an algorithm searches for the configuration corresponding to the minimum of the total energy. The sequence of generated configurations does not correspond to the real time evolution of the system, and there is no problem of conversion from simulated time t_s to real time t_r. However, for the dynamic behavior (such as growth) of a system the correlation between t_s and t_r is important, and the configuration sequence has to mimic the real one. The appropriate tool for this is KMC. Here the dynamics of the system is obtained from the local microscopic transition rates in terms of Poisson processes.

In contrast to MD, in KMC the place and time of events are no longer deterministically obtained, but are chosen by statistical considerations. The task of the KMC is to build up an artificial chronological sequence of distinct events separated by certain *interevent* times. This chronology is based on the hierarchy dictated by the transition rates and may not correspond to reality. For example, during growth events may occur simultaneously, but KMC introduces an artificial time interval between them. At the other extreme, events may be separated by a very large interval that is not reflected in the KMC *interevent* time. Thus, KMC does not model the "deterministic" microscopic dynamics yielding the exact times of various processes (as MD does). Therefore, in KMC it is possible to avoid the explicit calculation of all unsuccessful attempts, but the chain of events and corresponding interevent times must be constructed from probability distributions weighting appropriately all possible outcomes.

To illustrate the strategy of a KMC simulation (for more details see Ref.[22]) let us consider a system of M particles that are capable of undergoing o total of m transition events (diffusions, deposition, etc.). The m transition events are described by a set of rates $\mathbf{\Gamma} \equiv \{\Gamma^{(1)}, \Gamma^{(2)}, ..., \Gamma^{(m)}\}$ where $\Gamma^{(j)}$ has the form given by Eq. (13). The M particles can be partitioned in classes among the various possible transition events as $\mathbf{M} \equiv \{n^{(1)}, n^{(2)}, ..., n^{(m)}\}$ where $n^{(j)}$ is the number of particles capable of undergoing a transition with a rate $\Gamma^{(j)}$. It should be pointed out that $\sum_j n^{(j)} \geq M$ since a particle may belong to several classes of $\mathbf{M}$ at the same time. Thus, a configuration of the system at a particular time can be labeled by the distribution of $\mathbf{M}$ over $\mathbf{\Gamma}$. A transition is determined by picking randomly among various possible events available at each time step, and the configuration is updated. The choice of the event is dictated by the dynamical hierarchy described by the set of relative transition probabilities $\mathbf{W} \equiv \{W^{(1)}, W^{(2)}, ..., W^{(m)}\}$, each of them defined as:

$$W^{(j)} = \frac{n^{(j)}\,\Gamma^{(j)}}{R} \quad , \tag{20}$$

where $R = \sum_j n^{(j)}\,\Gamma^{(j)}$ is the total rate. These probabilities are constructed such that detailed balance is achieved at thermal equilibrium. After each Monte Carlo step time should be updated with an increment $\Delta t = -ln(\rho)/R$ where ρ is a random number in the range $(0, 1]$. In this way, a direct and unambiguous relationship between KMC time and real time is established, since the KMC algorithm effectively simulates stochastic processes described by a Poisson distribution.

KMC simulations have been used to study crystal growth of semiconductors (e.g. [53,54,55]) and metals (e.g. [56,57,58,59]). However, most of these studies have been based on restrictive approximations. For example, the input parameters have been treated as effective parameters determined rather indirectly by fitting to experimental quantities, like intensity oscillations in helium atom scattering (HAS) measurements or reflection high energy electron diffraction (RHEED), or they were obtained from STM studies of island densities. Thus, the connection between these parameters and the microscopic nature of the processes may be somewhat uncertain. Often the correct surface structure was neglected and the simulation was done on a simple cubic lattice while the system of interest had an fcc or bcc structure. Despite these approximations such studies have provided qualitative and in some cases also quantitative insight into growth phenomena.

It is desirable to carry out KMC simulations with the proper geometry and microscopically well founded parameters. This has been done for example in Refs.[60,61,62] where semi-empirical calculations have been employed to evaluate the PES. The most accurate, but also most elaborate approach to obtain the PES employs DFT as described in Sections 2 and 3. It is interesting to note that the result of a KMC study will be the same as that of a MD simulation, provided that the underlying PES is the same. More details on the KMC method are contained in Ref.[22] and references therein.

An example of the possible combination of DFT results and KMC simulations is the study of island shape in functions of substrate temperature for Al on Al (111). This surface, as each (111) surface of an fcc crystal, is characterized by the presence of two types of close-packed steps, labeled as {100} and {111} facets, referring to the plane passing through the atoms of the step and the atom of the substrate (often these steps are labeled A and B). Using the DFT parameters for Al/Al (111) by Stumpf and Scheffler[63,64] one can carry out KMC simulations under typical growth conditions. The results of the *ab initio* KMC simulations where all relevant processes, including the different mechanisms for a particular event (i.e. hopping or exchange) are included, are shown in Fig. 5. They are obtained after 0.5 s with a deposition rate of 0.8 ML/s. When the growth temperature is 50 K the shape of the islands is highly irregular and indeed fractal. Adatoms which reach a step cannot leave it anymore and they cannot even diffuse along the step. Thus, at this temperature ramification takes place into random directions, and island formation can be understood in terms of the hit and stick model. At $T = 150$ K the island shapes are triangular with their sides being {100}-faceted steps, but increasing the temperature to $T = 200$ K a transition from triangular to hexagonal shape occurs. For $T = 250$ K the islands become triangular again, but they are rotated by π and are mainly bounded by {111}-faceted steps.

The transition can be understood in terms of different diffusivities along the step edges of the islands: The lower the migration probability along a given step edge, the higher is the step roughness and the faster is the speed of advancement of this step edge. As a consequence, this step edge shortens and eventually it may even disappears. Since diffusion along the densely packed steps on the (111) surface (the {100} and {111} facets) is faster than along steps with any other orientation, this criterion explains the presence of islands which are mainly bounded by {100}- or {111}-faceted steps. The same argument can

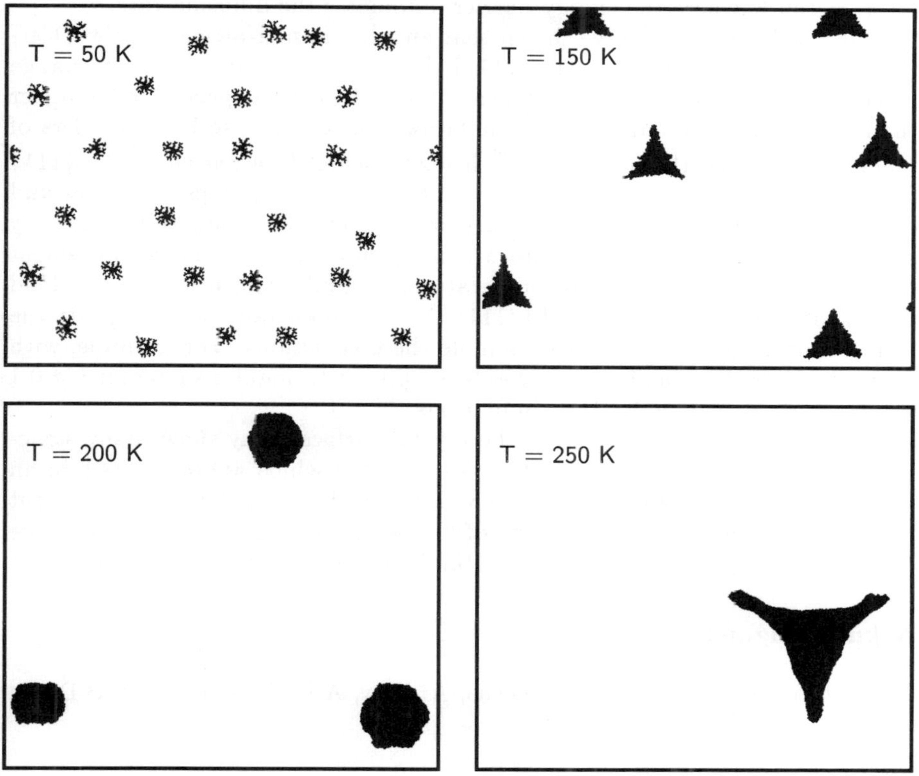

Figure 5: A surface area of (1718 × 1488) Å^2 (half of the simulation area) at four different substrate temperatures. The deposition rate was 0.08 ML/s and the coverage in each picture is Θ = 0.04 ML.

be extended to the diffusion along the two close-packed steps. By considering the energy barriers (0.32 eV for diffusion along the {100} facet vs. 0.42 eV along the {111}) we would expect only islands with {100} sides, until the temperature regime for the thermal equilibrium is reached. This is the case for $T = 150$ K where the energy barrier dominates the diffusion rate.

DFT calculations[63] have shown that an adatom migrates along the {100} facet by hopping, while along the {111} the diffusion occurs via the exchange mechanism. Since the attempt frequency for exchange is expected to be higher than the one for hopping (in this simulation it is assumed to be two orders of magnitude larger), the higher $\Gamma_0^{\{111\}}$ leads to faster diffusion along the {111} facet than along the {100} one at $T = 250$ K. The latter steps disappear and only triangles with {111}-faceted sides are present. The roughly hexagonally shaped islands at $T = 200$ K are a consequence of the equal advancement speed for the two steps at that temperature. A similar transition in the island shapes has been observed for Pt (111)[65]. An important point regards the time required by the system to assemble these structures. For example, with $F = 0.8$ ML/s the island at $T = 250$ K assumes a triangular shape for $t \geq 0.1$ s, a time not reachable by MD simulations.

In conclusion, during growth the crystal surfaces may show many structural details such as islands, clusters, steps, all of which are originated on an atomic length scale (microscopically well-founded description is needed) but may lead to cooperative phenomena of mesoscopic dimensions with time scale of the order of seconds (stochastic methods are necessary).

Acknowledgments

The authors are indebted to M. Fuchs, A. Kley, A.P. Seitsonen, and B.D. Yu for many valuable discussions.

References

1. P. Hohenberg and W. Kohn, Phys. Rev. **136**, B864 (1964).
2. M. Levy, Proc. Natl. Acad. Sci. (USA) **76**, 6062 (1979).
3. We limit our discussion in this paper to non-magnetic systems. However, it is straightforward to generalize DFT and to write the total-energy functional in terms of the electron density *and* the magnetization density (see for example R.M. Dreizler and E.K.U. Gross, *Density Functional Theory* (Springer Verlag, Berlin, Heidelberg, New York, 1990)).
4. All equations are noted in Hartree atomic units, i.e., the unit of length is 1 bohr = 0.5292 Å, the unit of energy is 1 hartree = 27.2116 eV, and

$\hbar = m = e = 1$.

5. W. Kohn and L.J. Sham, Phys. Rev. **140**, A1133 (1965).
6. J.P. Perdew and Y. Wang, Phys. Rev. B **45**, 13244 (1992); J.P. Perdew, in *Electronic Structure of Solids '91*, eds. P. Ziesche and H. Eschrig (Akademie Verlag, Berlin, 1991), p. 11; J.P. Perdew, J.A. Chevary, S.H. Vosko, K.A. Jackson, M.R. Pederson, D.J. Singh, and C. Fiolhais, Phys. Rev. B **46**, 6671 (1992).
7. J.P. Perdew and Y. Wang, Phys. Rev. B **33**, 8800 (1986); J.P. Perdew, Phys. Rev. B **33**, 8822 (1986); *ibid.* **34**, 7406(E) (1986).
8. Y. Wang and J.P. Perdew, Phys. Rev. B **43**, 8911 (1991); J.P. Perdew, Physica B **172**, 1 (1991); J.P. Perdew, K. Burke, Y. Wang, unpublished.
9. A.D. Becke, Phys. Rev. A **38**, 3098 (1988); J. Chem. Phys. **98**, 3892 (1993).
10. C. Lee, W. Yang, and R.G. Parr, Phys. Rev. B **37**, 785 (1988).
11. W. Kohn, A.D. Becke, R.G. Parr, J. Chem. Phys. **100**, 12974 (1996).
12. M. Ernzerhof, J.P. Perdew, and K. Burke, in *Topics in Current Chemistry* **180** (Springer Verlag, Berlin, Heidelberg, New York, 1996), p. 1.
13. L. Mitas and R.M. Martin, Phys. Rev. Lett. **72**, 2438 (1994); J.C. Grossman, L. Mitas, and K. Raghavachari, Phys. Rev. Lett. **75**, 3870 (1995).
14. C. Fillipi, X. Gonze, and C.J. Umrigar, to be published in *Recent Developments and Applications of Density Functional Theory*, ed. J.M. Seminario (Elsevier, Amsterdam, 1996).
15. B. Hammer, M. Scheffler, K.W. Jacobsen, and J.K. Norskov, Phys. Rev. Lett. **73**, 1400 (1994); B. Hammer and M. Scheffler, Phys. Rev. Lett. **74**, 3487 (1995).
16. B.D. Yu and M. Scheffler, Phys. Rev. Lett. **77**, 1095 (1996); Phys. Rev. B **55**, 13916 (1997).
17. C. Ratsch, A.P. Seitsonen, and M. Scheffler, Phys. Rev. B **55**, 6750 (1997).
18. M.C. Payne, M.P. Teter, D.C. Allan, T.A. Arias, and D.J. Joannopoulos, Rev. Mod. Phys. **64**, 1045 (1992).
19. R. Stumpf and M. Scheffler, Comput. Phys. Commun. **79**, 447 (1994), M. Bockstedte, A. Kley, J. Neugebauer, and M. Scheffler, to be published.
20. D.J. Singh, *Planewaves, Pseudopotentials, and the LAPW Method* (Kluwer Academic Publ., 1994).
21. H.L. Skriver, *The LMTO Method* (Springer Verlag, Berlin, Heidelberg, New York, 1983).
22. P. Ruggerone, C. Ratsch, and M. Scheffler, in *The Chemical Physics of Solid Surfaces* vol. 8, eds. D.A. King and D.P. Woodruff (Elsevier Science, Amsterdam, 1997), in press.

23. D.C. Senft and G. Ehrlich, Phys. Rev. Lett. **74**, 294 (1995).
24. T.R. Linderoth, S. Horch, E. Lægsgaard, I. Stensgaard, and F. Besenbacher, Phys. Rev. Lett. **78**, 4978 (1997).
25. S. Glasston, K.J Laidler, and H. Eyring, *The Theory of Rate Processes*, (McGraw-Hill, New York, 1941).
26. G.H. Vineyard, J. Phys. Chem. Solids **3**, 121 (1957).
27. G. Wahnström, in *Interaction of Atoms and Molecules with Solid Surfaces*, eds. V. Bortolani, N.H. March, and M.P. Tosi (Plenum Press, New York and London, 1990), p. 529.
28. R. Gomer, Rep. Prog. Phys. **53**, 917 (1990).
29. G. Boisvert, L.J. Lewis, and A. Yelon, Phys. Rev. Lett. **75**, 469 (1995).
30. A. Kley and M. Scheffler, in *The Physics of Semiconductors*, Eds. M. Scheffler and R. Zimmermann (World Scientific, Singapore, 1996), p. 1031.
31. D.W. Bassett and P.R. Webber, Surf. Sci. **70**, 520 (1978).
32. J.D. Wrigley and G. Ehrlich, Phys. Rev. Lett. **44**, 661 (1978).
33. K.C. Pandey, Phys. Rev. Lett. **57**, 2287 (1986).
34. G.L. Kellog and P.J. Feibelman, Phys. Rev. Lett. **64**, 3143 (1990); C. Chen and T.T. Tsong, Phys. Rev. Lett. **64**, 3147 (1990).
35. P.J. Feibelman, Phys. Rev. Lett. **65**, 729 (1990).
36. S. Stoyanov and D. Kashchiev, in: Vol. 7 of *Current Topics in Materials Science*, Ed. E. Kaldis (North-Holland, Amsterdam, 1981), p. 69.
37. C. Ratsch, A. Zangwill, P.Šmilauer, and D.D. Vvedensky, Phys. Rev. Lett. **72**, 3194 (1994).
38. J. Villain, A. Pimpenelli, L. Tang, and D.E. Wolf, J. Phys. I France **2**, 2107 (1992).
39. L. Kuipers and R.E. Palmer, Phys. Rev. B **53**, R7646 (1996).
40. G.S. Bales and D.C. Chrzan, Phys. Rev. B **50**, 6057 (1994).
41. H. Brune, K. Bromann, H. Röder, K. Kern, J. Jacobsen, P. Stolze, K.W. Jacobsen, J. Nørskov, Phys. Rev. B **52**, R14380 (1995).
42. H. Brune and K. Kern, in *Growth and Properties of Ultrathin Layers*, eds. D.A. King and D.P. Woodruff (Elsevier Science, Amsterdam, 1997), in press; and references therein.
43. C. Roland and G.H. Gilmer, Phys. Rev. B **46**, 13428 (1992).
44. C. Ratsch and M. Scheffler, unpublished.
45. J.J. Mortensen, B. Hammer, O.H. Nielsen, K.W. Jacobsen, and J.K. Nørskov, in *Elementary Processes in Excitations and Reactions*, eds. A. Okiji, H. Kasai, and K. Makoshi (Springer, Berlin 1996), p.173.
46. B.D. Yu and M. Scheffler, unpublished.

47. V. Fiorentini, M. Methfessel, and M. Scheffler, Phys. Rev. Lett. **71**, 1051 (1993).
48. C. Ratsch, V. Bostanov, and M. Scheffler, unpublished.
49. C.L. Liu, J.M. Cohen, J.B. Adams, and A.F. Voter, Surf. Sci. **253**, 334 (1991).
50. *The Monte Carlo Method in Condensed Matter Physics*, ed. K. Binder (Springer Verlag, Berlin, Heidelberg, New York, 1992), and references therein.
51. H.C. Kang and W.H. Weinberg, J. Chem. Phys. **90**, 2824 (1989); K.A. Fichthorn and W.H. Weinberg, J. Chem. Phys. **95**, 1090 (1991); H.C. Kang and W.H. Weinberg, Acc. Chem. Res. **25**, 253 (1992).
52. M. Metropolis, A.W. Rosenbluth, M.N. Rosenbluth, A.N. Teller, and E. Teller, J. Chem. Phys. **21**, 1087 (1953).
53. A. Madhukar, Surf. Sci. **132**, 344 (1983); A. Madhukar and S.V. Ghaisas, Appl. Phys. Lett. **47**, 247 (1985); S.V. Ghaisas and A. Madhukar, J. Vac. Sci. Technol. B **3**, 540 (1985); S.V. Ghaisas and A. Madhukar, Phys. Rev. Lett. **56**, 1066 (1986).
54. S. Clarke and D.D. Vvedensky, Phys. Rev. Lett. **58**, 2235 (1987); Phys. Rev. B **36**, 9312 (1987); Phys. Rev. B **37**, 6559 (1988); J. Appl. Phys. **63**, 2272 (1988); S. Clarke, M.R. Wilby, D.D. Vvedensky, T. Kawamura, K. Miki, and H. Tokumoto, Phys. Rev. B **41**, 10198 (1990); T. Shitara, D.D. Vvedensky, M.R. Wilby, J. Zhang, J.H. Neave, and B.A. Joyce, Phys. Rev. B **46**, 6815 (1992); Phys. Rev. B **46**, 6825 (1992).
55. H. Metiu, Y.-T. Lu, and Z. Zhang, Science **255**, 1088 (1992).
56. M.C. Bartelt and J.W. Evans, Phys. Rev. Lett. **75**, 4250 (1995);
57. Z. Zhang, X. Chen, and M. Lagally, Phys. Rev. Lett. **73**, 1829 (1994).
58. J.G. Amar and F. Family, Phys. Rev. Lett. **74**, 2066 (1995).
59. S.V. Khare, N.C. Bartelt, and T.L. Einstein, Phys. Rev. Lett. **75**, 2148 (1995).
60. S. Liu, Z. Zhang, G. Comsa, and H. Metiu, Phys. Rev. Lett. **71**, 2967 (1993).
61. J. Jacobsen, K.W. Jacobsen, P. Stolze, and J.K. Nørskov, Phys. Rev. Lett. **74**, 2295 (1995); J. Jacobsen, K.W. Jacobsen, and J.K. Nørskov, Surf. Sci. **359**, 37 (1996).
62. Z.-P. Shi, Z. Zhang, A.K. Swan, and J.F. Wendelken, Phys. Rev. Lett. **76**, 4927 (1996).
63. R. Stumpf and M. Scheffler, Phys. Rev. Lett. **72**, 254 (1994); *ibid.* **73**, 508 (1995)(E).
64. R. Stumpf and M. Scheffler, Phys. Rev. B **53**, 4958 (1996).
65. T. Michely, M. Hohage, M. Bott, and G. Comsa, Phys. Rev. Lett. **70**, 3943 (1993).

Atomistic Simulation Methods

Arthur F. Voter
Theoretical Division, Los Alamos National Laboratory, Los Alamos, NM 87545

Abstract

Atomistic methods for simulating surface morphology evolution are discussed. Although the time scale limitation of molecular dynamics precludes direct simulation of deposition at realistic growth rates, methods based on transition state state theory can extend the time significantly. For example, kinetic Monte Carlo simulations, using rates computed from transition state theory, offer coarse-grained dynamics on extremely long time scales while maintaining a direct connection with the underlying interatomic potential. If some reaction pathways are missing, however, the connection is not complete and the dynamics are approximate. The hyperdynamics method, a recently developed continuous-space trajectory approach, offers an alternative to kinetic Monte Carlo that will be useful in certain situations. I will discuss transition state theory, kinetic Monte Carlo, and hyperdynamics, from the point of view of their suitability for simulating surface morphology evolution.

1 Introduction

Molecular dynamics-based simulation approaches hold an appealing position in the spectrum of theoretical methods for studying cal methods for studying surface growth. Given a fast computer and an accurate model for the interatomic forces, a molecular dynamics (MD) simulation can, in principle, give exact predictions of the growth (or removal) dynamics. Experimental observables can be calculated for a range of length scales, and trajectories can be analyzed in complete detail to discover the important atomistic events underlying the larger-scale behavior. Two factors limit the realization of this ideal application of MD simulations: the quality of available interatomic potentials, and the time scale limitations of direct MD.

Interatomic potentials are continually improving [1,2]. Enhancements in computer speed allow consideration of more complicated forms and higher quality fits to large amounts of first-principles and experimental data. Improvements in electronic structure algorithms are making "first-principles potentials" more accessible, although they are still orders of magnitude more computationally expensive than fast "classical" potentials, such as the embedded atom method (EAM) [3] for metals or the Stillinger-Weber potential for silicon [4].

The time scale issue is perhaps more daunting. Using a fast interatomic potential on the fastest available computer workstation, an MD simulation of a 100-atom system (a very small model for a surface growth study) evolves at a rate of $\sim 10^{-7}$ s per week of processor time. Massively parallel computers, useful for simulating much larger systems, have almost no impact on this accessible time scale. Thus, even with the optimistic assumption that computer speeds double every 18 months, it will be 20 years before MD simulation of surface growth or etching at a realistic rate (e.g., 10^3 impacts/second) will run as fast as one impact per week of computer time.

One way around this time scale problem is to simulate artificially fast deposition rates, and simulations of this type are certainly valuable[5,6]. However, some of the most interesting behavior arises from the diffusion and reorganization processes that occur between deposition events. These processes can be accelerated by raising the temperature between depositions. However, because each of the microscopic rate constants has its own temperature dependence, raising the temperature upsets the delicate interplay of different rates that results in a particular morphology.

Although direct MD simulation is hopeless, there are alternative atomistic tools, based on transition state theory (TST), that can access the required time scales. The essence of the TST approximation is to focus on the interesting diffusive events by discarding, in a rigorous way, the irrelevant vibrational dynamics that occupy the time between those events. This chapter gives an introduction to TST and its use in computing accurate rate constants for elementary events. Two TST-based simulation methods relevant to surface growth are then discussed. The first is kinetic Monte Carlo, which takes TST rate constants as input, producing long-time dynamical behavior that is exact to the extent that the rate information is accurate and complete. The second is hyper-MD, a recently developed approach that combines the at combines the features of MD and TST to extend the MD time scale without any precalculation of rate constants. As interatomic potentials improve, it becomes increasingly important to have methods that can perform these accurate long-time dynamical simulations. The focus here will be on the characteristics of the methods for producing correct dynamics for a given interatomic potential, rather than on how well results using current potentials compare with experiment. The basis for most of the discussion will be silver surfaces modeled with an EAM potential[7], although most of the concepts carry over to other systems.

2 Transition State Theory

Consider the classical trajectory of an N-atom system representing a solid surface at some stage of growth. This system vibrates about the minimum of a basin in the $3N$-dimensional potential energy surface that we will call state A. State A thus corresponds to a particular pattern of terraces, step edges and adatoms, e.g., as shown in Fig. 1. After some length of time, one of the vibrational excursions takes the system over a ridge into an adjacent potential basin; a transition to a new state has occurred, perhaps via an adatom hop event. In transition state theory [8,9,10], the rate constant for this event is approximated as the flux through the dividing surface separating the two states. The beauty of TST is that this flux is an equilibrium property of the system, so once the dividing surface has been defined, the TST escape rate at some temperature T can be directly calculated without ever running a trajectory. It depends only on the properties of the partition function of state A and the partition function of the $3N-1$-dimensional dividing surface. If the dynamical connectivity in phase space is such that successive crossings of the dividing surfaces are uncorrelated (i.e., if a trajectory passing through a dividing surface always becomes thermalized in the new state and loses its memory before any subsequent crossings), then the TST rate constant is equivalent to the classically exact rate constant. If correlated events do exist, then the TST rate is an upper bound on the true rate. The methodology for correcting for the effect of correlated crossings in the condensed phase is well developed [11,12,13], but will not be considered here for the following reason: under the conditions typical for a solid surface between deposition events, the correlated dynamical events usually make a minor contribution, and certainly cause much less error than the inaccuracy of the interatomic potential. Thus, for our purposes, TST offers an excellent approximation to the exact dynamical transition rate, provided the dividing surface is chosen as the saddle plane, as discussed below.

Formally, the TST rate can be written as

$$k_{A\rightarrow}^{\mathrm{TST}} = \langle |dx/dt|\,\delta(x)\rangle_A \,. \tag{1}$$

Here, assuming a one-dimensional system for notational simplicity, the dividing surface is positioned at $x = 0$, $\delta()$ is a Dirac delta function and $\langle\cdots\rangle_A$ indicates a classical, canonical-ensemble average restricted to the configuration space of state A. Equation (1) can be computed exactly at a given temperature for the $3N$-dimensional system using importance sampling methods [14] implemented with either Metropolis Monte Carlo or MD. As an alternative, Eq. (1) can be evaluated analytically within the harmonic approximation, leading to the

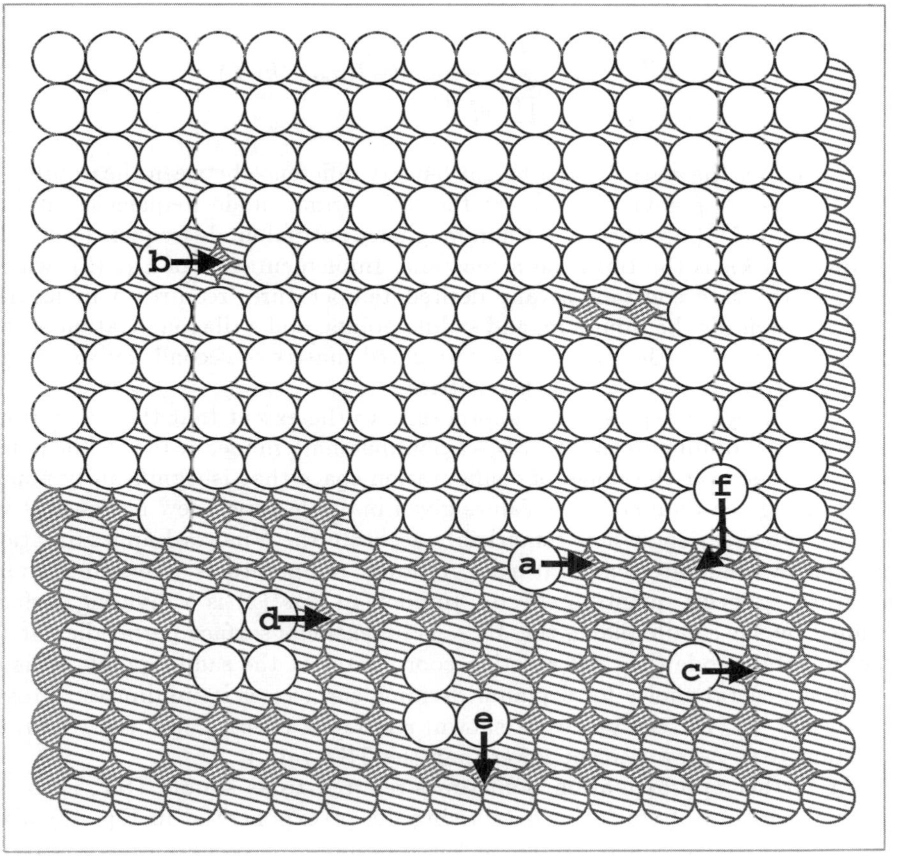

Figure 1: A particular state during growth on fcc(100) surface. Arrows indicate some of the possible hopping events that take the system to a different state, as discussed in text. The EAM static barriers for Ag(100) with one layer relaxed, from Ref. [7] are: (a) 0.26 eV, (b) 0.46 eV, (c) 0.49 eV, (d) 0.81, (e) 0.55 eV, and (f) 0.8 eV (present work).

so-called Vineyard expression [15]:

$$k^{HTST} = \frac{\prod_i^{3N} \nu_i^{min}}{\prod_i^{3N-1} \nu_i^{sad}} \exp(-E_{static}/k_B T). \tag{2}$$

Here E_{static} is the static barrier height (energy difference between the minimum and the saddle point), $\{\nu_i^{min}\}$ are the $3N$ normal mode frequencies at the minimum, $\{\nu_i^{sad}\}$ are the $3N-1$ nonimaginary normal mode frequencies at the saddle, and k_B is the Boltzmann constant. Implementation of Eq. (2), which predicts the rate constant at any desired temperature, requires a molecular statics search for the minimum and saddle points, and a diagonalization of the dynamical matrix (the inverse-mass-weighted matrix of second derivatives of the potential) at each of those geometries.

The Vineyard expression is appropriate to the extent that the $3N$ normal modes at the minimum and the $3N-1$ nonimaginary modes at the saddle point are harmonic over the range of configuration space that is significantly populated at a given temperature. Values given in Table I for a few representative surface processes show how remarkably accurate the Vineyard approximation is, even at moderate temperatures. At lower temperatures it becomes more accurate, ignoring quantum effects. Implicit in Eq. (2) is a dividing surface defined by the saddle plane, the $3N-1$-dimensional surface perpendicular to the unstable mode (i.e., the reaction coordinate) at the saddle point. This is the best general choice for a planar dividing surface, as it minimizes recrossing events [16]. (For the processes shown, recrossing effects cause the true rate constant to be a few percent lower than the TST rate.)

A further approximation, sometimes referred to as simple TST, is to replace the ratio of frequencies comprising the preexponential in Eq. (2) with the single frequency of the mode at the minimum corresponding to the reaction coordinate. Preexponentials calculated in this way are shown in Table I for comparison. This approximation is less reliable (e.g., see the Ag/Ag(100) surface vacancy diffusion result), but it saves constructing and diagonalizing the matrix of second derivatives. Simple TST also has the disadvantage that the mode along which the force constant should be calculated is not unambiguously defined for complicated reaction mechanisms such as the fcc(100) adatom exchange event. Taking a constant preexponential of, say, 5×10^{12} Hz may be preferable as a less biased approximation, although the true ion, although the true exponentials vary over a factor of ten or more.

The form of the Vineyard approximation merits further comment. Note that the only temperature dependence is in the exponential, and depends only on the static (i.e., T=0) barrier height [17]. No correction is needed, say, to account for the extra potential energy that the system has as it passes over the saddle at a finite temperature. This, and all the entropy effects, cancel out in the integration over the normal modes, leaving the simple form of Eq. (2). Also note that Planck's constant h does not appear in Eq. (2). The kT/h preexponential sometimes found in final TST expressions is an artifact of incomplete evaluation of the partition functions involved, or a dubious approximation made along the way. TST is a classical theory, so h cannot remain when the integrals are all evaluated properly. Confusion about this expression, which introduces the wrong temperature dependence and an inappropriate physical constant, has persisted because kT/h at T=300K (6.2×10^{12} Hz) is coincidentally similar to a typical preexponential factor. a typical preexponential factor.

3 kinetic Monte Carlo

If the rate constants $\{k_{i\to j}\}$ are known for all n possible escape paths from the present state (i) of the system (i.e., if all the saddle points bordering state i can be found), it is easy to show that the average time before the system escapes to one of these n adjacent states is given by

$$\tau_{i\to}^{esc} = \left[\sum_{j}^{n} k_{i\to j}\right]^{-1}, \tag{3}$$

and the probability that the escape event leads to a particular state j is

$$p(i \to j) = k_{i\to j}/[\sum_{j}^{n} k_{i\to j}] = k_{i\to j}\tau_{i\to}^{esc}. \tag{4}$$

These two equations form the basis of the kinetic Monte Carlo (kMC) approach, which can be described quite simply. Starting in a given state (i), a new state (j) is chosen randomly according to the probabilities in Eq. (4) [18], the simulation clock is incremented by $\tau_{i\to}^{esc}$, and the system is moved to the new state. The rate constants are then recalculated (or updated) to account for all the possible escape paths from this new state, and the procedure is repeated. If one is interested in time scales that are not long relative to the average escape time in the system, the simulation clock should be advanced at each step not by $\tau_{i\to}^{esc}$, but rather by

$$\Delta t_i^{kMC} = -\tau_{i\to}^{esc} ln(r) \tag{5}$$

Table 1: Harmonic and exact transition state theory results for some metallic surface diffusion events using embedded atom method interatomic potentials. ν^{STST} indicates the simple-TST preexponential frequency, $\nu^{Vineyard}$ is the full harmonic (Vineyard) preexponential, k^{HTST} is the Vineyard rate constant evaluated at the temperature shown, summed over all the equivalent escape paths (n_{path}), and k^{TST} is the exact TST rate at that temperature. The lattice in each case is expanded to the quasiharmonic lattice constant for the temperature shown (except the Ag/Ag(100) edge-runner case, which is at the T=0 lattice constant).

Event	Figure	T (K)	ν^{STST} (10^{12}s^{-1})	$\nu^{Vineyard}$ (10^{12}s^{-1})
Ag/Ag(100) hop[a]	1c	400	2.7	4.87
Ag/Ag(100) exchange[a]	3, 4a	400	(2.7)	4.48
Ag(100) surface vacancy hop[b]	1b	300	3	26.1
Ag/Ag(100) edge runner[b]	1a	400	2.24	2.36
Ni/Ni(100) terrace[c]	-	500	4.79	8.11

Event	E_{static} (eV)	n_{path}	k^{HTST} (s^{-1})	k^{TST} (s^{-1})
Ag/Ag(100) hop[a]	0.542	4	2.8x10^6	3.0±0.2x10^6
Ag/Ag(100) exchange[a]	0.555	4	1.8x10^6	
Ag(100) surface vacancy hop[b]	0.531	4		
Ag/Ag(100) edge runner[b]	0.277	2	1.53x10^9	1.7±0.3x10^9
Ni/Ni(100) terrace[c]	0.744	2	5.14x10^5	5.0±0.1x10^5

[a] Ref. [45].
[b] This work.
[c] Ref. [44].

which is designed to give the correct exponential distribution of decay times. Here r is a random number on the interval [0,1]. As shown by Gillespie [19], this simple stochastic procedure captures the exact dynamical state-to-state evolution for a given set of rate constants [20]. Moreover, we can make an even stronger statement. If TST is exact for the system under study, then kinetic Monte Carlo based on TST rate constants gives dynamical evolution that is indistinguishable from a long classical trajectory, provided we only inquire about the *state* of the system at any given time, and not the position within a state. Even though the classical equations of motion are deterministic, the intrinsic chaotic instability of a classical trajectory guarantees that all memory of the preceding state is lost within a few picoseconds of entering a new state;

the trajectory thus "chooses" the next transition event just as randomly as the kMC procedure does [21].

The real utility of kMC for studying surface growth comes after making two additional assumptions. The first is that the system can be mapped can be mapped onto a lattice; i.e., for each state of the system, the position of each atom can be indexed by its proximity to a particular atomic site on a lattice derived from the unperturbed bulk crystal. The second assumption is that transition events are local (e.g., the hop of an adatom), and that the rate constants depend only on the local environment. With these assumptions, a catalog of rate constants can be computed for all possible local environments influencing a reactive event [22]. Because a transition only changes the system locally, the rate list needed for Eqs. (3) and (4) can be updated quickly after each transition with a few lookups in the rate catalog, even if the total number of rates in the list is very large. The general problem of finding the escape paths from each state, (which is intractable for more than about 10 atoms) is thus reduced to a manageable amount of work, although precomputation of the rate catalog might be computationally intensive. Inclusion of deposition events is also straightforward, by placing new adatoms randomly on surface lattice sites at proper time intervals on the simulation clock. It is thus not surprising that TST-based kinetic Monte Carlo now enjoys considerable popularity as a simulation tool for surface growth processes (e.g., see [23,24,25,26,27,28,29]).

As an example, consider the dynamics of Ag adatoms on the Ag(100) surface. A catalog of rate constants for an adatom hop in the <110> direction can be constructed using the local environment shown in Fig. 2. In the Vineyard approximation to TST, this requires a search for 2^{10} saddle points and minima (some equivalent by symmetry), corresponding to the various occupations of the 10 sites neighboring the initial and final adatom binding sites. A few representative barrier heights from such a catalog [7] are given in Fig. 1. In this case, an embedded atom method (EAM) interatomic potential was employed, the top layer of substrate atoms were allowed to relax in the molecular statics searches (the barriers change slightly if more substrate layers are relaxed), and the crystal was held at the zero-temperature lattice constant. Qualitative aspects of the overlayer dynamics can be gleaned from these barriers. For example, the fastest event is "edge-running" of an atom along a step (Fig. 1a). (Edge-running accounts for most of the events in a kMC simulation with this rate catalog.) Vacancies (Fig. 1b) diffuse faster than adatoms (Fig. 1c), especially if the ~5x larger preexponential for vacancy diffusion is considered (see Table I), and small clusters that form stable blocks (e.g., Fig. 1d) diffuse more slowly than their counterparts with one more, or one fewer, atoms (e.g., Fig. 1e). Armed with this catalog, kinetic Monte Carlo simulations can be per-

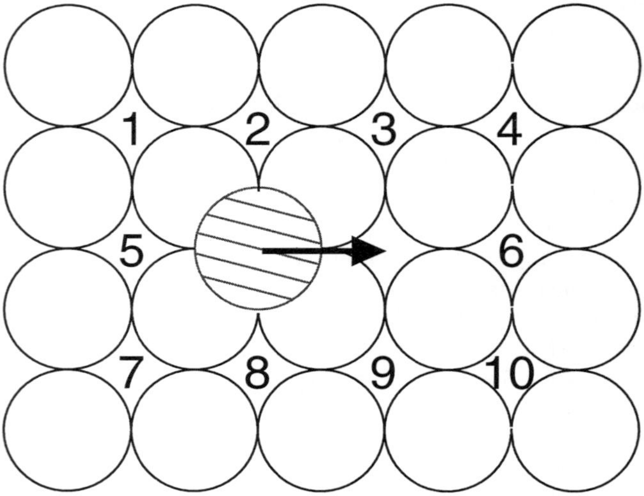

Figure 2: Local environment considered for computing the rate catalog for adatom hopping on the Ag(100) surface. Each of the sites is either empty or occupied, leading to 2^{10} rates total, ignoring symmetry.

formed ulations can be performed to study diffusion in an Ag(100) overlayer (e.g., this rate catalog shown was employed to study cluster diffusion constants as a function of cluster size[7].) To study surface growth, additional rates would need to be computed, corresponding to the Ehrlich-Schwoebel barrier [30] for descent of an adatom at a step, and other interlayer mechanisms. Specific results will not be discussed here, because a different point is of interest.

Implicit in the use of kinetic Monte Carlo is the assumption that all possible reactive events are known (or at least all those with low enough barriers to be important at the temperature of interest). If some rates are missing from the list, the predicted dynamics are incorrect. In recent years, a number of elementary processes have been discovered that cast doubt on the assumption that finding all the important processes is simply a matter of carefully applying physical intuition. For example, prior to 1990, adatom diffusion on the atomically smooth fcc(100) face was assumed to proceed solely by the hopping mechanism discussed above. roceed solely by the hopping mechanism discussed above. Feibelman's discovery of the substrate-exchange diffusion pathway [31] (shown in Fig. 3), verified experimentally to be the preferred event for Pt[32] and Ir[33], illustrates dramatically that surprises can occur even in systems thought

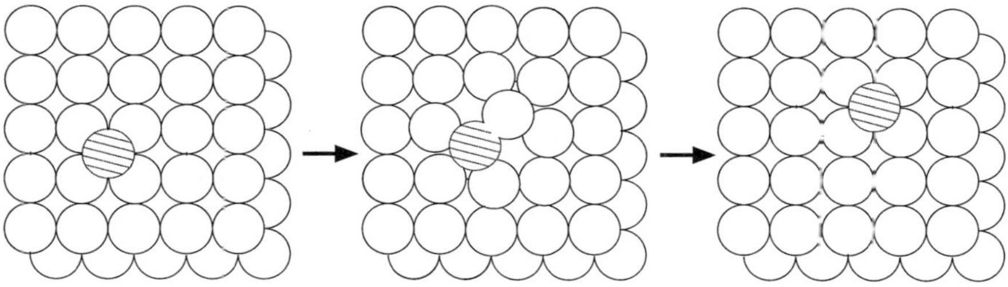

Figure 3: Adatom exchange event on fcc(100) surface. a) initial minimum; b) saddle point; c) final minimum. The initial adatom is shaded for clarity.

to be well understood. Exchange events are now known to be important in many situations.[34,35,36,25,37,38].

Thus, even for the relatively simple case of diffusion in a single Ag(100) overlayer, with no deposition or interlayer activity, key rate processes are missing from the catalog discussed above. Although the barrier for adatom exchange on Ag(100) is nominally higher than for hopping (0.60 eV versus 0.48 eV, fully relaxed at the T=0 K lattice constant [39]), exchange becomes competitive as the lattice constant increases with temperature. At the T=400K quasiharmonic lattice constant, the exchange barrier is 0.56 eV while the hopping barrier is 0.54 eV. One could augment the hopping rate catalog with an analogous set of rates for all the local environments of an exchange event (e.g., see Fig. 4, events a-e). Even then, however, the catalog would be incomplete. For example, long-distance exchange events have been observed [38,40], as indicated in Fig. 4f, and small, block clusters gain mobility via "dimer-shearing" moves [41], as shown in Fig. 4g. On fcc(111) surfaces, even more interesting mechanisms appear. Whole clusters can move as a unit in a concerted fashion [42], a mechanism also observed to occur spontaneously for larger clusters during deposition [6], and arbitrarily large clusters can move by passing dislocations [43]. (Examples of these events are shown in the next section.) Although, in principle, each of these mechanisms could be included in a kMC rate catalog (although with great difficulty in some cases), we must wonder what new mechanisms remain to be discovered. Also problematic for kinetic Monte Carlo are morphologies that defy simple mapping onto a lattice, such as if a bulk dislocation grows into the film, or if etching creates an amorphous region.

The conclusion is that although kinetic Monte Carlo is an extremely powerful approach, it should be applied with great care, and predictions should be extracted cautiously. In view of these limitations, it seems relevant to consider ways to extend the time scale of MD without requiring that mechanisms be known in advance. One such method is described next.

4 Hyperdynamics

The concept of the hyperdynamics, or hyper-MD, method [44,45] is to incorporate advantages of both MD and TST into a single simulation procedure. As in normal MD, a trajectory propagates in continuous space, with no propagates in continuous space, with no assumptions about the nature of transitions that might occur, simply "finding" a dividing surface occasionally, and passing through it to another state. However, the state-to-state transitions occur more rapidly than in normal MD because the trajectory is run on a potential surface that is biased to raise the energy within the basins without affecting the transition state dividing surfaces. This is illustrated schematically in Fig. 5. As discussed below, a bias potential ($\Delta V_b(x)$) with this characteristic can be constructed even if the escape paths are not known in advance. Assuming that the system obeys TST both before and after applying the bias potential (and assuming the bias potential does not "trap" the trajectory in a way that blocks rapid ergodic sampling within a state), this simulation procedure has some remarkable properties. Because the bias potential is zero along the dividing surfaces, the relative escape rate to different adjacent states is preserved. As a consequence, the trajectory propagates from state to state in a sequence that is indistinguishable from the exact dynamics. The time scale, which is accelerated in a nonlinear fashion (the bias potential will, in general, increase the escape rates more for some states than for others) can be recovered from an extremely simple procedure. At each integration step during the trajectory, the hyper-MD (or "boosted time") clock is advanced by the normal MD time step (Δt_{MD}) multiplied by the inverse Boltzmann factor for $\Delta V_b(x)$ at that point in configuration-space, i.e.,

$$\Delta t_b(t) = \Delta t_{MD}\ exp(\Delta V_b(x(t))/k_B T). \tag{6}$$

This definition for the boosted time converges on the exact time as the simulation runs, even though the system may escape from each state long before adequate sampling of the bias potential is obtained, and even if every state of the system is different [44].

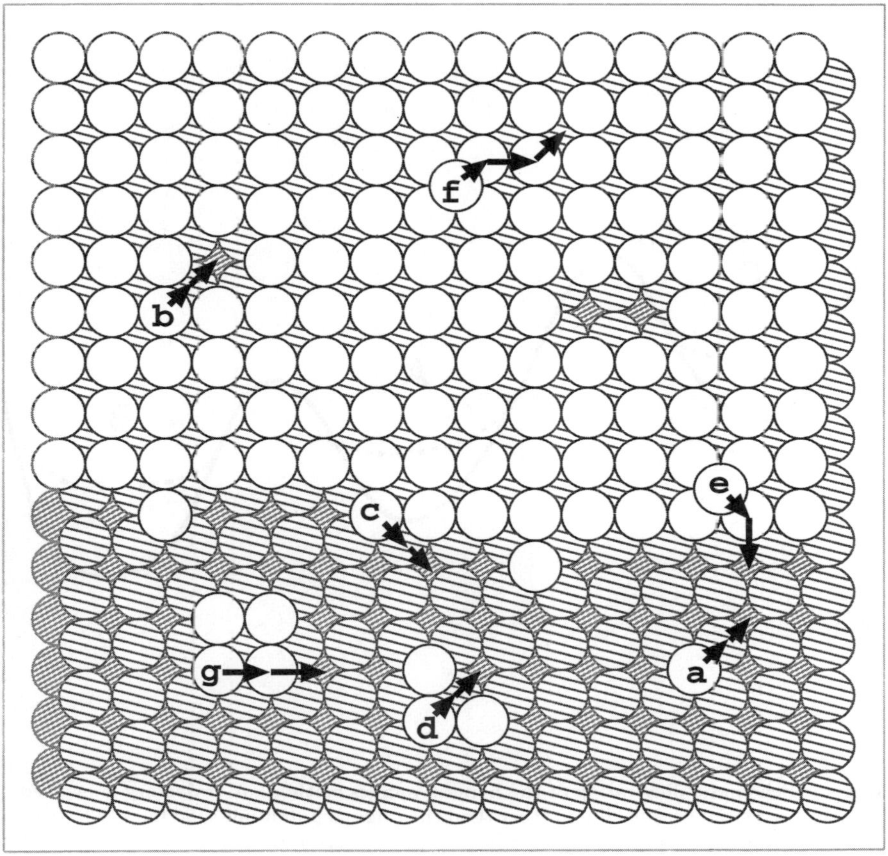

Figure 4: Fcc(100) surface during growth, indicating some of the concerted events omitted from the simple hopping picture in Fig. 1, as discussed in text. The substrate exchange event in (a) is shown in detail in Fig. 3. Some barriers at the T=0 K lattice constant are: (a) 0.60 eV, (d) 0.7 eV, (e) 0.55 eV, and (g) 0.69 eV.

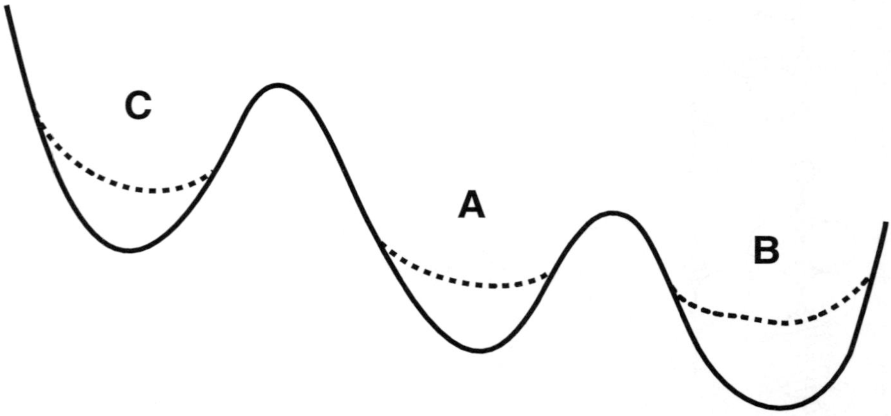

Figure 5: Schematic illustration of the hyperdynamics method. A trajectory is evolved on the biased potential (dashed line) rather than the original potential (solid line). The relative escape probabilities from a given state (e.g., $k_{A\to B}/k_{A\to C}$) are preserved because the bias potential is zero at the dividing surfaces. The nonlinear acceleration of the time scale is calculated as the simulation proceeds.

This formalism is only useful if it is possible to design a computationally efficient bias potential that is zero at the dividing surfaces without advanced knowledge of their location. For an N-atom system, a viable approach is to base $\Delta V_b(\mathbf{r})$ on local properties of the potential $V(\mathbf{r})$ via the gradient vector, $\mathbf{g}$ ($g_i \equiv \partial V(\mathbf{r})/\partial x_i$, where x_i is a component of the $3N$-dimensional configuration-space vector $\mathbf{r}$), and Hessian matrix, $\mathbf{H}$ ($H_{ij} \equiv \partial^2 V(\mathbf{r})/\partial x_i \partial x_j$). Although the definition of the best TST dividing surface between two states is inherently nonlocal (requiring steepest-descent trajectories to determine the state to which each $\mathbf{r}$-point relaxes), Sevick, Bell, and Theodorou[46] have proposed a local definition: the set of points in $\mathbf{r}$-space for which $\epsilon_1 < 0$ and $g_{1p} = 0$, where $g_{1p} = \mathbf{C}_1^\dagger \mathbf{g}$ is the projection of $\mathbf{g}$ onto the lowest eigenvector ($\mathbf{C_1}$) of $\mathbf{H}$. Near the saddle points this is an excellent approximation to the best TST surface, and we know from the accuracy of the Vineyard approximation that the saddles are the most important regions. These same quantities ($\epsilon_1(\mathbf{r})$ and $g_{1p}(\mathbf{r})$) can be used to construct a good definition for $\Delta V_b(\mathbf{r})$. The idea is to design a function that smoothly turns on in regions that do not satisfy the approximate dividing surface condition. An estimate of the lowest barrier in the system sets the scale for the maximum value the bias potential can reach (although nothing in the formalism prevents $\Delta V_b(\mathbf{r})$ from exceeding the barrier energy at some points points in $\mathbf{r}$-space, provided that the trajectory can find ways around these maxima to rapidly sample the whole state). The details of the form for $\Delta V_b(\mathbf{r})$ used in simulations so far are discussed elsewhere[45]. Although the bias potential in this approach depends formally on the second derivatives of $V(\mathbf{r})$ (and the forces neeeded for MD would thus depend on the third derivatives), an iterative algorithm[45] allows propagation of the hyper-MD trajectory using only first derivatives, just as for normal MD.

Although the hyperdynamics method is still in a developmental stage, preliminary results[44,45,47] have demonstrated that it can extend the MD time scale significantly, and gives the correct results if the bias potential is chosen properly. For example, in a hyperdynamics simulation of Ag adatom diffusion on Ag(100) at T=400K, both hopping and exchange events were observed, and the average boost factor (the final boosted time divided by the final MD time) was 1356. The extra work per computational step (compared to normal MD) was about 30 times. Thus, the same simulation (9.89 μs total boosted time) performed with direct MD would have required $1356/30 = 45$ times more computational work. The rate constants estimated from the observed number of events (23 hops, 16 exchanges) agreed with the exact values (see the first two lines of Table I). For that simulation, the known barrier heights were consulted to choose the strength of the bias potential[45].

Hyper-MD has also been applied to diffusion of a 10-atom Ag cluster on the Ag(111) surface [45]. Using a bias potential with parameters similar to that employed in the Ag/Ag(100) case, and allowing two substrate layers to move (for a total of 70 moving atoms), a boost factor of 8310 was obtained at a cost of about 20 times the work per integration step. In a few weeks of workstation time, the system evolved for t_b=221.2 μs. The nonlinear evolution of the boosted time is shown in Fig. 6. The cluster did not move far in this time (see Fig. 7), but a number of complicated, concerted mechanisms were observed. For example, At 9.5 μs, all 10 atoms jumped in the same direction to move the cluster from hcp to fcc registry (Figs. 7a and 7b), a mechanism mentioned in the last section. This occurred many times, with the jump direction essentially always perpendicular to the long axis of the cluster, for reasons discussed previously [45]. At 94.5 μs, the cluster broke out of its compact configuration, via a mechanism analogous to the dimer-shearing on fcc(100) [41], allowing adatom motion at the cluster periphery (Figs. 7c-7f). At 197.8 μs (Figs. 7g-7i), there was a dislocation event [43], resulting in a rotation of the cluster orientation and a change from hcp to fcc registry. Only one of these occurred in the simulation.

This example serves to illustrate the generality of the hyperdynamics approach and the time scales accessible with it. While for this system, all the observed events (and perhaps their relative frequencies) could have been predicted based on extrapolation downward from direct MD simulations at higher temperatures, such an extrapolation becomes impossible for a system as complex as a growing surface. Hyper-MD should be helpful for understanding these more complicated systems, on time scales approaching those that occur between deposition events. With further development, direct simulation of realistic deposition rates may be possible in certain cases.

The characteristics of a hyper-MD run depend intimately on the definition of the bias potential and its interaction with the system under study, and appear hard to predict *a priori*, although this should change as more experience is gathered. Properties of interest include the average boost factor, the change in boost factor as the system evolves through different states, and the possible inaccuracies caused by partial blocking of some pathways (i.e., if $\Delta V_b \neq 0$ along part of a TST surface) or the introduction of correlated crossing events. While correlated crossings violate the requirements of the derivation, a small fraction of recrossing events do not cause significant inaccuracies in the predicted dynamics [44]. Similarly, minor blocking of the transition state region may be acceptible, although for a given bias potential, the resulting dynamical error may increase as the temperature is lowered.

The average boost factor in a hyperdynamics simulation increases rapidly as the temperature is decreased, as expected from its exponential form. The

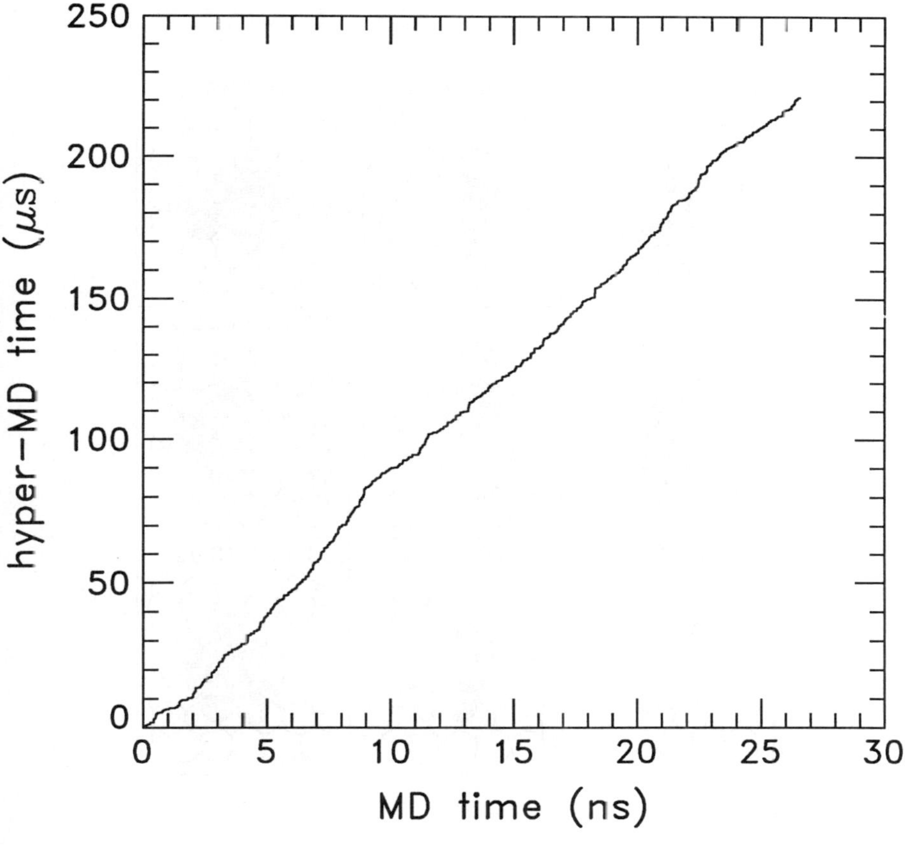

Figure 6: Evolution of the cumulative boosted time (t_b) during the hyper-MD simulation of a 10-atom Ag cluster on Ag(111). The average boost factor is 8310.

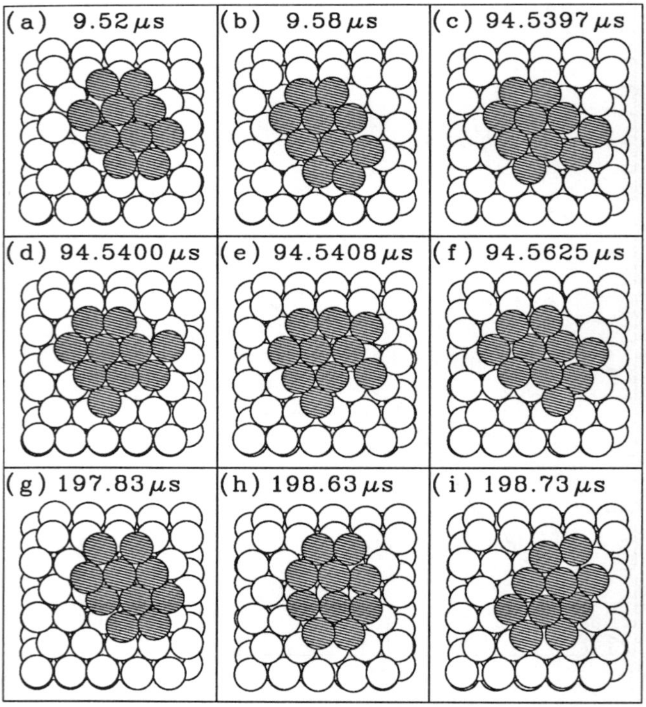

Figure 7: Snapshots from Ag_{10}/Ag(111) hyper-MD simulation, discussed in text. Boosted time values (t_b) are shown.

effective simulation rate (events per cpu time) scales as the true rate constant multiplied by the average boost factor, divided by the extra work of hyper-MD relative to normal MD per integration step. In test runs on Ag/Ag(100), the boost factor was found to increase faster than exponentially with $1/T$ (presumably because at lower T, the system spends more time in regions where the bias potential is high). Because rate constants decrease exponentially with $1/T$, this offers the tantalizing possibility that below a certain temperature, the effective simulation rate actually levels off, or even *increases* as the temperature is lowered further. For fast classical potentials, this temperature range should be accessible on computers available a few years from now.

5 Conclusions

Direct molecular dynamics has the virtue of yielding the exact dynamical evolution of a system. The burden is on the interatomic potential if results do not agree with experiment, and potentials keep getting better every year. However, for times longer than 10^{-7} s or so, direct MD is no longer possible. The jumping off point for longer time scales is transition state theory, in which the rate constant for an infrequent event is calculated directly, rather than waiting for it to occur in a long simulation.

TST rates can be used as the input for the lattice-based, kinetic Monte Carlo method, which accesses much longer time scales because each simulation step is a reactive event. The problem with kMC is that errors can result from not knowing all the possible transitions. The discovery of a variety of novel mechanisms in the last few years, both on metal and semiconductor surfaces, makes this a real concern.

The new hyper-MD approach may offer a useful alternative to both direct MD and kMC. Working in continuous space, making no assumptions about lattice mapping, and requiring no specific knowledge about the existing saddle points, hyper-MD simulations can extend the time scale of direct MD by a few orders of magnitude.

Both kMC and hyper-MD tend to be limited by the time scale of the fastest event (e.g., edge running on the periphery of a cluster, or diffusion of a free adatom on the (111) face). Overcoming this effect in a general way, without the possibility of corrupting the long-time dynamics, remains a problem for future research.

6 Acknowledgements

I am grateful for helpful discussions with J.D. Kress, F. Houle, and H. Jònsson, and for support from the Laboratory Directed Research and Development program at Los Alamos National Laboratory.

References

1. B.J. Garrison and D. Srivastava, Ann. Rev. Phys. Chem. **46**, 373 (1995).
2. A.F. Voter, *MRS Bulletin*, **21**, 17 (Materials Research Society, 1996).
3. M.S. Daw, S.M. Foiles, and M.I. Baskes, Mater. Sci. Reports **9**, 251 (1993).
4. F.A. Stillinger and T.A. Weber, Phys. Rev. B **31**, 5262 (1985).
5. C.M. Gilmore and J.A. Sprague, Phys. Rev. B **44**, 8950 (1991).
6. M. Villarba and H. Jónsson, Phys. Rev. B **49**, 2208 (1994).
7. A.F. Voter, in *Modeling of Optical Thin Films*, M.R. Jacobson, Ed., Proc. SPIE, **821**, 214 (1987).
8. R. Marcelin, Ann. Phys. **3**, 120 (1915).
9. E. Wigner, Z. Phys. Chem. **B 19** 203, (1932).
10. H. Eyring, J. Chem. Phys., **3**, 107 (1935).
11. C.H. Bennett, in *Algorithms for Chemical Computation*, edited by R.E. Christofferson (American Chemical Society, Washington, DC, 1977), p. 63.
12. D. Chandler, J. Chem. Phys., **68**, 2959 (1978); J.A. Montgomery, Jr., D. Chandler, and B.J. Berne, J. Chem. Phys. **70**, 4056 (1979).
13. A.F. Voter and J.D. Doll, J. Chem. Phys. **82**, 80 (1985).
14. A.F. Voter, J. Chem. Phys. **82**, 1890 (1985).
15. G.H. Vineyard, J. Phys. Chem. Solids **3**, 121 (1957).
16. M. Toller, G. Jacucci, G. DeLorenzi, and C.P. Flynn, Phys. Rev. B **32**, 2082 (1985).
17. Of course, an additional T dependence is introduced if the quasiharmonic method is employed to give a different lattice constant (and hence different barrier height and preexponential) at each temperature, but we are considering a fixed lattice constant case here.
18. A.B. Bortz, M.H. Kalos, and J.L. Lebowitz, J. Comp. Phys. **17**, (1975) 10.
19. D.T. Gillespie, J. Comp. Phys. **22**, 403 (1976).
20. Alternative methods have been explored for advancing the simulation clock at a faster rate by allowing multiple events per time step. For example, see P-L. Cao, Phys. Rev. Lett. **73**, 2595 (1994), and references therein. Care must be taken in this kind of approach to not allow attempted moves by two interacting events, or the dynamics are incorrect.)

21. Of course, we also require that this memory loss occurs between transitions, by stating that TST holds.
22. A.F. Voter, Phys. Rev. *B* 34, 6819 (1986).
23. A.P. Smith and J. Jónsson, Phys. Rev. Lett. **77**, 1326 (1996).
24. D.J. Dawnkaski, D. Srivastava, and B.J. Garrison, J. Chem. Phys., **104**, 5997 (1996).
25. J. Jacobsen, K.W. Jacobsen, P. Stoltze, and J.K. Norskov, Phys. Rev. Lett. **74**, 2295 (1995).
26. S. Liu, Z. Zhang, G. Comsa, and H. Metiu, Phys. Rev. Lett. **71**, 2967 (1993).
27. Y.T. Lu, Z.Y. Zhang, and H. Metiu, **257**, 199 (1991).
28. J.M. McCoy and J.P. LaFemina, Phys. Rev. B **50**, 17127 (1994).
29. M. Breeman, G.T. Barkema, M.H. Langelaar, and D.O. Boerma, Thin Solid Films **272**, 195 (1996).
30. G. Ehrlich and F.G. Hudda, J. Chem. Phys. **44**, 1039 (1966); R.L. Schwoebel J. Appl. Phys. **40**, 614 (1969).
31. P.J. Feibelman, Phys. Rev. Lett **65**, 729 (1990).
32. G.L. Kellogg and P.J. Feibelman, Phys. Rev. Lett. **64**, 3143 (1990).
33. C. Chen and T.T. Tsong, Phys. Rev. Lett.,**64**, 3147 (1990).
34. P. Stoltze and J.K. Norskov, Phys. Rev. B **48**, 5607 (1993).
35. R. Wang and K.A. Fichthorn, Molec. Sim. **11**, 105 (1993).
36. M. Villarba and H. Jónsson, Surf. Sci. **317**, 15 (1994).
37. G.L. Kellogg and A.F. Voter, Phys. Rev. Lett. **67**, 622 (1991).
38. J.M. Cohen, Surf. Sci. Lett. **306**, L545 (1994).
39. C.L. Liu, J.M. Cohen, J.B. Adams, and A.F. Voter, Surf. Sci., **253**, 334 (1991).
40. J.E. Black and Z-J. Tian, Phys. Rev. Lett. **71**, 2445 (1993).
41. Z-P. Shi, Z. Zhang, A.K. Swan, and J.F. Wendelken, Phys. Rev. Lett. **76**, 4927 (1996).
42. S. Liu, Z. Zhang, J. Norskov, and H. Metiu, Surf. Sci. **321**, 161 (1994).
43. J.C. Hamilton, M.S. Daw, and S.M. Foiles, Phys. Rev. Lett. **74**, 2760 (1995).
44. A.F. Voter, J. Chem. Phys. **106**, 4665 (1997).
45. A.F. Voter, Phys. Rev. Lett. **78**, 3908 (1997).
46. E.M. Sevick, A.T. Bell, and D.N. Theodorou, J. Chem. Phys. **98**, 3196 (1993).
47. A.F. Voter, in *Novel Materials*, edited by B. K. Rao and S. N. Behera (Nova Scientific, New York, 1997), in press.

SUBMONOLAYER NUCLEATION AND GROWTH OF 2D ISLANDS AND MULTILAYER MOUND FORMATION DURING HOMOEPITAXY

J.W. Evans[1] and M.C. Bartelt[2]

[1]*Department of Mathematics and Ames Laboratory, Iowa State University, Ames, IA 50011*

[2]*Computational Materials Science Department, Sandia National Laboratories, Livermore, CA 94550*

Abstract

Submonolayer nucleation theory is reviewed noting the limitations of traditional mean-field treatments. We describe recent advances in the characterization of island size distributions, the transition to reversible island formation, island shape selection, island coalescence and percolation, island spatial correlations and ramifications for diffraction profiles. We also describe the transition to multilayer growth in systems with step-edge barriers, relating multilayer "mound" structure to the submonolayer island morphology.

1. Introduction

A general feature of homoepitaxy is that the configuration of the growing film typically cannot relax to equilibrium on the time scale of deposition. Thus, a variety of far-from-equilibrium morphologies are produced. In the submonolayer regime, a distribution of "small" two-dimensional (2D) islands is formed, in contrast to the equilibrium state which consists of a single large region of a 2D condensed phase coexisting with a very dilute 2D "gas" phase (below a critical temperature for phase separation).[1] Also, the shapes of these individual islands can differ dramatically from their equilibrium forms. For multilayer growth, kinetic roughening of the growing film is observed even below the transition temperature for thermal roughening (where the equilibrium state corresponds to a smooth surface). Various film morphologies are possible including self-affine structures, and ensembles of "mounds".[2]

Submonolayer nucleation and growth of epitaxial islands during deposition has been studied for decades.[1] The key features are as follows. Atoms are deposited randomly with flux F, on the substrate which provides a periodic array of adsorption sites. Adatoms typically undergo diffusion between such sites at a rate determined by the substrate temperature, T, leading to nucleation and subsequent growth of islands. Experimental results for island density behavior have traditionally been obtained for heteroepitaxial systems via SEM and TEM, and were interpreted using mean-field rate equation theories. The dependence on flux provides information on the nucleation mechanism, and specifically on the critical size, i, above which islands are effectively stable. Arrhenius behavior then provides information on the energetics of the underlying atomistic processes.[1]

Recent STM and high-resolution diffraction studies have provided detailed data on island formation for both homoepitaxy[3] and heteroepitaxy.[4] The former provides an ideal testing ground for theory, which has received renewed scrutiny. Also, recently, kinetic Monte Carlo (KMC) techniques have been applied to analyze the behavior of various models of nucleation and growth. These models might prescribe the critical size or bond-breaking rules, as well as specifying island structure (or restructuring dynamics).[5-8] Alternatively, they might prescribe activation barriers for all relevant atomic hopping processes, either via simple rules,[9] or based on (approximate) values from suitable theories.[10,11] KMC not only allows determination of "exact" behavior of key quantities for these models, but also allows for examination of a variety of quantities characterizing the adlayer morphology that are not readily accessible in mean-field treatments.[6]

Clearly, the morphology of the growing multilayer film will to some degree reflect submonolayer island structure. This is most apparent in the presence of a significant additional step-edge barrier to downward transport at island edges.[12] Here multilayer growth is "unstable", and characterized by the formation of mounds which are built upon the "template" provided by submonolayer islands.[13] Analytic theories have recently been extended to estimate the population of higher layers near the onset of multilayer growth,[14,15] but KMC simulation is more effective in providing a detailed characterization of all aspects of the film morphology.[16-20] The latter investigations have also exploited ideas from generic theories of "kinetic roughening" developed over the last decade.[2]

In this contribution, we consider explicitly homoepitaxial growth, although many of the ideas also apply for heteroepitaxy. We primarily review recent developments in submonolayer nucleation theory, where comprehensive understanding of several basic issues is just emerging. We also consider more briefly the transition to multilayer growth for systems with a step-edge barrier, emphasizing the relationship between the morphology of submonolayer islands and multilayer "mounds". Most issues are illustrated with results from a "canonical model" for metal(100) homoepitaxy involving formation of near-square islands in each layer.[6,18] This was the first realistic model for which "exact" behavior was determined by comprehensive simulation analyses. We refer to experimental results throughout, contrasting metal(100) and metal(111) homoepitaxy.

2. Theory of Submonolayer Nucleation and Growth

Here, we consider only the formation of 2D islands in isotropic systems, assuming that stable islands are effectively immobile. Other cases are discussed in the Appendix.

Rate Equations

Let N_s denote the density of islands of s adatoms. Thus N_1 is the density of isolated diffusing adatoms, $N_{av}=\Sigma_{s>1}N_s$ is the average island density, and $\theta=A\Sigma_{s\geq 1}sNs$ is the coverage, where A is the area per site. Note that the average island separation and size satisfy $L_{av}=N_{av}^{-1/2}$ and $s_{av}=(\theta-N_1)/N_{av}\approx\theta/N_{av}$, respectively. The diffusion coefficient for isolated adatoms is written as $D=D_0\exp[-E_d/(k_BT)]$, where E_d is the activation barrier. Then the rate at which diffusing adatoms aggregate with an island of size s is denoted by[1] $R_{agg}(s)=D\sigma_sN_sN_1$, *defining* σ_s as the "capture number" for islands of size s. Alternatively, one might replace N_1 in $R_{agg}(s)$ by the local density of diffusing adatoms, $N_1{}^*$, thus rescaling the definition of σ_s. The rate at which an island of size s directly captures atoms deposited on top of or adjacent to itself is denoted by $F\kappa_sN_s$. [For compact islands, $\kappa_s\approx sA$ for large s, in the pre-coalescence regime and in the absence of a large step-edge barrier, where essentially all such deposited atoms are incorporated; also $N_1{}^*=N_1/(1-\theta)$.] Finally, the rate at which atoms escape from an island of size s is written as η_sN_s. Then one has[1]

$$dN_1/dt = F(1 - 2\kappa_1N_1 - \Sigma_{s>1}\kappa_sN_s) - 2R_{agg}(1) - \Sigma_{s>1}R_{agg}(s) + 2\eta_2N_2 + \Sigma_{s>2}\,\eta_sN_s,$$
$$dN_{s>1}/dt = F(\kappa_{s-1}N_{s-1} - \kappa_sN_s) + R_{agg}(s-1) - R_{agg}(s) - \eta_sN_s + \eta_{s+1}N_{s+1}. \qquad (1)$$

Note that for diffusion between nearest neighbor (NN) sites on a square lattice, one has $D=a^2h/4$, where h is the total hop rate out of a site, and 'a' is the surface lattice constant. One often specifies densities per site, so $A=a^2\rightarrow 1$, and F is in ML per unit time. We anticipate solutions to (1) satisfying a scaling ansatz, $N_s\sim\theta(s_{av})^{-2}f(s/s_{av},\theta)$, for large s_{av}, with 's' corresponding to stable islands.[5,6] This form implies *similarity* of island distributions for fixed θ, and various large s_{av}.

Critical Sizes and Classic Island Density Scaling

A simplified analysis of (1) neglects escape from islands above some critical size, i, and furthermore assumes quasi-equilibrium for islands of size $s\leq i$.[1] The latter implies the Walton relation,[1] $N_i\approx c(\theta)\exp[E_i/(k_BT)](N_1)^i$, where $E_i>0$ is the binding energy for "critical clusters" of i atoms. Retaining only dominant terms, one has for compact islands

$dN_1/dt \approx F(1 - \theta) - D\sigma_{av}N_{av}N_1$, and

$$dN_{av}/dt \approx D\sigma_i N_i N_1 \approx k(\theta)\, D\sigma_i \exp[E_i/(k_BT)](N_1)^{i+1}, \tag{2}$$

where $\sigma_{av} = \Sigma_{s>i}\sigma_s N_s/\Sigma_{s>i}N_s$ is the average capture number for stable islands. These equations can be integrated given the form of σ_{av} (and σ_i), which is discussed in detail below. One also anticipates an initial "transient regime" of increasing $N_1 \approx Ft$, followed by a "steady-state regime" in which the gain in N_1 due to deposition is roughly balanced by the loss due to aggregation. This balance yields the steady-state condition $F(1-\theta) \approx D\sigma_{av}N_{av}N_1$. Substituting into the N_{av}-equation, and assuming that σ_{av} (and σ_i) depend only on θ (as shown below), one obtains the central result of classic nucleation theory[1]

$$N_{av} \sim K(\theta)(F/D_0)^{\chi}\exp[E/(k_BT)], \text{ where } \chi = i/(i+2) \text{ and } E = \chi(E_d + E_i/i). \tag{3}$$

Thus the F-dependence of N_{av} yields χ, and thus i, and its Arrhenius behavior yields E (and in particular, one obtains E_d if i=1, where E_1=0).

Mean-field (MF) Theory of Capture Numbers

We first treat *irreversible* capture (i=1). The strategy here is to develop a diffusion equation for the density, $N_1(\mathbf{r})$, of adatoms at a displacement $\mathbf{r}$ from the center of a "typical" island of size s. Thus, one has N_1(perimeter)=0, since the island perimeter is a perfect sink for diffusing adatoms, and $N_1(\mathbf{r}) \to N_1$, as $r=|\mathbf{r}| \to \infty$. This analysis is only viable with the key MF assumption that the environment of the island is independent of its size, and with a simplistic "effective medium" treatment of the background of other islands via

$$dN_1(\mathbf{r})/dt = D\nabla^2 N_1(\mathbf{r}) + F(1 - \theta) - D\xi^{-2} N_1(\mathbf{r}). \tag{4}$$

Here, the characteristic diffusion length satisfies $\xi \approx \sigma_{av}^{-1/2}L_{av}$ [cf. (2)]. The steady-state solution of (4) reveals a *depletion zone* surrounding the island whose width scales like ξ, and shows that $\sigma_s = N_1^{-1}\int_{perim} dl\ \mathbf{n}\cdot\nabla N_1(\mathbf{r}) = \sigma(R_s/\xi)$ is a function only of the *ratio* of the linear island dimension, R_s, and ξ. Based on the explicit form of the solution for circular islands, one expects that in general $\sigma(y) \sim 1/|\ln(y)|$, as $y \to 0$, and $\sigma(y) \sim y$, for large y.

This result for σ_s allows determination of σ_{av}, which however is required as input to (4), so there is a need for *self-consistent* analysis.[1,21] For compact islands, one has $s = cR_s^2$ and $\sigma_s = k(\sigma_{av}\theta s/s_{av})$, where $k(y) = \sigma(c^{-1/2}y^{1/2})$. This implies $\sigma_{av} \approx \int dx\ k(\sigma_{av}\theta x) f(x,\theta)$ which should be solved self-consistently. One finds that $\sigma_{av} = \sigma_{av}(\theta)$ depends only on θ, as stated above, and also that $\sigma_{av}(\theta) \sim 1/|\ln(\theta)|$, for small θ. For an alternative treatment of capture based on the properties of random walks (RW's), see the Appendix.[5]

Recently, the above analysis has been extended to treat *reversible* island formation without any prescribed critical size, i.[22] The key idea is the appropriate description for the diffusion equation (4) of a boundary condition at island edges. Furthermore, one must distinguish between the rate of *detachment*, ω_s, from islands of size s, and the associated *escape* rate, η_s. The former is a microscopic rate controlled by an effective barrier for detachment of $E_{diss}=E_d+\delta E_b$, where δE_b is the energy change on dissociation. The latter involves statistical factors,[23] and satisfies $\eta_s \propto \sigma_{s-1}\omega_s$. See also Sec.3.

Exact Behavior of the Capture Numbers

Here, it is useful to consider (Voronoi) tessellations of the surface, in which each island is associated with a cell in the tessellation (see Fig.1).[1] It has long been recognized that islands with larger cells should grow faster, thus producing a correlation between the size and local environment of islands.[1,24] This non-MF behavior was observed experimentally,[25] and in simulations.[25] However, what has been lacking until recently[26] is a characterization of the non-MF dependence on island size, s, of the cell areas, A_s, and of capture numbers, σ_s, and an appreciation that this size-dependence controls the behavior of the N_s.

To isolate the non-MF component of this s-dependence, it is natural to consider a point-island model, wherein islands occupy a single site but carry a size label.[5] Clearly, here MF capture numbers are strictly s-independent, in contrast to the case of compact islands. For i=1 (irreversible capture), Fig.1 shows simulation results for A_s and σ_s indicating a MF-type plateau in these quantities for smaller $s<s_{av}$, followed by a dramatic non-MF increase for $s>s_{av}$. This reveals a correlation between island size and separation: larger islands have larger cell areas, and are thus further separated from their neighbors. This behavior can be understood as follows. Initially, $A_{av}=\Sigma_s A_s N_s/N_{av}=1/N_{av}$ decreases quickly with t, which tends to produce A_s increasing with s (since larger islands were typically nucleated earlier, when A_{av} was larger). However, as deposition continues, areas for smaller islands are transferred to bigger islands simply by virtue of island growth. This tends to produce a plateau in A_s at the value of the cell area, A_{nuc}, for *just-nucleated* islands. A_{nuc} is less than A_{av} since just-nucleated islands must "fit between" existing islands. The selection of the non-MF behavior for $s>s_{av}$ is non-trivial, and influenced by preferential nucleation in larger cells associated with larger islands (see Ref. 26).

Detailed analysis reveals that the scaling forms $A_s/A_{av}\approx a(s/s_{av})$ and $\sigma_s/\sigma_{av}\approx C(s/s_{av})$ are almost independent of θ and (large) h/F, as is the scaled island size distribution, $f(s/s_{av})$. Normalization requires that $\int_0^\infty dx\, a(x)f(x)=\int_0^\infty dx\, C(x)f(x)=1$. Note that if all atoms deposited in a cell were to aggregate with the associated island, then clearly σ_s/σ_{av} would equal A_s/A_{av}. However, we find that

$\sigma_s/\sigma_{av} \approx (1-\alpha)A_s/A_{av}+\alpha$, where $\alpha \approx 0.3$, so Voronoi cells do not correspond exactly to "capture zones" for diffusion-mediated aggregation of adatoms with islands. Also, we emphasize that while the above results are for the artificial case of point-islands, basically the same behavior is found for compact islands, i.e., the form of A_s and σ_s versus s is controlled by correlations between island size and separation.

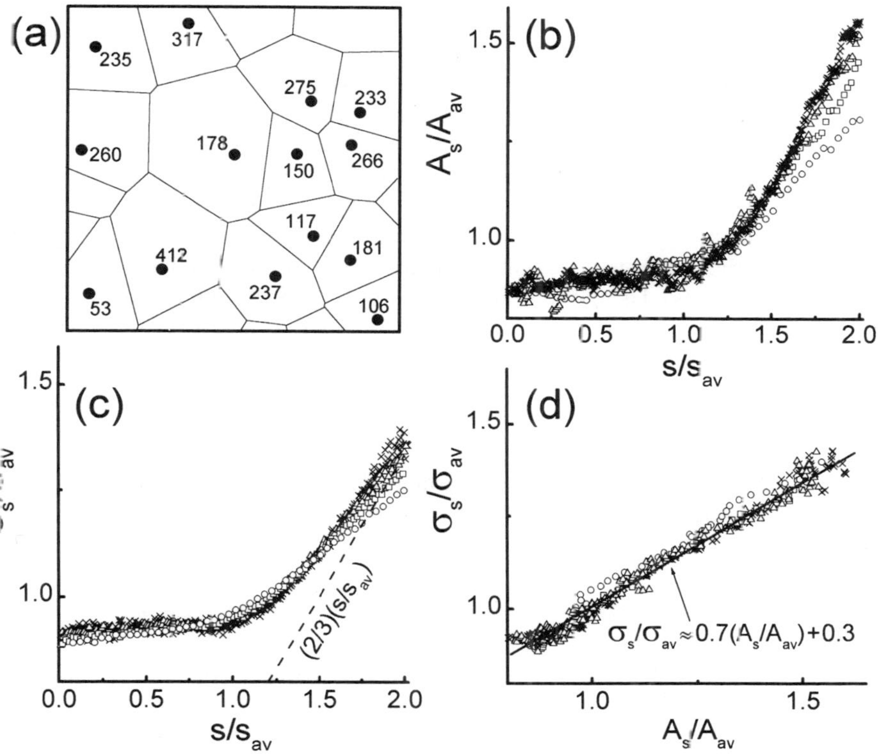

Figure 1: Simulation results for irreversible formation of point islands[5,26] at 0.2ML. (a) island distribution on a 120×120 lattice for $h/F=10^{10}$, and the associated Voronoi tessellation; (b) normalized Voronoi cell areas versus island size; (c) normalized capture numbers versus island size; and (d) quasi-linear capture number versus cell area relationship, for $h/F=10^6$ (circle), 10^7 (square), 10^8 (triangle), 10^9 (cross).

Bales[27] suggested developing rate equations for the A_s akin to those for the N_s. Consider the fractional area, A_sN_s, associated with islands of size s. Contributions to $d/dt(A_sN_s)$ come from the above mentioned transfer of areas due to island growth including, e.g., a gain term of the form $A_{s-1}R_{agg}(s-1)$. To conserve total area, $\Sigma_s A_sN_s=1$, one must also reduce the area associated with stable islands each time an area A_{nuc} is assigned to a just-nucleated island. We account for this effect with a loss term, $-A_{nuc}(dN_{av}/dt)P_s$, where P_s gives the probability that nucleation occurs in a cell of an island of size s, so $\Sigma_s P_s = 1$. Since P_s should scale with island population, we naturally write $P_s = Q_s N_s / N_{av}$. If P_s was simply proportional to the available area, one would set $Q_s = A_s / A_{av}$. Integration of these equations, assuming a suitable form for σ_s in terms of A_s, has some qualitative success (see below), but indicates that further refinements of the theory are needed.

Quasi-Hydrodynamic Analysis of the Scaling Regime

We apply a quasi-hydrodynamic approach of treating island size as a continuous variable[1] to elucidate the form of the island size distribution for large s_{av}. With the appropriate selection of σ_s, this technique reveals the exact form of the scaling function, f(x), for the island size distribution.[26] We restrict our attention to the low θ regime, and to irreversible capture. Then, the rate equations (1) become $dN_s/dt \approx -d/ds\, R_{agg}(s)$. Using the scaling forms of N_s and σ_s with the natural variable $x = s/s_{av}$, neglecting any explicit θ-dependence of C(x) and f(x), and noting that $F \approx D\sigma_{av} N_1 N_{av}$, one obtains[26]

$$(1-2\varpi) f(x) - \varpi x\, d/dx\, f(x) = -\, d/dx\, [C(x) f(x)], \quad (5)$$

where $\varpi = d(\ln s_{av})/d(\ln t)$ is assumed constant. (The effective value of ϖ ranges from 2/3 to unity, depending on the degree of saturation of N_{av}.) Integration yields the *key result*[26]

$$f(x) = f(0) \exp\{\textstyle\int_0^x dy\, [(2\varpi - 1) - dC(y)/dy]/[C(y) - \varpi y]\}. \quad (6)$$

In the MF theory, where C(y) is constant (for point-islands) or "slowly increasing" (for compact islands), there is an artificial singularity in f(x) at $x = x_s$, where $C(x_s) = \varpi x_s$. However, there is no singularity with the exact "rapidly increasing" C(y). See Sec.3.

The same type of analysis could be applied to the cell areas based on the rate equations $d/dt\,(A_s N_s) \approx -d/ds\,[A_s R_{agg}(s)] - A_{nuc}(dN_{av}/dt) P_s$. This produces a differential equation in the scaling variable, x, involving a(x), C(x), f(x), and a scaling function q(x), where we assume that $Q_s \approx q(s/s_{av})$. Note that $\int_0^\infty dx\, q(x) f(x) = 1$, and that $a(0) = A_{nuc}/A_{av}$ naturally appears below. Using (5) to eliminate f(x) from this equation, one finds that

$$[C(x) - \varpi x]\, d/dx\, a(x) = (3\varpi - 1) a(x) - (1 - \varpi) a(0) q(x). \quad (7)$$

Integration of (7), using reasonable forms for C(x) and q(x) in terms of a(x), yields the qualitative behavior of Fig.1, but only with unphysically high values of q(x).

3. Detailed Characterization of Adlayer Morphology

Island Density Behavior for Irreversible Island Formation

The rate equations (2) show that N_{av} will at first increase with coverage like θ^3 in the transient regime, which extends to $\theta^* \sim (D/F)^{-1/2}$. Neglecting the θ-dependence of σ_{av} (as is valid for point-islands), yields a slow increase of N_{av} in the steady-state regime with $K(\theta) \sim \theta^{1/3}$ in (3).[5] However, for compact or fractal islands, nucleation is quenched even more completely, so $K(\theta) \sim$ constant (for large D/F), as is clear from simulations.[5]

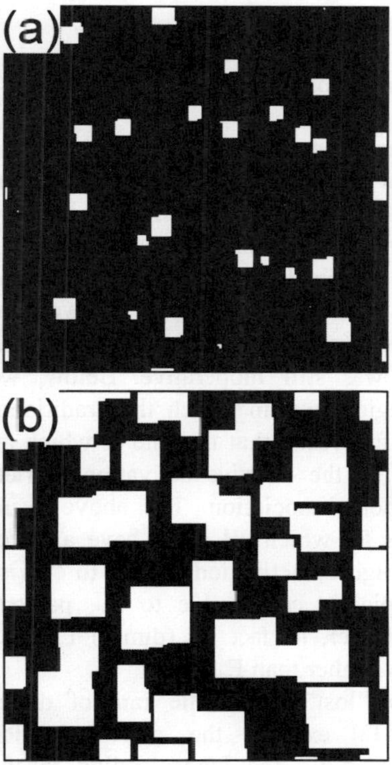

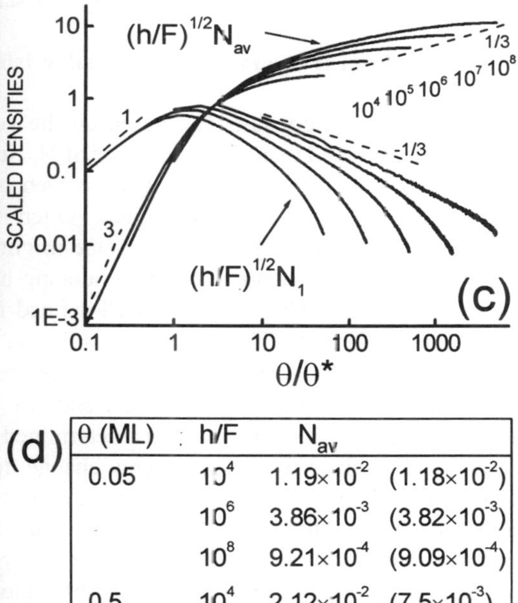

(d)

θ (ML)	h/F	N_{av}	
0.05	10^4	1.19×10^{-2}	(1.18×10^{-2})
	10^6	3.86×10^{-3}	(3.82×10^{-3})
	10^8	9.21×10^{-4}	(9.09×10^{-4})
0.5	10^4	2.12×10^{-2}	(7.5×10^{-3})
	10^6	5.27×10^{-3}	(2.2×10^{-3})
	10^8	1.18×10^{-3}	(5.1×10^{-4})

Figure 2: Simulation results for irreversible formation of square islands[6]: island configurations on a 170×170 site lattice for $h/F=10^8$ at (a) 0.05 ML, (b) 0.5 ML; (c) scaled densities per site of isolated adatoms (N_1) and nucleated islands (N_{av}), where the latter does not account for the reduction in density due to coalescence; (d) tabulated values of density per site of nucleated islands, and of the actual island density (in parenthesis) which is lower due to coalescence.

This feature is reproduced quantitatively for compact islands, at least up to 0.3ML, by incorporating the MF θ-dependence of σ_{av}.[21] For higher θ, a RW-based theory might better describe the capture of atoms deposited in the "slits" between islands.[6] For fixed θ (in the steady-state regime), equation (3) shows that N_{av} scales like $(D/F)^{-1/3}$, for large D/F, which allows determination of E_d from the Arrhenius behavior of $N_{av} \sim \exp[E_d/(3k_BT)]$ for large enough T. Realistic simulations, or rate equations incorporating precise σ_{av}, also allow possibility to extract E_d by matching observed N_{av} values at a single T (preferably for varying F or θ).[28] This is now a common strategy given accurate STM data, even in the transient regime.[29,30] However, for low-T, one must take into account post-deposition nucleation,[31] and any unusual adatom "clumping" processes.[18,32] In Fig.2, we show N_{av}-behavior in our model for irreversible formation of square-islands.[6]

Transition from Irreversible to Reversible Island Formation

The classic results (3), based on the generic concept of a critical size, are traditionally applied to analyze behavior of N_{av} for higher T in order to assess adatom bonding. The validity of such results, or even of the idea of a critical size, had not been tested, and was recently critiqued.[9] Often i=1 (irreversible island formation) is observed for a broad range of low T. One might anticipate that increasing T would allow at least single-bond scission, producing a transition to a regime of reversible island formation, with potentially well-defined i=2 (on a triangular lattice), or i=3 (on a square lattice), if double-bond scission was still inoperative. Below, we characterize in detail this "broad" transition region, within which the traditional analysis is not likely to apply. To this end, we first recall that the rate at which an atom dissociates from an island is determined by the effective activation barrier, $E_{diss}=E_d+\delta E_b$, where δE_b is the energy change upon dissociation. The above result applies even if some dissociation pathway exists for which all steps have a lower activation barrier than E_{diss}! For example, the highest activation barrier to "twist apart" a dimer on a metal(100) surface is typically comparable to E_d, perhaps suggesting an i=1 regime is not possible. However, in fact $E_{diss}(\text{dimer})=E_d+E_{NN}$, where E_{NN} is the NN bond energy, is significantly higher than E_d.[33]

One expects that the i=1 regime is "lost" when the rate of dimer dissociation, $H_{diss}(\text{dimer})=\nu \exp[-E_{diss}(\text{dimer})/(k_BT)]$ exceeds the rate at which monomers aggregate with dimers, $H_{agg}(\text{dimer})=h\sigma_2N_1$, the latter leading to the formation of larger more stable islands with doubly-bonded adatoms.[33] It is possible to show from (2) that the key ratio H_{diss}/H_{agg} is "universally" determined by the variable $Y=(\nu/F)\exp[-(E_d+3E_{NN}/2)/(k_BT)]$. Thus Y provides the natural parameter to characterize the transition out of i=1.[33] This is demonstrated in Fig.3 which shows the transition in the flux-scaling exponent from $\chi\approx1/3$ (i=1) for $Y\leq5$, to $\chi\approx3/5$ (i=3) for $Y\geq50$, for various models of nucleation and growth on a square-lattice with NN pair interactions.[23,34] Our study[33] also indicates that a well-defined i=3 regime requires large E_{NN}/E_d, or many-body interactions providing additional stabilization of tetramers. Fig.3 also shows the predictions for χ versus Y from traditional MF rate equations, which equate escape rates, η_s, with detachment or dissociation rates, ω_s, as determined above. The MF results predict the onset of reversible island formation too early. We attribute this inadequacy to the neglect of the possibility that adatoms in a dissociating dimer *recombine* rather than aggregate with other islands.[23] This possibility is highly likely due to the recurrence property of 2D random walks.

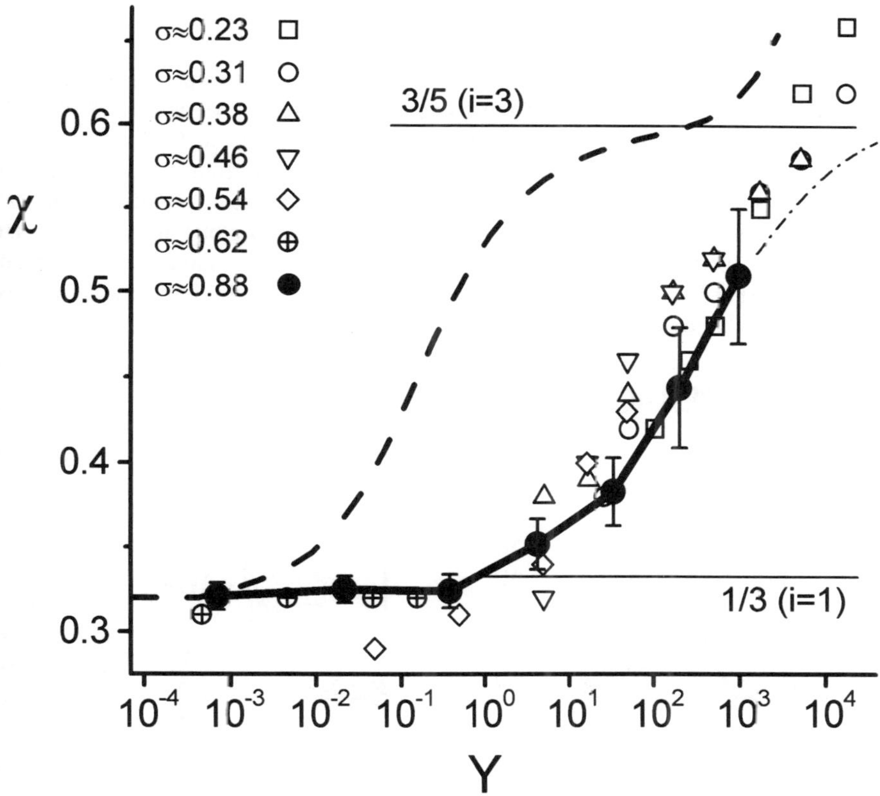

Figure 3: Effective exponent χ=d(ln N_{av})/d(ln F) versus the crossover variable Y=(ν/F)exp[-(E_d+3E_{NN}/2)/(k_BT)] for the transition from irreversible to reversible island formation for models with nearest-neighbor pairwise-additive adatom bonding of strength E_{NN}, for various σ=E_{NN}/E_d. Open symbols are simulation results from the model of Ratsch *et al.*,[9] and solid symbols are from our reversible square-island model.[34] Conventional mean-field rate equations (dashed line) predict an earlier onset of reversibility due to neglect of recombination of dimers.

The criterion Y≈5 for the demise of i=1 allows direct determination of E_{NN} from the observed transition temperature, T* (given E_d, etc.).[34] However, it has been suggested that values thus determined for metal(100) homoepitaxy are too high.[35] We find that a given T* can be preserved with a lower E_{NN} if one allows for longer range attractions between adatom pairs (enhancing the recombination probability).[23] For very low E_{NN}, additional attractive many-body interactions are needed to account for observed behavior.

Island Size Distributions (for Low Pre-Coalescence θ)

For i=1, the scaling function, f(x), for compact (or point) islands, as determined by (6), has intercept f(0)≈0.4 and maximum value of about 0.9 slightly above x=1. An almost complete independence on θ, at least up to 0.15 ML, derives from that of C(x). The form of f is qualitatively different from those previously postulated with f(0)=0,[8,25] and does not exhibit the artificial singularity of MF predictions.[26] Note that significant deviations from the asymptotic form of the size distribution occur for lower *D*/F, as shown in Fig.4 for our square-island model. These are observed in deposition experiments at lower T, with some modifications due to post-deposition effects.[31]

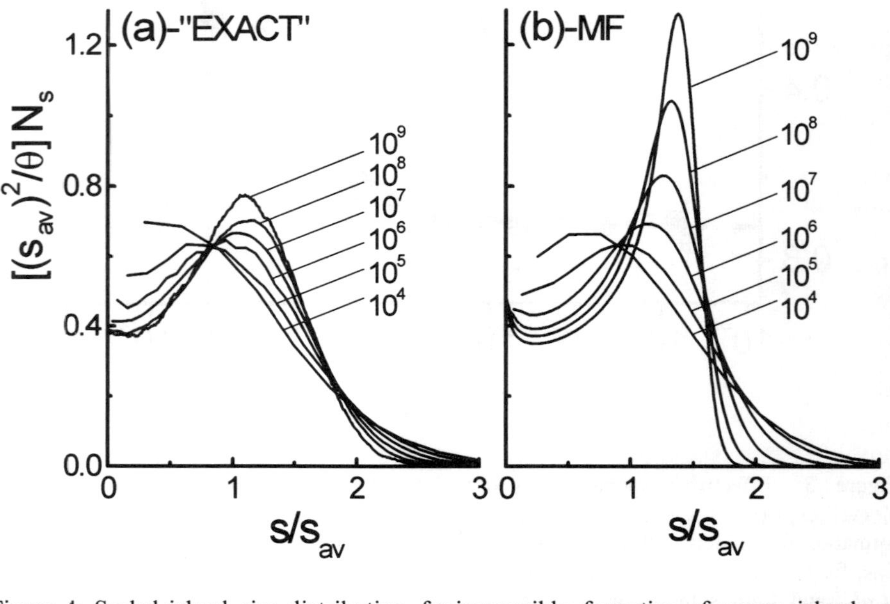

Figure 4: Scaled island size distribution for irreversible formation of square islands at 0.05 ML with h/F=10^4-10^9: (a) "exact" simulation results; (b) mean-field rate equation predictions using Bales-Chrzan[21] capture numbers.

Our studies for point-islands,[28] and later studies for more general island structures,[8,9] revealed a sharpening of the scaling function with the introduction of reversibility, or with an increase in critical size. The shape of the size distribution during the transition out of i=1 is parameterized by the crossover variable H_{diss}/H_{agg} (or Y).[9] Finally, it should be emphasized that f varies strongly with θ (at fixed T) once island coalescence becomes significant (see below).

Island Separation Distributions

Above we described the form of the adatom density surrounding an island of specified size, $N_1(\mathbf{r})$. In particular, we noted the existence of a depletion zone, the width of which scales like the average island separation in the steady-state regime. Here, we consider the adatom-island pair distribution obtained by averaging $N_1(\mathbf{r})$ over island sizes. For simplicity, we also denote the averaged quantity by $N_1(\mathbf{r})=N_1G(\mathbf{r})$, where $G\to1$, as $r\to\infty$. If $N_{av}(\mathbf{r})$ denotes the probability of finding an island with center a distance **r** from some island (the island-island pair distribution), then $N_{av}(\mathbf{r})=N_{av}C_{isl}(\mathbf{r})$ with $C_{isl}(\mathbf{r})\to1$, as $r\to\infty$. From (2), one expects that $d/dt\, N_{av}(\mathbf{r}) \approx d/dt\, N_{av}\, G(\mathbf{r})^{i+1}$, for a well defined critical size i. Integrating this equation yields $C_{isl}(\mathbf{r})\approx G(\mathbf{r})^{i+1}$, if one neglects the time-dependence of $G(\mathbf{r})$.[1,6] Thus, depletion in the adatom density near islands leads to enhanced depletion of pairs of nearby islands. This non-randomness in the island distribution significantly influences the diffraction profiles (see below). Simulation results[6] shown in Fig.5a indicate that the above provides a fairly accurate description at least for lower θ. These studies also confirm the scaling behavior, $G(\mathbf{r})\approx g(\mathbf{r}/L_{av},\theta)$ and $C_{isl}(\mathbf{r})\approx c(\mathbf{r}/L_{av},\theta)$, suggested by the above analysis. We also note that the non-randomness of the island distribution affects the two-point correlation function for the adlayer, $C(\mathbf{r})$, which does not decrease monotonically to zero, but has negative minimum due to depletion.[6]

Island Shapes

Diffusion-mediated growth of islands naturally produces a Mullins-Sekerka type shape instability which can lead to fractal or dendritic islands.[36-38] This instability can be quenched if island restructuring processes are sufficiently active on the time-scale of aggregation. The dominant restructuring process will usually involve diffusion of adatoms along the island edge, which potentially allows aggregating atoms to reach highly coordinated (HC) sites thereby producing compact island shapes. One expects[39] that islands will grow compact until a critical size, L_c, above which atoms cannot reach HC sites before being caught by other aggregating atoms, thus initiating the shape instability. In the simplest scenario of a single edge hopping rate, h_e, we find that $L_c\sim(h_e/I)^{1/2}$, where $I=F/N_{av}$ is the rate at which atoms aggregate with an island.[40] The above picture is overly simplistic. Typically, one must at least distinguish between diffusion along straight edges, and rounding of "global" corners and of "local" corners (or "kinks"). Rapid diffusion along straight edges is *not* sufficient to obtain compact islands. One must also have at least rounding of kinks. Otherwise adatoms get trapped "on top" of 1D terraces, unable to reach HC sites, thus leading to a shape instability.[40,41]

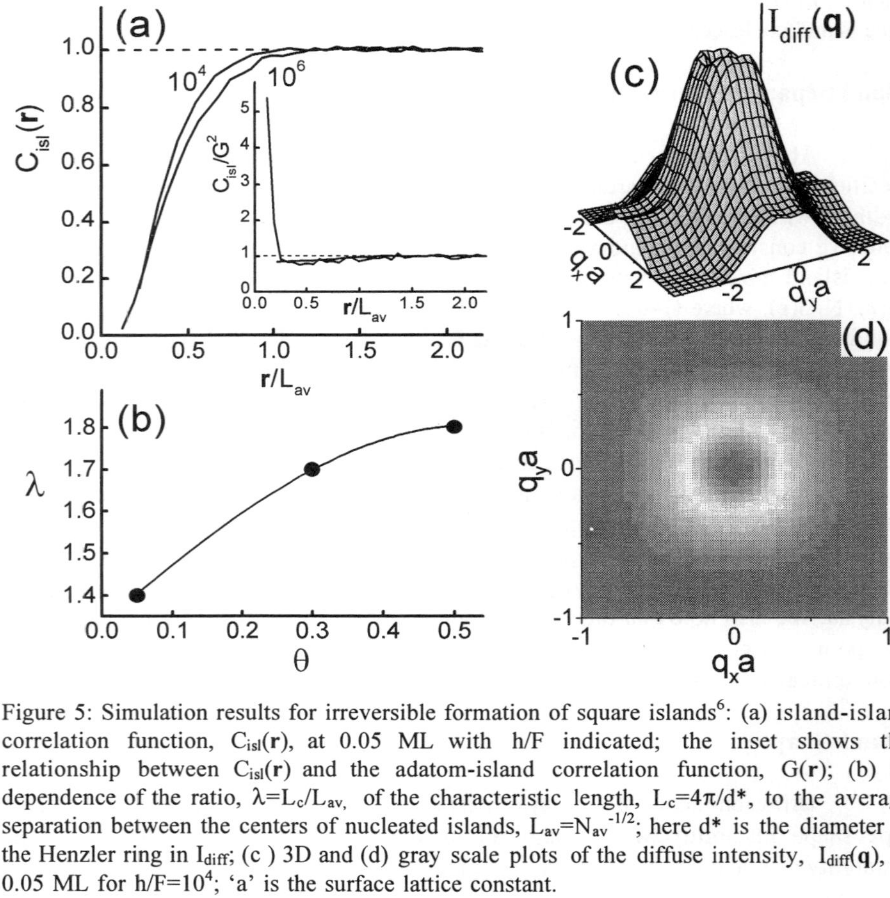

Figure 5: Simulation results for irreversible formation of square islands[6]: (a) island-island correlation function, $C_{isl}(\mathbf{r})$, at 0.05 ML with h/F indicated; the inset shows the relationship between $C_{isl}(\mathbf{r})$ and the adatom-island correlation function, $G(\mathbf{r})$; (b) θ-dependence of the ratio, $\lambda=L_c/L_{av}$, of the characteristic length, $L_c=4\pi/d^*$, to the average separation between the centers of nucleated islands, $L_{av}=N_{av}^{-1/2}$; here d* is the diameter of the Henzler ring in I_{diff}; (c) 3D and (d) gray scale plots of the diffuse intensity, $I_{diff}(\mathbf{q})$, at 0.05 ML for $h/F=10^4$; 'a' is the surface lattice constant.

On metal(100) surfaces, diffusion along straight edges is generally rapid, and the characteristic appearance of square islands indicates that kink rounding is also active. We note that the local environment at kinks is more favorable to an "easy" exchange pathway for corner rounding than it is at global corners. We also note that a near-square equilibrium shape applies for attractive NN interactions,[42] but this shape is modified by the presence of next NN or many-body attractive interactions. Also, it is not generally recognized that the island "growth shape" during deposition is more square than the equilibrium shape! This feature further justifies models with prescribed square island shape.

For metal(111) surfaces, shape instability due to limited edge diffusion is often manifested in the appearance of fractal or dendritic islands.[37,38] Three distinct generic regimes have been proposed:[43] no edge diffusion, producing DLA type fractals; only hopping from singly coordinated corner sites to doubly coordinated edge sites, producing limited arm thickening; more extensive edge diffusion allowing compactification. The actual behavior in these systems can be even more complex. For example, inclusion of anisotropy in corner rounding is sometimes necessary to reproduce details of exotic shape transitions with varying T.[44-46]

While edge diffusion is the only pathway for island shape equilibration when i=1, the onset of reversibility (or i>1) leads to an additional "evaporation-recondensation pathway".[9,33] This pathway is particularly effective because of the high probability of recondensation due to the recurrence property of 2D random walks. One can confirm this by switching off edge diffusion and observing compactification with the onset of reversibility. We noted above that kink rounding is required for compactification, so if its barrier is comparable to that for bond-scission, then compactification should coincide with the transition to reversible island formation. This may be the case for Rh/Rh(111).[47]

Island Coalescence and Percolation

For systems with compact, convex islands, coalescence of individual islands due to growth becomes significant at coverages above 0.25ML. Simulations with near-square islands reveal that coalescence produces a broadening of the island size distribution.[6] Eventually, coalesced clusters of islands percolate at a critical threshold value of $\theta=\theta_c$ (for an infinite system), which depends on model details. However, basic behavior near this threshold is controlled by random percolation universality.[6] For models with compact islands, increasing T from 0K, where adatoms are randomly distributed, produces an initial decrease in θ_c, followed by a dramatic increase as the linear island dimension increases to many lattice constants, and one crosses over to a continuum percolation regime. In fact, θ_c rises above the 0.7ML value typical for continuum percolation of randomly distributed islands. This is due to depletion effects (producing an "effective repulsion" between islands).[6] Finally, we note that non-convex island configurations created by coalescence tend to restructure to form larger convex islands, but often on a time scale slower than that of deposition.

Kinematic Diffraction Profiles

For interlayer spacing b, the kinematic approximation for the diffracted intensity at a lateral (vertical) momentum transfer **q** (q_z) is given around **q**=0 by[6]

$$I(\mathbf{q}, q_z) \propto (2\pi)^2[1-2(\theta+\theta^2)\{1-\cos(q_z b)\}]\,\delta(\mathbf{q}) + 2[1-\cos(q_z b)]\, I_{diff}(\mathbf{q}), \tag{8}$$

where the first term is the Bragg delta-function intensity, and the second term is the diffuse intensity. Here, $I_{diff}(\mathbf{q})=\Sigma_\mathbf{r} \exp(i\mathbf{q}\cdot\mathbf{r})C(\mathbf{r})$ is the diffuse intensity for the overlayer, which exhibits a "ring" due to the weak oscillation in $C(\mathbf{r})$ resulting from depletion.[6]

We naturally recast the expression for I_{diff} in an "island representation" as the sum $I_{diff}=I_0+I_{int}$, where the "random phase approximation", I_0, neglects interference between scattering from different islands, and I_{int} accounts for this interference. One has[6,48]

$$I_0(\mathbf{q}) = \Sigma_s N_s |\mathcal{A}_s(\mathbf{q})|^2, \text{ and } I_{int}(\mathbf{q}) \approx |\Sigma_s N_s \mathcal{A}_s(\mathbf{q})|^2\, \Sigma_\mathbf{r} \exp(i\mathbf{q}\cdot\mathbf{r})[C_{isl}(\mathbf{r})-1], \tag{9}$$

where $\mathcal{A}_s(\mathbf{q})$ is the average scattering amplitude for islands of size s, and $C_{isl}(\mathbf{r})$ is the island-island correlation function introduced above. We emphasize that this expression for I_{int} neglects correlations between island size and separation,[6] as is reasonable for sharp island size distributions. Its general validity, even for low coverages where island separation far exceeds typical linear size, is less clear (cf. Sec.2). However, (9) provides at least a semi-quantitative description of intensity: I_0 is monomodal with width reflecting the mean island size, and symmetry reflecting island shape; I_{int} has a negative value at $\mathbf{q}=0$, with a magnitude measuring the "total amount of depletion", $\Sigma_\mathbf{r} [1-C_{isl}(\mathbf{r})]$, of nearby island pairs. When combined with I_0, this produces a ring feature to I_{diff} mentioned above, the ring diameter, d*, being controlled by the mean island separation. See Figs. 5b-d.

4. Transition to Multilayer Mounding and Roughening

During homoepitaxial growth, kinetic limitations on interlayer transport produce rough films. Indeed, for a model without interlayer diffusion in a simple cubic (SC) geometry, one finds very rough Poisson growth. This result is well known in the absence of all diffusion,[49] but it applies for any amount of intralayer diffusion and islanding.[18,50] Limited interlayer transport often results from an *additional* step-edge (SE) barrier, E_{SE}, to downward transport at island edges, which produces unstable growth characterized by the formation of mounds.[13] Particularly for large SE barriers and i=1, the multilayer mound morphology reflects that of the submonolayer island distribution, as these islands form a "template" for the subsequent growth.[19] Thus, the above characterization of submonolayer morphology is directly relevant to the analysis of multilayer growth.

Onset of Multilayer Growth

The population of the second layer at the onset of multilayer growth is very sensitive to the magnitude of the SE barrier, E_{SE}, and also dependent on the details of the submonolayer island morphology. Various analytic MF treatments of nucleation on top of compact islands have been developed.[14,15] These lead to estimates of the SE barrier by matching observed layer populations. A more precise approach utilizes simulation models tailored to match submonolayer structure, and adjusts the SE barrier until matching observed behavior.[18,20] This gives at best an effective SE barrier, since its value likely depends on the local step configuration, and the result depends on the assumed degree of island restructuring upon coalescence.[18] This approach is more viable for metal(100) homoepitaxy where islands are near-square with somewhat straight step edges, than for metal(111) homoepitaxy where islands can be fractal with a more complex step edge structure. In the latter case, account of various SE barriers, and of transitions in island structure, is generally necessary to realistically describe multilayer growth.[40,51]

General Features of Unstable Growth

The key observation of Villain[13] is that the presence of an additional SE barrier to downward transport produces biased reflection of adatoms from descending steps and incorporation at ascending steps. This corresponds to a lateral "Schwoebel" mass current, $\mathbf{J}_{Sch}$, in the *uphill* direction, resulting in unstable growth and the formation of "mounds". Let $\mathbf{m}$ denote the local slope of the surface. Then, for a large SE barrier, one expects the limiting forms $\mathbf{J}_{Sch} \sim F\mathbf{m}/m^2$ in the step flow regime for narrow terraces (at larger m), and $J_{Sch} \sim FL_{av}{}^2\mathbf{m}$ for broad terraces (at smaller m) where only adatoms deposited within a distance $O(L_{av})$ of ascending steps are captured by these. Interpolating, and also accounting for a finite SE barrier, it was proposed that $\mathbf{J}_{Sch} \sim FSL_{av}{}^2\mathbf{m}/[1+(BL_{av}m)^2]$, where $S = 1\text{-}\exp[-E_{SE}/(k_BT)]$.[16] We emphasize, however, that it is the development of additional counterbalancing *downhill* currents which is responsible for the often observed stabilization or selection of mound slopes, Θ. Typically, these currents are determined by detailed features of the dynamics of depositing atoms, and of the adsorption site geometry.[52] For metal(100) homoepitaxy, with four-fold hollow (4FH) adsorption sites, the dominant process is likely *downward funneling* of atoms depositing at step edges and other microprotrusions to lower 4FH adsorption sites.[18,52] The downward funneling current roughly satisfies $\mathbf{J}_{DF} \approx -CF\mathbf{m}$, where C is $O(1)$.[52]

We now discuss the roughening of the film, as quantified by its interface width, W(t), the coarsening of the linear mound size, L(t) (which is initially strongly dependent on L_{av}), and the evolution of mound slopes, S(t). Note that if $W(t)\sim t^{\beta}$ and $L(t)\sim t^{n}$, then one has $S(t)\sim t^{\lambda}$ with $\lambda=\beta\text{-}n$. It is simplest to first consider the regime of *large* island sizes. Here various dynamic processes producing downhill currents will initially be insignificant, being related to step edge density. Thus, for an insurmountable SE barrier, one has $\beta=1/2$ as for an SC geometry.[18,50] The uphill current produces formation of mounds with slope initially increasing in time from its submonolayer value of order $1/L_{av}$. Eventually the increased slope and step edge density will result in an increase in the counterbalancing downhill currents. Examples are provided by metal(111) homoepitaxy, where for Ag it was found that $\beta=1/2$ and $\lambda=1/2$-$2/3$ at lower T.[53,54] For *smaller* islands or smaller SE barriers, the stabilization and selection of slope can be achieved much more quickly, e.g., within a few ML as observed for metal(100) homoepitaxy.[18,55]

Finally, we comment briefly on the coarsening of mounds. This process relies on "filling the gap" between the 2D island bases of individual mounds. This will be easier for smaller SE barriers, and for greater degree of reversibility of island formation.[19] We note that a competition between the coarsening and steepening of mounds is often observed, which is consistent with maintaining a fixed $\beta=n+\lambda$.[19] Thus, e.g., for large islands and significant SE barriers, one expects the initial slow coarsening to be enhanced once the slope is stabilized, and a significant downhill current has developed.

Detailed Behavior for Metal(100) Homoepitaxy

Despite "low" SE barriers for metal(100) homoepitaxy, these systems generally exhibit kinetic roughening with $\beta>0$; they also display mound formation with rapid slope selection.[55-57] A detailed analysis of growth models for metal(100) homoepitaxy suggests that film height distributions deviate from Gaussian forms, and specifically have a skewness $\kappa<0$ (i.e., the tops of mounds are flatter than the valleys between them).[18] The height-height correlation function, $H(\mathbf{r})$, which measures the mean-square height difference between two surface atoms separated laterally by $\mathbf{r}$, develops *strong* oscillations [contrasting the submonolayer $C(\mathbf{r})$] with increasing film thickness.[18,19,55] These oscillations reflect the development of a fairly ordered array of mounds, and are a generic feature of unstable growth, distinguishing it from self-affine growth. The full distribution function, $G_n(\mathbf{r})$, for the height-difference 'n' at separation $\mathbf{r}$ has the form $G_n(\mathbf{r})\sim H(\mathbf{r})^{-1/2}g[n/H(\mathbf{r})^{-1/2}]$, where $g()$ exhibits significant deviations from a commonly assumed (normalized) Gaussian form.[18] Fig.6 summarizes these features.

These models further predict that lowering the growth temperature (from some "high" value) initially produces rougher films as the SE barrier becomes more difficult to surmount. However, eventually the reduced island size and increased step density ensure that downward funneling becomes dominant, producing "re-entrant" smooth growth.[18] The selected slope, determined by balancing the above expressions for $\mathbf{J}_{Sch}$ and $\mathbf{J}_{DF}$, is generally observed to increase with decreasing T.[20,34,50] As T is lowered, eventually the $\mathbf{J}_{Sch}$ versus $\mathbf{m}$ curve crosses below the quasi-linear $\mathbf{J}_{DF}$ versus $\mathbf{m}$. At this critical T (where $\mathbf{J}_{DF}$ always dominates $\mathbf{J}_{Sch}$), one has a "kinetic phase transition" from mounding to smooth Edwards-Wilkinson growth.[50] There are many complications to the above picture particularly at lower T where L_{av} is small, and local slopes are larger: $\mathbf{J}_{DF}$ depends on film morphology when multiple steps occur,[50] downward funneling can break down leading to defect formation,[50] adatom clumping[18,50] and step attraction[20] can modify growth, etc..

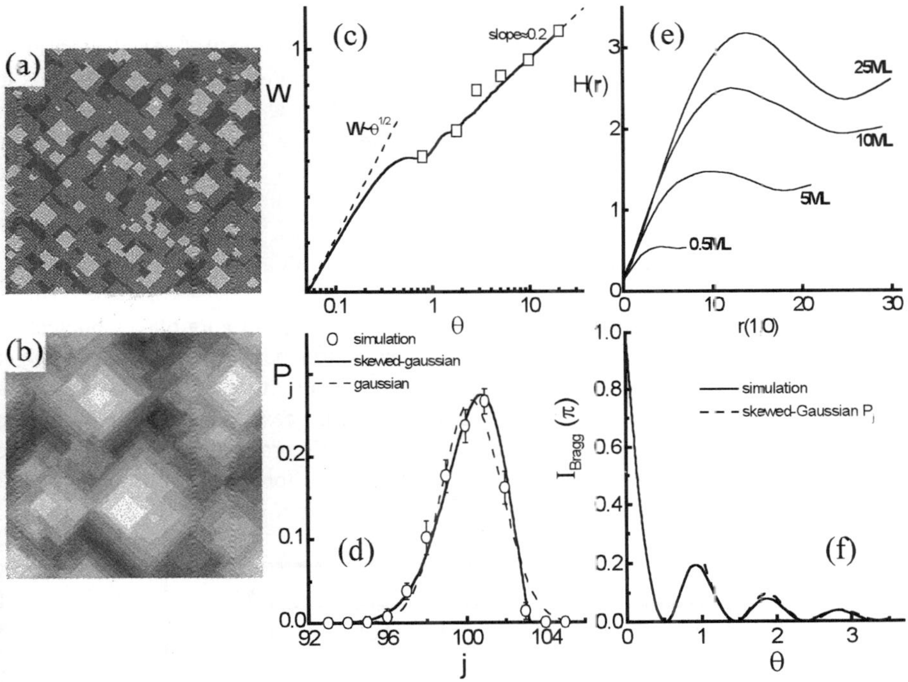

Figure 6: Simulation results for a model of Fe/Fe(100) multilayer growth at 295K and F=1/70 ML/s, with $\nu=10^{12}$, E_d=450meV, and E_{SE}=45meV, involving irreversible formation of square islands in each layer[18]: film morphology at (a) 1 ML and (b) 50 ML in a $170a/\sqrt{2}\times170a/\sqrt{2}$ region with darker regions indicating lower film height; (c) interface roughness, W (in units of b), versus θ; symbols are experimental data[55]; (d) film height distribution $P_j=\theta_j-\theta_{j+1}$ versus layer j, at 100 ML (skewness $\kappa=-2.5$); (e) height-height correlation function, H(**r**), in the (110) direction; (f) decay of the anti-phase Bragg intensity, $I_{Bragg}(\pi)$.

Kinematic Diffraction Theory

By analogy with the submonolayer regime, near **q**=0, and for $|q_z|\leq\pi b$, one has[18,58]

$$I(\mathbf{q}, q_z) \propto (2\pi)^2 I_{Bragg}(q_z/b)\,\delta(\mathbf{q}) + I_{diff}(\mathbf{q}, q_z), \text{ where}$$
$$I_{Bragg}(\pi) \approx 4\cos^2[\pi\theta\text{-}\kappa\pi^3W^3/3!]\exp[-\pi^2W^2] \text{ and } I_{diff}(\mathbf{q}, q_z) = \Sigma_{\mathbf{r}} \exp(i\mathbf{q}\cdot\mathbf{r})C(q_z,\mathbf{r}). \quad (10)$$

Here $C(q_z,\mathbf{r}) = \Sigma_n \exp(iq_z nb)\,[G_n(r)\text{-}G_n(\infty)] \approx \wp[q_z H(r)^{1/2}]$, if q_z is not too small, where $\wp$ is the Fourier Transform of g, and thus vanishes for large arguments.[18] These results reveal a characteristic oscillatory decay of the Bragg oscillations at the anti-phase condition, with minima shifted below half-monolayers for $\kappa<0$.[18] At the onset of multilayer growth, I_{diff} will exhibit a ring structure associated with the oscillations in $H(\mathbf{r})$ which begin to amplify those in $C(\mathbf{r})$. However, as the film roughens and characteristic values for W and $H(\mathbf{r})$ become larger, the oscillations in $H(\mathbf{r})$ are pushed into the "tail" of $\wp[q_z H(\mathbf{r})^{1/2}]$. Thus they have less influence on I_{diff}, leading to a disappearance of the ring structure.[18]

5. Summary and Applications

A rather comprehensive picture is emerging of the submonolayer nucleation process for metal homoepitaxy, and of the connection between the submonolayer island morphology and the subsequent multilayer film structure. A test of the success of this effort is the ability to elucidate the nucleation process, and to reliably extract key energetic parameters, as well as to predict growth properties. In Table 1, we summarize some estimates and predictions of this type for metal homoepitaxy. At this point, estimates for E_d are substantially more reliable than those for E_{NN} and E_{sch}.

SYSTEM	E_d	T*	E_{NN}	E_{SE}	β, Θ
Fe/Fe(100)	0.45eV[59]	500K[59]	0.5eV	40meV[18]	[a]0.16, 13°
Ag/Ag(100)	0.33eV[23]	310K[23]	0.3eV	30meV[23]	[b]0.17, 6°
Cu/Cu(100)	0.36eV[60,34]	240K[60]	0.2eV		[c]0.45, 6°
Ag/Ag(111)	0.10eV[61]	120K[62]	0.13eV	120meV[61]	[d]0.5, incr.
Pt/Pt(111)	0.26eV[30,40]	265K[30]	0.35eV	80meV[51]	[e]complex
Cu/Cu(111)	0.03eV?[63]	120K?[63]	0.2eV		[f]0.5, incr.?
Rh/Rh(111)	0.18eV[47]	600K[47]	0.8eV		[g]0.33, 6°

Table 1: E_d values estimated from nucleation theory; measured temperature, T*, of transition to reversible island formation; *effective* NN bond-energies from T* using $Y\approx5$; estimated *effective* values of E_{SE}, except that for Pt(111) is the low value at kinks; effective β and Θ values below 100ML; (a) measured[55] and simulated[18,20,34] at 300K; (b) measured[54] and simulated[23,34] at 300K; $\beta\approx0.28$ at 200K[23,34,54]; (c) measured at 300K[57]; $\beta\approx0.56$ (0.26) and $\Theta\approx16°$ (25°) measured at 200K (160K)[56]; (d) measured below 400K[53,54]; (e,f) see Ref.3; (g) measured at 725K, Θ increasing for lower T.[47]

Appendix: Modified Theories of Nucleation and Growth

Random Walk (RW) Based Theory of Capture

Deposited adatoms typically aggregate with "nearby" islands or adatoms due to the space filling nature of random walks. Thus, the aggregation rates can be rewritten as[5,6] $R_{agg}(s)=N_1P_s/\tau_s$, where $P_s \approx N_s/N_{av}$ gives the probability of being deposited near an island of size $s \geq 1$, and τ_s denotes the lifetime of such an adatom. τ_s is determined by the theory of RW's in the presence of traps. For $s=O(1)$, or for point-islands, one has $D\tau_s \sim N_{av}^{-1}\ln(N_{av}^{-1})$ for isotropic diffusion, and $D\tau_s \sim N_{av}^{-2}$ for 1D diffusion. We note that this approach is quite versatile allowing treatment of moderately anisotropic systems[64], anomalous diffusion due to repulsive defects,[65] and capture in high θ regimes.[6]

Anisotropic Systems

The above RW formalism was applied to extend traditional nucleation theory for isotropic diffusion to treat strongly anisotropic diffusion[5] (motivated by recent studies of Si/Si(100) homoepitaxy[66]). The different form of the aggregation rates modifies the fundamental exponent, χ, for the flux scaling of N_{av} from $i/(i+2)$ to $i/(2i+2)$.[28] The results have been applied to analyze metal epitaxy on anisotropic reconstructed (100) surfaces,[67] and on (110) surfaces,[68] where island shape anisotropy is also of interest.

Small Cluster Mobility

"Significant" diffusion of dimers (with diffusion coefficient D') modifies the scaling of the island density[69] for $i=1$ from the classic $D'=0$ form (3) to $N_{av} \sim (DD'/F^2)^{-1/5}$. Classic scaling is "lost" when the rate of dimer loss due to aggregation with existing islands, $H_{loss}(\text{dimer}) \propto D'N_{av}$, compares with the rate at which monomers aggregate with, and thus immobilize dimers, $H_{agg}(\text{dimer})$. The key ratio H_{loss}/H_{agg} is "universally" determined by $Z=(\nu/F)\exp[-(3E_d'-2E_d)/(k_BT)]$, where E_d' is the activation barrier for dimer diffusion, so Z provides a natural crossover variable (cf. Sec.3).[70] Modified scaling has not yet been observed, but small cluster mobility no doubt perturbs island size distribution shapes.[70]

Acknowledgements: This work was supported by NSF Grant CHE-9700592. It was performed at Ames Laboratory, which is operated for the US Department of Energy by Iowa State University under contract No. W-7405-Eng-82. At Sandia, it was supported by the Office of Basic Energy Sciences, Division of Materials Science, of the US Department of Energy under contract No. DE-AC04-94AL85000.

References

1. J.A. Venables, Philos. Mag. **27**, 697 (1973); S. Stoyanov and K. Kashchiev, Curr. Topics Mater. Sci. **7**, 69 (1981).
2. A.-L. Barabási and H.E. Stanley, "Fractal Concepts in Surface Growth" (Cambridge UP, Cambridge, 1995).
3. G. Rosenfeld, B. Poelsema, and G. Comsa, in "Growth Properties of Ultrathin Epitaxial Layers", The Chemical Physics of Solid Surfaces, Vol.8, Ch.3.
4. R.Q. Hwang and M.C. Bartelt, Chem. Rev. **97**, 1063 (1997); C. Günther, S. Günther, E. Kopatzki, R. Q. Hwang, J. Schröder, J. Vrijmoeth, and R.J. Behm, Ber. Buns. Phys. Chem. **97**, 522 (1993).
5. M.C. Bartelt and J.W. Evans, Phys. Rev. B **46**, 12675 (1992); Surf. Sci. **284**, L437 (1993); M.C. Bartelt, M.C. Tringides, and J.W. Evans, Phys. Rev. B **47**, 13891 (1993); MRS Proc. **280**, 363 (1993).
6. M.C. Bartelt and J.W. Evans, Surf. Sci. **298**, 421 (1993); MRS Proc. **312**, 255 (1993).
7. L.-H. Tang, J. Physique I **3**, 935 (1993).
8. J.G. Amar and F. Family, Phys. Rev. Lett. **74**, 2066 (1995).
9. C. Ratsch, A. Zangwill, P. Smilauer, and D.D. Vvedensky, Phys. Rev. Lett. **72**, 3194 (1994); Surf. Sci. **329**, L599 (1995).
10. A.F. Voter, SPIE **821**, 214 (1987).
11. Z. Zhang, S. Liu, and H. Metiu, this Volume; G. Barkema, O. Biham, M. Breeman, D. Boerma, and G. Vidali, Surf. Sci. **306**, L569 (1994).
12. G. Ehrlich and F.G. Hudda, J. Chem. Phys. **44**, 1039 (1966); R.L. Schwoebel and E.J. Shipsey, J. Appl. Phys. **37**, 3682 (1966).
13. J. Villain, J. Physique I **1**, 19 (1991).
14. J. Tersoff, A.W. Denier van der Gon, and R.M. Tromp, Phys. Rev. Lett. **72**, 266 (1994).
15. J.A. Meyer, J. Vrijmoeth, H.A. van der Vegt, E. Vlieg, and R.J. Behm, Phys. Rev. B **51**, 14790 (1995); P. Smilauer and S. Harris, *ibid*, **51**, 14798 (1995).
16. M.D. Johnson, C. Orme, A.W. Hunt, D. Graff, J. Sudijono, L.M. Sander, and B.G. Orr, Phys. Rev. Lett. **72**, 116 (1994).
17. M. Siegert and M. Plischke, Phys. Rev. Lett. **73**, 1517 (1994); Phys. Rev. E **53**, 307 (1996); J. Krug, M. Plischke, and M. Siegert, Phys. Rev. Lett. **70**, 3271 (1993).
18. M.C. Bartelt and J.W. Evans, Phys. Rev. Lett. **75**, 4250 (1995); MRS Proc. **399**, 89 (1996); Langmuir **12**, 217 (1996).
19. P. Smilauer and D.D. Vvedensky, Phys. Rev. B **52**, 14263 (1995).
20. J. G. Amar and F. Family, Phys. Rev. B **52**, 13801 (1995); **54**, 14071, 14742 (1996); Phys. Rev. Lett. **77**, 4584 (1996).
21. G.S. Bales and D.C. Chrzan, Phys. Rev. B **50**, 6057 (1994).
22. G.S. Bales and A. Zangwill, Phys. Rev. B **55**, 1973 (1997).

23. C.-M. Zhang, M.C. Bartelt, J.-M. Wen, C.J. Jenks, J.W. Evans, and P.A. Thiel, J. Crystal Growth **174**, 851 (1997); Surf. Sci. **406**, 178 (1998).
24. J.A. Venables and D.J. Ball, Proc. R. Soc. Lond., Ser. A **322**, 331 (1971).
25. P.A. Mulheran and J.A. Blackman, Phil. Mag. Lett. **72**, 55 (1995); Phys. Rev. B **53**, 10261 (1996).
26. M.C. Bartelt and J.W. Evans, Phys. Rev. B **54**, R17359 (1996); MRS Proc. **440**, 247 (1997).
27. G.S. Bales, presentation at ACCG-10/ICVGE-9 (Vail, Colorado, August 1996).
28. J.W. Evans and M.C. Bartelt, J. Vac. Sci. Technol. A **12**, 1800 (1994).
29. H. Brune, H. Röder, C. Boragno, and K. Kern, Phys. Rev. Lett. **73**, 1955 (1994).
30. M. Bott, M. Hohage, M. Morgenstern, T. Michely, and G. Comsa, Phys. Rev. Lett. **76**, 1304 (1996).
31. B. Müller, L. Nedelmann, B. Fischer, H. Brune, and K. Kern, Phys. Rev. B **54**, 17858 (1996).
32. M. Breeman, G. Rosenfeld, and G. Comsa, Phys. Rev. B **54**, 16440 (1996).
33. M.C. Bartelt, L.S. Perkins, and J.W. Evans, Surf. Sci. **344**, L1193 (1995).
34. J.W. Evans and M.C. Bartelt, in "Surface Diffusion: Atomistic and Collective Processes", edited by M.C. Tringides (Plenum, New York, 1997).
35. P.J. Feibelman, Phys. Rev. B **52**, 12447 (1995).
36. R.Q. Hwang, J. Schröder, C. Günther, and R.J. Behm, Phys. Rev. Lett. **67**, 3279 (1991).
37. T. Michely, M. Hohage, M. Bott, and G. Comsa, Phys. Rev. Lett. **70**, 3943 (1993).
38. H. Brune, C. Romainczyk, H. Röder, and K. Kern, Nature **369**, 469 (1994).
39. A. Pimpinelli, J. Villain, and D.E. Wolf, J. Physique **13**, 447 (1993).
40. M.C. Bartelt and J.W. Evans, Surf. Sci. **314**, L829, L835 (1994).
41. Z. Zhang, NATO ASI presentation (Rhodes, Greece, August, 1996).
42. C. Rottman and M. Wortis, Phys. Rev. B **24**, 6274 (1981).
43. Z. Zhang, X. Chen, and M.G. Lagally, Phys. Rev. Lett. **73**, 1829 (1994).
44. H. Brune, H. Holger, K. Bromann, K. Kern, J. Jacobsen, P. Stoltze, K. Jacobsen, and J. Nørskov, Surf. Sci. **349**, L115 (1996).
45. M. Hohage, M. Bott, M. Morgenstern, Z. Zhang, T. Michely, and G. Comsa, Phys. Rev. Lett. **76** (1996).
46. G. Rosenfeld and G. Comsa, in "Surface Diffusion" (Ref. 34).
47. F. Tsui, J. Wellman, C. Uhler, and R. Clarke, Phys. Rev. Lett. **76**, 3164 (1996).
48. T.-M. Lu, L.-H. Zhao, M.G. Lagally, G.-C. Wang, and J.E. Houston, Surf. Sci. **122**, 519 (1982); R.S. Nord and J.W. Evans, Phys. Rev. B **35**, 6004 (1987).
49. J.D. Weeks, G.H. Gilmer, and K.A. Jackson, J. Chem. Phys. **65**, 712 (1976).
50. M.C. Bartelt and J.W. Evans, Surf. Sci., accepted (1997).

51. J. Jacobsen, K.W. Jacobsen, P. Stoltze, and J. Nørskov, Phys. Rev. Lett. **74**, 2295 (1995).
52. J.W. Evans, D.E. Sanders, P.A. Thiel, and A.E. DePristo, Phys. Rev. B **41**, 479 (1990); J.W. Evans, *ibid*, **43**, 3897 (1991); D.E. Sanders and J.W. Evans, in "Structure of Surfaces III" (Springer, Berlin, 1991); H.C. Kang and J.W. Evans, Surf. Sci. **271**, 321 (1992).
53. E.Z. Luo, J. Wollschläger, and M. Henzler, Appl. Phys. A **60**, 19 (1995).
54. W.C. Elliot, P.F. Miceli, T. Tse, and P.W. Stephens, Phys. Rev. B **54**, 17938 (1996).
55. J.A. Stroscio, D.T. Pierce, M. Stiles, A. Zangwill, and L.M. Sander, Phys. Rev. Lett. **75**, 4246 (1995).
56. H.-J. Ernst, F. Fabre, R. Folkerts, and J. Lapujoulade, Phys. Rev. Lett. **72**, 112 (1994).
57. J.-K. Zuo and J.F. Wendelken, Phys. Rev. Lett. **78**, 2791 (1997); L.C. Jorritsma, M. Bijnagte, G. Rosenfeld, and B. Poelsema, *ibid*, **78**, 911 (1997).
58. H.-N. Yang, G.-C. Wang, and T.-M. Lu, "Diffraction from Rough Surfaces and Dynamic Growth Fronts" (World Scientific, Singapore, 1993).
59. J.A. Stroscio, D.T. Pierce, and R.A. Dragoset, Phys. Rev. Lett. **70**, 3615 (1993).
60. H. Dürr, J.F. Wendelken, and J.-K. Zuo, Surf. Sci. **328**, L527 (1995).
61. K. Bromann, H. Brune, H. Roder, and K. Kern, Phys. Rev. Lett. **75**, 677 (1995).
62. G. Rosenfeld, N.N. Lipkin, W. Wulfhekel, J. Kliewer, K. Morgenstern, B. Poelsema, and G. Comsa, Appl. Phys. A **61**, 455 (1995).
63. W. Wulfhekel, N.N. Lipkin, J. Kliewer, G. Rosenfeld, L.C. Jorritsma, B. Poelsema, and G. Comsa, Surf. Sci. **348**, 227 (1996); M. Henzler, T. Schmidt, and E.Z. Luo, in "Structure of Surfaces IV" (World Scientific, Singapore, 1994).
64. M.C. Bartelt and J.W. Evans, Europhys. Lett. **21**, 99 (1993).
65. S. Liu, L. Bönig, J. Detch, and H. Metiu, Phys. Rev. Lett. **74**, 4495 (1995).
66. Y.W. Mo, J. Kleiner, M.B. Webb, and M.G. Lagally, Phys. Rev. Lett. **66**, 1998 (1991).
67. S. Günther, E. Kopatzki, M.C. Bartelt, J.W. Evans, and R.J. Behm, Phys. Rev. Lett. **73**, 553 (1994); T.R. Linderoth, J.J. Mortensen, K.W. Jacobsen, E. Lægsgaard, I. Stensgard, and F. Besenbacher, *ibid*, **77**, 87 (1996).
68. J.-P. Bucher, E. Hahn, P. Fernandez, C. Massobrio, and K. Kern, Europhys. Lett. **27**, 473 (1994); Y. Li, M.C. Bartelt, J.W. Evans, N. Waelchli, E. Kampshoff, and K. Kern, Phys. Rev. **56**, 12539 (1997).
69. J. Villain, A. Pimpinelli, L. Tang, and D. Wolf, J. Physique I **2**, 2107 (1992); S. Liu, L. Bönig, and H. Metiu, Phys. Rev. B **52**, 2907 (1995).
70. M.C. Bartelt, S. Günther, E. Kopatzki, R.J. Behm, and J.W. Evans, Phys. Rev. B **53**, 4099 (1996).

Equilibrium shape of a coherent epitaxial cluster

Christophe Duport[1], Catherine Priester[2], Jacques Villain[1]
[1] *CEA, Département de Recherche Fondamentale sur la Matière Condensée, SPSMS, 38054 Grenoble Cedex 9, France*

[2] *IEMN, Département ISEN. Av. Poincaré BP 69 F-59652 VILLENEUVE D' ASCQ CEDEX - France*

Abstract

A qualitative investigation of the equilibrium shape of coherent, isolated, adsorbed clusters is presented, using analytic formulae which are correct for small thickness, and appropriate qualitative evaluations for thick islands. Qualitative (and partly new) evaluations are obtained for: i) the energy of a cluster having the shape of a truncated pyramid, ii) the volume at which the transition from a flat shape to a three-dimensional cluster takes place, iii) the corresponding activation energy, iv) the optimal size of an assembly of flat clusters. Near the transition, the shape is, within our approximation, a weakly truncated pyramid.

1 Introduction.

In this article, we address the equilibrium properties and metastable states of an adsorbate deposited *coherently* (i.e. without misfit dislocations) on a substrate. Such coherent structures of two different semiconductors are commonly used in microelectronics. We consider here *free* surfaces (as they are during growth) rather than buried adsorbates commonly used in technology. Indeed, due to growth conditions and burying process, the shape of buried systems is closer to equilibrium shape of free islands than to that of buried clusters. The interface will be assumed to be parallel to a high-symmetry direction. In technological applications, it has generally the (001) orientation.

The stability or metastability of the coherent state depends very much on the "misfit" $\delta a/a$, which is the relative difference betwen the atomic distances of the two materials when they are pure. If $\delta a/a$ is small, say less than 1 or 2%, layer by layer growth is possible and quantum wells can be realized. The adsorbate is constituted by completely filled layers. When the number of adsorbed layers increases beyond a certain "critical thickness" n_c, the system can in principle lower its energy by forming misfit dislocations, but n_c is large and, even for a thick adsorbate, the commensurate state can be metastable. Misfit dislocations are beyond the scope of the present article, but one can also consider the stability of the smooth adsorbate surface with respect to formation of bumps. Bump formation implies overcoming an activation energy W_{ac} which, as will be seen in section 6, is proportional [1] to $(\delta a/a)^{-4}$ and therefore very large for small misfits. For larger misfits, roughly until 2 or 3%, depending on growth conditions, W_{ac} can still be too high for bumps to form, and misfit dislocations appear for thick enough layers. In this paper, we have chosen to focus on the case of large misfits, when W_{ac} is so low that bumps can appear before misfit dislocations. Such bumps have a fairly regular, pyramidal shape and their size distribution is fairly narrow, so that they can be used as quantum dots. The InAs/GaAs [2,3,4,5,6] and Ge/Si [7,8,9,10,11] systems have been studied in a particularly extensive way. The reason of the narrow size distribution is still controversial [12,13,14] . An important feature in these systems is related to the effective *elastic* interaction between adsorbed atoms: this interaction results from the strain produced by adsorbed atoms. The strain produced by a given atom on a flat surface decays as the power -3 of the distance to this atom, so that the interaction energy between two atoms at points $\vec{r}$ and $\vec{r'}$ on a planar surface is [15]

$$W_{int}(r) = \frac{\text{Const}}{|\vec{r} - \vec{r'}|^3} \tag{1}$$

The constant turns out to be positive. Thus, the elastic interaction is long-ranged and repulsive. As will be seen in section 7, this does favour a narrow size distribution in a particular circumstance [12], namely, when the adsorbate forms one monolayer thick clusters. Such clusters will be called *platelets* in the following. Surprisingly the stable form of a cluster of given volume V, when the strain is taken into account, turns out to be a platelet [12,14] if V is not too large. This will be seen in section 6. For larger volumes, platelets can still be metastable. As will be seen in section 7, a population of platelets of fixed coverage has a preferred size [12]. This is a quite remarkable feature, which completely modifies Ostwald's theory of coarsening [16,17,18,19].

As well known, the equilibrium shape of a free crystal, as given by Wulff's construction [20,15], is independent of its volume. In that case, the energy is just the surface energy, or *capillary* energy, which is essentially proportional to the number of chemical bonds broken by the surface. In the present case, this contribution is still there, but an elastic contribution should be added, which in the simplest case is just the sum of the contributions (1) for all pairs.

The knowledge of the total energy of the system, i.e. the sum of the capillary and elastic contributions, is necessary to calculate the equilibrium shape, the equilibrium size and the activation energies. This energy can be calculated numerically [21,22,23] for any given configuration of the adsorbate, e.g. a collection of platelets of given size, but a systematic study would imply the knowledge of the energy of all configurations for all values of the misfit and of the elastic constants. Alternatively, what we shall do is an approximate analytic calculation of the energy of a *single* cluster in a simplified model, thus ignoring interactions between clusters. The assumptions about the shape of the cluster are explained in the next section, and those concerning the energy in sections 3 and 4.

2 Basic parameters for a pyramidal cluster

Our study will be restricted to a cluster having the shape of

i) either a square of thickness one monolayer and side-length R.

ii) or a full or truncated, regular pyramid with a square basis of side R and height h, with its flanks making an angle θ with the substrate surface (Fig. 1). Although a truncated pyramid is fully determined by these 3 parameters, it is of interest to introduce the side

$$R' = R - 2h \cot\theta \tag{2}$$

of the upper side, as well as the height

$$H = \frac{R}{2} \tan\theta \tag{3}$$

of the full pyramid which would result from prolonging the edges (Fig. 1).

The parameter

$$x = \frac{R'}{R} = 1 - \frac{2h}{R} \cot\theta \tag{4}$$

will turn out to be very convenient.

The volume of the truncated pyramid is

$$V = \frac{4h^3}{3} \frac{1 - x^3}{(1 - x)^3} \cot^2\theta \tag{5}$$

The equilibrium shape for a given volume V is obtained by minimizing the energy with respect to x, h and θ with the constraint (5).

In the case of a one monolayer thick, square cluster of side R, (5) is to be replaced by $V = cR^2$, where c is the thickness of a monolayer. In the case of the (001) surface of a binary semiconductor like InAs, a monolayer is a pair of one In layer and one As layer.

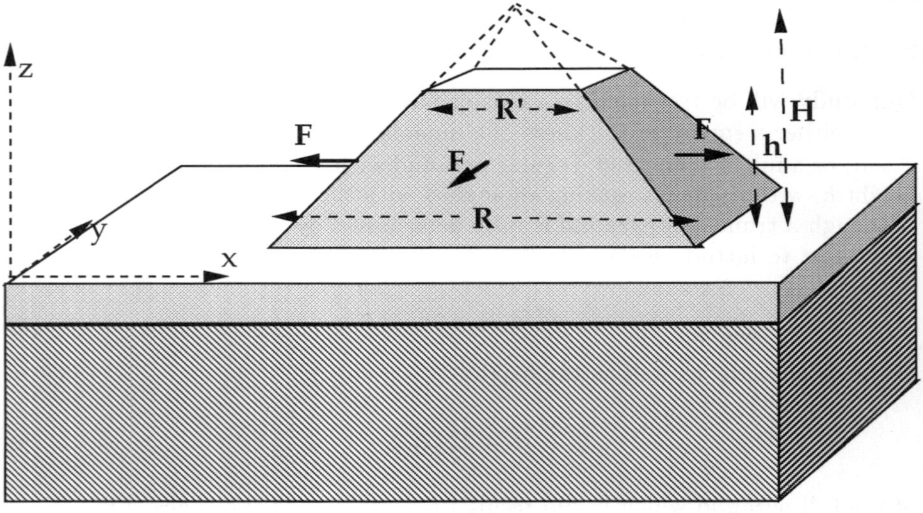

Figure 1: A cluster forming a truncated pyramid. The force F acting on the sides is proportional to the misfit $\delta a/a$.

3 Surface energy

The energy of any configuration with N_a adsorbed atoms will be counted from a reference state with the same number N_a of adsorbed atoms, where the adsorbate forms a layer of uniform thickness. Therefore, if the elastic strain is neglected, the bulk energy density vanishes and the energy reduces to the surface contribution. In terms of the parameters introduced in the previous section, the surface energy of a truncated pyramid contains 3 parts:

i) The first part, proportional to the total area of the trapezoidal faces and to their surface tension $\sigma(\theta)$, is

$$W_s' = 4\sigma(\theta)\frac{h}{\sin\theta}\frac{R+R'}{2} = 4\sigma(\theta)\frac{h}{\sin\theta}(R - h\cot\theta) \tag{6}$$

ii) The second part, proportional to the area R'^2 of the upper face and to its surface tension σ_0, is

$$W_s'' = \sigma_0 R'^2 \tag{7}$$

iii) The third part, proportional to the area R^2 of the lower face can be written as

$$W_s''' = (\gamma - \sigma_0)R^2 \tag{8}$$

where γ depends on how many full layers are adsorbed between the substrate and the pyramid. If there is none, γ is the interface tension of the adsorbate-substrate interface. In InAs/GaAs, γ is strongly negative and the pyramid is unstable. One should subtract σ_0 because of the energy origin which has been chosen. In the case of a pyramid deposited on a thick, smooth adsorbate layer, $\gamma = 0$ because the basis of the pyramid separates two identical media and has therefore no interface energy. In the following, γ will be neglected. In the case of InAs/GaAs(001), the pyramids are deposited on a single adsorbate monolayer, and we have numerically calculated the decay of the interface tension for a 2D InAs layer deposited on a (001) GaAs substrate as the layer becomes thicker and thicker. This calculation suggests that neglecting γ is a good approximation.

Taking into account that $(R^2 - R'^2)$ is the area of the projection of the trapezoidal faces, the total capillary energy reads

$$W_s = W_s' + W_s'' + W_s''' = 4\Gamma h\,(R - h\cot\theta) \tag{9}$$

where

$$\Gamma = \frac{\sigma(\theta) - \sigma_0 \cos\theta}{\sin\theta} \tag{10}$$

depends, in principle, on θ. However, if θ is small, the trapezoidal (or triangular) faces of the pyramid can be considered as formed by non-interacting steps and, since $4h(R - h\cot\theta)/c$ in (9) is the total step length, Γc can be regarded as the step energy per unit length and assumed independent of the orientation. *This approximation will be used throughout this article.* It is only qualitatively acceptable in practice, but it should give the correct trends.

As a function of x, h and θ, (9) reads

$$W_s = 4\Gamma h^2 \frac{1+x}{1-x} \cot\theta \tag{11}$$

Since the energy will have to be minimized at constant volume, it is of interest to use the variables V, x and θ, so that

$$W_s = (6V)^{2/3} \Gamma \mathcal{U}(x) \tan^{1/3}\theta \tag{12}$$

where

$$\mathcal{U}(x) = \frac{1-x^2}{(1-x^3)^{2/3}} \tag{13}$$

Finally, for a one-monolayer-thick, square cluster, formula (9) should obviously be replaced by

$$W_s = 4\Gamma c R = 4\Gamma\sqrt{Vc} \quad . \tag{14}$$

We shall now discuss the elastic energy resulting from the misfit.

4 Elastic energy: the Tersoff approximation

If the surface of the adsorbate were planar and smooth, as well as the adsorbate-substrate interface, the unit cell size in the direction parallel to the interface would be the same everywhere, the unit cell size in the direction perpendicular to the interface would be the same everywhere in the substrate and everywhere in the adsorbate, apart from a "relaxation" in the very vicinity of interfaces, which is beyond the scope of continuous elasticity theory, and will be disregarded in the present work. The presence of clusters disturbs the atomic distances in an inhomogeneous way, and the energy resulting from this perturbation will be called the elastic energy. The elastic energy in the substrate, assumed to be an isotropic medium with Lamé coefficients λ and μ, is

$$\mathcal{F}_s = \frac{\lambda}{2} \int_s d^3r \left[\epsilon_{xx} + \epsilon_{yy} + \epsilon_{zz}\right]^2 + \mu \int_s d^3r \left[\sum_{\alpha,\gamma} \epsilon^2_{\alpha,\gamma}\right] \tag{15}$$

where the subscript s in the integral indicates an integration on the substrate, and $\epsilon(x, y, z)$ designates the local strain.

In the adsorbate, the formula would be the same if the strains were counted from the free adsorbate. Since we count them from the uniformly constrained, coherent adsorbate, ϵ_{xx}, ϵ_{yy} and ϵ_{zz} must be respectively replaced by $\epsilon_{xx} + \epsilon_0$, $\epsilon_{yy} + \epsilon_0$ and $\epsilon_{zz} - 2\lambda\epsilon_0/(\lambda + 2\mu)$, where $\epsilon_0 = \delta a/a$. Moreover, an appropriate constant should be added to the free energy, so that it vanishes in the uniformly constrained, coherent adsorbate (but not in the free material). Finally, the free energy of the adsorbate is found after a straightforward calculation as

$$\begin{aligned} \mathcal{F}_a = & \frac{\lambda_a}{2} \int_a d^3r \left[\epsilon_{xx} + \epsilon_{yy} + \epsilon_{zz}\right]^2 + \mu_a \int_a d^3r \left[\sum_{\alpha,\gamma} \epsilon^2_{\alpha,\gamma}\right] \\ & + \frac{E_a \epsilon_0}{1 - \zeta_a} \int_a d^3r \left[\epsilon_{xx}(x, y, z) + \epsilon_{yy}(x, y, z)\right] \end{aligned} \tag{16}$$

where the subscript a in the integral indicates an integration on the adsorbate, λ_a, μ_a, E_a and ζ_a are the Lamé coefficients, Young modulus and Poisson ratio of the adsorbate.

Since the strain is related to the displacement $\vec{u}$ by $\epsilon_{xx} = \partial u_x/\partial x$ and $\epsilon_{yy} = \partial u_y/\partial y$, the last term of (16) can be integrated. Thus, the existence of the misfit ϵ_0 is equivalent to forces acting on the sides of the clusters as shown on figure 1. It may be as convenient not to make the integration, and to state that the misfit is equivalent to a uniform volume distribution of force dipoles in the adsorbate, with strength $E_a\epsilon_0/(1 - \zeta_a)$ per unit volume. This is exact in the harmonic approximation, since (16) is exact. Thus, the elastic energy would be known if we were able to calculate the linear response of the system *substrate+adsorbate* to the system of forces or force dipoles. This problem is quite difficult to treat analytically, and therefore we shall replace any force or force dipole acting at point (x, y, z) in the adsorbate by the same force or force dipole acting at the surface of the substrate at point $(x, y, 0)$. This approximation has been used by Tersoff and his coworkers [24,1] and we propose to call it the *Tersoff approximation*. It can be expected to give the total elastic energy with an acceptable accuracy if

i) the elastic constants of the both materials are not too different,

ii) the slope $\tan\theta$ is small.

Condition (ii) is necessary since the strains at step edges are obviously not accurately given, so that it is necessary to have few step edges. Nevertheless, our hope (to be tested by future numerical calculations) is that the Tersoff approximation yields the total energy with the correct order of magnitude even for $\tan\theta$ no too far from 1. Condition (i) results from the fact that the elastic energy is shared by the substrate and the adsorbate, except for a very thin adsorbate. A test of the Tersoff approximation for a thin adsorbate can be obtained from Hu's careful treatment [25].

We can now calculate the elastic energy. Within the Tersoff approximation, it is the same as if each atom of the adsorbate is assumed to be at the substrate-adsorbate interface. Therefore, the elastic energy is the sum of pair interactions given by (1). It is appropriate to call $w_{el,i,j}$ the partial sum for $\vec{r}$ and $\vec{r}\,'$ situated respectively in the i'th and j'th monolayer, so that the elastic energy is

$$W_{el} = 2\sum_{j}\sum_{i<j} w_{el,i,j} + \sum_{i} w_{el,i,i} \tag{17}$$

The successive floors are classified by increasing side length L_i, so that $i<j$ means $L_i < L_j$. A long algebraic calculation [26] yields for $i<j$

$$\begin{aligned} w_{el,i,j} = {} & -8L_j c^2 B\left[\sqrt{2} - \ln\left(\sqrt{2}+1\right)\right] + 8c^2 B \\ & \left\{\sqrt{2(L_i^2+L_j^2)} - \frac{L_i+L_j}{2}\ln\frac{\sqrt{2\left(L_j^2+L_i^2\right)}+L_j+L_i}{L_j-L_i}\right. \\ & \left. -\frac{L_j-L_i}{2}\ln\frac{\sqrt{2\left(L_j^2+L_i^2\right)}+L_j-L_i}{L_j+L_i}\right\} \end{aligned} \tag{18}$$

while, for $i=j$,

$$w_{el,i,i} \simeq -8c^2 B L_i\left[\sqrt{2} - \ln\left(\sqrt{2}+1\right) - 2 + \ln\frac{4L_i}{a}\right] \tag{19}$$

where the length a is of the order of the atomic distance, and

$$B = \frac{1+\zeta}{1-\zeta}\frac{E}{2\pi}\left(\frac{\delta a}{a}\right)^2 \tag{20}$$

where E is the Young modulus and ζ the Poisson ratio.

Formula (19) has a logarithmic divergence for long L_i. The factor which multiplies the logarithm is exact, but the additive constant (e.g. the factor 4 in the fraction) is only approximate because it depends on the short distance behaviour of the elastic interaction, which is not given by (1). For the same reason, (18) is only correct for $L_j - L_i \gg a$.

Particular cases: platelet, truncated and full pyramids

Formulae (17), (20) and (19) yield the elastic energy of a one-monolayer-thick, square platelet of radius R, namely[24]

$$W_{el} = -8c^2 \frac{1+\zeta}{1-\zeta} \frac{E}{2\pi} \left(\frac{\delta a}{a}\right)^2 R \ln \frac{R\eta}{a} \tag{21}$$

where

$$\ln \eta \approx 2\ln 2 - 2 + \sqrt{2} - \ln\left(\sqrt{2}+1\right) = -0.0809 \tag{22}$$

For the truncated pyramid described in section 2, $L_j = 2jc \cot\theta$. In the case $h >> c$, the second term of (17) can be neglected, the first one can be replaced by a double integral which can be calculated after a new piece of lengthy algebra. Finally, the elastic energy is found to be

$$W_{el} = -12 \frac{1+\zeta}{1-\zeta} \frac{E}{2\pi} \left(\frac{\delta a}{a}\right)^2 V G(x) \tan\theta \tag{23}$$

where x is defined by (4) and

$$\begin{aligned} G(x) = \quad & \frac{1}{1-x^3}\left\{\frac{1}{3}\left[\sqrt{2} - \ln\left(\sqrt{2}+1\right)\right](2+x)(1-x)^2 \right. \\ & \left. +\frac{2}{3}\left[\sqrt{2}\ln\left(1+\sqrt{2}\right) - 1\right](1+x^3) - \Xi(x)\right\} \end{aligned} \tag{24}$$

and

$$\Xi(x) = \quad -\frac{2x}{3}\sqrt{2+2x^2} + \frac{2x^3\sqrt{2}}{3} \ln \frac{1+\sqrt{1+x^2}}{x}$$

$$+\frac{2\sqrt{2}}{3}\ln\left(x+\sqrt{1+x^2}\right)$$
$$-\frac{1-x-x^2+x^3}{2}\ln\frac{1+x+\sqrt{2(1+x^2)}}{1-x}$$
$$+\frac{1+x-x^2-x^3}{2}\ln\frac{1-x+\sqrt{2(1+x^2)}}{1+x} \tag{25}$$

Equation (23) is remarkably simple except for the form of the function $G(x)$. For a flat truncated pyramid ($h << (R/2)\tan\theta$, or $x \approx 1$) formulae (23), (24) and (25) yield, after some algebra, the result of Tersoff and LeGoues [1]:

$$W_{el} = -4\frac{1+\zeta}{1-\zeta}\frac{E}{2\pi}\left(\frac{\delta a}{a}\right)^2 h^2(R+R')\ln\frac{(R+R')\eta'}{R-R'} \tag{26}$$

except that, instead of $\eta' = e^{3/2} = 4.48$ as found by Tersoff and LeGoues, we obtain

$$\ln\eta' = \sqrt{2}-\frac{1}{2}-\ln\left(\sqrt{2}+1\right)+\ln 2 = 0.726 \tag{27}$$

and $\eta' = 2.07$.

In the case of a full pyramid ($h = H$, $R' = 0$, $x = 0$) formulae (23), (24) and (25) yield

$$W_{el} = -8\left(\sqrt{2}-1\right)\left[1+\ln(1+\sqrt{2})\right]\frac{1+\zeta}{1-\zeta}$$
$$\frac{E}{2\pi}\left(\frac{\delta a}{a}\right)^2 V\tan\theta \tag{28}$$

The predictions of the Tersoff approximation have been compared with numerical calculations based on a Keating potential [12] for platelets and for pyramids with (104) orientation (i.e. $\tan\theta \approx 1/4$). The elastic energy variation law can be fitted from formulae (21) and (28) respectively, by making use of an effective Young modulus E_{eff}. Both formulae lead to approximately the same value of $E_{eff} \approx 5$ eV/atom, of the same order of magnitude, even if smaller, than the actual Young modulus.

The value of Γ deduced from (12) with $x = 0$ is

$$\Gamma c^2 = 140\text{meV}$$

It is more difficult to deduce Γ from formula (14) valid for platelets because a contribution from (21) should be added, which depends on η which is not very well known. Neglecting this correction, one finds $\Gamma c^2 \approx 80$ meV. The difference between both values, as well as the discrepancy between E_{eff} and E can be imputed to limitations of the Tersoff approximation such as a large interaction between steps, which is not unexpected in semiconductors.

5 Elastic energy: beyond the Tersoff approximation

We will now show that the Tersoff approximation completely fails for a *full* pyramid with $\tan\theta >> 1$, i.e. $h >> R$.

Inside the substrate, the strain created by a periodic perturbation of wave vector q is known to decay as $\exp(-q|z|)$ with the depth $|z|$. In the case of an isolated cluster, the profile $z(x, y)$ can be written as a Fourier integral dominated by Fourier components of wavelength of order R and the strain is expected to penetrate inside the substrate to a depth of order R and to decay as the reciprocal cube of the depth at larger depth [15]. Similarly, *inside* a steep adsorbed pyramid, the strain (counted from the free adsorbate) should penetrate until a height of order R above the interface. Therefore, the part of the pyramid whose height is larger than R is almost completely relaxed. The free energy density of that relaxed part is obtained from (16), with $\epsilon_{xx} = \epsilon_{yy} = -\epsilon_0$ and $\epsilon_{zz} = 2\lambda\epsilon_0/(\lambda + 2\mu)$, and is equal to $3K$, with

$$K = \frac{1}{3}\frac{E}{1-\zeta}\left(\frac{\delta a}{a}\right)^2 . \tag{29}$$

The total elastic energy W_{el} is obtained by multiplying the relaxed energy density by the volume $hR^2/3$ of the pyramid, and adding a positive correction proportional to the volume R^3 of the constrained zone (which extends both in the substrate and the adsorbate). Thus

$$W_{el} \approx -KhR^2 + QKR^3 \tag{30}$$

where Q is a positive constant, presumably of order unity.

6 Equilibrium shape of an epitaxial cluster of volume V

a) Optimal h at fixed $\tan\theta << 1$

The total energy W of a truncated pyramid, to be minimized, is the sum of (12) and (23) and has the form

$$W = CV\left[\alpha V^{-1/3}\mathcal{U}(x) - G(x)\right] \tag{31}$$

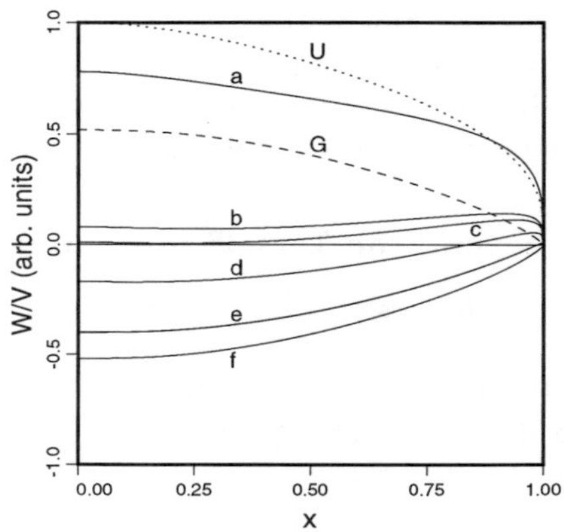

Figure 2: The functions $\mathcal{U}(x)$ (dotted line) et $G(x)$ (dashed line) defined by (13) and (24). The full curves show the total energy for several values of $\alpha V^{-1/3}$. For $\alpha V^{-1/3} > 0.53$ (1.3 for curve a and 0.6 for curve b), the energy mi nimum corresponds to $x = 1$ and platelets. For $0.12 \leq \alpha V^{-1/3} \leq 0.53$ (.53 for curve c, .35 for curve d and 0.12 for curve e), the weakly marked minimum corresponds to $0 \leq x \leq 0.25$ and slightly truncated pyramids. For $\alpha V^{-1/3} \leq 0.12$ (curve f), the energy minimum corresponds to $x = 0$ and full pyramids.

where $\mathcal{U}(x)$ and $G(x)$ are defined by (13) and (25) and

$$C = 12 \frac{1+\zeta}{1-\zeta} \frac{E}{2\pi} \left(\frac{\delta a}{a} \right)^2 \tan\theta \tag{32}$$

and

$$\alpha = \frac{1-\zeta}{1+\zeta} \frac{\pi\Gamma}{6^{1/3} E} \left(\frac{\delta a}{a} \right)^{-2} \tan^{-2/3}\theta \tag{33}$$

depending only on θ, $\delta a/a$, E, ζ and Γ.

The function $W(x)$ defined by (31) is displayed for various values of $\alpha V^{-1/3}$ by Fig. 2. For $\alpha V^{-1/3} > 0.53$ (i.e. for small volumes) $W(x)$ is minimum for $x = x_{min} = 1$, i.e. the equilibrium shape is a platelet. For larger volumes, there is a discontinuous jump to $x_{min} = x_c \simeq 0.25$, and then a continuous decrease of x_{min} towards the value 0. Thus, the pyramid tends to become full. Anyway it is never strongly truncated, according to the Tersoff approximations and to our additional approximations.

The transition from the platelet to the pyramidal shape occurs at a volume

$$V_c = \left[\frac{\alpha \mathcal{U}(x_c)}{G(x_c)} \right]^3 \simeq \left[\frac{\alpha}{G(0)} \right]^3 \tag{34}$$

The second form, where x_c has been replaced by 0, results from the inspection of Fig. 2. The above results are correct for $\tan\theta \ll 1$, but can be expected to be qualitatively acceptable even for $\tan\theta \approx 1$.

Numerical calculations based on a Keating potential have been made[12] for InAs/GaAs. Both platelets and pyramids with (104) orientation (i.e. $\tan\theta \approx 1/4$) which correspond to the experimental observation, have been treated. The latter were found to be more stable than the former for clusters of more than 16 000 atoms, while platelets are stable more stable than pyramids for smaller sizes. Experimentally, the smallest pyramids which have been observed contain 50 000[2] or 20 000[4] atoms. It is interesting to note that applying Fig. 2 to InAs/GaAs confirms that the observed clusters are slightly truncated (the optimal shape is got for $x = 0.165$ for 20 000 atoms clusters and $x = 0.100$ for 50 000 atoms clusters. Moreover (33) and Fig. 2 both together explain why, for lower misfits, the range of cluster sizes which lead to clearly truncated pyramids is widened. This has been observed for InAs/InP [27], InP/GaAs [28] and InGaAs/GaAs [29].

b) Optimal θ

The optimal θ, is obtained by minimizing W, defined by (31), (32) and (33), with respect to θ. We shall assume a full, non-truncated pyramid i.e. $x = 0$. This is reasonable because, as seen in the beginning of this section, the pyramid is never strongly truncated and, when the truncation is largest (25%) the energy is almost the same as that of a full pyramid (see Fig. 2). These statements hold for any value of θ.

Let us start with a very small volume V. Then, W is an increasing function of θ as far as (31) is valid. For $\tan\theta_M > 1$, one should apply (30). Adding the capillary energy (12), the total energy of a pyramid is obtained as

$$W \approx (6V)^{2/3}\Gamma \tan^{1/3}\theta - KV + 2QKV\cot\theta \tag{35}$$

and again W is an increasing function of θ. There is therefore a single minimum at $\theta = 0$ (Fig. 3, curve a), which corresponds to very flat clusters which are actually platelets.

If V increases, but not too much, (31) is an increasing function of θ for small $\tan\theta$ until it reaches a maximum for a value $\tan\theta_M$ smaller than 1. Then W decreases until $\tan\theta$ reaches a value of order 1. Then, one should apply (35), which is an increasing function of θ for large θ. One concludes that the energy has a minimum (Fig. 3, curves b and c) for a value $\tan\theta_m$ of order 1. Platelets are stable for small V (Fig. 3, curve b), as long as $W(\theta_m)$ is greater than the energy of a platelet, given by the sum of (14) and (21),

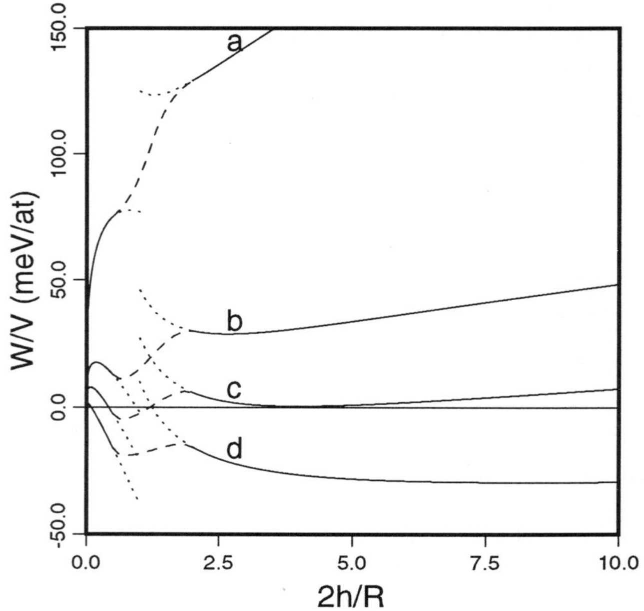

Figure 3: Approximate behaviour of the energy variations of a full pyramid of given volume as a function of $\tan\theta = 2h/R$. For small values of $2h/R$ (full lines for $2h/R < 0.5$ and dotted lines for $0.5 < 2h/R < 1$) one uses (31) , whereas, for sharper pyramids (full lines for $2h/R > 2$ and dotted lines for $1 < 2h/R < 2$) one uses (35). The dashed line is a spline interpolation between both full lines (in the area where $2h/R$ can neither be considered as $\ll 1$ nor as $\gg 1$). Numerical application has been done for InAs/GaAs, and four pyramid volumes have been considered : 50 (a), 1000 (b), 5000 (c) and 100 000 (d) atoms . As discussed in the text, for small volumes, the absolute minimum corresponds to a pyramid with $\tan\theta \approx 1$, whereas for large volumes the absolute minimum corresponds to a sharp pyramid.

$$W = 4\Gamma\sqrt{Vc} - 8c^2 \frac{1+\zeta}{1-\zeta} \frac{E}{2\pi} \left(\frac{\delta a}{a}\right)^2 \sqrt{V/c} \ln \frac{\eta\sqrt{V/c}}{a} \qquad (36)$$

For larger volumes, the pyramidal shape becomes stable (Fig. 3, curve c). The existence of a potential barrier at θ_M is in agreement with the first order nature of the transition, demonstrated in the previous subsection.

The result $\tan\theta_m \approx 1$, or lower, as can be seen in Fig. 3(b,c), is in qualitative agreement with the experimental fact [2,3,4,5,6] that observed pyramids in InAs/GaAs(001) are rather flat. The generally observed $\tan\theta_m \approx 1/4$ is somewhat lower the $\tan\theta_m$ one can deduce from Fig. 3. This disagreement can be due to the approximations made both on W_s and on W_{el}. Moreover, one cannot forget that the exact facets orientation of the pyramid is strongly related to surface reconstruction (often 2x4). As emphasized in section III, this has not been taken into account within our surface energy description.

For very large volumes (Fig. 3, curve d), the optimal pyramid is given by the minimum of (35) and corresponds to $\tan\theta \simeq (KV/\Gamma)^{1/4}$. However, in this case where the elastic energy overwhelms the surface tension, adsorbed clusters have no reason to be pyramids. They rather tend to expel themselves from the substrate and to form presumably bubbles similar to those observed in the Vollmer-Weber growth. The difference, however, is that the contact angle is not fixed. For very large coherent bubbles, it can be shown that the size of the interface tends to a finite limit, so that the contact angle tends to the value π. Whether very large clusters can still be coherent is questionable. The evolution of the stable size of an isolated coherent cluster with its volume V is displayed by Fig. 4.

The equations of continuous elasticity have been solved numerically (after discretization) by Johnson and Freund [30] for a dome-shaped coherent cluster. They find that, at the top of the cluster, the adsorbate strain (counted from the free adsorbate) can be negative. This possibility is not included in our approximation.

c) Activation energy

To determine the activation energy, one has to find how to transform continuously a platelet into a pyramid at the minimal energy cost. It can be argued that the best way is to make a pyramid on the platelet and to increase its volume progressively, keeping $x \approx 0$ and $\tan\theta \approx 1$ constant. The energy (31) of the pyramid is maximum for $V' = V_M = [2\alpha/(3G(0))]^3$ and its value at the maximum is, in qualitative agreement with Tersoff and LeGoues [1]

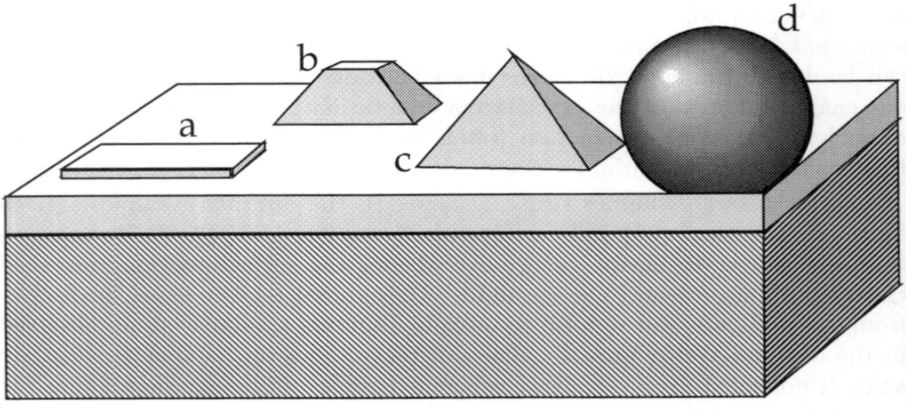

Figure 4: Successive shapes of an isolated, coherent cluster for increasing volum e: a) platelet. b) weakly truncated pyramid. c) nearly full pyramid. d) Bubble (with facets which have not been represented)

$$W_{ac} \simeq \frac{4C\alpha^3}{27G^2(0)} \tag{37}$$

This is the activation energy, provided the energy change due to the change of the platelet size can be neglected, as it can for large sizes or for small misfits. Indeed, the value (37) is proportional to $(\delta a/a)^{-4}$ according to (32) and (33), and the platelet energy is of order $(\delta a/a)^{-3}$ according to (36). For the large misfits of interest here, (37) is expected to give the right order of magnitude.

7 Optimal size of a collection of clusters.

In the previous section, the optimal shape of a coherent cluster of fixed volume has been investigated. It is actually reasonable that, in the absence of growth, the clusters have a fixed volume, because atoms presumably do not diffuse easily from a cluster to another one. In the present section, it is assumed that atoms can diffuse between clusters. When elasticity is disregarded, it is known that atoms preferably go from small to big clusters [16,17,18,19], so that the average cluster size increases to infinity. The effect of elasticity on kinetics has been addressed by Chen and Washburn [14], by Dobbs et al [31] and in the numerical work of Ratsch et al [32]. Here, only the equilibrium distribution will be discussed: is there, in principle, a huge, single cluster? Or are there many clusters with a well-defined size? In the forthcoming study, interactions between clusters are neglected, as correct only if their distance is very large.

The answer is given by minimizing W/V with respect to x, θ, *and* V. To simplify this operation, we first assume a pyramidal shape. It follows from the previous section that, for large enough volumes, the optimal shape corresponds to $\tan\theta > 1$, and in that case formula (35) is qualitatively applicable. It yields

$$W/V \approx (6)^{2/3} V^{-1/3} \Gamma \tan^{1/3}\theta - K + 2QK\cot\theta \tag{38}$$

This is minimum for a single, very sharp pyramid. As argued in the preceding section, this pyramid is unstable and transforms into a bubble, but the above argument is sufficient to show that the stable form of a sufficiently large coherent adsorbate is a single cluster. Taking into account interactions between pyramids would not alter this conclusion. Therefore, the evolution of a population of coherent clusters is expected to increase the average size, although perhaps less rapidly [14,31] than in the absence of elastic effects.

Platelets

The energy density of a collection of non interacting platelets is given by (36), namely

$$W/V = \frac{4\Gamma}{R} - 4K'\frac{c}{R}\ln\frac{R\eta}{a} \tag{39}$$

where

$$K' = 2\frac{1+\zeta}{1-\zeta}\frac{E}{2\pi}\left(\frac{\delta a}{a}\right)^2 \tag{40}$$

The energy density (39) is minimum when R is equal to

$$R_0 = a\eta^{-1}\exp\frac{\Gamma + K'c}{K'c} \tag{41}$$

Insertion into (39) yields the minimum of energy density

$$(W/V)_{min} = \frac{-4K'c\eta}{a}\exp\frac{-(\Gamma + K'c)}{K'c} \tag{42}$$

The existence of an optimal radius R_0 shows that a collection of coherently adsorbed platelets evolves, as first noticed by Priester and Lannoo, towards a situation where all platelets have the same radius R_0. As demonstrated in ([12]), taking into account interactions between platelets just shifts the optimal radius R_0 towards larger sizes. If the misfit is small, K' is large, R_0 is unphysically large and (42) is extremely small in absolute value.

In the case of large misfits addressed here, it is not so. For InAs/GaAs(001), R_0 corresponds to 26 000 atoms and (42) is of the order of magnitude of the maximum energy density gain K. However, it is smaller than K in absolute value, so that the platelet population is in principle unstable. It can, however, be metastable if the activation barrier evaluated in the preceding section is high.

8 Conclusion.

We have discussed the equilibrium shape of coherent, adsorbed clusters within an approximation which yields analytical formulae for an *isotropic* solid and, even in that case, would only be quantitatively correct for very thin clusters. The isotropic approximation would be quantitatively correct for a cubic material in which the ratio $A = 2c_{44}/2(c_{11} - c_{12})$ would be close to 1. In GaAs and InAs, it is of order 0.5 as seen from the following table, where the elastic constants are given in units of 10^{11} ergs/cm^3.

	c_{11}	c_{12}	c_{44}	A
GaAs	11,85	5,35	5,94	0,45
InAs	8,53	4,53	3,96	0,52

In the case of thick clusters, our approximation is expected to give the correct energy if the slope θ is weak, provided the elastic constants are not too different.

We have neglected the surface stress as well as the dependence of the surface energy on strain. These approximations are probably not too serious for a pyramid, but are questionable for a platelet.

In the frame of our approximation, we have confirmed the stability of the platelet shape for an isolated island of small volume, and of the pyramidal shape for larger volumes. This was already known [12,14] but we have evaluated the transition volume (34) and stressed the importance of the activation energy (37) which can impede the transition for weak misfits, thus making the formation of misfit dislocations unavoidable. We also have given the expression of the optimal size (41) of platelets, whose existence was derived by Priester and Lannoo [12].

The formation of three-dimensional clusters is analogous to the Grinfeld instability [33,34,20] which arises in a crystal subject to an external, uniaxial stress. However, the fact that these clusters develop on a singular surface has specific effects, for instance the existence of the activation energy (37). The investigation of the latest stage of the instability is also different when the size of the clusters is not limited by the adsorbate thickness, but by the steric interactions with other bubbles [35,36,37,38].

In growth problem, the equilibrium shape is clearly not the only relevant factor. We have completely left aside kinetic mechanisms [14,31,32] which are also important.

References

1. J. Tersoff and F.K. LeGoues, Phys. Rev. Lett. **72**, 3570 (1994).
2. D. Leonard, M. Krishnamurthy, C.M. Reaves, S.P. DenBaars, P.M. Petroff, Appl. Phys. Lett. **63** 3203 (1993)
3. R. Nötzel, K. Ploog, Adv. Materials **5** 22 (1993)
4. J.M. Moison, F. Houzay, F. Barthe, L. Leprince, E. André, O. Vatel Appl. Phys. Lett. **64** 196 (1994)
5. J.Y Marzin, J.M. Gérard, A. Izraël, D. Barrier, G. Bastard, Phys. Rev. Lett. **73** 716 (1994)
6. Z. Liliental-Weber et al, App. Phys. Lett. 68, 111 (1996)
7. D.J. Eaglesham et M. Cerullo, Phys. Rev. Lett. 64, 1943(1990)

8. B. Voigtländer et A. Zinner, Appl. Phys. Lett. **63** 3055 (1993)
9. Z. Liliental-Weber et al.,Phys. Rev. B 52, 16581 (1995)
10. S. Theiss et al Appl. Phys. Lett. **66** 448(1995)
11. D.E. Jesson, K.M. Chen, S.J. Pennycook , MRS bulletin ,31 (April 1996) and references therein
12. C. Priester, M. Lannoo, Phys. Rev. Lett. **75** 93 (1995)
13. V.A. Shchukin et al., Phys. Rev. Lett. **75** 2968 (1995)
14. Y.Chen and J. Washburn Phys. Rev. Lett. 77, 4046 (1996)
15. J. Villain and A. Pimpinelli, Physique de la Croissance Cristalline, Eyrolles-Alea-Saclay, Paris (1994) and english version (Cambridge University Press), to be published.
16. Ostwald, W. (1908) "Foundation of Analytical Chemistry", p. 22 (Mc Millan, London)
17. Lifshitz, I.M., Slyozov, V.V. (1961) J. Phys. Chem. Solids **19**, 35.
18. Chakraverti, B.K. J. Phys. Chem. Sol. **28** 2401 et 2413 (1967). Avignon, M., Chakraverti, B.K. (1969) Proc. Roy. Soc. **A 310**, 277. Avignon, M. (1971) J. Cryst. Growth **11**, 265.
19. Villain, J. (1986) Europhys. Lett. **2**, 531.
20. P. Nozières, Lectures at the Beg Rohu Summer School (1989), in *Solids far from equilibrium*, C.Godrèche Ed. (Cambridge Univ. Press,1991).
21. C. Priester, I. Lefebvre, G. Allan, M. Lannoo, Mat. Res. Soc. Symp. Proc. **317** 131 (1994).
22. Y. Androussi, P. François, A. Lefebvre, C. Priester, I. Lefebvre, G. Allan, M. Lannoo, J.M. Moison, N. Lebouche, F. Barthe, Mat. Res. Soc. Symp. Proc. **355** 569 (1995).
23. E. Pehlke, N. Moll and M. Scheffler, Proc. 23rd Int. Conf. on the Physics of Semiconductors, Eds M. Scheffler and R. Zimmermann (World Scientific, Singapore) 1301 (1996).
24. J. Tersoff and R.M. Tromp, Phys. Rev. Lett. **70**, 2782 (1993).
25. S.M. Hu, Appl. Phys. Lett. **50**, 4661 (1979)
26. C. Duport, thèse (Université de Grenoble, juin 1996)
27. A. Ponchet, A. Le Corre, H. L'Haridon, B. Lambert, S. Salaün, Appl. Phys. Lett. **67**, 1850 (1995).
28. M.E. Pistol et al, Proc. 23rd Int. Conf. on the Physics of Semiconductors, Eds M. Scheffler and R. Zimmermann (World Scientific, Singapore) 1317 (1996).
29. N. Grandjean, J. Massioes, O. Tottereau, Phys. Rev. B. **55**, 10189 (1997)
30. H.T. Johnson, L.B. Freund, preprint (1996); L.B. Freund, H.T. Johnson, R. V. Kukta, Mat. Res. Soc. Symp. Proc. **399** 359 (1996)

31. H.T. Dobbs, A. Zangwill, D.D. Vvedensky, Phys. Rev. Lett. **70**, 897 (1997)
32. C. Ratsch, P. Smilauer, D.D. Vvedensky, A. Zangwill, J. Physique I **6** 575 (1996)
33. R.J. Asaro, W.A. Tiller, Metall.Trans. **3**, 1789 (1972).
34. M.Ya. Grinfeld, Dokl. Akad. Nauk SSSR **283**, 1139 (1985) [Sov. Phys. Dokl. **31**, 831 (1986)] and J. Nonlinear Sci. **3**, 35 (1993)
35. P. Nozières, J. de Physique I **3**, 681 (1993).
36. J. Grilhé, Europhys. Lett. **23**, 141 (1993).
37. Bena, C. Misbah and A. Valance, Phys. Rev. B **47**, 7408 (1993)
38. K. Kassner and C. Misbah, Europhys. Lett. **28**, 245 (1994)

Dynamic Scaling in Epitaxial Growth

S. Das Sarma
Physics Department, University of Maryland, College Park, Maryland 20742-4111

1 Outline

In this article I provide a brief theoretical perspective on our current understanding of the dynamic scaling phenomena in nonequilibrium epitaxial growth with the emphasis on the extensively studied ultrahigh vacuum thin film deposition growth, such as MBE growth, where the main growth front smoothening mechanism competing against the surface kinetic roughening induced by inherent shot noise fluctuations in the random deposition process is the adatom mobility associated with surface diffusion at the growth front. The Introduction sets the tone, and in subsequent sections I expand upon the key themes arising in the Introduction. In particular, the issues of universality, continuum growth equations, atomistic growth models, and unstable growth as in smooth growth and mound formation are discussed in the four sections following the Introduction.

2 Introduction

Epitaxy is usually thought of as an extremely smooth growth process leading to atomically sharp surfaces/interfaces with little ($\sim$ one atomic monolayer) interface roughness. In particular, layer by layer epitaxial growth on a singular surface takes place via two dimensional island formation where the surface roughness oscillates between zero and one monolayer as each new atomic layer is formed by the deposited atoms from the incident flux with no more than a few (~ 1) incomplete atomic monolayers being active at the growth front. The reason for this extremely smooth epitaxial growth morphology is the high adatom mobility at the growth front enabling the incident atoms to be "quickly" (i.e. before the arrival of the next incident atom in its neighborhood) incorporated at epitaxial growth sites on the surface resulting in the atomically flat growth morphology. The somewhat loosely defined "epitaxial growth temperature" (T_e) is, in fact, thought to be the temperature which is high enough for fast adatom diffusion to cause layer by layer (step flow) growth on a singular (vicinal) surface, but low enough for atomic desorption from the growth front to be unimportant. This scenario for epitaxial growth is well established in the literature through *in situ* RHEED analysis (and other similar experimental techniques) of the growth front and through extensive computer kinetic growth simulations.

It turns out that this well-accepted paradigm for morphologically smooth epitaxial growth is conceptually flawed from a statistical mechanical perspective — the smooth epitaxial growth mode is actually unstable. The layer by layer growth mode on a singular surface (or the step flow growth mode on a vicinal surface) is, in fact, always an initial transient which asymptotically must lead to kinetically rough growth [1,2,3,4] at long times and large distances. The random fluctuations (i.e. the shot noise) inherent in the incident beam dominate the adatom surface diffusion at long times (i.e. after the initial layer by layer growth transient), and the interface roughness (as measured, for example, by the root mean square fluctuation in the interface width, or equivalently, by the number of incomplete layers dynamically active at the growth front) always increases monotonically (until it reaches saturation at a steady state determined by the substrate size) as a function of time after the initial transient (during the initial transient the interface roughness oscillates between zero and one monolayer indicating smooth epitaxial growth with an essentially flat atomically sharp growth morphology). It is indeed true that for fast enough adatom diffusion the (unstable) layer by layer growth transient may last for many deposited layers, and for many practical purposes this may suffice. Also, for fast diffusion (i.e. high temperatures) and small enough substrate sizes, the steady state saturated roughness may be small[5] producing a smooth surface.

In addition to unstable layer by layer and asymptotically "stable" kinetically rough growth modes, there has been much recent interest in a third surface growth scenario where mounds or pyramids dominate the growth morphology, and these mounds/pyramids may coarsen or steepen with time. Such a surface growth process via mound formation is thought to arise [6] from the so-called Ehrlich-Schwoebel (ES) barrier existing at surface step edges which hinders the downward adatom motion by introducing an additional step-edge diffusion bias. The additional step-edge diffusion bias makes it less likely for an adatom to attach to a down-step than to an up-step, and therefore gives rise to characterstic "wedding-cake" structures on the growing surface as adatoms on upper terraces are prevented by the Ehrlich-Schwobel barrier from coming down. The net result is a rough morphology (albeit of a particular type involving mounds/pyramids or similar "wedding-cake" structures) with the roughness increasing monotonically in time as growth progresses.

The kinetic surface roughening phenomenon is experimentally well established[1,2,3,4,5,6,7]. Layer by layer epitaxial growth on a flat singular surface manifests itself in RHEED intensity oscillations from the growing surface — each oscillation corresponds to the completion of a single layer. The RHEED intensity oscillation phenomenon associated with the layer by layer growth process is so well-established that it is routinely used as a characterization tool in MBE growth monitoring the number of grown layers. Experimentally it is known that at low temperatures growth is kinetically rough or three-dimensional (because the mobility of the incident atoms is low) as indicated by there being no RHEED intensity oscillations (the intensity drops monotonically with time indicating progressively rougher growth) whereas at high temperatures (when the adatoms are "sufficiently" mobile) growth occurs via the two dimensional layer by layer mechanism as indicated by the existence of pronounced RHEED intensity oscillations. The important conceptual point is, however, the generic observation that these pronounced RHEED intensity oscillations always decay and eventually disappear, indicating that the high-temperature smooth two-dimensional layer by layer growth is an unstable transient, which eventually turns into kinetically rough three dimensional growth. How long the transient layer by layer growth mode persists depends on the adatom mobility (which in turn depends exponentially on the growth temperature), and at high enough temperatures it may be possible to grow many layers without appreciable kinetic roughening. It should be noted, however, that the growth temperature cannot be arbitrarily high so as to avoid desorption from the growth front. In practice, the growth temperature should be as high as possible (within the restriction of avoiding significant desorption) to obtain the best layer by layer epitaxial growth. (I should mention here that most real surfaces being necessarily slightly vicinal, RHEED intensity oscillations actually disappear at high temperatures as the diffusion length becomes comparable to the terrace size and two-dimensional layer by layer growth mode gives way to the step flow growth mode — in a truly high symmetry singular surface this would not happen and the high temperature limit then is set by the no-desorption constraint.)

One of the key recent developments in epitaxial growth phenomena has been the realization that the coarse-grained kinetically rough growth morphology follows a rather general dynamical scaling behavior. In particular, the root mean square fluctuation (W) in the evolving surface height, $h(\boldsymbol{r},t)$ where h is the height of the interface at time t for the substrate position $\boldsymbol{r}$, obeys the following dynamic scaling law (a similar dynamic scaling law holds for the height-height correlation function as well [1,2,3,4]):

$$W(L,t) \sim L^{\alpha} f(L/\xi(t)), \tag{1}$$

where

$$W^2 = \langle (h(\boldsymbol{r},t) - < h >)^2 \rangle, \tag{2}$$

is the *mean* squared interface width ($< h >$ being the average film thickness, $< h >= Ft$ where F is the average growth rate), L is the substrate width, and the correlation length $\xi(t)$ grows with time as

$$\xi(t) \sim t^{1/z}, \tag{3}$$

with z being the dynamical exponent (which describes how correlations spread laterally along the substrate over time), and α the roughness exponent. At long times, when $\xi \sim t^{1/z} \gg L$, lateral correlations have essentially spread over the whole substrate, and the system reaches a saturated steady state, indicating that the scaling function $f(x)$ must have the long-time asymptotic form:

$$f(x \gg 1) \sim 1. \tag{4}$$

In the short-time situation ($\xi(t) \ll L$), however, the surface roughness W should be independent of the lateral system size L because lateral correlations have not yet spread over the whole substrate, and therefore the short-time asymptotic form for the scaling function $f(x)$ must be:

$$f(x \gg 1) \sim x^{-\alpha}. \tag{5}$$

Combining the above equations one obtains the following limiting behavior of the dynamical surface roughness $W(L,t)$:

$$\begin{aligned} W(L,t) \quad & \sim t^{\beta} \text{ for } t \ll L^z \\ & \sim L^{\alpha} \text{ for } t \gg L^z, \end{aligned} \tag{6}$$

where $\beta = \alpha/z$ is the growth exponent.

The physical picture underlying the dynamic scaling description is the following. The shot noise associated with the random deposition of the incident beam produces kinetic roughness in the evolving growth morphology which leads to monotonic increase in the interface width (roughness) W as a function of time. In the limit of very large substrates ($L \to \infty$) there is no natural limiting length scale in the problem and therefore the kinetic roughness $W \sim t^{\beta}$ increases forever. This increase follows a power law, $W \sim t^{\beta}$, because there is no natural time scale in the problem other than that defined by the growth rate which defines the unit of time. Note that the average thickness of the growing film is, by definition, proportional to the growth time, $< h >= Ft$, and therefore in all these scaling relations one could replace t by $< h >$. For a finite substrate, a steady state is reached when the lateral correlations spread over the whole system ($\xi \geq L$) and the interface width saturates to a steady state value, $W(L, t \gg L^z) = W_0 \sim L^{\beta z} \equiv L^{\alpha}$, which naturally scales as a power of the system size. The key point is the existence of a lateral (dynamical) correlation length, $\xi(t) \sim t^{1/z}$, which scales with time according to the dynamical exponent z. This is, of course, the standard scenario for dynamic scaling in physical phenomena. Note that the kinetically rough surface morphology is a self-affine fractal object which exhibits anisotropic generic scale invariance where the anisotropy arises from the inequivalence between growth and substrate directions [1,2,3,4].

It should be noted that there exists no rigorous proof that a dynamic scaling scenario must, in general, apply to the epitaxial growth morphology. But there is a great deal of empirical evidence, based on both theoretical and simulational work as well as rapidly accumulating experimental support, for the applicability of dynamic scaling to epitaxial growth. The essential idea underlying the dynamic scaling hypothesis is that there being no characteristic length and time scales in the problem, dynamic scaling must hold at large length scales and long times. I note that in all real epitaxial growth processes there obviously are several characteristic length scales (eg. the lattice size, the diffusion length, etc.) which determine the "critical regime", i.e. the asymptotic dynamic scaling sets in at distances (and times) much longer than the characteristic lengths (and times) naturally appearing in the problem. (Calculation [8] of these "short" distance cutoffs, eg. the diffusion length in MBE growth, could be quite tricky for specific experimental situations.) Below these "short" distance (time) cutoffs dynamic scaling does not occur, and (non-universal) transient behavior dominates. One example of such "small-scale" transient behavior is the smooth layer by layer growth regime with oscillatory surface roughness occuring at length scales smaller than $\tilde{\ell}$, where $\tilde{\ell}$ (determined by the adatom diffusion length) is the appropriate "coherence length" [9] for epitaxial growth. (Equivalently one could define a coherence time [9].)

The usual scale invariant dynamical scaling behavior is the stable long distance (time) asymptotic behavior of the system at large scales. It should be emphasized that in many practical situations the characteristic "short" distance or time cutoff (eg. the coherence length or the coherence time for layer by layer growth) may be so large that the stable asymptotic behavior is not of particular experimental or practical relevance, and the transient unstable growth regime completely dominates the experimental observations. For the layer by layer growth transient the "short distance" cutoff grows exponentially (or faster) with temperature, for example, the diffusion length and therefore the coherence length for the smooth layer by layer growth regime increases exponentially with temperature making it possible to grow very smooth large area thin films without any appreciable kinetic surface roughness. It is to be noted, however, that the layer by layer growth regime is always a fnite size (time) phenomenon with kinetic surface roughening dominating at large scales.

3 Universality in Kinetic Surface Roughening

The set of exponents $\alpha, \beta, z = \alpha/\beta$ for a particular growth process defines a dynamic universality class which characterizes the long wavelength asymptotic properties of kinetic surface roughening associated with that particular growth process. Motivated by the central role played by the concept of universality in our understanding of equilibrium critical phenomena, one naturally asks whether a few universality classes determined by symmetry, conservation laws, and other factors controlling growth dynamics could describe seemingly many different surface growth processes. In particular, the important question is whether there are only a few possible independent sets of exponents (α, β, z) describing epitaxial growth in different materials and under different growth conditions. We would of course also want to know what fundamental aspects of a growth process uniquely determine its universality class. A great deal of attention has recently been focused [1,2,3,4,5,6,7] on the issue of the possible universality class(es) of epitaxial growth (in particular, MBE type growth). Although the subject is very much in flux and no firm conclusion has yet been reached, some consensus has emerged in the last few years. I summarize below the currently existing consensus about the four different possible dynamic universality classes for kinetic surface roughening in epitaxial growth, and discuss the continuum equations appropriate for each of these four universality classes in the next section.

1. *KPZ universality* If the growth process is nonconserved (i.e. if desorption and/or formation of surface overhang and bulk vacancies is dynamically significant at the growth front), then the asymptotic universality class of the growth process is thought to belong to the Kardar-Parisi-Zhang (KPZ) universality class[10] whose exact exponents are known only in one substrate dimension ($d = 1 + 1$) with $\alpha = 1/2$, $\beta = 1/3$, $z = 3/2$ in $d = 1 + 1$. The KPZ exponents for the actual surface growth problem, where the substrate dimension is two ($d = 2+1$), are known only approximately (from extensive numerical simulations) to be $\alpha \simeq 0.39$, $\beta \simeq 0.24$, $z \simeq 1.61$. I note that as a matter of principle all epitaxial growth processes must be nonconserved at the largest scales of length and time because some desorption as well as some overhang/vacancy formation is unavoidable in any real growth problem. Thus, the current consensus is that the KPZ universality class is, in fact, the *generic* growth universality class which must apply to all growth problems at long enough scales of lengths and times. On the other hand, the KPZ exponents are almost

never found in any real epitaxial surface growth experiments where the measured roughness (α) and dynamical (z) exponents tend to be much larger than the KPZ values ($\alpha \simeq 0.39$; $z \simeq 1.61$). The KPZ universality class therefore has this dubious dichotomy of being theoretically generic for the large scale asymptotic growth behavior and, at the same time, being empirically inapplicable to almost all of the currently existing experimental data on surface growth. Part of the reason for this dichotomy obviously lies in the fact that in most epitaxial growth situations, desorption from the growth front as well as vacancy/overhang formation during growth are dynamically insignificant in the experimental time and length scales, and therefore the asymptotic generic universality simply does not manifest itself. After all, in epitaxial growth experiments, eg. MBE growth, one takes great care in avoiding evaporation and vacancy/overhang formation during growth, and it is therefore perhaps not surprising that the KPZ exponents are never seen in epitaxial growth experiments. Although numerical simulations of MBE growth allowing for defect (eg. overhangs and vacancies) formation rather clearly demonstrate [11,12,13] the expected asymptotic crossover to the KPZ universality, one of the most serious open issues in this subject has been the persistent lack [14] of clear and convincing experimental support for the existence of the generic KPZ universality in real epitaxial growth.

2. *EW universality* For conserved (i.e. no desorption or vacancy/overhang formation) solid-on-solid (SOS) type growth processes the KPZ universality is inapplicable, and at least three different possible conserved SOS growth universality classes have extensively been discussed in the literature during the last five years. The most obvious one is the so-called Edwards-Wilkinson (EW) growth universality class [15] which applies whenever the growth process involves some "downward funneling" type relaxation mechanism which pushes the growth front atoms down to the local height minima [16]. Any tendency (however small) for the atoms to preferentially attach to down-steps at terrace edges rather than to up-steps during growth automatically leads to the EW universality class. (Note that this tendency of preferential attachment to down-steps over up-steps is precisely opposite to what happens for growth under an Ehrlich-Schwoebel step-edge diffusion bias where atoms preferentially attach to up-steps rather than down-steps — this explains why one occasionally encounters in the literature statements to the effect that a "negative" Ehrlich-Schwoebel barrier leads to EW universality.) The EW growth universality class is characterized by the exactly known critical exponents $\alpha = 0(1/2)$, $\beta = 0(1/4)$, $z = 2(2)$ in $d = 2 + 1(1 + 1)$

dimensions. (The EW critical exponents for kinetic roughening are exactly known in arbitrary dimensions.) An important point is that EW universality predicts "smooth" growth ($\alpha = \beta = 0$) in the physical $2+1$-dimensional kinetic surface roughening process, implying only logarithmic increase in the kinetic roughness with time and system size. (Note that vanishing critical exponents implies logarithmic increase in kinetic roughening and *not* vanishing roughening as one may naively assume.) Of all the theoretically possible growth universality classes, the EW universality class has the smoothest evolving growth morphology because it has the smallest possible values of the growth (β) and the roughness (α) exponents in any dimension. This is significant from a technological perspective because large area smooth thin films and surfaces are required for most electronic materials applications. Understanding growth techniques and physical mechanisms which may induce EW universality in epitaxial growth would be immensely valuable technologically. It has recently been suggested[17] that surfactants induce smooth epitaxial growth by causing preferential attachment of deposited adatoms at down-steps and thereby introducing the EW universality behavior ($\alpha = \beta = 0$) in a natural manner. A systematic experimental search for various techniques which induce EW growth could be quite useful.

3. *MH universality* In conserved SOS growth situations where "surface diffusion" is dynamically significant in the absence of any EW relaxation process (i.e. no downward funneling or preferential attachment to down-steps), the growth process may belong to the linear surface diffusion (the so-called Mullins-Herring, MH) universality class[13,19]. The critical exponents for the MH growth universality class are exactly known theoretically: $\alpha = 1(3/2)$, $\beta = 1/4(3/8)$, $z = 4(4)$ in $d = 2+1(1+1)$ dimensions. Although several experimental claims in the literature report measurements of kinetic surface roughening exponents, $\beta \approx 0.25$ and $\alpha \approx 0.9 - 1$, consistent with the MH universality predictions, it is unclear at the present time whether the MH universality could ever be a true asymptotic universality class (at long times and distances) rather than being a short distance/time crossover phenomenon with the asymptotic universality being given by the EW universality defined in (2) above or the MBE universality defined in (4) below. The current theoretical consensus is that the MH universality can only be a crossover and *not* an asymptotic universality class. I note that the MH universality (sometimes also referred to as the "linear surface diffusion" universality) has the largest critical exponents α and β among the four epitaxially relevant universality classes being discussed here, implying that MH universality

produces the maximum large-scale kinetic surface roughness in the evolving growth morphology. Note that $\alpha = 1$ for the MH universality in the surface growth problem, which means that the saturated steady state interface width is proportional to the substrate size, $W(L, t \to \infty) \propto L$ in $d = 2 + 1$. In one dimension ($d = 1 + 1$), the roughness exponent, $\alpha = 1.5$, exceeds unity, which is a manifestly peculiar situation implying that the large scale steady state morphology of the growing surface is not self-affine in $d = 1 + 1$ because $W(L, t \to \infty)/L$ actually diverges as $\sqrt{L}$ in the thermodynamic limit. This extreme pathological roughness in the evolving growth morphology of the MH universality class effectively rules it out as a true asymptotic description for a physical surface growth problem even though it may very well dominate the short-time (non-asymptotic) kinetic surface roughening in many experimental situations where surface diffusion plays an important role. I mention that the dynamic exponent, $z = 4$, is very large for the MH universality class, making the evolution of lateral correlations extremely slow, $\xi(t) \sim t^{1/4}$, and this slow evolution of correlations considerably complicates studying crossover properties as well as steady state behavior in the MH universality problem.

4. *MBE universality* The most relevant universality class in the context of conserved epitaxial growth is the molecular beam epitaxy (MBE) universality, which somewhat confusingly goes by several other names in the literature as well: conserved epitaxial growth, conserved KPZ, nonlilnear surface diffusion, and Villain-Lai-Das Sarma (VLD) or equivalently Lai-Das Sarma-Villain (LDV) universality [20,6] all describe this same universality I refer to as the MBE universality in this article. This universality class is believed by many to be the correct large scale asymptotic description for the kinetic surface roughening properties of epitaxial growth morphology (which explains my nomenclature for this universality) provided that the growth process is conserved SOS (no desorption and overhang/vacancy formation) and that the "downward funneling" type EW relaxation processes are absent or are dynamically unimportant. The critical exponents for the MBE universality class are calculated to be [20] $\alpha = 2/3(1)$, $\beta = 1/5(1/3)$, $z = 10/3(3)$ in $d = 2 + 1(1 + 1)$ dimensions. These calculated exponents were thought to be exact until very recently when some questions [21] have been raised about their exactness. Any corrections (if there are any, the issue is by no means settled) to these exponent values are, however, estimated [21,22] to be well below 1%, and are therefore of no practical significance. There are many reports in the literature of measuring critical exponents $\beta \approx 0.2 - 0.3$ and $\alpha \approx 0.6 - 0.9$

in surface growth experiments, which support the applicability of the MBE universality class to real epitaxial growth morphology. Existing realistic MBE growth simulations based on kinetic Monte Carlo studies[4,23] also lead to critical exponents consistent with the MBE universality class exponents. It should be noted that often these experimentally measured or numerically simulated exponents appear to be somewhat larger than the MBE universality exponents ($\beta = 0.2$, $\alpha = 0.67$), indicating that real epitaxial growth morphology may actually be exhibiting a crossover from the MH universality ($\beta = 0.25$, $\alpha = 1.0$) to the MBE universality. One definitive way of distinguishing between these two universality classes would be a measurement of the surface skewness s

$$s = \langle (h- < h >)^3 \rangle / \langle (h- < h >)^2 \rangle^{3/2}, \tag{7}$$

associated with the growth morphology. Both the EW and the MH universality (being linear universality classes) predict a vanishing value of s whereas the MBE universality class [4] predicts $s \neq 0$ in the steady state saturated morphology (as well as in the pre-asymptotic transient regime). There have recently been reports [24] of non-zero surface skewness measurements in epitaxial growth morphology, again supporting the viewpoint that the MBE universality class indeed provides the appropriate dynamic scaling description of epitaxial growth. (In the context of surface skewness, I point out that the non-conserved KPZ universality also has a nonzero value of s in the preasymptotic transient regime in both $d = 1+1$ and $2+1$ dimensions, but in the asymptotic steady state situation the skewness magically vanishes, $s = 0$, in $d = 1+1$ dimensions by virtue of the validity of a fluctuation-dissipation theorem which is also responsible for producing the exact values of KPZ exponents in one substrate dimension[25].)

4 Continuum Theory

Theoretical understanding of dynamic scaling in kinetic surface roughening has been exclusively based on stochastic partial differential equations describing the time evolution of the coarse-grained surface height variable $h(\boldsymbol{r}, t)$ ever since the pioneering work[10] of Kardar, Parisi and Zhang (more than ten years ago) who showed that the well-known (in the context of fluid dynamics) noisy Burgers equation is the appropriate long wavelength asymptotic dynamical description for nonconserved generic growth under non-SOS ballistic deposition conditions (which allow the formation of overhangs and vacancies at the growth front). In this section I will briefly discuss the continuum growth equations which define the four universality classes described in the last section.

It is notationally more convenient to write the dynamical equations in terms of the height fluctuation, $h(\boldsymbol{r},t) - < h >$, which is equivalent to a transformation to a moving frame of reference which takes the average interface height $< h >= Ft$ as the reference position rather than the original substrate $< h >= 0$. This eliminates the trivial constant term, given by the average growth rate F, explicitly from the equation. From now on h implies the height fluctuation (around the average) rather than the height variable itself. The celebrated KPZ equation [10] is then the deceptively simple-looking nonlinear stochastic equation

$$\text{KPZ}: \ \frac{\partial h}{\partial t} = \nu_2 \nabla^2 h + \lambda_2 (\boldsymbol{\nabla} h)^2 + \eta, \tag{8}$$

where the spatial derivative $\boldsymbol{\nabla} \equiv \frac{\partial}{\partial \boldsymbol{r}}$ is in the substrate plane, and η is the stochastic shot noise associated with the random fluctuations in the incident beam. Usually $\eta(\boldsymbol{r},t)$ is taken to be a Gaussian white noise of zero mean $< \eta >$ with the noise correlator

$$\langle \eta(\boldsymbol{r}_1, t_1) \eta(\boldsymbol{r}_2, t_2) \rangle = D \delta(\boldsymbol{r}_1 - \boldsymbol{r}_2) \delta(t_1 - t_2)$$

where D is the strength of the bare noise.

I do not discuss here the KPZ equation at all because, as stated in the last section, it is not particularly relevant for understanding epitaxial growth morphology in experimental length and time scales. (Several excellent reviews [1,2,25] of the KPZ equation exist in the literature.) Two particular features of Eq. 4.1 should be noted: (a) It is the lowest order growth equation consistent with the symmetries of the problem, and (b) the nonlinear $(\nabla h)^2$ term in the equation, a manifestly nonequilibrium growth term as it breaks the up-down symmetry in the problem, is a nonconserved term associated with vacancy/overhang formation (or desorption) at the growth front.

Before proceeding any further I write down below the three continuum growth equations which correspond respectively to the three other universality classes introduced in the last section:

$$\text{EW}: \ \frac{\partial h}{\partial t} = \nu_2 \nabla^2 h + \eta \qquad , \tag{9}$$

$$\text{MH}: \ \frac{\partial h}{\partial t} = -\nu_4 \nabla^4 h + \eta \qquad , \tag{10}$$

$$\text{MBE}: \ \frac{\partial h}{\partial t} = -\nu_4 \nabla^4 h + \lambda_{22} \nabla^2 (\boldsymbol{\nabla} h)^2 + \eta, \tag{11}$$

Several features of these four growth equations, Eqs. 4.1-4.4, corresponding to the universality classes 1-4 respectively in section II, should be noted: the EW and the MH equations are linear and are therefore very simple; the KPZ (MBE) equation is simply the nonlinear version of the EW (MH) equation with the appropriate 2nd (4th) order nonlinearity being included in the equation; the MBE equation is effectively a conserved KPZ equation with a nonconserved external noise rather than a conserved noise[26] $\frac{\partial h}{\partial t} = -\nu_4\nabla^4 h + \lambda_{22}\nabla^2(\boldsymbol{\nabla} h)^2 + \eta = \nabla^2(-\nu_4\nabla^2 h + \lambda_{22}(\boldsymbol{\nabla} h)^2) + \eta$; EW and KPZ equations are 2nd order equations while MH and MBE equations are fourth order and therefore a growth equation containing all of these terms will be dominated by the KPZ nonlinearity (if it is non-zero) or by the EW term (if $\lambda_2 = 0$) with the $\nu_4\nabla^4 h$ and the $\lambda_{22}\nabla^2(\boldsymbol{\nabla} h)^2$ terms being irrelevant in the renormalization group sense. I emphasize, however, that even in such a situation the $\nabla^4 h$ and the $\nabla^2(\boldsymbol{\nabla} h)^2$ terms could significantly affect crossover properties and dominate the critical properties at "short" length and time scales.

Restricting to epitaxialy relevant conserved SOS growth (with the nonconserved noise arising from external beam fluctuations) it is easy to see[6,20] that the coarse-grained continuum growth equation must obey a continuity equation consistent with the conservation of the surface current in the problem:

$$\frac{\partial h}{\partial t} + \boldsymbol{\nabla}\cdot\boldsymbol{j} = \eta, \tag{12}$$

which follows simply from the mass and volume conservation in epitaxy associated with the no overhang/vacancy/desorption hypothesis. Thus the growth equation must have the form $\frac{\partial h}{\partial t} = -\boldsymbol{\nabla}\cdot\boldsymbol{j} + \eta$ where the surface current density $\boldsymbol{j}$ can only be of a form which is consistent with the symmetries in the growth problem. (It should be noted that the KPZ equation, Eq. 4.1, is explicitly ruled out because the KPZ nonlinearity, $(\nabla h)^2$, cannot be expressed as the divergence of a current.) The only general symmetries in the epitaxial growth problem are the translational invariance along the growth direction (i.e. it does not matter where the substrate is placed) and the rotational invariance in the substrate $\boldsymbol{r}$-plane. Taking into account these general symmetries and the conserved current constraint, the conserved epitaxial growth equation could be written as[20]

$$\frac{\partial h}{\partial t} = \nu_2\nabla^2 h - \nu_4\nabla^4 h + \lambda_{22}\nabla^2(\boldsymbol{\nabla} h)^2 + \lambda_{13}\boldsymbol{\nabla}(\boldsymbol{\nabla} h)^3 + \eta, \tag{13}$$

keeping terms only up to the fourth order. This is the most general possible low order conserved epitaxial continuum growth equation. The conserved universality classes discussed in the last section are all special limiting cases of Eq. 4.6, as is obvious by comparing Eq. 4.6 with Eqs. 4.2-4.

A simple power counting analysis shows that in Eq. 4.6 the most relevant long wavelength term is the EW $\nabla^2 h$ term, followed respectively by the

$\nabla(\nabla h)^3$ term, the $\nabla^2(\nabla h)^2$ term, and the $\nabla^4 h$ term in order of most relevant to the least relevant growth processes at the largest scales.

Since asymptotic critical exponents are always determined by fluctuations at the largest scales, one concludes that the EW universality, if present (i.e. if $\nu_2 \neq 0$), is the most *relevant* universality in the problem. The other terms may still be quite important in determining the *effective* exponents at shorter scales (and thus affect crossover behavior), but the asymptotic universality class is always EW if $\nu_2 \neq 0$ in Eq. 4.6. An additional notable point is that the most relevant fourth order term, the $\lambda_{13}\nabla(\nabla h)^3$ term, originally introduced in ref. [20] through symmetry considerations and recently rediscovered [27] in the context of unstable growth under an ES barrier, generates the EW $\nabla^2 h$ term upon renormalization [28] and therefore also leads to EW universality although it is formally a higher order term. Thus even if $\nu_2 = 0$, but $\lambda_{13} \neq 0$ in Eq. 4.6, the growth universality class is still EW! (I note that the most relevant conserved term in each order, namely the $\nabla(\nabla h)^{2n+1}$ term with $2n = 2, 4, 6$, etc., always generates EW universality upon renormalization — it is, however, difficult to imagine physical processes which could lead to $\nu_2 = 0$ but these nonlinear terms non-zero.)

The fourth order linear (Eq. 4.3) or nonlinear (Eq. 4.4) continuum growth equations become applicable to conserved epitaxial growth only in the absence of all EW relaxation processes (i.e. when $\nu_2 = \lambda_{13} = 0$), or as crossover effects at short distance/time scales before the EW growth terms in Eq. 4.6 become dynamically significant. As mentioned in section III, there is a large body of experimental data [1,2,4,7] on the surface kinetic roughening in epitaxial growth reporting a roughness exponent $\alpha \sim 0.6 - 0.9$ and a growth exponent $\beta \sim 0.2 - 0.3$, which are totally inconsistent with the EW exponents ($\alpha = \beta = 0$), but are consistent with the MH growth equation ($\alpha = 1.0$, $\beta = 0.25$) and the MBE growth equation ($\alpha = 0.67$, $\beta = 0.2$). Empirically, therefore, it seems that epitaxial growth belongs to the MBE universality class as defined by Eq. 4.4 (which incorporates as a special case the MH equation, Eq. 4.3, in the special situation of the vanishing nonlinearity $\lambda_{22} = 0$).

Obtaining the critical exponents from the continuum growth equation is simple and straightforward for the linear equations (Eqs. 4.2 and 4.3) because the linear equations can be directly solved via Fourier transformation. The nonlinear MBE growth equation, Eq. 4.4, is thoroughly nontrivial, however, and has been analyzed using the dynamical renormalization group technique [20] and the direct numerical integration technique [28,29,22]. I summarize below the theoretical exponent values for Eqs. 4.2-4:

$$\text{EW}: \; \alpha = (3-d)/2, \; \beta = (3-d)/4, \; z = 2, \tag{14}$$

$$\text{MH}: \ \alpha = (5-d)/2, \ \beta = (5-d)/4, \ z = 4, \tag{15}$$

$$\text{MBE}: \ \alpha = (5-d)/3, \ \beta = (5-d)/(7+d), \ z = (7+d)/3. \tag{16}$$

In Eqs. 4.7-9, d is the total spatial dimension, and obviously the physically relevant dimension for kinetic surface roughening in epitaxial growth is $d = 2 + 1 = 3$. The one dimensional substrate problem, $d = 1 + 1$, is, in fact, extensively studied theoretically (mainly because of convenience and ease in numerical simulations), and may have considerable potential experimental relevance in terms of kinetic roughening of step edges [14] on vicinal surfaces. While the above results (Eqs. 4.7 and 8) for the linear EW and MH universality classes are obviously exact, the MBE universality results given in Eq. 4.9 are obtained from a one-loop dynamical renormalization group analysis which is leading order in the ϵ-expansion where $\epsilon = 5 - d$. Given that the expansion parameters $\epsilon = 2(3)$ in $d = 3(2)$ are obviously not small, one may legitimately question[21] the validity of the calculated exponents in Eq. 4.9. Direct numerical integration of Eq. 4.4, however, gives[28,29,22] the same (within the numerical accuracy $\sim$ 1-2%) exponents in $d = 2$ and 3 dimensions as those of Eq. 4.9, providing support to the one-loop dynamical renormalization group result [20]. It is possible that the exponents of Eq. 4.9 are, in fact, exact for the MBE universality class due to some hidden symmetry in the problem[20,26], but the issue has to be considered open at this time[21].

Combining the current conservation condition with the general conserved SOS growth equation (Eq. 4.6) one gets the following expression for the surface current in epitaxial growth

$$\boldsymbol{j} = -(\boldsymbol{\nabla} h)\{\nu_2 + \lambda_{13}(\boldsymbol{\nabla} h)^2 + \cdots\} + \boldsymbol{\nabla}\{\nu_4(\nabla^2 h) - \lambda_{22}(\boldsymbol{\nabla} h)^2 + \cdots\}, \tag{17}$$

where the first term, a nonequilibrium surface current j_{NE}, is proportional to the surface slope $\boldsymbol{\nabla} h$ and the second term can be thought of [6] as the contribution from a generalized chemical potential μ, where $\mu = \mu_0 + \mu_{\text{NE}}$ with $\mu_0 \propto \nabla^2 h$ is the standard "equilibrium" chemical potential leading to the Mullins-Herring equation[30] and the so-called nonequilibrium chemical potential μ_{NE} is given by $\mu_{\text{NE}} \propto \lambda_{22}(\boldsymbol{\nabla} h)^2 + O((\boldsymbol{\nabla} h)^4) + \cdots$. Thus, Eq. 4.10 can be rewritten [31] as $\boldsymbol{j} = \boldsymbol{j}_{\text{NE}} + \boldsymbol{\nabla}\mu$ where $\boldsymbol{j}_{\text{NE}} \propto (\boldsymbol{\nabla} h)$. Note that the chemical potential contribution to the current $\boldsymbol{j}$ does not explicitly depend on $\boldsymbol{\nabla} h$ whereas the nonequillibrium current contribution, $\boldsymbol{j}_{\text{NE}}$, which leads to the EW universality is proportional to $\boldsymbol{\nabla} h$. This provides[31] a simple and effective technique of detecting the possible existence of the EW universality in a particular growth model. If there is a net inclination dependent surface current on a tilted substrate in a growth model, then $j_{\text{NE}} \neq 0$, indicating the presence of the EW universality in the model (this is true only if the current is downward and therefore stabilizing — if the inclination dependent current is upward then that indicates the existence of an ES barrier in the problem with the associated growth instability). On the other hand, if the measured inclination dependent current vanishes on a tilted substrate, then obviously the EW universality is absent in the model. This simple technique of measuring the inclination dependent surface current on a tilted substrate has turned out to be a powerful tool for detecting the presence or absence of the EW universality in various growth models, and is operationally very effective in determining the EW universality long before the measured critical exponents show a crossover to EW behavior [2,31]. Application of this technique of measuring inclination dependent surface current on tilted substrates enables one to effectively distinguish among unstable ES barrier growth (upward current), EW universality (downward current), and the fourth order MBE growth universality (vanishing inclination dependent current) in various discrete epitaxial growth models (some of which are discussed in the next section).

I conclude this section by mentioning that the actual values of the coefficients ν_2, ν_4, λ_{22}, λ_{13}, etc. entering the continuum growth equations are not important in determining the respective asymptotic universality classes except in the trivial sense that a particular coefficient must be non-zero for that particular universality class to show up. The explicit values of these coefficients are, however, important in determining the crossover behavior in the growth problem, for example, if only the (most relevant) ν_2 and the (most irrelevant) ν_4 terms are non-zero in Eq. 4.6 then the growth universality class is obviously EW since $\nu_2 \nabla^2 h$ is more relevant than $\nu_4 \nabla^4 h$, but a simple dimensional analysis shows that the EW universality can show up only for lateral length scales larger than $(|\nu_4|/|\nu_2|)^{1/2}$ or equivalently for time scales larger than $(|\nu_4|/|\nu_2|)^2$, and therefore in a real finite size/time experiment the EW universality may simply not manifest itself, leading to effective MH universality exponents. This is a situation where an inclination dependent current measurement on a tilted substrate may lead to the "correct" (i.e. EW) asymptotic universality even though the "effective" exponents are still in the MH crossover regime. Finally, I point out that the coefficients ν_2, λ_{13}, and λ_{22} are expected to be proportional to the external flux rate F in the leading order because these terms are manifestly nonequilibrium contributions (ν_2 and λ_{13} to $\boldsymbol{j}_{\mathrm{NE}}$, and λ_{22} to μ_{NE}) to the growth process, and therefore should vanish in the absence of the external flux. The ν_4 term, on the other hand, is the usual equilibrium surface diffusion contribution [30] and its presence in the nonequilibrium growth problem is still somewhat mysterious [6,19,2].

5 Atomistic Growth Models

All real epitaxial growth takes place in discrete systems with atoms and lattices whereas the coarse-grained long wavelength theory discussed in the last section is, by definition, a continuum theory. A question naturally arises about whether such a coarse-graining procedure leading to a continuum dynamical growth equation which describes the long wavelength asymptotic dynamic scaling properties of the growth problem is, in general, allowed for all discrete expitaxial growth phenomena. The answer to this question is not known, and only in a few problems a rigorous connection between a discrete growth model and its coarse-grained continuum description has been established. One hopes, with considerable physical justification, that such a coarse-graining prescription leading from a discrete problem to a continuum description should, in fact, be possible at "sufficiently long" wavelengths. (Whether real experimental growth studies are always carried out in this "sufficiently long" wavelength regime or not is, of course, a totally different issue which is extremely difficult to answer.) Recently, some subtle and serious questions have been raised [32,33,34] by several different groups regarding the validity of the continuum descriptions for discrete growth problems, particularly in cases involving nonlinear (eg. KPZ, MBE, etc.) growth terms. Although the details of these doubts vary considerably from case to case, it is fair to say that one of the tentative conclusions emerging from these analyses [32,33,34] is that in some situations involving nonlinear growth processes discrete growth problems and their coarse-grained continuum descriptions may belong to different universality classes. (Not surprisingly no such problem arises in linear growth problems, eg., EW universality.) There is thus sufficient reason to be critical and skeptical about blindly applying continuum theoretical results to discrete growth problems.

The most effective technique to study atomistic epitaxial growth is via the direct numerical simulation of the discrete growth process. There have been numerous such studies using a variety of techniques and models in the context of dynamic scaling in epitaxy — see, for example, refs. [4,5,11,12,13,16,18,19,22,23,34] and the review articles in refs. [1,2,3]. In the context of epitaxy, where adatom mobility at the growth front is the dominant smoothening mechanism and the shot noise fluctuations inherent in the deposition beam is the roughening mechanism, the kinetic (or, the stochastic) Monte Carlo simulation [4,5,11,12,13,23] has been the most realistic direct numerical technique for studying MBE growth. In this technique both deposition and diffusion are taken to be stochastic processes simulated by various random number generators (and hence the name, Monte Carlo simulation) with the atomic diffusion at the growth front taken to be an Arrhenius activated hopping process with the activation energy (consistent with the principle of detailed balance) determined by the nature of local bonding for the hopping atom. Depending on the local bonding of the diffusing atom, several different activation energies (and consequently several different diffusion rates) participate in the hopping process, leading to the possibility of rather strong finite size and crossover effects in the simulation. Thus, the full activated diffusion stochastic Monte Carlo simulation of epitaxial growth, while being reasonably realistic in capturing the evolving surface morphology, is not always the ideal method for obtaining the asymptotic growth exponents.

Purely nonequilibrium zero temperature growth models, with instantaneous relaxation of the deposited adatoms according to some physically[18,19] or mathematically[16,22] motivated local rules, have been very popular and extremely successful in characterizing the asymptotic growth universality class and the dynamic scaling behavior of epitaxy. I will discuss one such conserved SOS nonequilibrium growth model (called the DT model by Krug[2]), introduced in ref. [19] and extensively studied in refs. [4,5,35,36,37]. Before discussing the model I mention two significant salient features of the model which make it particularly relevant for studying dynamic scaling in epitaxial growth: (1) the scaling exponents α, β calculated for this simple nonequilibrium model agree *quantitatively* with the corresponding exponents for the full activated diffusion stochastic Monte Carlo simulation results in both $d = 1 + 1$ and $d = 2 + 1$ dimensions[4,37]; (2) due to a simple symmetry[2] in the growth rules of this model, the inclination dependent surface current vanishes exactly for tilted substrates, implying that this model most certainly does not have the most relevant EW term in its continuum description (i.e. $\nu_2, \lambda_{13} = 0$ in this growth model) — this particular feature makes this model unique among the existing nonequilibrium SOS growth models, the other two such models introduced in ref. [16] and in ref. [18] are known to belong asymptotically to the EW universality class.

In the growth model of ref. [19], atoms are deposited randomly and sequentially (i.e. one at a time at a randomly chosen spatial position on the surface) on a cubic substrate according to an average rate and within the SOS constraint, after deposition each atom could relax within a lateral diffusion length (which is most commonly taken to be just one lateral lattice spacing) subject to the following conditions: only deposited atoms with no lateral bonds to other surface atoms are allowed to move, and the atom moves (within the lateral diffusion length) only to increase its local coordination number (i.e. lateral bonding) — otherwise it does not move and the next atom is deposited. Note that these local diffusion rules allow only downward relaxation of the adatoms (to kink and trap sites, but *not* necessarily to height minima[16] or to sites of maximum local bonding[18]), and in case of ties (i.e. more than one final site satisfying the relaxation condition) the atom moves randomly with an equal probability to any of the allowed final sites. Each deposited atom is allowed to relax only once immediately following deposition, but more elaborate local relaxation rules have also been considered [38] with concommitant increase in the dynamic scaling possibilities. Note that the particularly simple local relaxation rules of this manifestly nonequilibrium growth model are motivated by the actual atomistic hopping processes at the MBE growth front in the low to intermediate growth temperatures. The idea is that only singly bonded atoms (to their neighbors underneath them, an essential ingredient of conserved SOS growth) have appreciable surface mobility at low growth temperatures, the hopping probability of adatoms with higher coordinations being suppressed exponentially. The nonequilibrium growth model of ref. [19] is a typical example of limited mobility diffusion models which have been successful in modeling epitaxial growth.

Extensive computer simulations (using up to 10^{12} deposited atoms) of this nonequilibrium growth model in $d = 2+1$ dimensions show [37] a clear crossover in the growth exponent β from $\beta \approx 0.25$ at short time scales to $\beta \approx 0.20$ at longer times (the estimated roughness exponent $\alpha \approx 0.6 \pm 0.1$), indicating that this model truly belongs to the MBE growth universality, and the crossover is from the MH universality, dominated by the $\nabla^4 h$ term, to the MBE universality, dominated by the nonlinear $\nabla^2(\boldsymbol{\nabla} h)^2$ term, in Eq. 4.4. Note that the crossover time scale to see the MBE universality is approximately $\sim |\nu_4/\lambda_{22}|^4$ and it should show up only for substrate sizes larger than $\sim |\nu_4/\lambda_{22}|$. Consistent with these expectations, the observed crossover to MBE universality manifest itself [37] only for system sizes larger than 100×100 — in smaller systems saturation occurs before the crossover and one sees only the MH universality ($\beta \approx 0.25$, $\alpha \approx 0.9$). These findings [37] in this simple limited mobility nonequilibrium growth model [19] are in agreement with many experimental reports on dynamic scaling in epitaxy with the experimental $\beta \approx 0.2 - 0.3$ and the experimental $\alpha \approx 0.6 - 0.9$. I speculate that the dynamic scaling in most epitaxial kinetic roughening experiments is essentially in the crossover regime between the MH universality and the MBE universality class. It is certainly not possible to rule out the asymptotic existence of an EW universality in experimental systems at still larger scales; all one can say is that such an eventual crossover to the EW universality ($\beta = \alpha = 0$) has *not* been reported in the literature in the context of dynamic scaling in epitaxial growth experiments. Based on all of these observations, it is not unreasonable to conclude that the simple limited mobility nonequilibrium growth model of refs. [19,37] captures the essence of the dynamic scaling behavior in epitaxial growth.

Before concluding this section I mention several features of the epitaxial atomistic growth model of ref. [19,37] which are of potential importance:

1. Explicit calculation of the inclination dependent surface current on tilted substrates shows a vanishing current in both $d = 1+1$ and $2+1$ dimensions, implying the non-existence of any EW growth term in the model of ref. [19,37].

2. Very recent theoretical work[39,40] using the master equation approach find that the model of ref. [19,37] actually follows the MBE growth equation, Eq. 4.4 of this paper. (The model of Wolf and Villain[18], which is closely related to the model of ref. [19], is however found [39,40] to contain the EW $\nabla^2 h$ term, which is consistent with the earlier finding[31] of a tilt dependent surface current in the Wolf-Villain model.)

3. The model of ref. [19,37] shows considerable evolving skewness in the growing surface (implying a breaking of the up-down symmetry under the nonequilibrium growth condition) with the best estimate for the steady state skewness in the surface morphology being $s \approx -0.5$ [36,37,4].

4. For reasons not completely understood at the present time, the model exhibits an extremely long crossover regime[19,5,4] in $d = 1+1$ dimensions, with $\beta \approx 0.37$ for upto 10^8 monolayers of growth on a $L = 10^4$ substrate (the value of α for smaller system sizes, upto $L = 200$, is $\alpha \approx 1.3$) — thus the expected crossover to the MBE universality ($\beta = 0.33$, $\alpha = 1.0$) is extremely slow in one substrate dimension where the MH exponents ($\beta = 0.375$, $\alpha = 1.5$) seem to dominate for a long time except that the existence of a finite skewness definitively rules out a linear equation as the underlying continuum description.

5. Finally, the discrete model of ref. [19,37] shows an extremely intriguing anomalous scaling[35] and multiaffine scaling[36] behavior where each moment of the height-height correlation function seems to have its own roughness exponent, indicating the growth problem to be similar to the intermittency phenomenon in fluid turbulence[36]. A recent detailed study [34] of the multiscaling phenomena[36] indicates that the continuum equation for the discrete nonequilibrium growth model of ref. [19] may actually be

$$\frac{\partial h}{\partial t} = [-\nu_4 \nabla^4 h + \lambda_{22} \nabla^2 (\boldsymbol{\nabla} h)^2 + \eta] + \sum_{n=2,3,\cdots} \lambda_{2n} \nabla^2 (\boldsymbol{\nabla} h)^{2n}, \quad (18)$$

where the terms within the square brackets comprise the MBE growth equation, Eq. 4.4, and the infinite order series of $\nabla^2(\boldsymbol{\nabla} h)^{2n}$ terms with

$2n = 4, 6, 8, \cdots$ defines a rather complex correction to scaling, arising from an infinite series of marginally relevant (irrelevant) terms in $d = 1 + 1(2 + 1)$ dimensions. The multiscaling phenomenon seems to be transient, lasting only up to 10^4-10^5 layers of growth in $d = 2 + 1$ dimensions[37], but is a very long lasting transient (at least up to 10^9 layers) in $d = 1 + 1$ dimensions. A direct experimental measurement of the scaling properties of the higher moments of the height-height correlation functions in epitaxial growth will be very interesting in this context. Any expeirmental evidence for multliaffine growth, where different moments of the height-height correlation function scale with different exponents and where the step height distribution of the evolvilng morphology shows an intermittent stretched exponential distribution [36,2,4,34,37], would not only be interesting from the kinetic surface roughening perspective, but will be of broad general interest as it will be an example of a rather simple stochastic nonequilibrium model exhibiting intermittency and turbulence.

6 Related Topics

I will very briefly discuss two topics in this section without really doing much justice to their importance and significance. (Each of these topics deserves their own theoretical chapters.) The first topic is the role of growth temperature in dynamic scaling of epitaxial kinetic surface roughening. The second topic is the issue of dynamic scaling in "unstable" growth under ES step edge diffusion bias. In contrast to the standard lore in the literature, where temperature induced smooth layer by layer growth and growth under ES diffusion bias are thought of as phenomena quite distinct from kinetic surface roughening, I tend to take the view that the powerful dynamic scaling ideas describe all these growth scenarios with appropriate modifications and qualifications.

1. *Temperature effects* Note that the growth temperature (T) plays no direct or explicit role in the dynamic scaling properties of kinetic surface roughening discussed so far in this article. This is in sharp contrast to equilibrium critical phenomena where T has to be tuned close to the critical temperature in order to observe the critical behavior and scaling. Kinetic surface roughening is thus generically scale invariant and the dynamic scaling behavior is present at long wavelengths without any tuning of growth parameters. The growth temperature does, however, play an extremely important role in controlling the various crossover fegimes, finite size effects, and in determining the size of the transient layer by layer smooth growth regime. Growth temperature sets the dis-

tance/time scales which separate the transient and the dynamic scaling regimes. Temperature actually determines the important short distance cutoff in the growth problem by controlling[8] the diffusion length ℓ which scales as

$$\ell \sim (D/F)^{\gamma} \tag{19}$$

where the exponent $\gamma \sim 1/2 - 1/6$ depends explicitly on the temperature dependent minimum stable island size and can, in principle, be calculated [8] by stochastic Monte Carlo simulations or kinetic rate-theoretic arguments, and D, F are the diffusion rate and the deposition rate respectively. The diffusion rate D depends exponentially on the temperature via the Arrhenius activation law

$$D \sim k_B T e^{-E_A/k_B T}, \tag{20}$$

and therefore $\ell \sim \frac{T^{\gamma} e^{-\gamma E_A/k_B T}}{F^{\gamma}}$ with the activation energy E_A depending explicitly on the size of the stable islands on the surface (i.e. how many bonds are being cut in the hopping process). The important point to note is that the characteristic diffusion length varies superexponentially ($\sim T^{\gamma} e^{-\gamma E_A/k_B T}$) with temperature, and therefore the crossover properties of dynamic scaling are expected to be strong functions of temperature in the appropriate temperature windows[9,5,41,42].

It has recently been shown[9] that the short distance cutoff in MBE growth is not the diffusion length itself, but the closely related coherence length $\tilde{\ell}$ which varies as a power of the diffusion length ℓ:

$$\tilde{\ell} \sim \ell^{\delta} \tag{21}$$

with $\delta = 4/(5-d)$ for the MBE growth equation. Putting $d = 2(3)$ one concludes therefore that the finite size scaling behavior of the critical exponents α, β, etc. in the presence of the short distance cutoff defined by the coherence length $\tilde{\ell}$ should show a characteristic $L/\ell^2 (d = 2+1)$ or $L/\ell^{4/3} (d = 1+1)$ dependence. Such a finite size scaling behavior, in particular, an approximate $L/\ell^{4/3}$ scaling of the effective growth exponent β on system size and diffusion length, has earlier been reported in $d = 1+1$ [5,41]. Since the diffusion length ℓ itself depends strongly on temperature, this implies a potentially strong temperature dependence of the critical exponents arising just from temperature dependent finite size effects. In particular, the dynamical exponent β and the roughness exponent α should vary between their asymptotic values as defined by the appropriate universality for $L/\tilde{\ell} \to \infty$ to effectively zero as $L/\tilde{\ell} \to 0$.

For Arrhenius activiated full diffusion MBE model, however, the zero temperature ($L/\tilde{\ell} = \infty$) limit of β is trivially 1/2, corresponding to pure random deposition with no relaxation because at $T = 0$ the adatoms do not diffuse [5,19]. Such a behavior of β, varying strongly with temperature and diffusion length, has been seen in computer simulations [5,19,41,42], and can easily be understood as manifestations of crossover/finite size scaling behavior in the presence of a characteristic short-distance cutoff $\tilde{\ell}$ over which the growing surface morphology is essentially smooth. In particular, for $\tilde{\ell} \geq L$ the layer by layer growth regime persists indefinitely [9], and $\beta, \alpha \equiv 0$ due to finite size effect. One can therefore think of the smooth layer by layer growth regime as a trivial dynamic scaling regime where finite size effects push down the growth and roughness exponents to zero values. Such an idea has been used in the literature to define and calculate the epitaxial growth temperature [5,42].

The basic point of physics is that kinetic roughening (for example, coarse-grained continuum descriptions) applies only at length scales above $\tilde{\ell}$, and below $\tilde{\ell}$ one has smooth layer by layer growth transient. Even at large scales the coefficients ν_2, ν_4, λ_{22} etc. appearing in the growth Eqs. 4.2-4 depend on the growth temperature (and deposition rate), which could produce complex temperature dependent crossover behavior. Careful experimental work obtaining $\beta(T)$ as a function of temperature in epitaxial growth will be useful in the context of understanding finite size effects in dynamic scaling of epitaxy. Such systematic temperature dependent experimental information is lacking at the present time. I note that the so-called epitaxial growth temperature, T_e, is defined by the condition $\tilde{\ell}(T) = L$ in this picture, and the superexponential temperature dependence of $\tilde{\ell}$ implies a weak (sub-logarithmic) system size dependence of T_e [5,42].

2. *Unstable growth under ES barrier* In nonequilibrium growth under an ES step edge diffusion bias, where atoms preferentially attach to ascending over descending steps, a characteristic wedding cake morpology evolves on the surface with the formation of mounds/pyramids. These mounds coarsen and/or steepen with time — in the pre-saturation regime the typical mound radius (i.e. the lateral mound size), $R(t)$, increases with time according to a coarsening exponent n, and the mound slope, $m(t)$, increases with time according to a steepening exponent λ:

$$R(t) \sim t^n; \; m(t) \sim t^\lambda. \tag{22}$$

A crucial point (which is not always emphasized in the literature on the subject) is that the dynamic scaling law applies to this situation exactly in the same form as it does in the generically scale invariant kinetic surface roughening problem, $W \sim L^\alpha f(L/\xi(t))$ with $W(t, L \to \infty) \sim t^\beta$, except here the "surface roughness" $W(t)$ is essentially the same as the typical mound height, $H(t)$. Using the fact that the mound slope $m \sim H/R$ (i.e. mound height/mound size), one obtains the exponent identity

$$\beta = \lambda + n \tag{23}$$

in nonequilibrium growth under an ES barrier. (Alternative exponent definitions are, in principle, possible which may deviate somewhat from the above exponent identity, but in my view the definitions of β, λ, and n used here are the natural ones both for the experimental and the numerical simulation purpose — an important point to note is that coarsening and steepening compete, as is directly implied by Eq. 6.5.) The dynamical exponent z is essentially $1/n$ for most reasonable definitions of the mound size, implying that $R(t)$ for all practical purposes is the dynamical correlation length $\xi(t)$.

This leads to the conclusion that nonequilibrium growth on flat singular surfaces under an ES step edge diffusion bias follows the usual two exponent (eg. β and n) dynamical scaling behavior exactly in the same manner as in the kinetic roughening problem. There has been considerable earlier confusion in this subject based on erroneous claims in the theoretical literature of single exponent scaling in this problem. Note that the evolving mounds/pyramids etc. manifestly break generic scale invariance in this growth problem because obviously there are characteristic length scales (eg. the mound size) dominating the growth morphology. But the dynamic scaling concept applies here at the same level of generality as it does in the generically scale invariant kinetic surface roughening phenomenon.

Incorporation of an ES barrier in nonequilibrium surface growth has turned out to be a rather formidable theoretical challenge. There is no suitable coarse-grained continuum growth equation which catches all aspects of the interesting mound formation and its subsequent coarsening. Although there are a number of proposed growth equations, and in some cases the exponents β, n, λ have been calculated, there is no existing theoretical consensus on a continuum description for nonequillibrium surface growth under ES barriers. Surprisingly, numerical simulations of discrete nonequilibrium surface growth models under ES barriers have also turned out to be quite tricky, with the results depending strongly on the details of how the ES barrier is incorporated[43,44,45,46,47,48,49,50,51,27]. In fact, there is no consensus even on the basic nature of the instability[6] associated with surface growth under an ES bias — in particular, much debate and speculation can be found in the literature on whether or not there is slope selection (the exponent $\lambda \to 0$ if there is slope selection) during the coarsening of the mounds. Experimental studies seem to indicate that there may or may not be slope selection depending on the systems and growth temperatures one uses. I believe that a good part of the theoretical problem arises from our ignorance about how to incorporate the effect of ES barrier in the coarse-grained continuum description as well as in atomistic numerical simulations. In the stochastic Monte Carlo growth simulation, an ES bias can be included[46] either as a reflection barrier [46,47,48] or as an edge barrier [46,44]. In the reflection barrier case, an atom approaching a descending step is reflected back onto the upper terrace and is probabilistically prevented from reaching the top of the step edge. If it is already at the step edge, it may, however, go down and attach itself to the descending step without encountering any additional barrier. In the edge barrier case, there is a barrier for the atom to go down to the step edge, but there is no barrier to reach the top of the down-step (the ES bias only hinders it from going down once it is already at the top of the step). Both of these ways of incorporating ES barrier are simplifications of the realistic atomic potential near terrace edges [52] which have been calculated in a few cases. The correct situation is a complicated combination of edge and reflection barriers, neither by itself is adequate. Comparing with the realistic surface potential contours [52], however, it is quite obvious that the edge barrier model is a much better representation of the actual ES bias than the reflection barrier model. For reasons not very clear to me, most of the numerical growth simulations[47,48] utilize the reflection barrier approach (and make additional assumptions regarding the nature of surface relaxation to obtain slope selection). I myself believe that the edge barrier model is better suited to study surface growth under an ES bias than the reflection barrier model. Below I discuss our recent results [43] for nonequilibrium surface growth under an edge surface diffusion bias.

I conclude by providing the results of some very recent numerical calculations[43] of the exponents β, λ, and n in the nonequilibrium growth under an ES barrier in the $1+1$ and $2+1$ dimensional SOS growth model. Our finding [43] is that the growth exponent β is $1/2$ for this problem (both in $d = 2+1$ and $1+1$ dimensions), independent of the strength of the ES edge barrier. It has earlier been pointed out [38,44,45,46] that for a strong ES barrier, the growth exponent tends toward 1/2, but our new finding is that $\beta = 1/2$ always under an edge ES barrier, except the crossover time to observe this asymptotic $\beta(= 0.5)$ is extremely long for weak ES barriers which is why the existing ES barrier simulations have not always unambiguously observed this asymptotic regime. Note that $\beta = 1/2$ here does not imply uncorrelated random Poisson growth because $n = z^{-1} \approx 0.2 - 0.1$. The other question of considerable interest [47,48] to both theory and experiment is the issue of slope selection in nonequilibrium growth under an ES barrier, i.e. whether $\lambda = 0$ (i.e. $\beta = n$) asymptotically in this growth problem, and some magic slope is selected by the mounds which does not change with time and remains fixed. We find that within an SOS model, there is no slope selection $\lambda \neq 0$ [43] in the edge barrier model, and any experimental observation of slope selection must derive either from the physics of crystallographic orientations [31] or from some other processes [47,48] which are not essential ingredients of the ES barrier physics in the SOS model. My own belief is that, although slope selection may happen at extremely long times when the typical mound slope $m(t)$ is very large, it is not a generic phenomenon at small slopes. We find $\lambda \approx 0.2 - 0.4$, $n \approx 0.2 - 0.1$, and $\beta \approx 0.5$ in all our edge barrier simulations without encountering any slope selection in either $1 + 1$ or $2 + 1$ dimensions.

7 Concluding Remarks

In this article I have provided a theoretical review from my own perspective of our current understanding of dynamic scaling in nonequilibrium epitaxial growth on flat singular substrates in the high symmetry directions. The underlying theme has been the successful application of the dynamic scaling hypothesis, defined by two independent critical exponents, to a wide variety of epitaxial growth phenomena. I have tried to emphasize how local discrete rules for deposition and relaxation under nonequilibrium conditions lead at large scales to complex growth morphologies which obey dynamic scaling according to coarse-grained continuum equations. The subject is vast and necessarily my perspective is based on the work I am most familiar with. The list of references is by no means comprehensive. The cited review articles [1,2,3,7,25] as well as refs. [4,5] provide more references to the literature.

8 Acknowledgement

I am most grateful to the United States Office of Naval Research for their steadfast support of my research in this area. In particular, I wish to take this opportunity to thank Dr. Larry Cooper (ONR) for sponsoring my research on MBE growth.

References

1. A.L. Barabasi and H.E. Stanley, *Fractal Concepts in Surface Growth* (Cambridge, New York, 1995).
2. J. Krug, Adv. Phys. **46**, 139 (1997).
3. S. Das Sarma, J. Vac. Sci. Technol. A **8**, 2174 (1990); J. Vac. Sci. Technol. B **10**, 1695 (1992); Fractals **1**, 784 (1993); Comput. Mater. Sci. **6**, 149 (1996).
4. S. Das Sarma, C.J. Lanczycki, R. Kotlyar, and S.V. Ghaisas, Phys. Rev. E **53**, 359 (1996).
5. P.I. Tamborenea and S. Das Sarma, Phys. Rev. E **48**, 2575 (1993).
6. J. Villain, J. Phys. (France) I **1**, 19 (1991).
7. J. Krim and G. Palasantzas, Int. J. Mod. Phys. B **9**, 599 (1995).
8. S.V. Ghaisas and S. Das Sarma, Phys. Rev. B **46**, 7308 (1992).
9. H. Kallabis, L. Brendel, J. Krug, and D.E. Wolf, Int. J. Mod. Phys. B (1997).
10. M. Kardar, G. Parisi, and Y.C. Zhang, Phys. Rev. Lett. **56**, 889 (1986).
11. S. Das Sarma, C.J. Lanczycki, S.V. Ghaisas, and J.M. Kim, Phys. Rev. B **49**, 10693 (1994).
12. M. Schimschak and J. Krug, Phys. Rev. B **52**, 8550 (1995).
13. C.J. Lanczycki, V. Jejjala, and S. Das Sarma, Phys. Rev. E **54**, 4755 (1996).
14. F. Wu, S.G. Jaloviar, D.E. Savage, and M.G. Lagally, Phys. Rev. Lett **71**, 4190 (1993). A theoretical explanation of this experimental work is given in C.J. Lanczycki and S. Das Sarma, Phys. Rev. B **51**, 4579 (1995).
15. S.F. Edwards and D.R. Wilkinson, Proc. R. Soc. London, Ser. A **381**, 17 (1982).
16. F. Family, J. Phys. A **19**, L441 (1986).
17. I. Markov, Phys. Rev. B **50** 11271 (1994).
18. D.E. Wolf and J. Villain, Europhys. Lett. **13**, 389 (1990).
19. S. Das Sarma and P.I. Tamborenea, Phys. Rev. Lett. **66**, 325 (1991).
20. Z.W. Lai and S. Das Sarma, Phys. Rev. Lett. **66**, 2348 (1991).
21. H.K. Janssen, Phys. Rev. Lett. **78**, 1082 (1997).

22. J.M. Kim and S. Das Sarma, Phys. Rev. Lett. **72**, 2903 (1994).
23. D. Wilby, D.D. Vvedensky, and A. Zangwill, Phys. Rev. B **46**, 12896 (1992); M. Kotrla and P. Smilauer, Phys. Rev. B **53**, 13777 (1996).
24. C.Y. Mou and J.W.P. Hsu, Phys. Rev. B (1996).
25. T. Halpin-Healy and Y.C. Zhang, Phys. Rep. **254**, 215 (1995).
26. T. Sun, H. Guo, and M. Grant, Phys. Rev. A **40**, 6763 (1989).
27. J.A. Stroscio, D.T. Pierce, M. Stiles, A. Zangwill, and L.M. Sander, Phys. Rev. Lett. **75**, 4246 (1995).
28. J.M. Kim and S. Das Sarma, Phys. Rev. E. **51**, 1889 (1995); S. Das Sarma and R. Kotlyar, Phys. Rev. E **56**, R4275 (1994); A.K. Kshirsagar and S.V. Ghaisas, Phys. Rev. E **53**, R1325 (1996).
29. Y. Tu, Phys. Rev. A **46**, R729 (1992).
30. W.W. Mullins, J. Appl. Phys. **28**, 333 (1957); C. Herring, J. Appl. Phys. **21**, 301 (1950).
31. J. Krug, M. Plischke, and M. Siegert, Phys. Rev. Lett. **70**, 3271 (1993).
32. M. Marsili, A. Maritan, F. Toigo, and J.R. Banavar, Rev. Mod. Phys. **68**, 963 (1996).
33. M. Marsili and A.J. Bray, Phys. Rev. Lett. **73**, 2750 (1996); T.J. Newman and A.J. Bray, J. Phys. A **29**, 7917 (1996).
34. C. Dasgupta, J.M. Kim, M. Dutta, and S. Das Sarma, Phys. Rev. E **55**, 2235 (1997); C. Dasgupta, J.M. Kim and S. Das Sarma, Phys. Rev. E **54**, R4552 (1996).
35. S. Das Sarma, S.V. Ghaisas, and J.M. Kim, Phys. Rev. E **49**, 122 (1994).
36. J. Krug, Phys. Rev. Lett. **72**, 2903 (1994).
37. S. Das Sarma and P. Punyindu, Phys. Rev. E **55**, 5361 (1997).
38. S. Das Sarma and S.V. Ghaisas, Phys. Rev. Lett. **69**, 3762 (1992).
39. M. Predota and M. Kotrla, Phys. Rev. E **54**, 3933 (1996).
40. Z.F. Huang and B.L. Gu, Phys. Rev. E **54**, 5935 (1996).
41. S. Das Sarma, Z.W. Lai, and P.I. Tamborenea, Surf. Sci. Lett. **268**, L311 (1992).
42. S. Das Sarma and P.I. Tamborenea, Phys. Rev. B **46**, 1925 (1992).
43. S. Das Sarma and P. Punyindu, to be published (preprint, 1997).
44. Z. Zhang, J. Detch, and H. Metiu, Phys. Rev. B **48**, 4972 (1993).
45. C.J. Lanczycki and S. Das Sarma, Phys. Rev. E **50**, 213 (1994).
46. C.J. Lanczycki and S. Das Sarma, Phys. Rev. Lett. **76**, 780 (1996).
47. J. Amar and F. Family, Phys. Rev. B **54**, 14742 (1996).
48. P. Smilauer and D.D. Vvedensky, Phys. Rev. B **52** 14263 (1995).
49. A.W. Hunt, C. Orme, D.R.M. Williams, B.G. Orr, and L.M. Sander, Europhys. Lett. **27**, 611 (1994).
50. P. Politi and J. Villain, Phys. Rev. B **54**, 5114 (1996).
51. M. Rost and J. Krug, Phys. Rev. E **55**, 3952 (1997).
52. S.Kodiyalam, K.E. Khor, and S. Das Sarma, Phys. Rev. B **53**, 9913 (1996).

Scaling and crossovers in models for thin film growth

Alberto Pimpinelli[1], Pablo Jensen[2], Hernán Larralde[3], Philippe Peyla[4]
[1] *Institut Laue-Langevin,*
B.P. 156X, F-38042 Grenoble Cedex 9, France

[2] *Département de Physique des Matériaux,*
Université Claude Bernard Lyon-1, 69622 Villeurbanne Cedex, France

[3] *Instituto de Física, Lab. de Cuernavaca,*
Apdo. Postal 48-3, C.P. 62251, Cuernavaca, Mor., Mexico

[4] *Laboratoire de Physique et de Modélisation des Milieux Condensés*
CNRS/Université Joseph Fourier,
B.P. 166, F-38042 Grenoble Cedex 9, France

Abstract

A self-contained review of the scaling arguments used for computing the maximum island density in adatom aggregation on a substrate during thin-film growth, is presented. We show how a general argument can be formulated, which applies to all physical situations, and allows to re-derive known results, as well as generalizations and new ones. When several processes, such as island aggregation, capture by pre-existing steps, adatom desorption, are in competition, this scaling argument allows to determine the appropriate scaling parameters, and yields approximate analytic forms of the crossover scaling functions. The scaling form of the growth velocity of a film when adatom desorption is present, is investigated. Surprisingly, a strict analogy between layer-by-layer growth on a high-symmetry substrate and step flow on a vicinal substrate, is found.

1 Introduction

Since the pioneering works by Zinsmeister [1,2], Stoyanov and Kaschiev [3], and Venables et al. [4], it is known that the saturation density of clusters, or islands, formed by atoms diffusing on a high-symmetry substrate where they have been deposited from a vapour or a beam, varies as a power law

$$N \approx (F/D)^{\phi} \tag{1}$$

as a function of the ratio F/D of the atom deposition rate F to the atom diffusion constant D. The exponent ϕ depends on the characteristics of the process of atom aggregation, in particular it is a function of the size s^* of the largest dissociable, or mobile, cluster. The size s^* depends in turn on D and F, so that the range of validity of (1) must be carefully investigated [6,7] (see also [5] for more references, including experimental ones).

Formula (1) is an example of a *scaling relation*. The island density is invariant if the flux and the diffusion constant are multiplied by the same factor. Indeed, Villain et al. [8] set out to find a simple scaling argument, as usual in statistical physics, which allows to rederive (1) in the most straightforward way. This kind of argument is reviewed here, as well as its generalizations to more complicated situations, where crossovers between different scaling regimes are observed.

Of course, the scaling analysis of a problem like atom deposition, diffusion and aggregation rarely leads to a quantitative description, contrasting in that with computer simulations. However, the qualitative understanding one may gain from scaling relations is unvaluable, most of all when it comes to make sense out of simulations!

Scaling such as in (1) holds because of the presence in the problem of two time scales: the time for depositing one layer, $1/(Fa^2)$, and the hopping time a^2/D, where a is the atomic size. Their ratio allows to construct a length, $\ell_s/a \approx [D/(Fa^4)]^{\phi'}$, which is the natural lengthscale of the aggregation process. This means that all other lengths appearing in the problem, e.g. the distance between islands, their size, the adatom diffusion length, and so on, are expected to be functions of F and D through ℓ_s only.

In general, more than one lengthscale appears in the problem. Examples are (i) deposition and growth on a vicinal substrate, which consists of an ordered array of steps at average distance d; (ii) growth with adatom desorption at a rate $1/\tau_v$ (per unit surface and time), so that the adatoms diffuse over a length $x_s = \sqrt{D\tau_v}/a$ before they leave the surface, if they are not captured by an island. In such situations, the island density N is expected to be a function of all lengthscales $\{\ell\}$, $N = N(F/D, \{\ell\})$. A crossover is then expected between different regimes, depending on which length happens to be the smallest. These regimes are characterized by power laws such as (1), with different exponents ϕ depending on the dominant lengthscale. Several physical situations are discussed in sections 3 and 4.

2 Rate equations

In general, in the submonolayer regime (i.e. for coverages $\theta = Ft \leq 0.25$, say) the densities of adatoms, n, and of islands of size s, N_s, may be obtained by solving a set of coupled rate equations of the form (we will let $a = 1$ from now or)

$$\dot{n} = F - 2Dn^2 - \sum_{s \geq 2} [n\sigma_s - \frac{1}{\tau_s}] N_s - \frac{n}{\tau} , \tag{2}$$

$$\dot{N}_s = n\sigma_{s-1} N_{s-1} - n\sigma_s N_s - \frac{1}{\tau_s} N_s + \frac{1}{\tau_{s+1}} N_{s+1} \quad s \geq 2 , \tag{3}$$

where the coefficients σ_s are the adatom-capture cross-sections, and the τ_ss are the atom emission rates, of islands of size s. The rate $1/\tau$ of adatom disappearence stands for other processes than capture by islands, such as evaporation, capture by defects, steps and so on. It is understood that $\sigma_1 N_1 = 2Dn$ in (3).

For computational convenience, it is assumed that a critical size s^* exists, such that $1/\tau_s = 0$ if $s > s^*$. Strictly speaking, this definition of the critical size makes sense only for $s^* = 1$ (see the Discussion section). It is nevertheless usually assumed to hold true also for larger s^*. In this way, it is possible to write Eq. (2) as

$$\dot{n} = F - 2Dn^2 - \sum_{s=2}^{s^*} [n\sigma_s - \frac{1}{\tau_s}] N_s - n\sigma_{\text{av}} N - \frac{n}{\tau} , \tag{4}$$

where $N = \sum_{s > s^*} N_s$, and $\sigma_{\text{av}} = (1/N) \sum_{s > s^*} \sigma_s N_s$. Eq. (3) can be split into three formulae (according to s being less than, equal to or larger than s^*), and the third one summed over $s > s^*$, so that one has

$$\dot{N}_s = n\sigma_{s-1} N_{s-1} - n\sigma_s N_s - \frac{1}{\tau_s} N_s + \frac{1}{\tau_{s+1}} N_{s+1} \quad s < s^* , \tag{5}$$

$$\dot{N}_{s^*} = n\sigma_{s^*-1} N_{s^*-1} - n\sigma_{s^*} N_{s^*} - \frac{1}{\tau_{s^*}} N_{s^*} , \tag{6}$$

$$\dot{N} = n\sigma_{s^*} N_{s^*} . \tag{7}$$

The usual assumption at this point is that the densities of islands whose size is smaller than s^* satisfy the quasi-equilibrium ($\dot{N}_s \approx 0$) law of mass action (letting $\sigma_1 = 2D$)

$$N_s \approx \left(\prod_{k=2}^{s} \sigma_{k-1} \tau_k \right) n^s . \tag{8}$$

One can now dispose of Eqs. (5) and (6), while keeping Eqs. (4) and (7). Defining the adatom diffusion length ℓ through $D\ell^{-2} = 2n + \sigma_{\text{av}} N + 1/\tau$, formulae (4) and (7) read

$$\dot{n} = F - D\frac{n}{\ell^2} , \tag{9}$$

$$\dot{N} = n\sigma_{s^*} N_{s^*} \ . \tag{10}$$

The right-hand side of (10) is the nucleation rate of stable islands. The density of critical nuclei, N_{s^*}, is given by (8). The cross section σ_{s^*} is equal to $2D$ for $s^* = 1$, and it is proportional to D otherwise, with a proportionality coefficient which is of order unity[3,4]. It will be taken as unity in the following, that is, $\sigma_{s^*} = D$ for all s^*.

It can be shown (see[7] and the Discussion) that σ_{av}/D is a function of the coverage θ only, provided the island size distribution satisfy the scaling law $N_s = \theta/s_{\text{av}}^2 F(s/s_{\text{av}})$. Since we are interested in finding the maximum island density, i.e. the island density at the coverage θ_c where coalescence takes place, we can take σ_{av}/D as a constant. For simplicity, it will be assumed that the constant is equal to 1 and $\sigma_{\text{av}} = D$.

3 Scaling arguments

In this section we will present the simple scaling arguments that allow to arrive in the different cases, at the different forms of Eq. (1). These arguments were originally formulated in ref. [8] for the special case of deposition without evaporation on a high-symmetry substrate. Here, a slightly different, more general formulation is given. It allows to compute N with simple arguments in any situation. In all that follows the atomic size is taken as the length unit.

The first stage requires the determination of the nucleation rate per unit surface and time, $1/\tau_{\text{nuc}}$. This is given by (10), so that

$$\frac{1}{\tau_{\text{nuc}}} = DnN_{s^*} \ . \tag{11}$$

The density of critical nuclei, N_{s^*}, is given by (8). Since we are only concerned here with scaling laws, and order-of-magnitude relations, we shall further simplify formula (8) to

$$N_{s^*} \approx n^{s^*} \ . \tag{12}$$

From (11) and (12), one finds

$$\frac{1}{\tau_{\text{nuc}}} \approx Dn^{s^*+1} \ . \tag{13}$$

Another, independent equation can be written down to relate the nucleation rate and the density of stable islands $N = \sum_{s>s^*} N_s$. In Villain et al. [8], it was stated that in the area $\ell_s^2 = 1/N$ occupied by an island, only one (on average) nucleation event takes place, during the time $1/F$ needed to fill a layer. The present modified argument simply states that *in the area* $\ell_s^2 = 1/N$ *occupied by an island, only one (on average) nucleation event takes place, during the time* t_c *needed for the growing islands to come into contact.* This reads $\ell_s^2 t_c/\tau_{\rm nuc} \approx 1$, and thus

$$\frac{1}{\tau_{\rm nuc}} \approx \frac{N}{t_c} \, . \tag{14}$$

In other words, coalescence after a time of order t_c is explicitely included in equation (10):

$$\dot{N} = DnN_{s^\bullet} - \frac{N}{t_c} \, . \tag{15}$$

The time t_c is readily computed by knowing the growth velocity of an island of size R. This is found by equating the rate of increase of the mass of the island, $\mathrm{d}\mathcal{M}/\mathrm{d}t$, to the current of atoms impinging onto the island, j_n. The latter is approximately

$$j_n \approx Dn \, , \tag{16}$$

if the island growth is diffusion-limited, and

$$j_n \approx FR^{d_b} x_s \tag{17}$$

if the island growth is evaporation-limited (remember $x_s = \sqrt{D\tau_v}$). Here d_b is the dimensionality of the island boundary. It is $d_b = 1$ for 2-D compact islands, $d_b = 2$ for isotropically growing 3-D islands, and $1 < d_b < 2$ for compact 2-D island with fractal boundary.

Once $\mathrm{d}\mathcal{M}/\mathrm{d}t$ is known, t_c is defined as the time at which the island size $R \approx \ell_s$, since ℓ_s is by definition the typical distance between island centres.

We shall consider in the following a number of physically relevant situations.

• *High-symmetry substrate, no evaporation, compact 2-D islands.*

The adatom density results in this case from a balance between deposition at a rate F and capture by the stable islands at a rate DnN, all other processes being negligible, so that letting $\ell = \ell_s = 1/N^{1/2}$ in (9), and making $\dot{n} \approx 0$,

$$n \approx F/(DN) \approx F\ell_s^2/D \, . \tag{18}$$

The rate of growth of the area of an island of linear size R is diffusion-limited in this case, so that $R^2 \approx Dnt \approx Ft/N$. At $t = t_c$, $R \approx \ell_s$, and thus

$$t_c \approx N\ell_s^2/F \approx 1/F \, . \tag{19}$$

Using (13) and (18), one finds

$$\frac{1}{\tau_{\rm nuc}} \approx D[F/(DN)]^{s^*+1} . \tag{20}$$

From (14) and (19),

$$\frac{1}{\tau_{\rm nuc}} \approx FN . \tag{21}$$

Finally, (20) and (21) yield [3,4,8]

$$N \approx \left(\frac{F}{D}\right)^{s^*/(s^*+2)} . \tag{22}$$

Of course, equation (19) can be arrived at without calculations, since $1/F$ is the time needed to fill a layer. However, it is not always so trivial, as we will see next.

- *High-symmetry substrate, no evaporation, fractal islands.*

In this case, island growth is still diffusion-limited, but the mass-radius relation is $\mathcal{M} \approx R^{d_f}$ for fractal islands of fractal dimension d_f, so that at time t, $R^{d_f} \approx Dnt \approx Ft/N$. At $t = t_c$, $R \approx \ell_s$, and thus

$$t_c \approx N\ell_s^{d_f}/F \approx \ell_s^{d_f-2}/F . \tag{23}$$

Using (14) one has

$$\frac{1}{\tau_{\rm nuc}} \approx FN^{d_f/2} . \tag{24}$$

From (13) (note that only $s^* = 1$ is compatible with the assumption of fractal islands) and (18), one finds

$$\frac{1}{\tau_{\rm nuc}} \approx Dn^2 \approx D[F/(DN)]^2 . \tag{25}$$

Finally, (24) and (25) yield [9,10]

$$N \approx \left(\frac{F}{D}\right)^{2/(4+d_f)} . \tag{26}$$

See however [7] and [11].

- *High-symmetry substrate, no evaporation, 1-D adatom diffusion.*

If adatom diffusion happens to be strongly anisotropic, one could rightfully consider it to be 1-dimensional. In this case, adatom encounters take place along a line, so that adatom clusters are found in an area of size ℓ_s (since the atomic distance is equal to 1 in our units). Thus,

$$N = 1/\ell_s \, . \tag{27}$$

During its lifetime τ, that is, before being captured by an island, a diffusing adatom explores, in 1 dimension, a number ν of distinct sites given by $\nu \approx (D\tau)^{1/2}$. By definition of the adatom diffusion length, $\ell_s \approx \nu$, so that $\tau \approx \ell_s^2/D$. The adatom density is then given by

$$n \approx F\tau \approx F\ell_s^2/D \approx FN^{-2}/D \, . \tag{28}$$

The growth of the length L of an island is diffusion-limited, so that $L \approx Dnt \approx Ft/N^2$. At $t = t_c$, $L \approx \ell_s$, and thus

$$t_c \approx N^2 \ell_s / F \approx N/F \, . \tag{29}$$

Using (13) (which is still valid, since one nucleation event is expected on a length ℓ_s during a time of order t_c) and (28), one finds

$$\frac{1}{\tau_{\rm nuc}} \approx D[F/(DN^2)]^{s^*+1} \, . \tag{30}$$

From (29) and (14),

$$\frac{1}{\tau_{\rm nuc}} \approx F \, . \tag{31}$$

Finally, (20) and (31) yield [12,13]

$$N \approx \left(\frac{F}{D}\right)^{s^*/(2s^*+2)} \, . \tag{32}$$

• *High-symmetry substrate, no evaporation, restricted adatom diffusion.*

Besides being isotropic or strongly anisotropic, adatom diffusion might take place in the presence of adsorbed impurities acting as obstacles. In particular, these impurities might be strongly repulsive and restrict the diffusion path of an adatom, in such a way that it is only able to explore a number ν of distinct sites given by $\nu \approx (D\tau)^\alpha$, during its lifetime τ [14]. The exponent α is an (a priori unknown) function of the impurity concentration and of the temperature. In the absence of impurities, $\alpha = 1$ (for isotropic diffusion); otherwise $\alpha < 1$. Only isotropic diffusion will be considered in the following.

By definition of the adatom diffusion length, $\ell_s^2 \approx \nu$, so that $\tau \approx \ell_s^{2/\alpha}/D$. The adatom density is then given by

$$n \approx F\tau \approx F\ell_s^{2/\alpha}/D \approx FN^{-1/\alpha}/D \,. \tag{33}$$

The growth of the area of an island is diffusion-limited, so that $R^2 \approx Dnt \approx FtN^{-1/\alpha}$. At $t = t_c$, $R \approx \ell_s$, and thus

$$t_c \approx N^{1/\alpha}\ell_s^2/F \approx N^{(1-\alpha)/\alpha}/F \,. \tag{34}$$

Using (13) and (33), one finds

$$\frac{1}{\tau_{\rm nuc}} \approx D[FN^{-1/\alpha}/D]^{s^*+1} \,. \tag{35}$$

From (34) and (14),

$$\frac{1}{\tau_{\rm nuc}} \approx FN^{(2\alpha-1)/\alpha} \,. \tag{36}$$

Finally, (20) and (36) yield

$$N \approx \left(\frac{F}{D}\right)^{\alpha s^*/(s^*+2\alpha)} \,. \tag{37}$$

This formula generalizes the result that Liu et al. [14] have obtained for the special case $s^* = 1$.

• *High-symmetry substrate, no evaporation, diffusing islands I.*

The possibility that stable islands are not immobile will be considered here. However, if dimers, trimers, and so on, diffuse and at the same time are *not* stable, the validity of Eq. (12) should be discussed. Instead, it will be assumed here that $s^* = 1$, that is, even dimers are stable. Let D_N be the (average) cluster diffusion coefficient. Assuming, as reasonable, that $D \gg D_N$, the adatom density will still be given by (18).

The rate of growth of the area of an island of linear size R is diffusion-limited in this case. Since islands also diffuse, $R^2 \approx (Dn + D_N N)t$. The situation of interest here is $Dn \ll D_N N$, so that $R^2 \approx D_N N t$. At $t = t_c$, $R \approx \ell_s$, and thus

$$t_c \approx 1/(D_N N^2) \,. \tag{38}$$

Using (13) with $s^* = 1$ and (18), one finds

$$\frac{1}{\tau_{\rm nuc}} \approx D[F/(DN)]^2 \,. \tag{39}$$

From (38) and (14),

$$\frac{1}{\tau_{\text{nuc}}} \approx D_N N^3 . \tag{40}$$

Finally, (39) and (40) yield[8]

$$N \approx \left(\frac{F}{D}\right)^{2/5} \left(\frac{D}{D_N}\right)^{1/5} = \left(\frac{F^2}{DD_N}\right)^{1/5} . \tag{41}$$

The condition for the validity of (41) is $Dn \ll D_N N$. Using (18), the condition reads $F/N \ll D_N N$, or $N^2 \gg F/D_N$. From (41), $F^2/(DD_N) \gg (F/D_N)^{5/2}$, or

$$\frac{F}{D} \ll \left(\frac{D_N}{D}\right)^3 . \tag{42}$$

Equation (41) has been checked with Kinetic Monte Carlo simulations by Jensen et al. [15] and by Liu et al. [16].

• *High-symmetry substrate, no evaporation, diffusing islands II.*

Another possibility is that the critical nucleus size is not fixed by the detachment rate of atoms from an island, as in (2) and (3), but rather by the diffusion of small clusters towards big ones. This means $1/\tau_s \ll D_s N$, if D_s is the diffusion coefficient of a cluster of size s. In this sense, we say that s^* is such that clusters of size $s > s^*$ do not diffuse or decay, while smaller ones do not decay but diffuse. Making the further simplifing assumption $D_s = D$, the density of critical nuclei N_{s^*} can be found by replacing all $D\tau_s$ in Walton's formula (8) by $1/N$. One finds

$$N_{s^*} \approx \frac{n^{s^*}}{N^{s^*-1}} . \tag{43}$$

The rate of growth of the area of an island of linear size R is diffusion-limited in this case. Since we assume that cluster diffusion is faster than atom detachment, we can expect growing islands to be fractal. Although small clusters diffuse, the main contribution to island growth comes from adatoms, so that $R^{d_f} \approx Dnt \approx Ft/N$. At $t = t_c$, $R \approx \ell_s$, and thus

$$t_c \approx N\ell_s^{d_f}/F \approx \ell_s^{d_f-2}/F . \tag{44}$$

Using (43) with and (18), one finds

$$\frac{1}{\tau_{\text{nuc}}} \approx D\frac{n^{s^*+1}}{N^{s^*-1}} \approx D\left(\frac{F}{D}\right)^{s^*+1} \frac{1}{N^{2s^*}} . \tag{45}$$

From (14) and (44),

$$\frac{1}{\tau_{\rm nuc}} \approx F N^{d_f/2} \,. \tag{46}$$

Finally, (45) and (46) yield [8,11,17]

$$N \approx \left(\frac{F}{D}\right)^{2s^*/(4s^*+d_f)} \,. \tag{47}$$

• *High-symmetry substrate, strong evaporation, 2-D islands.*

Strong evaporation means the adatoms are more likely to disappear due to desorption, with probability $1/\tau_v$ per unit time and site, than to be captured by an island. In other words, the adatom diffusion length before desorption, $x_s = \sqrt{D\tau_v}$, is shorter than the average island-island distance, ℓ_s. In this case, the adatom density results from a balance between deposition and desorption at a rate n/τ_v, so that

$$n \approx F\tau_v \approx F x_s^2/D \,. \tag{48}$$

A growing island is only able to capture the adatoms falling at a distance smaller than the adatom diffusion length before desorption, x_s. If x_s is smaller than the island size R, the island will capture the adatoms falling inside an annular region of width $2x_s$ around its border, as well as those directly impinging on its edge. The area of the annular region is proportional to $(R + x_2)^2 - (R - x_s)^2 = 4Rx_s$. Including direct impingement, $\mathrm{d}(R^2)/\mathrm{d}t \approx FR(1+x_s)$, and at time t, $R \approx F(1+x_s)t$. At $t = t_c$, $R \approx \ell_s$, and thus

$$t_c \approx \ell_s/[F(1+x_s)] \approx 1/[FN^{1/2}(1+x_s)] \,. \tag{49}$$

Direct impingement is important for nucleation, too, so that (13) becomes

$$\frac{1}{\tau_{\rm nuc}} \approx (F + Dn)n^{s^*} \approx F(1+x_s^2)n^{s^*} \approx F(1+x_s)^2 n^{s^*} \,. \tag{50}$$

Using (50) and (48) one finds

$$\frac{1}{\tau_{\rm nuc}} \approx F(1+x_s)^2 (F\tau_v)^{s^*} \,. \tag{51}$$

From (14) and (49),

$$\frac{1}{\tau_{\rm nuc}} \approx F(1+x_s)N^{3/2} \,. \tag{52}$$

Finally, (51) and (52) yield

$$N \approx (F\tau_v)^{2s^*/3} (1+x_s)^{2/3} \,. \tag{53}$$

This scaling equation is new. It differs from previously published results by Stoyanov and Kaschiev[3] and by Venables et al. [4]. The reason for this difference is discussed in Ref. [18]. Formula (53) is, on the other hand, in agreement with computer simulations by Jensen et al. for $s^* = 1$ [18]. We shall compute in section 4 the crossover scaling function between the Eqs. (32) and (53).

- *Vicinal substrate, no evaporation, 2-D islands.*

On a vicinal surface, the arriving adatoms diffuse between parallel steps at an average distance d. If d is much larger than ℓ_s, then the presence of these steps is not expected to affect much the adatom density, nor the probability of nucleation, and Eq. (22) is obtained. In the opposite case, $d \ll \ell_s$, adatoms are predominantly captured at steps. The adatom density is then given by the balance between deposition and capture by steps,

$$n \approx Fd^2/D \,. \tag{54}$$

The rate of growth of the area of an island of linear size R is diffusion-limited in this case, so that $R^2 \approx Dnt$. At $t = t_c$, $R \approx \ell_s$, and thus

$$t_c \approx \ell_s^2/(Fd^2) \approx 1/(Fd^2N) \,. \tag{55}$$

Using (13) and (54), one finds

$$\frac{1}{\tau_{\text{nuc}}} \approx D(Fd^2/D)^{s^*+1} \,. \tag{56}$$

From (14) and (55),

$$\frac{1}{\tau_{\text{nuc}}} \approx Fd^2N^2 \,. \tag{57}$$

Finally, (56) and (57) yield [19]

$$N \approx \left(\frac{Fd^2}{D}\right)^{s^*/2} \,. \tag{58}$$

The crossover scaling function between Eqs. (22) and (58) will be computed in section 4. Note, in passing, that one could also monitor the island density at shorter times than t_c, which means for coverages $\theta = Ft \ll Ft_c$. If this is the case, the preceding calculation still holds; one has simply to replace t_c with θ/F in eq. (14), and one finds $Dn^{s^*+1} \approx D(Fd^2/D)^{s^*+1} \approx FN/\theta$, or

$$N \approx \left(\frac{F}{D}\right)^{s^*} d^{2(s^*+1)}\theta \,. \tag{59}$$

The result (58) has been checked with Kinetic Monte Carlo simulations by Pimpinelli and Peyla for $s^* = 3$ [19]. Formula (59) has also been obtained and checked with simulations by Bales [20], in the case $s^* = 1$.

• *High-symmetry substrate, no evaporation, 3-D islands.*

The adatom density is given by (18).

The rate of growth of the volume of an island of linear size R is diffusion-limited in this case, so that $R^3 \approx Dnt \approx Ft/N$. At $t = t_c$, $R \approx \ell_s$, and thus

$$t_c \approx N\ell_s^3/F \approx 1/(FN^{1/2}) \,. \tag{60}$$

Using (13) and (18), one finds

$$\frac{1}{\tau_{\text{nuc}}} \approx D[F/(DN)]^{s^*+1} \,. \tag{61}$$

From (14) and (60),

$$\frac{1}{\tau_{\text{nuc}}} \approx FN^{3/2} \,. \tag{62}$$

Finally, (61) and (62) yield [3,4]

$$N \approx \left(\frac{F}{D}\right)^{2s^*/(2s^*+5)} \,. \tag{63}$$

• *High-symmetry substrate, strong evaporation, 3-D islands.*

Strong evaporation means, as before, that $\ell_s \geq R > x_s$. The adatom density is given by (48).

A 3-D island grows by two mechanisms in the case of strong evaporation: (i) by capture of the adatoms falling on the surface at a distance smaller than x_s; (ii) by direct capture of all adatoms falling on it. Thus, $\mathrm{d}R^3/\mathrm{d}t \approx F(x_sR + x_s^2 + R^2) \approx F(R^2 + x_sR)$. Solving for R with the condition $R = 0$ at $t = 0$, one gets $R - x_s \ln(1 + R/x_s) = Ft$. At $t = t_c$, $R \approx \ell_s > x_s$, which means that direct capture always dominates. It follows $Ft_c \approx \ell_s$, or

$$t_c \approx 1/[FN^{1/2}] \,. \tag{64}$$

Using (50) and (48), one finds

$$\frac{1}{\tau_{\text{nuc}}} \approx (F + Dn)n^{s^*} \approx F(1 + x_s^2)n^{s^*} \approx F(1 + x_s^2)(F\tau_v)^{s^*} \,. \tag{65}$$

From (14) and (64),

$$\frac{1}{\tau_{\text{nuc}}} \approx F N^{3/2} . \tag{66}$$

Finally, (65) and (66) yield

$$N \approx (F\tau_v)^{2s^*/3} \left(1 + x_s^2\right)^{2/3} . \tag{67}$$

When $x_s > 1$, this scaling relation agrees with Stoyanov and Kaschiev [3] and with Venables et al. [4]. However, formula (67) extends to another regime, namely when $x_s < 1$. In this case, $N \approx (F\tau_v)^{2s^*/3}$, a result which does not seem to appear in the literature.

- *Vicinal substrate, no evaporation, 3-D islands.*
The adatom density is still given by (54).
The rate of growth of the volume of an island of linear size R is diffusion-limited in this case, so that $R^3 \approx Dnt$. At $t = t_c$, $R \approx \ell_s$, and thus

$$t_c \approx \ell_s^3/(Fd^2) \approx 1/(Fd^2 N^{3/2}) . \tag{68}$$

Using (13) and (54), one finds

$$\frac{1}{\tau_{\text{nuc}}} \approx D(Fd^2/D)^{s^*+1} . \tag{69}$$

From (14) and (68),

$$\frac{1}{\tau_{\text{nuc}}} \approx Fd^2 N^{5/2} . \tag{70}$$

Finally, (69) and (70) yield

$$N \approx \left(\frac{Fd^2}{D}\right)^{2s^*/5} . \tag{71}$$

The crossover between Eqs. (63) and (71) will be computed in section 4. Note, in passing, that one could also monitor the island density at shorter times than t_c, for instance at times of the order of $1/F$. If this is the case, one recovers Eq. (59), the same as for 2-D islands.

4 Crossover scaling functions

We shall show in this section that it is possible to derive interpolation formulae describing the crossovers between, for instance, the no-evaporation and the strong-evaporation regimes, or the flat and vicinal regimes, of the preceding section.

To do this, we shall use Burton, Cabrera and Frank's theory of step flow [26]. The adatom density will be computed on a terrace bounded by two parallel steps, at a distance ℓ. They may be the steps in the orderd array of a vicinal surface; in this case, we shall let $\ell = d$. Otherwise, they may represent the edges of a big enough island; in the latter case we shall let $\ell = \ell_s = 1/\sqrt{N}$.

The adatom density obeys the equation

$$\dot{n} = F + D\nabla^2 n - \frac{n}{\tau} , \tag{72}$$

with the conditions $n(\pm\ell/2) = 0$ at boundary steps (choosing the origin $x = 0$ at the terrace centre).

The time τ is the average lifetime of an adatom. Since adatoms disappear either by capture by islands, or by desorption, we shall alternatively let $1/\tau = DN = D/\ell_s^2$ or $1/\tau = 1/\tau_v = D/x_s^2$. In both cases, the notation $\kappa^2 = 1/(D\tau)$ will be used.

In the quasi-stationary approximation [26], $\dot{n} \approx 0$, and equation (72) can be solved. The solution reads

$$n(x) = F\tau \left[1 - \frac{\cosh(\kappa x)}{\cosh(\kappa\ell/2)}\right] . \tag{73}$$

This formula will be needed to compute the nucleation rate (11). The latter is a mean field quantity, independent of x. We let thus $x = 0$ in (73), since the higher nucleation probability is at the terrace centre, given the symmetry of our problem (equivalently, the average value over the terrace could be used). One finds

$$n(0) = \bar{n} = F\tau \left[1 - \frac{1}{\cosh(\kappa\ell/2)}\right] = 2F\tau \frac{\sinh^2(\kappa\ell/4)}{\cosh(\kappa\ell/2)} , \tag{74}$$

where we used the identity $\cosh(x) - 1 = 2\sinh^2(x/2)$.

The next task is the determination of the total island density N. Its time variation is simple: N increases each time a new island is nucleated, so that $(\dot{N})_1 = DnN_{s^*}$. On the other hand, N decreases when two islands touch and coalesce. Following the authorities [3,4], we write $(\dot{N})_2 = -(\mathrm{d}\mathcal{A}/\mathrm{d}t)N^2$, where $\mathcal{A} \approx R^2$ is the average area of an island of linear size R. This means that coalescence results from binary encounters of immobile islands, whose area increases at a rate $\mathrm{d}\mathcal{A}/\mathrm{d}t$. Collecting $(\dot{N})_1$ and $(\dot{N})_2$ yields

$$\dot{N} = DnN_{s^*} - \frac{\mathrm{d}\mathcal{A}}{\mathrm{d}t} N^2 . \tag{75}$$

At stationarity, which is what we are interested in, $\dot{N} = 0$. Note that, by definition, coalescence takes place when $\mathcal{A} \approx \ell_s^2$. One can thus write

$$DnN_{s^*} \approx \left(\frac{\mathrm{d}\mathcal{A}}{\mathrm{d}t}\right)_{\mathcal{A}\approx\ell_s^2} N^2 \; . \tag{76}$$

The final task concerns therefore the evaluation of the growth rate of an island. This can be done by noting that the mass $\mathcal{M}$ of an island increases either by direct capture of atoms from the beam, or by diffusion of adatoms on the surface. Since the surface diffusion current of adatoms is $-D\nabla n$, assuming circular ($d = 2$) or hemispherical ($d = 3$) islands of radius R one can write

$$\frac{\mathrm{d}\mathcal{M}}{\mathrm{d}t} \approx FR^{d-1} - DR\left(\frac{\mathrm{d}n}{\mathrm{d}r}\right)_R \; . \tag{77}$$

The result can be found in a number of papers [26,3,4,18]. It makes use of Bessel functions, which are not easy to manipulate. Approximate results, which have the merit of reproducing the correct limiting regimes (see below), and of allowing analytical expressions to be written down, will be used. They will be derived case by case for a few specific situations: (i) 2-D islands on a vicinal surface without evaporation; (ii) 2-D islands on a high-symmetry substrate, with evaporation; (iii) 3-D islands on a high-symmetry substrate, with evaporation.

• *Vicinal surface, no evaporation, 2-D islands.*

We consider here nucleation of islands on a vicinal surface. The temperature is high enough to neglect extra potential barriers at steps (Ehrlich-Schwoebel effect [27]), but low enough to neglect evaporation. Adatoms can only disappear at the step edges, or at island edges.

Thus, we let $\ell = d$, and $1/\tau = DN$, so that $\kappa = 1/\ell_s$ and Eq. (74) becomes

$$\bar{n} = 2F\tau\frac{\sinh^2(\kappa d/4)}{\cosh(\kappa d/2)} \; , \tag{78}$$

Direct impingement to the step edge is always negligible in this case, and islands grow by capture of diffusing adatoms. We simplify the problem with the assumption $(\mathrm{d}n/\mathrm{d}r)|_{R\approx\ell_s} \approx -\bar{n}/\ell_s$. The meaning of this may be clarified by considering the limit $\tau \to \infty$ of (74). In that case, $n(x) = F/(8D)(\ell_s^2 - 4x^2)$, and $(\mathrm{d}n/\mathrm{d}x)|_{x=\ell_s} = -F\ell_s/D \approx -n(0)/\ell_s$. Therefore from (77) we find

$$\left.\frac{\mathrm{d}\mathcal{M}}{\mathrm{d}t}\right|_{R\approx\ell_s} \approx D\bar{n} \approx 2FD\tau\frac{\sinh^2(\kappa d/4)}{\cosh(\kappa d/2)} \; . \tag{79}$$

Since $d\mathcal{M}/dt \approx d\mathcal{A}/dt$, inserting (79) in (76) yields

$$D\bar{n}N_{s^*} \approx D\bar{n}N^2 , \tag{80}$$

and using (78) together with $\kappa = N^{1/2}$,

$$N \approx \left[\frac{F}{D}\frac{\sinh^2(N^{1/2}d/4)}{\cosh(N^{1/2}d/2)}\right]^{s^*/(2+s^*)} . \tag{81}$$

It is trivial to check that the limit $N^{1/2}d \to \infty$ yields (22), and that $N^{1/2}d \to 0$ yields (58).

Formulae such as (81) allow to produce *scaling plots.* In a scaling plot, several data sets of, say, the island density as a function of flux, temperature or whatever, are made to collapse on a single, 'universal' curve by an appropriate choice of the 're-scaled' plot variables. The chosen variables are called *scaling variables*, and the 'universal' curve is said to be the graph of the *scaling function.*

In the present case, letting $N_{\text{hisym}} = (F/D)^{s^*/(2+s^*)}$, it can be seen that formula (81) has the form

$$d^2N \approx d^2N_{\text{hisym}} f_{\text{vic}}(d^2N) \tag{82}$$

which, inverting $f_{\text{vic}}(x)$, can be rewritten as

$$d^2N \approx g_{\text{vic}}(d^2N_{\text{hisym}}) \tag{83}$$

which shows that d^2N and d^2N_{hisym} are an appropriate choice of scaling variables. The scaling function $g_{\text{vic}}(x)$ behaves as follows:

$$g_{\text{vic}}(x) \sim \begin{cases} x & \text{for } x \to \infty, \\ x^{(2+s^*)/2} & \text{for } x \to 0. \end{cases} \tag{84}$$

Results of Monte Carlo simulations with $s^* = 3$ [19] are shown in figure 1. The simulations are described in detail in section 6. In the figure, the variables are $n_{\text{step}} = \sqrt{N}$ and $n_\infty = \sqrt{N_{\text{hisym}}}$. It is easy to check that the slopes in a log-log plot are 1 for $dn_\infty > 3$ and 5/2 for $dn_\infty < 3$ (for the limit value $dn_\infty = 3$ see [22]). Indeed, the solid line is the plot of (81) for $s^* = 3$.

• *High-symmetry surface, non-negligible evaporation, 2-D islands.* We consider here adatoms diffusing between islands, on a high-symmetry substrate, and

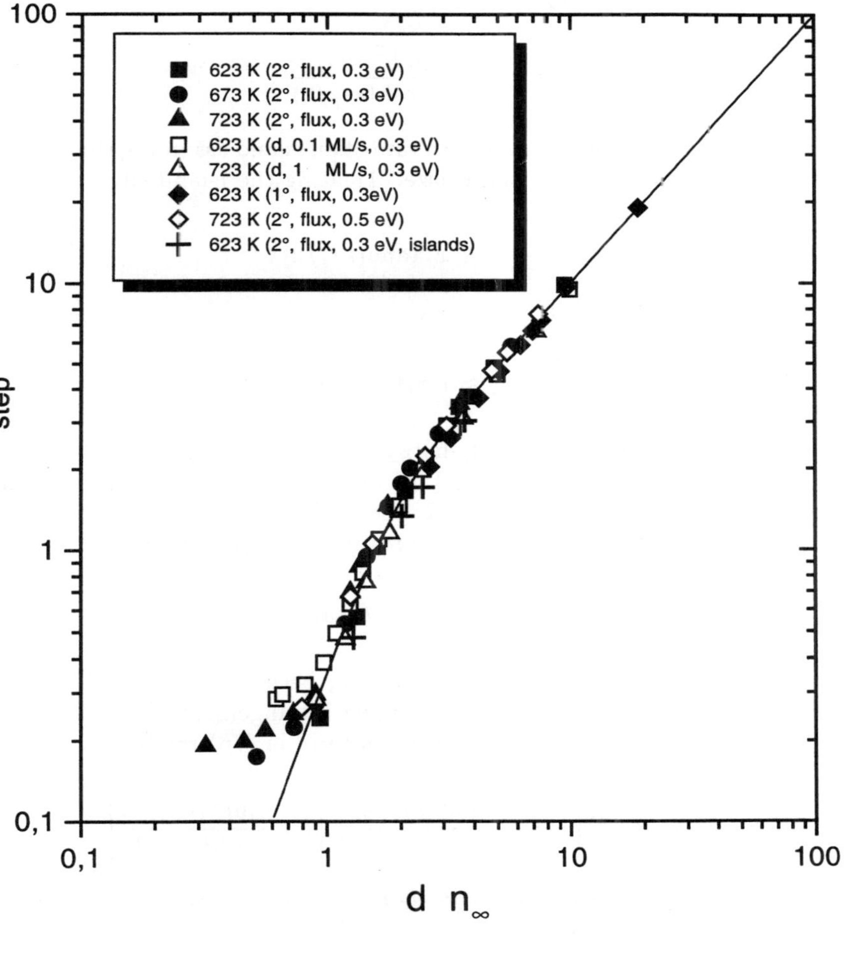

Figure 1: Scaling plot of simulation data in for growth on various vicinal surfaces, for different temperatures, fluxes and bond energies E_a (see text). The diffusion barrier is $E_D = 1$ eV everywhere. Symbols as in the inset. The continuous line is the scaling function derived from equation (81) with $i^* = 3$.

desorbing at a rate $1/\tau_v$ per atom. This implies that $\ell = \ell_s$, $\tau = \tau_v$ and $\kappa^{-2} = D\tau_v$. Eq. (74) becomes

$$\bar{n} = 2F\tau_v \frac{\sinh^2(\kappa\ell_s/4)}{\cosh(\kappa\ell_s/2)} . \tag{85}$$

We shall compute now the growth rate of an island assuming that its radius is large with respect to the atomic size. Treating the island edges as straight steps yields

$$\frac{\mathrm{d}R}{\mathrm{d}t} \approx Fx_s \tanh(\kappa\ell_s/2) . \tag{86}$$

Since $\mathrm{d}\mathcal{A}/\mathrm{d}t \approx R\mathrm{d}R/\mathrm{d}t$, one finds

$$\frac{\mathrm{d}\mathcal{A}}{\mathrm{d}t}|_{R=\ell_s} \approx Fx_s\ell_s \tanh(\kappa\ell_s/2) . \tag{87}$$

Note that the two limits $(\mathrm{d}\mathcal{A}/\mathrm{d}t)|_{R=\ell_s} \approx F\ell_s^2$ and $(\mathrm{d}\mathcal{A}/\mathrm{d}t)|_{R=\ell_s} \approx F\ell_s x_s$ are well reproduced for $\kappa\ell_s \ll 1$ and for $\kappa\ell_s \gg 1$, respectively. In fact, in the limit $x_s \to 0$, the area increase is FR, due to direct impingement. In order to interpolate all the way through to this regime, we finally write

$$\frac{\mathrm{d}\mathcal{A}}{\mathrm{d}t}|_{R=\ell_s} \approx F(1+x_s)\ell_s \tanh(\kappa\ell_s/2) . \tag{88}$$

In the limit where the diffusion length x_s becomes very small, the nucleation probability is also determined by direct impingement of a beam atom on a nucleus. Then, using (85), the nucleation term in (76) reads

$$(F + D\bar{n})N_{s^*} \approx F\left[1 + x_s^2\frac{\sinh^2(\kappa\ell_s/4)}{\cosh(\kappa\ell_s/2)}\right]\left[F\tau_v\frac{\sinh^2(\kappa\ell_s/4)}{\cosh(\kappa\ell_s/2)}\right]^{s^*} . \tag{89}$$

In fact, a simpler approximate expression will be used, which reproduces the correct limiting behaviour at large and small x_s:

$$(F + D\bar{n})N_{s^*} \approx \frac{(1+x_s)^2}{\tau_v}\left[F\tau_v\frac{\sinh^2(\kappa\ell_s/4)}{\cosh(\kappa\ell_s/2)}\right]^{s^*+1} . \tag{90}$$

Finally, (88) and (90) yield

$$\frac{(1+x_s)^2}{\tau_v}\left[F\tau_v\frac{\sinh^2(\kappa\ell_s/4)}{\cosh(\kappa\ell_s/2)}\right]^{s^*+1} \approx F(1+x_s)\ell_s \tanh(\kappa\ell_s/2)N^2 , \tag{91}$$

or,

$$(1+x_s)(F\tau_v)^{s^*} \approx \left[\frac{\cosh(\kappa N^{-1/2}/2)}{\sinh^2(\kappa N^{-1/2}/4)}\right]^{s^*+1} \tanh(\kappa N^{-1/2}/2)N^{3/2} \,. \qquad (92)$$

The crossover scaling formula (92) may be cast in the form:

$$(1+x_s)x_s^3(F\tau_v)^{s^*} \approx \left[\frac{\cosh(\kappa N^{-1/2}/2)}{\sinh^2(\kappa N^{-1/2}/4)}\right]^{s^*+1} \tanh(\kappa N^{-1/2}/2)(x_s^2 N)^{3/2} \,, \qquad (93)$$

or,

$$(1+x_s)x_s^{2s^*+3}\left(\frac{F}{D}\right)^{s^*} = f_{\mathrm{evap2D}}(x_s^2 N) \,, \qquad (94)$$

where

$$f_{\mathrm{evap2D}}(x) = \left[\frac{\cosh\left(\frac{1}{2x}\right)}{\sinh^2\left(\frac{1}{4x}\right)}\right]^{s^*+1} \tanh\left(\frac{1}{2x}\right)x^3 \,. \qquad (95)$$

Letting $N_{\mathrm{evap2D}} = (1+x_s)^{2/3}(F\tau_v)^{2s^*/3}$ (cf. Eq. (53)), and inverting f, one finds

$$x_s^2 N = g_{\mathrm{evap2D}}(x_s^2 N_{\mathrm{evap2D}}) \,, \qquad (96)$$

where

$$g_{\mathrm{evap2D}}(x) \sim \begin{cases} x^{3/[2(2+s^*)]} & \text{for } x \to \infty, \\ x & \text{for } x \to 0. \end{cases} \qquad (97)$$

Equations (94) and (97) agree with the Kinetic Monte Carlo results for $s^* = 1$ [18] shown in figure 2, where the solid line is equation (94) for $s^* = 1$ (note that also $D = 1$ has been assumed).

• *High-symmetry surface, with evaporation, 3-D islands.* The adatom density is still given by Eq. (74), and the nucleation rate by Eq. (90). The growth rate of the (projected) area of a 3-D island has two contributions, as for 2-D islands: a diffusion-limited contribution, which is given by

$$\frac{\mathrm{d}\mathcal{A}}{\mathrm{d}t}|_{R=\ell_s} \approx F x_s \ell_s \tanh(\kappa\ell_s/2) \,; \qquad (98)$$

and a direct capture contribution,

$$\frac{\mathrm{d}\mathcal{A}}{\mathrm{d}t}|_{R=\ell_s} \approx F\ell_s \,. \qquad (99)$$

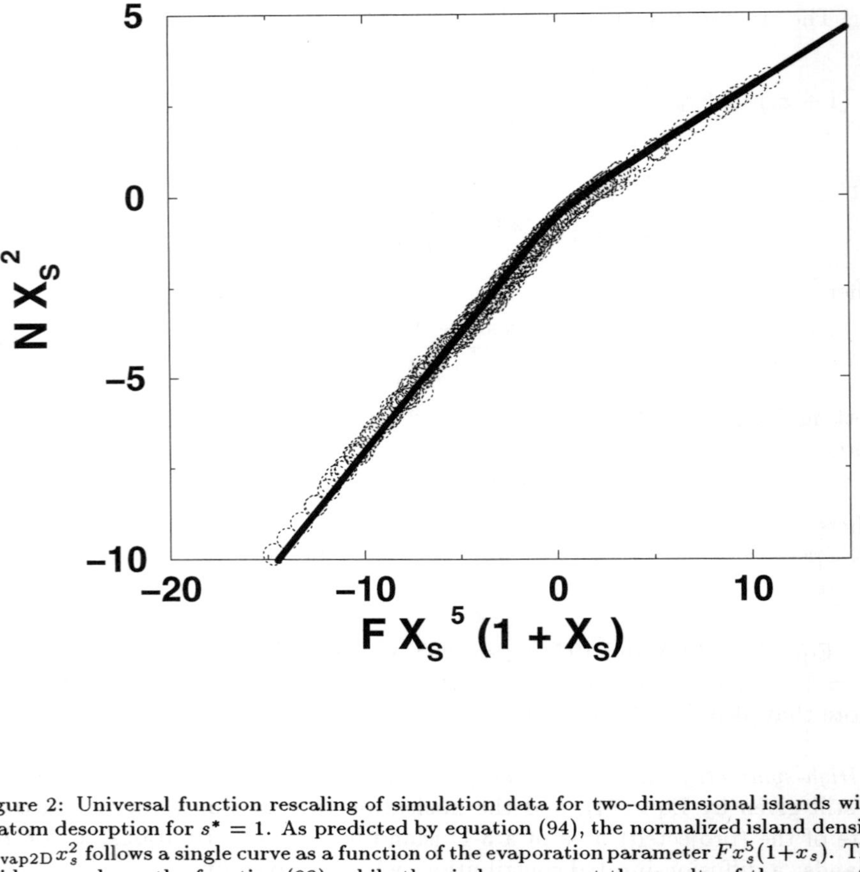

Figure 2: Universal function rescaling of simulation data for two-dimensional islands with adatom desorption for $s^* = 1$. As predicted by equation (94), the normalized island density $N_{\mathrm{evap2D}} x_s^2$ follows a single curve as a function of the evaporation parameter $F x_s^5(1+x_s)$. The solid curve shows the function (93), while the circles represent the results of the computer simulations.

A useful interpolation formula between (98) and (99), analogous to (88), is

$$\frac{d\mathcal{A}}{dt}|_{R=\ell_s} \approx F\ell_s \left(1 + \frac{x_s}{\ell_s}\right) \tanh(\kappa\ell_s/2) \ . \tag{100}$$

Finally, (90) and (100) yield

$$\frac{(1+x_s)^2}{\tau_v}\left[F\tau_v \frac{\sinh^2(\kappa\ell_s/4)}{\cosh(\kappa\ell_s/2)}\right]^{s^*+1} \approx F\ell_s\left(1+\frac{x_s}{\ell_s}\right)\tanh(\kappa\ell_s/2)N^2 \ , \tag{101}$$

or,

$$(1+x_s)^2(F\tau_v)^{s^*} \approx \left[\frac{\cosh(\kappa N^{-1/2}/2)}{\sinh^2(\kappa N^{-1/2}/4)}\right]^{s^*+1} \tanh(\kappa N^{-1/2}/2)(1+x_s N^{1/2})N^{3/2} \ . \tag{102}$$

Multiplying both sides by x_s^3 yields

$$x_s^{2s^*+3}(1+x_s^2)(F/D)^{s^*} \approx f(x_s N^{1/2}) \ , \tag{103}$$

where

$$f(x) = \left[\frac{\cosh\left(\frac{1}{2x}\right)}{\sinh^2\left(\frac{1}{4x}\right)}\right]^{s^*+1} \tanh\left(\frac{1}{2x}\right)(1+x)x^3 \ . \tag{104}$$

Letting

$$N_{\text{evap3D}} = \left(Fx_s^2/D\right)^{2s^*/3}\left(1+x_s^2\right)^{2/3} \tag{105}$$

(cf. Eq. 67), formula (103) can be cast in the form

$$\tilde{f}_{\text{evap3D}}(x_s^2 N) = x_s^2 N_{\text{evap3D}} \ , \tag{106}$$

where $\tilde{f}_{\text{evap3D}}(x) = [f(x^{1/2})]^{2/3}$. Inverting $\tilde{f}(x)$ yields the crossover scaling function

$$x_s^2 N = g_{\text{evap3D}}(x_s^2 N_{\text{evap3D}}) \ . \tag{107}$$

The function $g_{\text{evap3D}}(x)$ has the following properties:

$$g_{\text{evap3D}}(x) \sim \begin{cases} x^{3/(5+2s^*)} & \text{for} \quad x \to \infty, \\ x & \text{for} \quad x \to 0. \end{cases} \tag{108}$$

Results of simulations together with (108) are shown in figure 3. The critical size has been assumed $s^* = 1$, and thus the slopes (of a log-log plot) are predicted to be 3/7 and 1 in the large and small x limit, respectively, where x is the scaling variable $x_s^2 N_{\text{evap3D}} = x_s^{10/3} F^{2/3}\left(1+x_s^2\right)^{2/3}$ $(D=1)$ defined by (105).

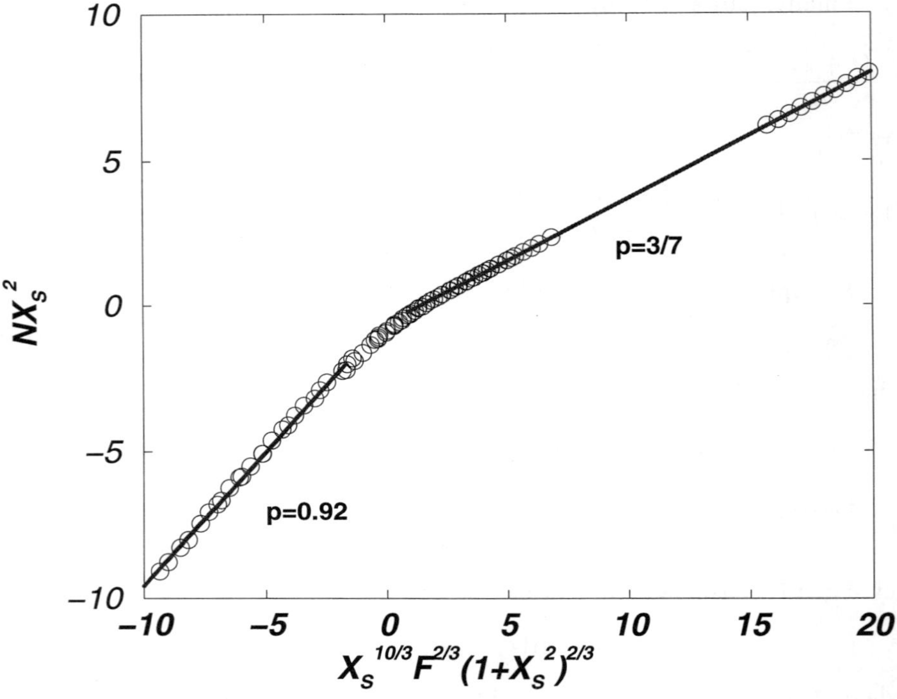

Figure 3: Universal function rescaling of simulation data for three-dimensional islands with adatom desorption for $s^* = 1$. As predicted by equation (103), the normalized island density $N_{\text{evap3D}}\, x_s^2$ follows a single curve as a function of the evaporation parameter $x_s^{10/3} F^{2/3} \left(1 + x_s^2\right)^{2/3}$. The circles represent the results of the computer simulations.

5 Scaling form of the growth velocity of a thin film when evaporation is present.

In this section we will compute the growth velocity of a thin film in the presence of strong adatom desorption, assuming perfect layer-by-layer growth.

The growth velocity of a film grown by UHV deposition at low temperature is simply equal to the incoming particle flux. At high enough T, adatoms start to desorb at an appreciable rate, and the growth velocity starts differing from F. It should be noted that the growth velocity is affected by desorption at a lower temperature, than the one where crystal sublimation is important. In fact, the crystal sublimation rate depends on the *equilibrium* adatom density, n_{eq}, while the rate of adatom desorption during growth, n/τ_v, depends on the actual adatom density in growth conditions, which is in general much larger than n_{eq}. For instance, in the case of a CdTe substrate with (100) orientation, growing under a stoechiomatric flux of 0.57 ML/s, the growth velocity is appreciably reduced already at 520 K, while the crystal sublimation becomes appreciable only above 620 K [21].

If growth is layer-by-layer, the island nucleation-spreading-coalescence mechanism leads to an oscillating surface step density. The period of the oscillations, approximately equal to the reciprocal of the coalescence time t_c, defines the growth velocity, V_{G}. When the diffusion length of the desorbing adatoms, x_s, becomes smaller than the average interisland distance, ℓ_s, each step spreads laterally at a constant velocity, $v_{\mathrm{step}} \approx Fx_s$. Coalescence takes place when each step has gone a distance of order ℓ_s, i.e. after a time $t_c \approx \ell_s/v_{\mathrm{step}}$. Thus,

$$V_{\mathrm{G}} \approx 1/t_c \approx Fx_s/\ell_s \ . \tag{109}$$

In the limit of negligibly small desorption, $\ell_s \ll x_s$, we expect the average adatom density to be given by

$$\bar{n} \approx \frac{F\ell_s^2}{3D} \ . \tag{110}$$

The growth rate of the crystal, V_{G}, is the balance between the incoming flux F, and the average desorbing matter flux $\bar{n}/\tau_v$,

$$V_{\mathrm{G}} = F - \frac{\bar{n}}{\tau_v} \ . \tag{111}$$

Using (110) yields

$$V_{\mathrm{G}} \approx F - F\frac{\ell_s^2}{x_s^2} \ . \tag{112}$$

In fact, we are clearly cheating to some extent, because the adatom density which should appear into (111) is the time-averaged density, while (110) is the space-averaged adatom density in the stationary state, and there is no *a priori* reason why the two should be equal. Indeed, the Monte Carlo simulations that we discuss in the following section support the following statement: in the layer-by-layer mode, a flat surface grows at the same velocity as a vicinal surface growing in the step flow mode, provided that the step density $1/d$ on the vicinal surface is replaced by the *time-averaged* step density $\bar{p}_{\text{step}} = 1/\ell_s$ on the initially flat surface. We can thus write (112) as a universal function of the step density $\bar{p}_{\text{step}}$ (total step length per unit surface):

$$V_{\text{G}} \approx F x_s \bar{p}_{\text{step}} \tanh\left(\frac{1}{x_s \bar{p}_{\text{step}}}\right) . \tag{113}$$

Indeed, the simulations show that the step density $\bar{p}_{\text{step}}$ can be computed using the rate equations for the island density and the stationarity condition [19,23].

6 Kinetic Monte Carlo simulations of growth with desorption

Performing Monte Carlo simulations allows us to test equation (113) directly. To do this, we use the activated rate model of Clarke and Vvedensky [24]. In this model, particles are randomly deposited on the sites of a square lattice at a rate F. Then, they diffuse at a rate $D = D_0 \exp[-E/(k_B T)]$, where the diffusion barrier E is written as a function of the n lateral neighbours of the diffusing particle: $E = E_S + nE_a$. Here, E_S is the substrate contribution, while E_a is the lateral bond energy. As a consequence, the critical size for aggregate formation is not fixed *a priori* in the model, but it is 'chosen' by the kinetics as a function of all the parameters. We extend the model in a natural way to include desorption by writing that each diffusing particle is tested for desorption at a rate $1/\tau_v = 1/\tau_0 \exp[-E_v/(k_B T)]$. If the move is accepted, the particle leaves the system and it is lost. The desorption rate Δ, i.e. the number of particles leaving the surface per unit surface and time, can thus be easily computed. The growth velocity is then given by

$$V_G = F - \Delta . \tag{114}$$

During the simulations we record the average step density $\bar{p}_{\text{step}}$ appearing in (113). Since the diffusion length $x_s = \sqrt{D\tau_v}$ is known, we can plot V_G as function of $x_s \bar{p}_{\text{step}}$. The result is shown in Fig.4 for two sets of parameter values. Open circles: $F = 1$ ML/sec, $E_D = 1$ eV, $E_a = 0.3$ eV and $E_v = 1.3$ eV. Filled squares: $F = 0.5$ ML/sec, $E_D = 1$ eV, $E_a = 0.5$ eV and $E_v = 0.9$ eV. It is clear that V_G/F is indeed a universal function of $x_s \bar{p}_{\text{step}}$, and that

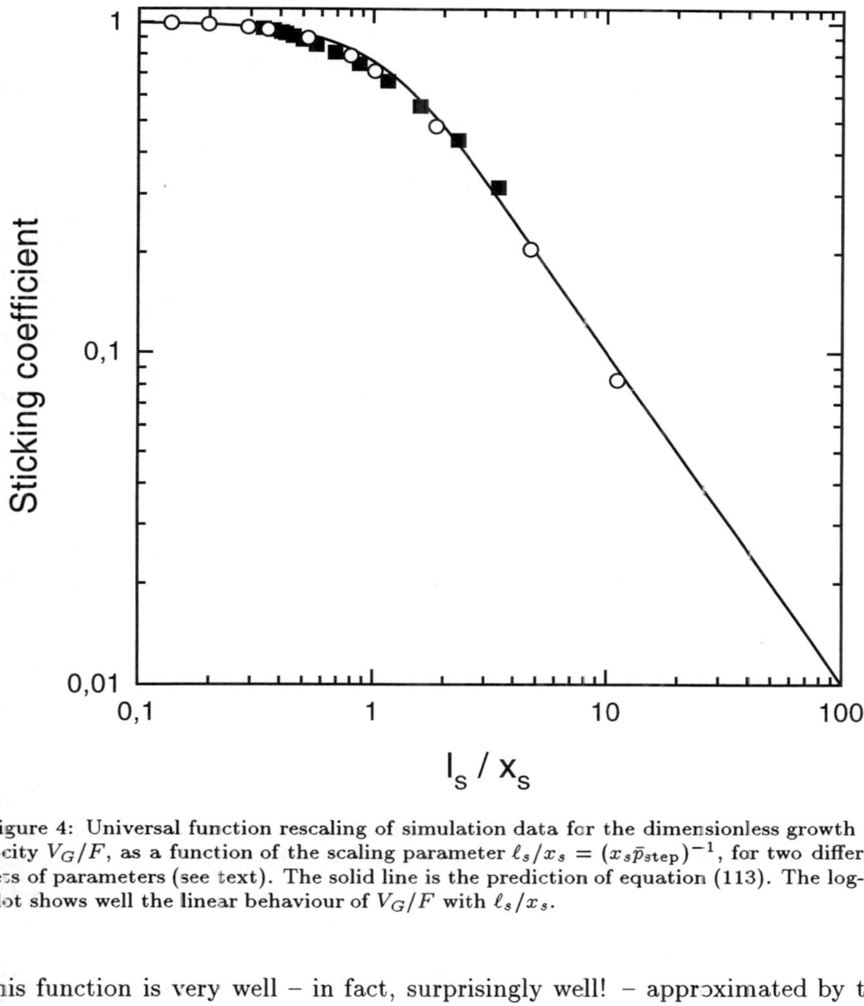

Figure 4: Universal function rescaling of simulation data for the dimensionless growth velocity V_G/F, as a function of the scaling parameter $\ell_s/x_s = (x_s\bar{p}_{\rm step})^{-1}$, for two different sets of parameters (see text). The solid line is the prediction of equation (113). The log-log plot shows well the linear behaviour of V_G/F with ℓ_s/x_s.

this function is very well – in fact, surprisingly well! – approximated by the BCF expression (113), with no ajustable parameter.

It is difficult to justify formula (113) from *a priori* considerations. It is a known experimental [25] and numerical [11,19] fact that a layer-by-layer-growing surface tends to a steady-state configuration which is the analog of step flow, in the sense that the step density becomes time-independent, regardless of the continuing nucleation and growth of islands. This is true if growth is not impeded by step-edge barriers (Schwoebel barrier [27]) which tend to produce mound-like islands, and hinder layer-by-layer growth [28,29]. A more thorough investigation of the 'pseudo step flow' regime of growth is certainly worth consideration.

7 Conclusions

We have shown in this paper that very simple scaling arguments yield the power-law behaviour of the island density as a function of the physical parameters (flux, diffusion constant, adatom residence time, step-step distance, etc.). The main hypotheses are the existence of a critical nucleus of (microscopic) size s^*, and the validity of the law of mass action Eq. (8).

A microscopic definition of the critical nucleus relies on a discussion of the stability of interatomic bonds as a function of the temperature and of the flux [30]. At very low temperature, and for strong enough fluxes, an atomic bond will be unbreakable during the time $1/F$ needed to fill a complete layer. Any cluster is thus stable, and $s^* = 1$ is a well defined occurrence. Problems start at higher temperatures, when a single atomic bond has a finite probability to break during the time $1/F$. In fact, it is reasonable to assume that a temperature interval exists where *doubly-bonded* atoms do not part, so that on a square lattice, say, dimers and trimers are unstable, while tetramers are stable. The point is that the pentamers *cannot* be stable, since the fifth atom is only single-bonded to the cluster [30]. Defining the critical nucleus as the largest unstable cluster seems therefore meaningless. However, we realize that a cluster will turn out to be *effectively* stable, if the probability of losing an atom is smaller than the probability of capturing a diffusing adatom: $N_s/\tau_s < n\sigma_s N_s$. One should also note that the lifetime τ_s should be strongly size-dependent: an atom which has detached from a cluster, has a high probability to stick back if the cluster is large. It is not surprising then, that critical cluster sizes larger than 1 are indeed observed, e.g. in simulations [19,30]. In any case, the validity of Eq. (12) has never, to our knowledge, been tested e.g. by simulations.

A point which is worth spending some time on is the dependence of the average capture cross section $\sigma_{\rm av}$ on the growth parameters for fractal islands [7,11]. In the case of two-dimensional islands on a high-symmetry substrate, Bales and Chrzan [7] have shown that, if one assumes a scaling form $\sigma_s = \sigma(s/\ell^2)$ for the cross sections, their average $\sigma_{\rm av}$ turns out to have the form

$$\sigma_{\rm av} = f(\theta \bar{s}^{2/d_f - 1}), \tag{115}$$

where d_f is the (possibly fractal) dimensionality of an island, and $\bar{s}$ is the average island size. For compact islands, $d_f = 2$ and $\sigma_{\rm av}$ is a function of the coverage θ only.

For fractal islands this is no longer true, in general. In the case of fractal islands, the mass of the island is proportional to s, while its *size*, defined as its gyration radius R, is $R \sim s^{1/d_f}$. When islands coalesce, the average island size R is by definition of order ℓ_s, so that we find the relation $\ell_s \approx \bar{s}^{1/d_f}$. Inserting the latter into formula (23) yields

$$Ft_c = \theta_c \approx \bar{s}^{1-2/d_f}, \tag{116}$$

so that according to (115), only at the maximum of the island density $\sigma_{\rm av}$ is a constant, independent of F/D, while it is not independent of it at lower coverages [7].

8 References

1. G. Zinsmeister, Thin Solid Films **2**, 497 (1968)
2. G. Zinsmeister, Thin Solid Films **7**, 51 (1971)
3. S. Stoyanov and D. Kaschiev, *Current topics in material science*, E. Kaldis Ed., Vol.7, p.69 (North Holland, 1951)
4. J. A. Venables, G. D. Spiller and M. Hanbücken, Rep. Prog. Phys. **47**, 399 (1984)
5. J. Villain and A. Pimpinelli *Physique de la Croissance Cristalline* (Eyrolles, 1995), and A. Pimpinelli and J. Villain, *Physics of Crystal Growth* (Cambridge University Press, expected 1997)
6. C. Ratsch, A. Zangwill, P. Šmilauer and D.D. Vvedensky, Phys. Rev. Lett. **72**, 3194 (1994)
7. G.S. Bales and D.C. Chrzan, Phys. Rev. B **50**, 6057 (1994)
8. J. Villain, A. Pimpinelli and D.E. Wolf, Comments Cond. Mat. Phys. **16**,1 (1992)
9. A. Pimpinelli, J. Villain, D.E. Wolf, J. de Physique I **3**, 447 (1993)
10. L.-H. Tang, J. de Physique I **3**, 935 (1993)
11. D.E. Wolf, in *Scale Invariance, Interfaces, and Non-Equilibrium Dynamics*, Vol. 344 of *NATO Advanced Study Institut, Series B: Physics*, M. Droz, A.J. McKane, J. Vannimenus and D.E. Wolf eds. (Plenum Press, New York 1996)
12. A. Pimpinelli, J. Villain, D.E. Wolf, Phys. Rev. Lett. **69**, 985 (1992)
13. M.C. Bartelt, J.W. Evans, Europhys. Lett. **21**, 99 (1993)
14. S. Liu, L. Bönig, J. Detch and H. Metiu, Phys. Rev. Lett. **74**, 4495 (1995)
15. P. Jensen, A. Barabási, H. Larralde, S. Havlin and H.E. Stanley, Phys. Rev. B **50**, 15316 (1994)
16. S. Liu, L. Bönig and H. Metiu, Phys. Rev. B **52**, 2907 (1995)
17. I. Furman and O. Biham, Phys. Rev. B **55**, 7917 (1997)
18. P. Jensen, H. Larralde and A. Pimpinelli, Phys. Rev. B **55**, 2556 (1997), and P. Jensen and H. Larralde, International J. of Mod. Physics B, **11**, 3635 (1997)

19. A. Pimpinelli and P. Peyla, International J. of Mod. Physics B, **11**, 3647 (1997)
20. G.S. Bales, Surface Sci. **356**, L439 (1996)
21. A. Arnoult and J. Cibert, Appl. Phys. Lett. **66**, 2397 (1995)
22. J. A. Venables, Surf. Sci. **299/300**, 798 (1994)
23. A. Pimpinelli and P. Peyla, **183**, 311 (1998)
24. S. Clarke, and D. Vvedensky, Phys. Rev. Lett. **58**, 2235 (1987)
25. J. Sudijono, M.D. Johnson, C.W. Snyder, M.B. Elowitz, and B.G. Orr, Phys. Rev. Lett. **69**, 2811 (1992)
26. W. K. Burton, N. Cabrera and F. Frank, Phil. Trans. Roy. Soc. **243**, 299 (1951)
27. R.L. Schwoebel, J. Appl. Phys. **40**, 614 (1968)
28. J. Villain, J. de Physique I **1**, 19 (1991)
29. P. Šmilauer and D.D. Vvedensky, Phys. Rev. B **52**, 14263 (1995)
30. M.C. Bartelt, L.S. Perkins and J.W. Evans, Surface Sci. Lett. **344**, L1193 (1995)

Quantum Effect in Metal Overlayers on Semiconductor Substrates

Jun-Hyung Cho
Solid State Division, Oak Ridge National Laboratory, Oak Ridge, Tennessee 37831-6032

Qian Niu and Chih-Kang Shih
Department of Physics, University of Texas at Austin, Austin, Texas 78712

Zhigang Suo
Mechanical and Aerospace Engineering Department and Princeton Materials Institute, Princeton University, NJ 08544

Zhenyu Zhang
Solid State Division, Oak Ridge National Laboratory, Oak Ridge, Tennessee 37831-6032 and Department of Physics and Astronomy, University of Tennessee, Knoxville, Tennessee 37996

In this article, we present the main results from our recent studies of metal overlayer growth on semiconductor substrates. We show that a variety of novel phenomena can exist in such systems, resulting from several competing interactions. The confined motion of the conduction electrons within the metal overlayer can mediate a surprisingly long-range repulsive force between the metal-semiconductor interface and the growth front, acting to stabilize the overlayer. Electron transfer from the overlayer to the substrate leads to an attractive force between the two interfaces, acting to destabilize the overlayer. Interface-induced Friedel oscillations in electron density can further impose an oscillatory modulation onto the two previous interactions. These three competing factors, of all electronic nature, can make a flat metal overlayer critically, marginally, or magically stable, or totally unstable against roughening. We further show that, for many systems, these electronic effects can easily win over the effect of stress. First-principles studies of a few representative systems support the main features of the present "electronic growth" concept.

1 Introduction

For important scientific and technological reasons, it is desirable to grow metal thin films with atomically flat interface and growth front [1]. This has not been possible in many heteroepitaxial systems, however, because of fundamental obstacles such as stress effects [2,3,4] and kinetic limitations [5,6]. It has been a major focus of intensive studies to identify and overcome such obstacles in the quest for technologically important overlayers.

One example is the growth of Ag on GaAs(110), where a rough surface would results in typical growth conditions[7]. However, by using a two-step process (low-temperature deposition followed by room-temperature annealing), Smith *et al.* [8] were able to obtain atomically flat silver films. One startling observation was the existence of a critical thickness: A flat silver overlayer can be formed only if the total Ag coverage deposited exceeds a minimum value of about 15 Å. Similar observations have also been confirmed independently for the same system [9], and on two other III-V semiconductors, GaP(110) and GaSb(110) [10]. The underlining physical reason for the success of this approach is not known to date, and it is unlikely due to stress effect, because one should then expect exactly the opposite behavior: A flat film becomes unstable at sufficiently large thickness, where the adsorbed material tend to form three-dimensional islands or cracks to release the strain energy [2,3,4].

In a recent study, we developed what we termed the "electronic growth" model for formation of metallic overlayers on semiconductor substrates [11]. Unlike commonly recognized growth mechanisms based on considerations of stress effects [2,3,4] or kinetics of individual atoms [5,6], we placed primary emphasis on the energetics of the conduction electrons which are extended throughout a metal overlayer. Earlier studies have established that the electronic states within an ultrathin metallic overlayer can be quantized due to confinement in the direction of film thickness [12,13]. For metal overlayers on semiconductor substrates, we showed that these discrete electronic levels can play a crucial role in defining the overall stability of the overlayers. Depending on a delicate energy balance between an energy loss due to confinement and a gain due to charge spilling from the film to the substrate, the overlayer can be either stable above a critical thickness of typically a few monolayers (ML), or unstable for any thickness greater than 1 ML. For those systems in which the interface-induced Friedel oscillations in electron density within the overlayer are sufficiently strong, additional magic thicknesses for stable film growth can be defined. Our theory not only confirms the existence of the critical thickness for Ag growth on GaAs [8,9], but also explains the well-known fact that only the first layer of alkali metals can be smooth on semiconductors [14]. It further points to new directions for achieving smooth growth in many other systems where magic thicknesses may exist.

The paper is organized as follows. In Sec. 2 we present the simple phenomenological model used to convey the main ideas of the electronic growth concept, as well as the application of the concept to the growth of metals belonging to several different columns of the periodic table [11]. In Sec. 3, we compare directly the relative importance of these purely electronic effects with stress effects, and show that, contrary to traditional belief, the electronic effects can easily win over stress effects in stabilizing ultrathin films on semiconductors. We also derive the asymptotic form of the effective repulsive potential between the metal-semiconductor interface and the growth front, mediated by the conduction electrons [15]. In Sec. 4 we present a first-principles study of a model metal/semiconductor system, Sb/GaAs(110), showing that the main features of the electronic growth concept persist to show up in several dramatic ways[16]. Finally in Sec. 5 we summarize the main findings, together with a brief discussion of the application of the electronic growth concept to the formation of two-dimensional metal islands on surfaces [17].

2 The "Electronic Growth" model and its applications

We present our theory within the framework of a general thermodynamic stability analysis, which in spirit is similar to the shell model for the existence of magic atomic numbers in metallic clusters [18]. Let $U_t(L)$ be the total energy of the system with a flat film of thickness L. $U_t(L)$ also plays the role of the Hemholtz free energy at low temperatures. The film is stable if the "compressibility" is positive, i.e., $\partial^2 U_t(L)/\partial L^2 \geq 0$. Under this condition, any small roughness in the film tends to be suppressed when sufficient atomic mobility is provided. The film is unstable if $\partial^2 U_t(L)/\partial L^2 < 0$. In this case, the system can achieve a lower total-energy state by developing a mixed phase of different film thicknesses. A critical thickness, L_c, can be defined if the film is stable for $L \geq L_c$ but unstable for $L < L_c$ (or the other way around). Furthermore, a magic thickness, L_m, can be defined if $U_t(L)$ has a downward cusp at L_m, namely, the film is unstable on both sides of L_m. Our main task is to show when and why there can exist critical and magic thicknesses in the formation of metal overlayers.

The energy of the system can be evaluated by referring to the ideal situation where the film is isolated from the semiconductor substrate, with charge neutrality maintained on both sides of the interface, and with the interface atoms neither reconstructed nor relaxed. The energy of the actual system, U_t, is then estimated as $U_t = U_0 - U_c$, where U_0 is the energy of the film in the ideal situation, and U_c is the energy decrease due to charge spilling as the delta potential is lowered. Figure 1 illustrates the band diagrams before and after charge spilling takes place. The vacuum levels of the film and the substrate should be equal before charge spilling. The Fermi energy and the energy of the film $U_0(L)$ in the ideal situation are estimated by a model of a free electron gas confined by a barrier step $(W_m + E_F)$ on the outer surface and by an infinite hard wall at the interface, where W_m and E_F are the work function and Fermi energy (relative to the bottom of the conduction band) of the metal in bulk form. After subtracting a term linear in the film thickness, which does not change the conclusion about film stability, the function $U_0(L)$

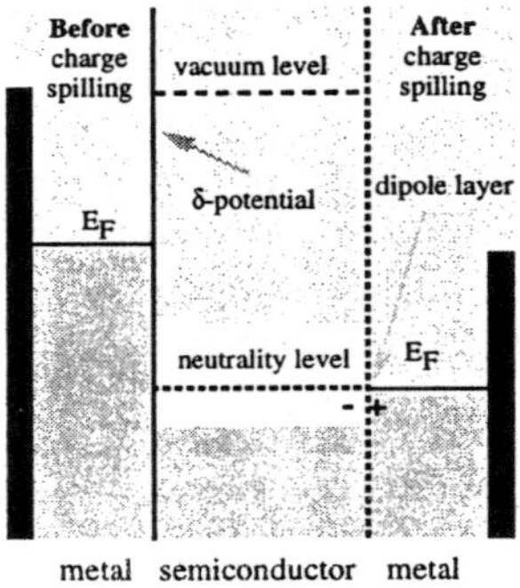

Figure 1: Schematic energy diagrams of a metal-semiconductor interface before charge spilling (the left two panels), showing the relatively high Fermi level of the electrons in the metal film; and after charge spilling (the right two panels), showing the alignment of the Fermi levels and the establishment of a dipole layer.

curves up as the film thickness becomes small, a quantum-size effect in the film thickness direction (see Fig. 2).

The Fermi energy of the semiconductor substrate is taken to be at the charge neutrality level in the gap [19]. Once the Fermi level difference $\Delta E_{\rm F}$ between the metal film and the substrate is calculated in the ideal situation, the energy lowering due to charge spilling can be expressed as $U_{\rm c} = 0.5C(\Delta E_{\rm F}/e)^2$, where e is the electron charge, $C = \varepsilon_0/(\ell_{\rm m} + \ell_{\rm s}/\kappa)$ is the interface capacitance, ε_0 is the vacuum dielectric constant, $\ell_{\rm m}$ and $\ell_{\rm s}$ are the length scales for charge redistribution on the metal and semiconductor side of the interface, respectively, and $\kappa \approx 2$ is the effective dielectric constant of the semiconductor near the interface [20]. The length scale for charge redistribution on the metal side can be estimated by the Thomas-Fermi screening length (0.59 Å for Ag) [21], while the length scale on the semiconductor side is taken as the tunneling distance at the charge neutrality level (about 2.8 Å for GaAs) [22]. The resulting energy lowering $U_{\rm c}$ also curves up as the thickness becomes small, because the Fermi energy of the film is squeezed higher by the quantum confinement (see Fig. 2).

The total energy $U_{\rm t} = U_0 - U_{\rm c}$ for Ag/GaAs(110), shown in Fig. 2, has a shape predicting the existence of a critical thickness. The energy lowering due to charge spilling has a steeper thickness dependence at smaller L, pulling the total-energy curve down. This makes the curvature of $U_{\rm t}(L)$ negative in that region, rendering flat films unstable. The curvature changes sign at $L = 3$ ML,

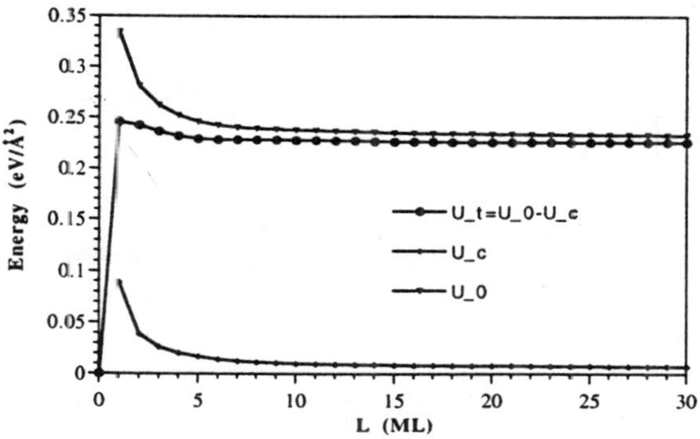

Figure 2: Film thickness dependence of the film energies for Ag on GaAs(110). The cusp-like feature at $L=$ 5 ML defines the critical thickness for flat film growth.

and has a cusp-like dip at $L = 5$ ML, indicating that a flat film at this thickness is particularly stable. This cusp is mainly caused by the sharp energy features of the quantum well states and the Fermi surface. The curve is practically flat beyond $L = 5$ ML, showing that thicker flat films are all marginally stable. We therefore identify $L_c = 5$ ML as the critical thickness, which is of the same magnitude as the experimental finding of $L_{c,\text{expt}} \sim 7$ ML[8,9]. The agreement is satisfactory, considering the fact that the simple model does not contain any adjustable parameters, and that there is some uncertainty in the determination of the absolute coverage of the film in the experiments[8,9]. Finally, we notice that if we follow Ref. 23 by using the empirical formula $\kappa = (1 + \kappa_s)/2 = 7$, where $\kappa_s = 13$ is the dielectric constant in bulk GaAs, then the enhanced energy lowering due to charge transfer would cause the change of curvature right at $L_c = 5$ ML.

The above calculations can be easily repeated for other metal-substrate systems. In these calculations, we assume that the metallic overlayer always prefers to grow in close-packed forms in the film thickness direction, as is the case for Ag[8,9,10]. Here we limit our discussions to different metals on the same GaAs(110) substrate, with detailed comparisons of the effects of different substrates presented elsewhere[10]. The stabiliy of Cu and Au films resembles very closely that of Ag, each with the same critical thickness of ~ 5 ML. This type of stability of an ultrathin metallic film is reproduced and designated as Type A in Fig. 3. In contrast, a qualitatively different type (Type B) is obtained for the alkali metals (Li, Na, K, Rb, and Cs), as represented by the

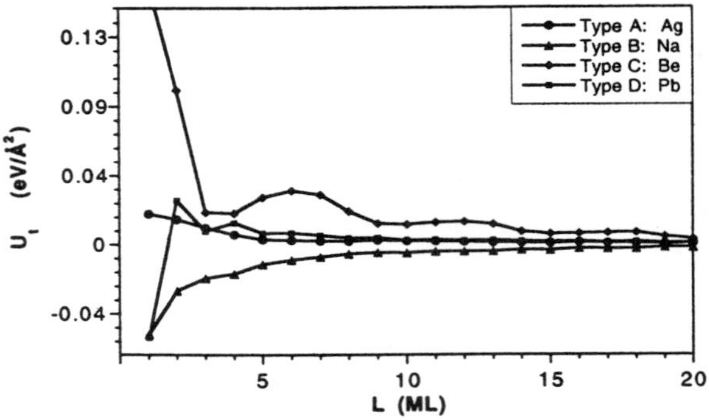

Figure 3: Comparison of four representative types of film stability for different metals on GaAs, as defined in the text. Notice that an element-specific constant term has been subtracted from each curve to make the total energy equal to zero at large film thickness.

case of Na in Fig. 3. Here, because of its small work function, the energy gain due to charge spilling dominates, leading to downward curving in U_t, with $\partial^2 U_t(L)/\partial L^2 < 0$. Therefore, a flat film thicker than 1 ML is unstable, because it is always possible to reduce its energy by phase separating into a rough film with smaller and larger thicknesses. This explains the well known experimental fact that one cannot grow a smooth film of alkali metals of more than 1 ML on GaAs(110) [14].

We have also investigated the quantum-size effects for overlayer growth of alkaline earth metals (Be, Mg, Ca, Sr, and Ba). Unlike the alkali metals which qualitatively all follow the same curve (Type B in Fig. 3), the alkaline earth metals have very different behaviors among themselves. Ca, Sr, and Ba are similar to the alkali metals (Type B), but Be and Mg belong to a new type (Type C), as represented by the case of Be in Fig. 3. Here, the film is magically stable around $L_m = 3$ ML; furthermore, there still exists a critical thickness at $L_c = 9$ ML above which the film is marginally stable. Calculations show that Zn and Cd also belong to Type C.

A fourth type of behavior is shown as Type D in Fig. 3, followed by both Al and Pb on GaAs. Here, the energy dependence is a damped oscillatory one, with a period of oscillation equal to 2 ML. Such oscillatory behaviors have been predicted in earlier studies of quantum-size effects in metal/metal systems [13].

The oscillatory thickness dependence of $U_t(L)$ for Types C and D is caused by the interface-induced Friedel oscillations. To illustrate this, it is sufficient

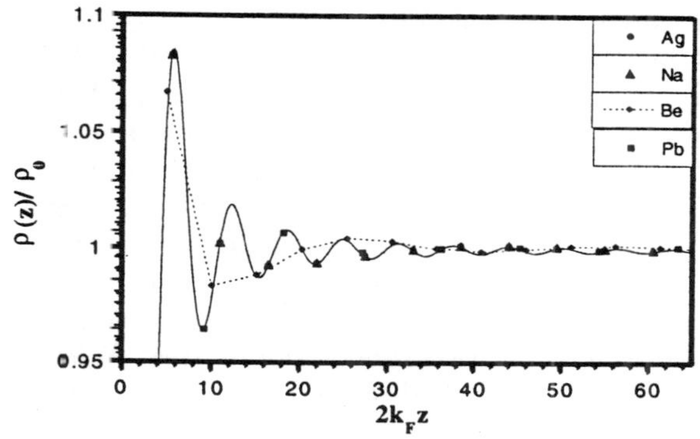

Figure 4: Friedel oscillations in electron density within a semi-infinite metal caused by a hard wall located at the metal-semiconductor interface. The different metal elements sample different sets of points on the same curve.

to plot in Fig. 4 the density of a semi-infinite electron gas confined by a hard wall (the interface):

$$\rho(u)/\rho_0 = 1 + 3[\cos(u) - \sin(u)/u]/u^2, \tag{1}$$

where $u = 2k_Fz$, z is the distance from the interface, and k_F is the Fermi wave vector [21]. For a given metal (characterized by its own k_F and interlayer spacing, d), the electron density at different layer thicknesses ($z = Ld$, with $L = 1,2,3, \ldots$) sample different points on the same curve. If a given film thickness coincides with a minimum of the Friedel oscillations, then there is an additional energy gain, because fewer electrons need to be pushed up in energy by the confinement of the outer surface. For Na and Ag, the first (and also the deepest) minimum in the density oscillations located between 3 and 4 ML is too shallow to induce an additional magic thickness. In contrast, the first minimum for Be is large, which results in the existence of a magic thickness. For Pb, the position of the first minimum located at 1 ML coincides with the first minimum of the universal Friedel oscillation curve; this perfect phase matching in the Friedel oscillations and that due to lattice periodicity explains why, as seen in Fig. 3, the total energy at $L = 1$ ML is so low for Pb.

The above semi-quantitative analysis has demonstrated a very plausible and important concept in film growth, that the electronic energy of an ultrathin metallic overlayer can play a crucial role in determining the overall stability of the overlayer and the morphological evolution during its growth. The three central and competing components of the theory, namely, the quantization of the electronic states, the charge spilling, and the Friedel oscillations, should show up in any reasonably accurate treatment of the problem. In contrast, some earlier treatments of the problem using free-standing films would miss the important effects due to charge transfer [13], without which it would be impossible to define the critical thickness for Ag/GaAs or to explain why only the first layer of an alkali metal can be grown smoothly. Also, the essential predictions of the present theory remain valid if all other factors can only give rise to a smooth modification to the total-energy curve, including possible shifts in the locations of the critical/magic thicknesses. What remains to be explored is how the electronic effects discussed here compete with stress effects, the subject of the next section.

3 Direct comparison between the quantum size effects and stress effects

An epitaxial film on a substrate of a different material is often under stress. The elastic energy may motivate atoms to diffuse on the surface and change the film morphology. Experimentally it has been known that whether a film remains flat or forms islands depends on its thickness. On a Si substrate, for example, a flat Ge film is stable up to three monolayers; above this thickness, islands form [24]. Similar behavior, sometimes with thicker wetting layers, has been reported for some other inorganic and organic semiconductor films [25]. The recent observation of the inverse critical thicknesses for smooth growth on Ag on III-V semiconductors adds another kind of thickness dependence for metallic films [8,9,10]. In the previous section, we showed that the long-ranged interactions associated with the electronic energy of a metallic overlayer can lead to various types of thickness dependence. In this section, we consider explicitly the competitions between surface tension, stress, and effective long-ranged forces mediated by the conduction electrons, and show that the latter can win over stress in stabilizing metal films of many monolayers [15].

Before introducing long-range force effects, we recall that the existing model addressing the stability of epitaxial films invokes surface tension and elasticity [3,26]. Figure 5 illustrates a film of thickness L on a semi-infinite substrate. The surface tension of the film, γ, is taken to be isotropic [27]. The stress in the film, σ, results from the difference of the film and the substrate in, for example, crystal structure, lattice constant or thermal expansion coefficient. When the surface is flat, the stress is uniform in the film. Consider a two-dimensional problem, representing the substrate by a semi-infinite plane, and the film by an overlaying strip; atoms can diffuse along the curve representing

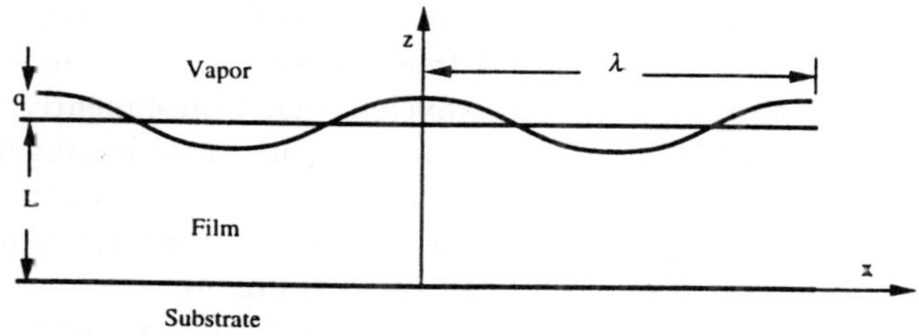

Figure 5: An epitaxial film on a substrate. The film surface is perturbed into a wavy shape of wavelength l and amplitude q.

the film surface. The model is based on a stability analysis. Atomic diffusion on the surface conserves the mass of the film. Consequently, one can perturb the surface into a wavy shape above and below the average film thickness, L, namely,

$$z = L + q\cos(2\pi x/\lambda) \tag{2}$$

where z is the perturbed film thickness, q the wave amplitude, and λ the wavelength. The x-axis coincides with the film-substrate interface.

To highlight the thermodynamic nature of this instability, we focus on energetics and avoid details of the mass transport process. All energies are computed for one period of the system, per unit thickness in the direction normal to the plane, to the leading order in the perturbation amplitude q. The surface energy, U_{S}, is γ times the length of the curve that represents the surface. One can readily show that the perturbation increases the surface energy by

$$\Delta U_{\mathrm{S}} = \pi^2 \gamma q^2/\lambda \tag{3}$$

Elementary considerations dictate that, when the surface undulates, the elastic energy stored in the system, U_{E}, should decrease, and its change should take the form

$$\Delta U_{\mathrm{E}} = -\beta\sigma^2 q^2/Y \tag{4}$$

Here Y is Young's modulus of the film, and β a positive dimensionless number, which has been calculated by solving the boundary value problem of a strained, perturbed film on a substrate[3,26]. If the film and the substrate have identical elastic constants, $\beta = \pi$. The total free energy, $U_S + U_E$, increases for short wavelengths, but decreases for long wavelengths. Consequently, the flat film of any thickness is unstable. For the film and the substrate having different elastic constants, β depends on λ/L and ratios of the elastic constants. Nonetheless the conclusion remains essentially unchanged: except for a film on a rigid substrate, the stressed film of any thickness is unstable.

This conclusion clearly disagrees with experimental observations cited in the beginning of this section. One may settle with the thought that the continuum model fails for thin films. On the other hand, thickness effects of various kinds have been observed in many systems, some of which have stable films of many monolayers[8,9,10,25]. Consequently, it is imperative to have a model with a wider applicability.

In a recent study, we have proposed to include additional thermodynamic forces acting over longer ranges than atomic length[15]. In the presence of a long-ranged force, the change in the interaction energy associated with the surface undulation is

$$\Delta U_L = \int_0^\lambda U(z)\,dx - U(L)\lambda \tag{5}$$

This expression is reasonable when the wavelength of the perturbation is larger than the film thickness, and the amplitude of the perturbation is small. To the leading order in q, the change is

$$\Delta U_L = \frac{\lambda}{4}\frac{\partial^2 U}{\partial L^2} q^2 \tag{6}$$

As described in Sec. 2, when the function $U(L)$ is concave up, i.e., $\partial^2 U/\partial L^2 > 0$, the long-range force tends to stabilize a flat film. When $\partial^2 U/\partial L^2 < 0$, the long-range force tends to destabilize the flat film.

Observe from (3), (4) and (6) that the surface tension is effective in stabilizing the film against perturbations of short wavelengths, the long range interaction (assuming $\partial^2 U/\partial L^2 > 0$) is effective in stabilizing the film against perturbations of long wavelengths, and the stress destabilizes the film for all wavelengths. Summing up (3), (4) and (6) leads to the conclusion that the net free energy increases for perturbations of all wavelengths only if

$$\partial^2 U/\partial L^2 > \sigma^4/Y^2\gamma \tag{7}$$

This establishes the condition under which the flat film is stable against any small perturbation. (We have taken $\beta = \pi$ because elastic constants are often not too dissimilar between the film and substrate.) An analogous condition has been derived under the assumption that the film surface tension is influenced by the presence of the substrate[28]. Because no physical origin of such dependence has been specified, to the best of our knowledge, it has never been established so far that a long-range force, of any kind, is strong enough to stabilize a film against stress.

In Ref. 15, we applied (7) to the case of dispersion forces and showed that even the van der Waals forces can be strong enough to stabilize films of a few monolayers. Here we focus on the long-ranged forces associated with the confined electrons in metal films. As discussed in Sec. 2, a recent model has highlighted forces of two origins: quantum confinement and charge transfer[11]. In a metallic film electronic states form discrete subbands[12,13]. Consequently, if insulated, the film has higher average electronic energy than the bulk. This difference results in an excess free energy of the film relative to the bulk. (As an approximation, ions in the film and in the bulk are taken to have identical free energy.) On the other hand, when the metallic film is brought in contact with a semiconductor substrate, electrons transfer between the two media to equalize the Fermi level. This lowers the free energy. For Ag on GaAs, the calculations given in Sec. 2 showed that the attraction due to charge transfer dominates for very thin films, and the repulsion due to quantum confinement dominates for thick films. Fig. 6 shows the qualitative shape of the combined interaction energy, U. The curve is concave down for thin films, but concave up for thick films; the small circle on the curve marks the inflection point, corresponding to the film thickness L_0, which is about a few monolayers. As pointed out in Sec. 2, if the effect of stress is negligible, such a long range interaction destabilizes a film thinner than L_0, but stabilizes a film thicker than L_0. This trend agrees with the experimental observations[8,9].

We now include the effects of the stress and surface tension. The bottom part of Fig. 6 shows the shape of $\partial^2 U/\partial L^2$ as a function of L. The quantity $\sigma^4/\gamma Y^2$ is a horizontal line. According to the stability condition (7), three situations exist. (a) When the horizontal line is too high to intersect with the curve, the flat film is unstable for any thickness. (b) When the horizontal line is tangent to the curve, the flat film is stable only for one particular thickness, and unstable for any other thicknesses. (c) When the horizontal line intersects with the curve at two points, corresponding to films of thickness L_1 and L_2, the flat film is stable if its thickness falls in between. Using $\gamma = 1$ J/m^2, $Y = 76$ GPa, and $\sigma = 500$ MPa (a relatively large stress in metallic films), we obtain $\sigma^4/\gamma Y^2 = 10^{13}$ J/m^4. Our calculations, including both quantum confinement and charge transfer, with either finite or infinite potential well, gave the magnitude of the maximum curvature, $(\partial^2 U/\partial L^2)_{\max} = 10^{18}$ J/m^4.

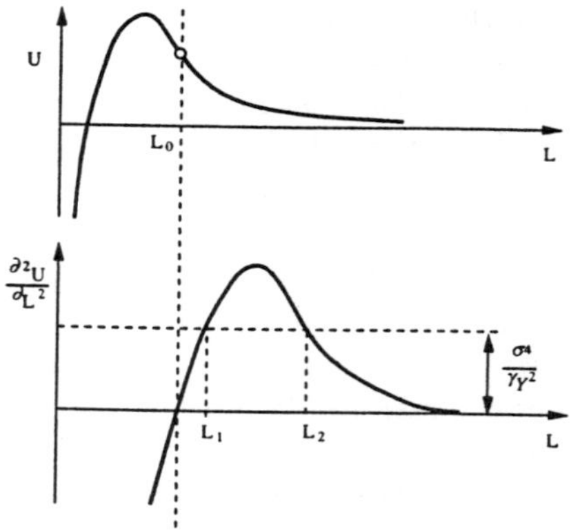

Figure 6: The top figure shows the qualitative shape of the function $U(L)$ for Ag on GaAs, where the inflection point is marked by a small circle. The bottom figure shows the qualitative shape of the curvature.

Note the huge difference between $\sigma^4/\gamma Y^2$ and $(\partial^2 U/\partial L^2)_{\max}$. Consequently, when the film is not too thick, the quantum confinement effect prevails over the stress by a large margin. Situation (c) is readily accessible experimentally: very thin films are destabilized by charge transfer, films of intermediate thickness are stabilized by quantum confinement, and thick films are destabilized by stress.

Because $(\partial^2 U/\partial L^2)_{\max} \gg \gamma Y^2$, from Fig. 6 we see that $L_1 \approx L_0$. However, L_2 must be estimated by using the long-range tail of the interaction energy. Everything else being equal, better confinement of electrons can stabilize thicker films. As an estimate of the magnitude of the long-range tail, consider electrons confined in a metallic film by infinite potentials on both sides. The energy levels are determined by the one-electron Schrödinger equation. The total free energy is estimated by the sum of energies over all electrons in the ground state of the film. As above, let $U(L)$ be the excess energy per unit area of the film relative to that of the bulk of the same thickness. Our analysis shows a surprisingly long ranging tail:

$$U(L) = \frac{3\pi^2\hbar^2\rho}{32mL} \equiv \frac{B}{L} \tag{8}$$

where $\hbar$ is the Planck constant, m the electron mass, and ρ the number of free electrons per unit volume. Figure 7 compares this asymptotic result with the exact numerical solution; they agree well beyond a few monolayers. A

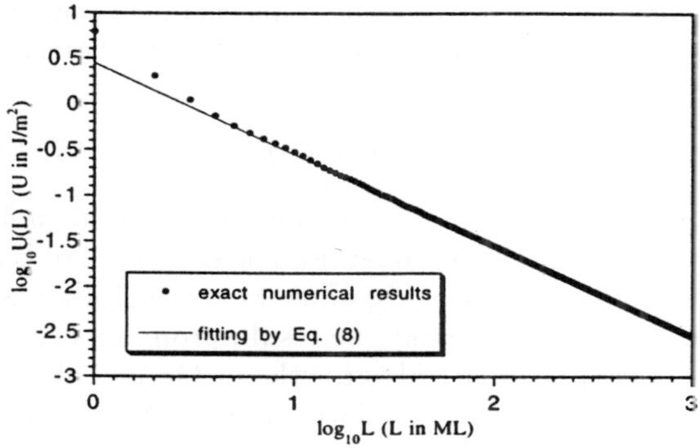

Figure 7: Comparison between the asymptotic long-range tail with numerical results, for a silver film confined by infinite potential on both sides.

combination of (7) and (9) gives

$$D_2 = \left(2B\gamma Y^2/\sigma^4\right)^{1/3} \tag{9}$$

For Ag, $n = 5.86 \times 10^{28}$ m^{-3} and $B = 6.62 \times 10^{-10}$ J/m. A stress of magnitude $\sigma = 500$ MPa leads to $L_2 = 496$ Å. The available experimental data do not permit a meaningful comparison. Equation (9) ignores fine oscillations that are invisible on the scale of Fig. 7. For finite confinement potentials, our numerical calculation shows that the interaction energies due to quantum confinement and charge transfer each has the $1/L$ tail, but with different proportionality constants. Consequently, these details do not change the qualitative behaviors at large L.

Our model predicts that the critical film thickness sensitively depends on the stress. This fact can be readily exploited in experiments. For example, the stresses in $In_xGa_{1-x}As$ films on a GaAs substrate depend on the composition x; the wetting layer thickness is known to be a strong function of the composition [29]. For a metallic film on a semiconductor substrate, thermal expansion misfit is large; for Ag on GaAs a temperature change can cause a stress change by 1.4 MPa/K. One expects that the critical thickness can be tuned by changing the temperature, as suggested in some experiments [9,25,30].

We are unaware of any experimental measurements of the long-range forces in crystalline films. The excess free energy U gives rise to a chemical potential of a thin film relative to the bulk:

$$\mu = \Omega \partial U/\partial L \tag{10}$$

where Ω is the volume per atom. Note that this chemical potential depends on the film thickness. It may be measured by suitable mass transfer experiments.

4 First-principles studies of a model system: Sb/GaAs(110)

So far all the discussions have been based on simple phenomenological models. The physical phenomena derived are novel and interesting, but will they survive if the systems are described by more rigorous, first-principles based interaction potentials? In this section, we present the results from a recent first-principles study of a representative system, Sb growth on GaAs(110) [16], in which we confirm many of the essential features derived using simple models.

As a prototype nondisruptive metal-semiconductor interface system, the growth of Sb on GaAs(110) has been investigated extensively[31,32,33,34,35,36,37]. Based on Auger electron spectroscopy [31] and scanning tunneling microscopy (STM) studies [32,33], the growth pattern of Sb has been found to follow a $(1+M)$ mode (or a modified Stranski-Krastonov mode): a monolayer followed by sets of multilayers of a well-defined thickness M (see Fig. 8). Moreover, $I-V$ measurements showed that the band gap at the Fermi energy decreases with increasing film thickness, suggesting a nonmetal-metal transition at a higher coverage[33]. Theoretically, most previous studies have concentrated on the adsorption of 1-ML Sb on GaAs(110) [34,35,36,37]. Our recent study this system in the multilayer regime aimed to provide the physical insights into the understanding of those growth phenomena.

In our study, the effects of quantum confinement and charge spilling are treated self-consistently, with inclusion of surface relaxation. We find strong manifestations of quantum size effects, in both known and unexpected ways. As the film thickness increases, the adsorption energy per layer oscillates, thereby defining the existence of magic thicknesses for smooth growth. This finding provides the microscopic basis for the $(1+M)$ growth mode. Furthermore, there exist corresponding oscillatory nonmetal-metal transitions, a surprising finding in contradiction with traditional belief. We will identify the underlying physical reasons for the existence of the oscillatory nonmetal-metal transitions, discuss the results in comparison with existing experiments, and suggest new ways to test some of the unique predictions made here.

In our calculations, we use norm-conserving separable pseudopotentials[38,39] together with the density-functional theory within the local-density approximation[40,41] (LDA). Partial-core corrections are included in the pseudopotentials of Ga [42]. We model the Sb/GaAs(110) system by a periodic slab geometry. Each slab contains seven GaAs substrate layers and a certain Sb overlayer on each side of the slab. The vacuum region between such slabs has a thickness of about 10 Å. The Sb overlayer is modeled by the so-called epitaxial continued layer structure[35,36,37], where the Sb atoms grow epitaxially in a form of zig-zag chains in the $[1\bar{1}0]$ direction on a nearly bulk-like GaAs(110) substrate (see Fig. 8). To optimize the atomic structure, atoms in the overlayer and the substrate are relaxed along the calculated forces until the remaining forces are all within 6 mRy/Å. We employ a plane-wave basis with an energy cutoff of 10 Ry and take a uniform grid of 24 **k** points within the (1×1) surface Brillouin-zone. The calculation scheme and its previous application to the Sb monolayer on GaAs(110) are described in detail elsewhere [37].

In order to examine the relative stability of Sb adsorption on GaAs(110) with increasing Sb coverage, we calculate the adsorption energy (E_{ad}) per Sb atom from

$$E_{\text{ad}} = (E(n-1) + 4E^a_{\text{Sb}} - E(n))/4 \tag{11}$$

where $E(n-1)$ and $E(n)$ is the total energy per unit cell for a slab with the Sb coverage, L, equal to $(n-1)$ and n ML, respectively, and E^a_{Sb} is the total energy of a free Sb atom. In the case of $n = 1$, $E(n-1)$ corresponds to the total energy of the clean GaAs(110) surface. The negative of the adsorption energy with respect to the Sb coverage is plotted in Fig. 9, showing strongly oscillatory size effects: The adsorption energies at $L = 1$, 3, and 5 ML are larger than those at 2 and 4 ML. This energetic information indicates that, the first Sb monolayer, at which the adsorption energy is the largest, binds most strongly to the GaAs substrate. A flat film at $L = 3$ or 5 ML is locally stable, but a flat film at 2 or 4 ML is unstable against roughening. Therefore, one expects Sb growth on GaAs(110) to follow the $(1+M)$ mode, with $M = 2$. This finding qualitatively explains the experimentally observed $(1+M)$ growth mode in this system. However, we note that on a quantitative level earlier experiments suggested M to be 3, estimated by using the constant bulk interlayer spacing for the Sb thin films[32,33]. As shown below, this discrepancy can be resolved by considering the large deviations of the interlayer spacings in the Sb thin films from the bulk value. Figure 9 also shows that the adsorption energy changes little above 6 ML of Sb; therefore, the double layer growth mode will no longer be favored at such higher coverages.

In Table 1, we summarize the calculated interlayer spacings of the Sb overlayers. It is worth to emphasize the following aspects. (i) the topmost Sb-Sb interlayer spacing varies with the film thickness in an oscillatory way, taking the values of 2.94, 2.77, 2.88, 2.79, and 2.81 Å as L increases from 2 to 6 ML. After 6 ML the oscillation disappears. Such variations in the topmost interlayer spacing should be observable, for example by measuring the height of sizeable monolayer-high islands formed at the growth front, as reported in

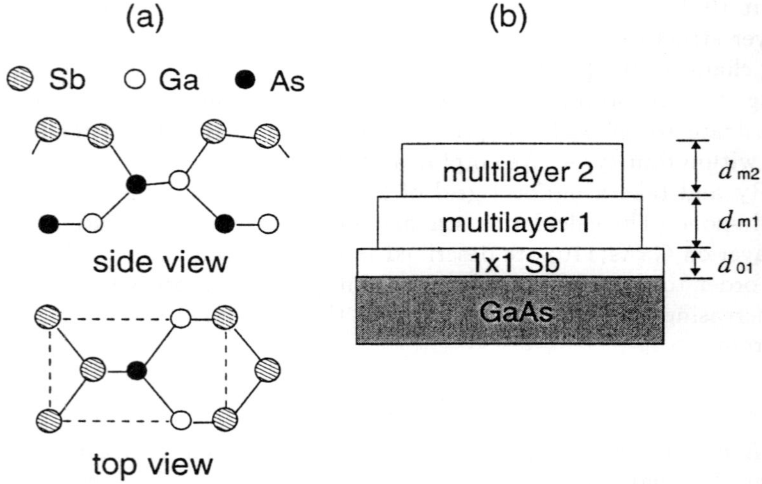

Figure 8: (a) Schematic diagrams of the side and top views of the epitaxial continued layer structure model for the 1ML-Sb/GaAs(110)-(1x1) system. (b) Schematic cross-section view depicting Sb multilayers on the GaAs(110) substrate.

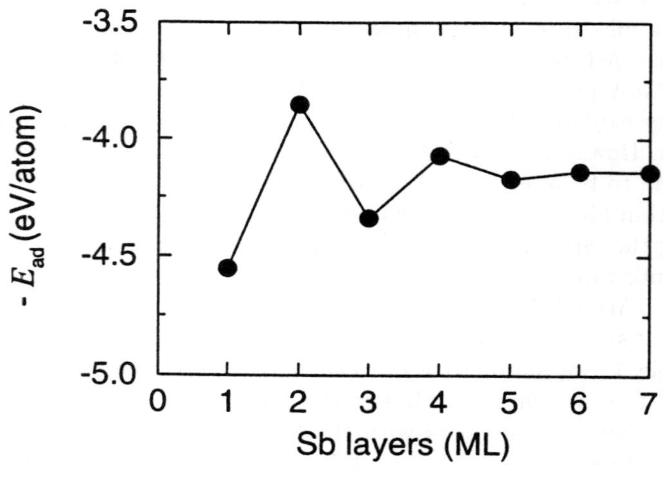

Figure 9: Negative adsorption energy as a function of the Sb coverage.

Table 1: Calculated interlayer spacings (in Å) for the Sb overlayers on GaAs(110).

	d_{01}	d_{12}	d_{23}	d_{34}	d_{45}	d_{56}	d_{67}
1 ML	2.39						
2 ML	2.54	2.94					
3 ML	2.41	3.44	2.77				
4 ML	2.44	3.28	2.99	2.88			
5 ML	2.43	3.25	2.86	3.16	2.79		
6 ML	2.44	3.28	2.87	2.97	3.02	2.81	
7 ML	2.43	3.24	2.89	3.00	2.94	3.08	2.81

a recent experimental study of Pb/Ge(100) [43]. (ii) for a given coverage, the interlayer spacing also oscillates from layer to layer.

The present results for the heights of double Sb layers (d_{m1} and d_{m2} in Fig. 8) are given in Table 2 together with those from experiments [32,33]. The values of $d_{m1} = 6.2$ Å and $d_{m2} = 6.0$ Å are in good agreement with the STM measurements of Shih, Feenstra, and Mårtensson [32] ($d_{m1} = 6.0 \pm 0.5$ Å and $d_{m2} = 6.0 \pm 0.5$ Å) and Patrin *et al.* [33] ($d_{m1} = 6.4 \pm 0.5$ Å and $d_{m2} = 6.4 \pm 0.5$ Å). However, both STM studies estimated the coverages of the first and the second multilayers to be 4 ML and 7 ML, respectively, by using the bulk interlayer spacing of about 2 Å. Our calculations show that the interlayer spacings in the thin Sb overlayers are significantly larger than the bulk value (see Table 1), a prediction to be confirmed in future experiments.

The oscillatory behavior of the interlayer spacing is a consequence of the quantum size effect. The electronic density in the quantum well has an oscillatory position dependence in the growth direction. It is natural to expect similar adjustment of the ions to minimize the electron-ion interaction energy[13]. Here we only like to emphasize the observation that the amplitude of oscillation is particularly large at $L = 3$ ML, when the system is in the nonmetallic state (see below). This can be explained qualitatively by the fact that in the nonmetallic state the charge density fluctuation is associated with one-dimensional screening, while in the metallic state the magnitude of the fluctuation associated with three-dimensional screening is smaller.

We have also obtained the band structures of the system at different Sb coverages, with those at 1 and 2 ML shown in Fig. 10. There are four subbands in the bulk gap at $L = 1$ ML[35,36]; the lower two subbands are fully occupied, and the higher ones are empty. Thus the 1-ML Sb overlayer is nonmetallic with a band gap of 0.9 eV. The 2-ML Sb overlayer has two additional subbands in the bulk gap, which overlap across the Fermi energy, leading to a metallic state.

Table 2: Calculated heights (in Å) of Sb multilayers in comparison with experimental results. For the denotations of d_{m1} and d_{m2}, see Fig. 8. The interlayer spacing between the Sb layer and the GaAs substrate (d_{01}) is given for comparison.

	d_{01}	d_{m1}	d_{m2}
present study	2.4	6.2	6.0
STM I[32]	2.5	6.0 ± 0.5	6.0 ± 0.5
STM II[33]	2.5	6.4 ± 0.5	6.4 ± 0.5

Surprisingly, these overlapping subbands appear oscillatorily with increasing Sb thickness: They disappear at 3 ML and reappear at 4 ML. As a result, the 3-ML Sb overlayer has a band gap of 0.1 eV, and the 4-ML Sb overlayer shows a metallic state. Above $L = 5$ ML the system is always metallic. The calculated band gap with respect to the Sb coverage is summarized in Fig. 11. It is well known that the LDA calculation underestimates the band gap. For example, the experimental band gap at 1 ML is about 1.3 eV[32,33], larger than the calculated value of 0.9 eV. Therefore, we expect that the real band gap at 3 ML is also larger than the present LDA value of 0.1 eV. Because the 2- and 4-ML Sb overlayers show metallic behavior with the presence of the two overlapping subbands at the Fermi energy, they are energetically unstable compared to the 1- and 3-ML Sb overlayers, consistent with the results shown in Fig. 9.

It is remarkable that the Sb overlayers on GaAs(110) show oscillatory nonmetal-metal transitions. For metal overlayers on semiconductor substrates, the typical picture for nonmetal-metal transition is as follows[44]: The overlayer is nonmetallic if the coverage is too low, becomes metallic at some critical coverage, and is expected to be more metallic if additional layers of metal are added. However, in the present study we find that a metallic overlayer at $L =$ 2 ML will turn into a nonmetallic state if one more layer of Sb is added.

The unusual oscillatory nonmetal-metal transitions can be explained by the classic Wilson rule[45], generalized to the present case, and constrained by the quantum size effect of the thin film. Because the dangling bonds of the topmost Sb layer atoms is fully occupied, we can regard our quantum well to really start at $L = 2$ ML. At $L \geq 2$ ML, because each Sb atom is tetrahedrally bonded in the (1x1) structure, each unit cell of a given Sb layer contains two nearly equivalent "free" electrons. Therefore, the total number of such electrons per unit cell of the thin film is always an even number, $2n$, with $n =$ 1, 2, 3, ... for $L =$ 2, 3, 4, ... ML. On the other hand, each additional Sb layer also contributes two overlapping subbands, due to the confinement in the

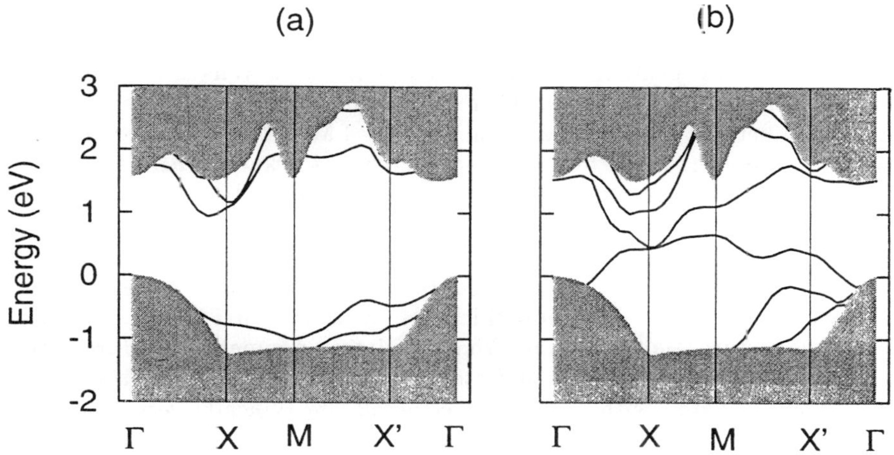

Figure 10: Surface band structures of Sb/GaAs(110)-(1x1) at two Sb coverages: (a) 1 ML; (b) 2 ML.

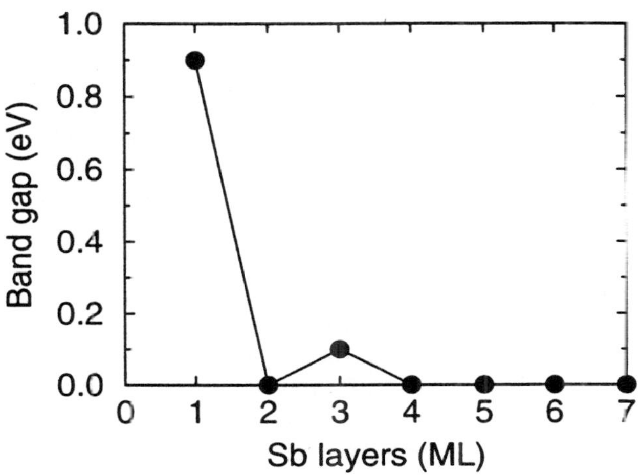

Figure 11: Band gap as a function of the Sb coverage.

vertical direction. For small enough film thicknesses at which different subbands originated from different Sb layers do not overlap in energy, we expect metal-nonmetal transitions according to a generalized Wilson rule: metal for odd n and nonmetal for even n. Therefore, we have a metallic system at $L = 2$ ML, a nonmetal at 3 ML, and a metal again at 4 ML. This oscillatory behavior will stop, however, if the film is above a critical thickness where the spacing between the subbands contributed by neighboring layers becomes smaller than the width of the subbands, because then the different subbands overlap in energy and only the metallic state prevails. Below the critical thickness, we expect that the gap between the filled and empty subbands of the nonmetallic state decreases with the film thickness, as shown in Fig. 11. Based on the above picture, we further infer that oscillatory nonmetal-metal transitions should *not* be expected in alkali metal films on semiconductors, because the subbands associated with those simple metals are typically very broad. On the other hand, for systems of metal overlayers with flatter subbands, oscillatory nonmetal-metal transitions can persist to even higher film coverages. At present, we are developing a phenomenological description of the oscillatory metal-nonmetal transitions aimed to explore the phase space in which such oscillations persist to higher film thicknesses [47].

5 Summary and discussions

This paper is centered around the presentation of the novel concept of "electronic growth". This concept contains three essential ingredients: quantum confinement, which leads to a repulsive force working to stabilize metal films of any thickness; charge spilling, which leads to an attractive force working to destabilize particularly thinner films; and interface-induced Friedel oscillations, which may introduce additional modulations in the effective long-ranged interaction potential. The theory provides new understanding on many existing observations in previous studies of ultrathin metallic overlayer growth on semiconductor substrates. For example, the observation of critical thicknesses for smooth growth of Ag on GaAs, GaP, and GaSb [8,9,10], and the fact that only the first layer of alkali metals can grow smoothly on GaAs [14] provide strong evidences for the validity of the theory. We also suspect that the peculiar features frequently observed during the growth of the first few monolayers of metals on semiconductors [46] may have their origin tied to the existence or absence of the critical/magic thicknesses in such systems. More importantly, one can devise new experiments to test systematically the unique predictions made here. In doing so, one should bear in mind two optimal conditions: the overlayer metal should be soft (as it is the case for Ag), so as to minimize the effect of the strain energy; and there should be minimal intermixing at the interface, so as to maximize the effect of a sharp interface. For many systems, the second requirement demands sufficiently low growth temperatures. Lower-temperature deposition is especially required to test the existence of the magic and critical thicknesses shown in Types C and D of Fig. 3, because one typically needs to have an initial film close to the stable flat film configuration. For systems in which such critical/magic thicknesses do exist, the morphology of the metallic overlayers can be controlled down to the atomic scale. It should also be possible to tune the values of the critical/magic thicknesses by tuning the band alignment. Therefore, the electronic growth mechanism in principle provides an important tool for quantum engineering of metallic thin films on semiconductor substrates.

We have also shown that suitable long-range interactions can stabilize epilayers against stresses. Substantial work is needed to study long-range interaction of various physical origins. In particular, forces prevalent in colloids and liquid films, such as that due to electrical double-layers, should be examined to determine their relevance to solid films. Long range interactions may play significant roles in other phenomena in nanostructures. It is hoped that experiments will soon succeed in confirming the predictions made here, and independently measuring the excess energy of solid films.

On a more rigorous level, our first-principles calculations have shown that, in the model growth system of Sb/GaAs(110), quantum size effects can prevail in several dramatic ways. The adsorption energy per layer has been found to oscillate with the overlayer thickness, making flat films at 1, 3, and 5 ML coverages magically stable, while films at 2 and 4 ML unstable. This finding qualitatively explains the $(1 + M)$ growth mode observed in previous experiments, and calls for more precise determination of M in future experiments. For films at different thicknesses, there exist strong oscillations in the topmost interlayer spacings; and for a film of a given coverage, the interlayer spacing within the film should also oscillate from layer to layer. There should also exist oscillatory nonmetal-metal transitions as the film thickness increases, a prediction to be confirmed in future experiments (for example by locally probing the band gaps using the scanning tunneling spectroscopy). All these oscillatory properties, in the stability, interlayer spacing, and transport, are correlated at the fundamental level.

Finally, we mention that the effects of quantum confinement, charge spilling, and Friedel oscillations are not limited to the cases of three-dimensional film growth. For example, in a recent study, these similar phenomena were found to play important roles in the formation of two-dimensional metal islands on surfaces, leading to the existence of magic length scales in metal submonolayer epitaxy [17]. We expect that the present line of study will stimulate intensive research activities, both experimental and theoretical, to further explore the importance and fascinating manifestations of quantum size effects in various systems of reduced dimensions.

We gratefully acknowledge Myung-Ho Kang, Jerry Mahan, Ward Plummer, Jerry Tersoff, John Weaver, and Hanno Weitering for helpful and informative discussions. This research was supported by Oak Ridge National Laboratory managed by Lockheed Martin Energy Research Corp. for the U.S. Department of Energy under Contract No. DE-AC05-96OR22464, and by the National Science Foundation (grant numbers MSS-9258115 and DMR-9705406).

References

1. See, for example, articles in Physics of Low-Dimensional Semiconductor Structures, P. Butcher, N. H. March, M. P. Tosi, Eds. (Plenum, New York, 1992).
2. J. H. van der Merwe, J. Appl. Phys. **34**, 117 (1963); *ibid.*, p123. J. W. Matthews and A. E. Blakeslee, J. Cryst. Growth **29**, 273 (1975); *ibid.* **32**, 265 (1976). J. Tersoff, Appl. Phys. Lett. **62**, 693 (1993). D. E. Jesson, K. M. Chen, S. J. Pennycook, T. Thundat, R. J. Warmack, Science **268**, 1161 (1995).
3. R. J. Asaro and W. A. Tiller, Metall. Trans. **3**, 1789 (1972). M. A. Grinfeld, J. Nonlinear Sci. **3**, 35 (1983). D. J. Srolovitz, Acta Metall. **37**, 621 (1989). B. J. Spencer, P. W. Voorhees, S. H. Davis, Phys. Rev. Lett. **67**, 3696 (1991). J. Tersoff and F. K. LeGoues, *ibid.* **72**, 3570 (1994).
4. See, for example, the review articles in the special issue on Heteroepitaxy and Strain, MRS Bulltin **22** (No. 4) (1996).
5. J. Villain, J. Phys. I (France) **1**, 19 (1991). J. A. Stroscio, D. T. Pierce, M. D. Stiles, A. Zangwill, L. M. Sander, Phys. Rev. Lett. **75**, 4246 (1995).
6. For a review on various growth mechanisms based on atomistic processes, see Z. Y. Zhang and M. G. Lagally, Science **276**, 377 (1997).
7. B. M. Trafas, Y.-N. Yang, R. L. Siefert, J. H. Weaver, Phys. Rev. B **43**, 14107 (1991).
8. A. R. Smith, K.-J. Chao, Q. Niu, and C. K. Shih, Science **273**, 226 (1996).
9. G. Neuhold, L. Bartels, J. J. Paggel, K. Horn, Surf. Sci. **376**, 1 (1997).
10. K.-J. Chao, Q. Niu, Z. Y. Zhang, C.K. Shih, (to be published).

11. Z. Y. Zhang, Q. Niu, and C.-K. Shih, Phys. Rev. Lett. (in press).
12. R. C. Jaklevic, J. Lambe, M. Mikkor, and W. C. Vassell, Phys. Rev. Lett. **26**, 89 (1971). M. Jalochowski and E. Bauer, Phys. Rev. B **38**, 5272 (1988). T. Miller, A. Samsavar, G. E. Franklin, T. C. Chiang, Phys. Rev. Lett. **61**, 1404 (1988). B. J. Hinch, C. Koziol, J. P. Toennies, and G. Zhang, Europhys. Lett. **10**, 341 (1989). E.Ortega and F. J. Himpsel, Phys. Rev. Lett. **69**, 844 (1992). D. A. Evans, M. Alonso, R. Cimino, K. Horn, Phys. Rev. Lett. **70**, 3483 (1993).
13. F. K. Schulte, Surf. Sci. **55**, 427 (1976). P. J. Feibelman, Phys. Rev. B **27**, 1991 (1983). P. J. Feibelman and D. R. Hamann, Phys. Rev. B **29**, 6463 (1984). I. P. Batra *et al.*, Phys. Rev. B **34**, 8246 (1986). N. Trivedi and N. Ashcroft, Phys. Rev. B **38**, 12298 (1988).
14. N. J. DiNardo, T. M. Wong, and E. W. Plummer, Phys. Rev. Lett. **65**, 2177 (1990). L. J. Whitman, J. A. Stroscio, R. A. Dragoset, and R. J. Celotta, Phys. Rev. Lett. **66**, 1338 (1991). O. Pankratov and M. Scheffler, Phys. Rev. Lett. **70**, 351 (1993).
15. Z. G. Suo and Z. Y. Zhang, Phys. Rev. Lett. (submitted, 1997).
16. J.-H. Cho, Q. Niu, and Z. Y. Zhang, Phys. Rev. Lett. **80**, 3582 (1998).
17. K.-J. Jin, G. D. Mahan, H. Metiu, and Z. Y. Zhang, Phys. Rev. Lett. **80**, 1026 (1998).
18. W. D. Knight, K. Clemenger, W. A. de Heer, W. A. Saunders, M. Y. Chou, M. L. Cohen, Phys. Rev. Lett. **52**, 2141 (1984).
19. J. Tersoff, Phys. Rev. Lett. **52**, 465 (1984).
20. S. G. Louie, J. R. Chelikowsky, and M. L. Cohen, Phys. Rev. B **15**, 154 (1977).
21. N. W. Ashcroft and N. D. Mermin, Solid State Physics (Saunders, Philadelphia, 1976).
22. A. M. Cowley and S. M. Sze, J. Appl. Phys. **36**, 3212 (1965).
23. S. B. Zhang and A. Zunger, Phys. Rev. B **53**, 1343 (1997).
24. D. J. Eaglesham and M. Cerullo, Phys. Rev. Lett. **64**, 1943 (1990). Y.-M. Mo, D. E. Savage, B. S. Swartzentruber, M. G. Lagally, Phys. Rev. Lett. **65**, 1020 (1990). F. K. LeGoues, M. Copel, and R. M. Tromp, Phys. Rev. B **42**, 11690 (1990). M. Asai, H. Ueba, and C. Tatsuyama, J. Appl. Phys. **58**, 2577 (1985).
25. W. J. Schaffer et al., J. Vac. Sci. Technol. B **1**, 688 (1983). C. W. Snyder, B. G. Orr, and H. Munekata, Appl. Phys. Lett. **62**, 46 (1992). D. Leonard, M. Krishnamurthy, S. Fafard, J. L. Merz, and P. M. Petroff, J. Vac. Sci. Technol. B **12**, 1063 (1994). R. Leon, S. Fafard, D. Leonard, J.L. Merz, and P.M. Petroff, Appl. Phys. Lett. **67**, 521 (1995). Q. K. Xue and T. Sakurai, (preprint, 1997). S.R. Forrest, Chem. Rev. **97**,

1793 (1997).
26. M. Grinfeld, Sov. Phys. Dokl. **31**, 831 (1986). H. Gao, Int. J. Solids Structure **28**, 703 (1991). L. B. Freund and F. Jonsdottir, J. Mech. Phys. Solids **41**, 1245 (1993).
27. Here we use the simplest version of the model to compare the relative magnitudes of various thermodynamic forces. The effect of surface tension anisotropy can be important. See J. Tersoff and F. K. LeGoues, Phys. Rev. Lett. **72**, 3570 (1994); D. E. Jesson, K.M. Chen, S. J. Pennycook, T. Thundat, R. J. Warmack, Phys. Rev. Lett. **77**, 1330 (1996). A quantitative comparison of effects of surface tension anisotropy and long-range interaction is beyond the scope of this paper.
28. C.-H. Chiu and H. Gao, Mat. Res. Soc. Proc. **356**, 33 (1995).
29. P. M. Petroff, and S. P. DenBaars, Superlattices and Microstructures **15**, 15 (1994).
30. A. Grossmann, W. Erley, J. Hannon, and H. Ibach, Phys. Rev. Lett. **77**, 127 (1996).
31. R. Strümpler and H. Lüth, Surf. Sci. **182**, 545 (1987).
32. C. K. Shih, R. M. Feenstra, and P. Mårtensson, J. Vac. Sci. Technol. A **8**, 3379 (1990).
33. J. C. Patrin, Y. Z. Li, M. Chander, and J. H. Weaver, Phys. Rev. B **46**, 10221 (1992).
34. C. B. Duke *et al.*, Phys. Rev. B **26**, 803 (1982). P. Mårtensson and R. M. Feenstra, Phys. Rev. B **39**, 7744 (1989). W. K. Ford *et al.*, Phys. Rev. B **42**, 8952 (1990). M. G. Betti *et al.*, Phys. Rev. B **50**, 14336 (1994). H. Ascolani *et al.*, Phys. Rev. Lett. **78**, 2604 (1997).
35. G. P. Srivastava, Phys. Rev. B **46**, 7300 (1992); *ibid.*, **47**, 16616 (1993).
36. W. G. Schmidt, B. Wenzien, and F. Bechstedt, Phys. Rev. B **49**, 4731 (1994).
37. J.-H. Cho, Z. Y. Zhang, S. H. Lee, and M. H. Kang, Phys. Rev. B **57**, 1352 (1998).
38. L. Kleinman and D. M. Bylander, Phys. Rev. Lett. **48**, 1425 (1982).
39. N. Troullier and J. L. Martins, Phys. Rev. B **43**, 1993 (1991).
40. P. Hohenberg and W. Kohn, Phys. Rev. **136**, B864 (1964); W. Kohn and L. J. Sham, Phys. Rev. **140**, A1133 (1965).
41. D. M. Ceperley and B. J. Alder, Phys. Rev. Lett. **45**, 566 (1980). J. P. Perdew and A. Zunger, Phys. Rev. B **23**, 5048 (1981).
42. S. G. Louie, S. Froyen, and M. L. Cohen, Phys. Rev. B **26**, 1738 (1982).
43. A. Crottini, D. Cvetko, L. Floreano, R. Gotter, A. Morgante, and F. Tommasini, Phys. Rev. Lett. **79**, 1527 (1997).
44. E. W. Plummer and P. A. Dowben, Progress in Surface Science, **42**, 201

(1993). P. A. Dowben, Rev. Modern Phys. (to appear).
45. A. H. Wilson, Proc. Roy. Soc. A **133**, 458 (1931).
46. V. A. Grazhulis, Prog. Surf. Sci. **36**, 89 (1991). M. Jalochowski, H. Knoppe, G. Lilienkamp, E. Bauer, Phys. Rev. B **46**, 4693 (1992). E.T. I. M. Bootsma and T. Hibma, Surf. Sci. **287/288**, 921 (1993).
47. M.-C. Chang, J.-H. Cho, Z. Y. Zhang, and Q. Niu, (preprint).

SEMICONDUCTOR-ON-SEMICONDUCTOR GROWTH

SELF-ORGANIZED ISLAND ARRAYS IN SiGe/Si MULTILAYERS

CHRISTIAN TEICHERT
Max-Planck-Institut für Mikrostrukturphysik,
Weinberg 2, D-06120 Halle, Germany
E-mail: teichert@mpi-halle.mpg.de

JERRY TERSOFF
IBM Thomas J. Watson Research Center,
Yorktown Heights, NY 10598, USA
E-mail: tersoff@us.ibm.com

MAX G. LAGALLY,
Department of Materials Science and Engineering,
University of Wisconsin-Madison, 1509 University Ave., Madison, WI 53706, USA
E-mail: lagally@neep.engr.wisc.edu

Abstract

In the Stranski-Krastanov growth of $Si_{1-x}Ge_x$ films on Si(001) the lattice strain results at a certain stage in the formation of three-dimensional, {105} faceted islands that usually show prism-like shapes. These so called "hut clusters" form an interlocked array resulting in a broad distribution of island sizes, shapes and separations. Upon growth of an alternation of SiGe layers and Si spacer layers the island arrangement transforms in the multilayer stack into an increasingly periodic array of uniform pyramid-like islands. This self-organized island growth in a semiconductor multilayer film is explained in the frame-work of continuum elasticity theory as a result of strain mediation through the spacer layers.

1 Introduction

Spatial self-organization within particle ensembles is a common phenomenon in nature [1], covering a wide range of length scales and leading to beautiful patterns such as ripples on dunes or snow fields and cloud structures. Surface reconstruction and pattern formation in homoepitaxial growth are self-organization processes on an atomic scale, driven by the system's desire to reduce its chemical free energy. In the Stranski-Krastanov growth mode of heteroepitaxy where after the formation of a wetting layer growth of three-dimensional crystallites is thermodynamically required [2], the relief of the built-up stress can be an additional driving force for self-organization. Besides stress-induced surface reconstruction stress can cause morphological changes at a large scale, such as one-dimensional (1D) ripple formation due to bunching of atomic-height steps [3-5] and the formation of – frequently faceted – 3D crystallites coherent with the substrate. These crystallites (conventionally and hereafter called "islands") may show a certain degree of size uniformity as it is the case for a variety of III-V semiconductor systems [6-10] and also for II-VI semiconductors [11]. Several explanations by total energy calculations [12,13] as well as kinetic considerations [14,15] have been suggested for the size uniformity. The situation is slightly different for growth of a single layer of pure Ge or SiGe alloy on Si(001), the main representative of IV-IV heteroepitaxy. In this case also coherently strained small islands are observed [16]. However, these islands form in the shape of elongated prism-like crystallites therefore called "hut clusters" [17]. These islands arrange in an interlocked array exhibiting a broad distribution in size and shape [18-20]. We will show in this review, that the morphological organization of stress-induced crystallites improves dramatically by growing a multilayer stack consisting of two individual layers that differ slightly in lattice constant. This will be demonstrated for the example of alternating layers of pure Si and SiGe alloy on Si(001), where the prism-like hut clusters organize into uniformly sized islands in a square pattern. It is shown experimentally that this self-organization is related to the presence of the Si spacer layers. In the frame work of continuum elasticity theory the self-organization is explained as strain mediation through the spacer layers [21].

2 Experimental

The samples that are used to demonstrate the self-organization of nanoscale islands in multilayer films are device-quality $Si_{0.25}Ge_{0.75}/Si$ superlattices. They were grown by solid-source molecular beam epitaxy at 550 °C on vicinal Si(001) wafers. The polar and azimuthal miscut angles of the substrate, θ and ϕ, respectively, as well as the Ge concentration x in the alloy layers, were measured directly using x-ray diffraction [22]. Figure 1 shows the schematic layer structure of these multilayer films. After depositing a 100 nm Si buffer layer up to n = 40 bilayers were grown at a deposition rate of 0.055 nm/s. If not otherwise specified, the individual layer thicknesses were 2.5 nm for the alloy layers and 10 nm for the Si spacer layers. All the samples investigated here belong to a series of multilayers grown under identical conditions on Si(001) substrates with an identical miscut of $\theta = 0.25°$ and $\phi = 25°$ with respect to $[\bar{1}10]$ direction. They differ only by the number of bilayers deposited and the multilayer termination (alloy or silicon). The surface morphology of these samples was investigated using *ex-situ* atomic-force microscopy (AFM), in contact mode with a 12 μm xyz-scanner. As the measurements were performed *ex-situ*, a native oxide of about 1 nm thickness is present on all samples. The samples were always oriented on the microscope in such a way that the scanning direction is parallel to the [110] direction within an uncertainty of a few degrees. Conventional cantilevers with silicon nitride tips with a 70° side wall profile and a tip radius ranging between

20 and 40 nm were used. All images were taken with a resolution of 512 x 512 pixels allowing to calculate good-quality 2D power spectra of the images.

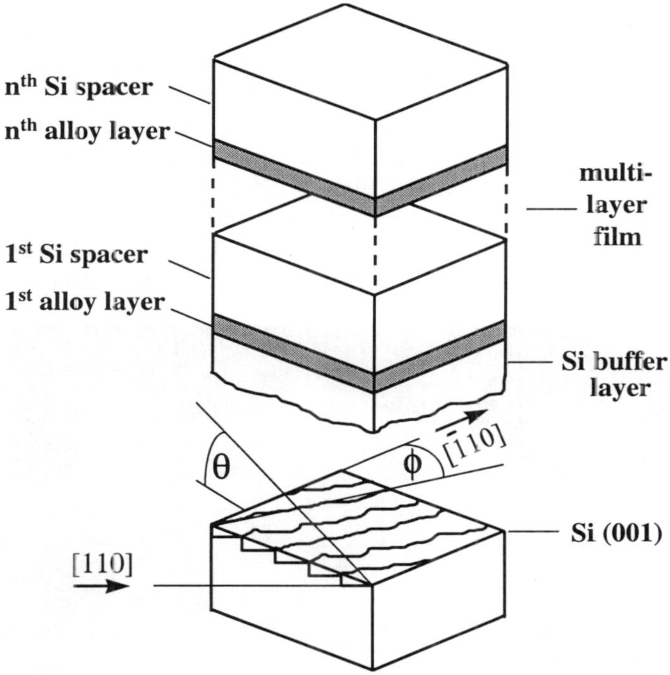

Figure 1: Illustration of the layer structure of the SiGe/Si multilayer samples. The layers are grown by solid-source MBE at 550 °C on vicinal Si(001) surfaces that are characterized by a polar miscut angle θ and an azimuthal miscut angle ϕ. Before depositing the multilayers a 100 nm Si buffer layer is grown. The bilayers of the multilayer stack consist of a $Si_{0.25}Ge_{0.75}$ alloy layer of 2.5 nm nominal thickness and, if not otherwise specified, a Si spacer layer of 10 nm. *Ex-situ* AFM is used to image the surface topography of identical samples that differ only by the number of bilayers n.

3 Strain relief in $Si_{1-x}Ge_x$ films grown on Si(001)

During $Si_{1-x}Ge_x$/Si(001) growth strain causes a series of morphological changes, whose existence and onset depend on the concentration, x, of Ge. The sequence of stress relief, at least in its initial stages, is well established: At submonolayer coverages a (2xn) surface reconstruction begins to form,[23-25] in which increasingly well-aligned rows of dimer vacancies [26] provide stress relief [27]. The (2xn) reconstruction modifies the relative roughness of steps [28]. The value of n depends on x [25]. The Si(001) surface stress tensor is anisotropic; deposition of Ge in sufficient amounts (>1ML) reverses this anisotropy through the creation of the (2xn) reconstruction [29]. For the deposition of pure Ge on Si(001), this stress tensor anisotropy leads to the breakdown of the epitaxial-layer structure into elongated two-dimensional islands at approximately three monolayers coverage [29]. For x<1, strain builds up less rapidly and this breakdown is delayed, although apparently at least one monolayer of Ge resides on the surface at all conventional growth conditions using molecular beam epitaxy, making the expected surface behavior quite similar in all cases [30]. For deposition of SiGe alloy on vicinal Si(001) substrates at appropriate growth conditions, the first manifestation of a stress-induced effect, after the formation of the 2xn reconstruction, is the bunching of preexisting substrate steps [4,5], first observed in $Si_{1-x}Ge_x$/Si multilayer films [22,31]. If the strain is large enough, then 3D islands with well-defined facets and coherent with the substrate lattice begin to form. The islands that initially form are metastable but with a well-defined faceted shape [17,18]. Figure 2 shows an atomically resolved scanning tunneling microscopy

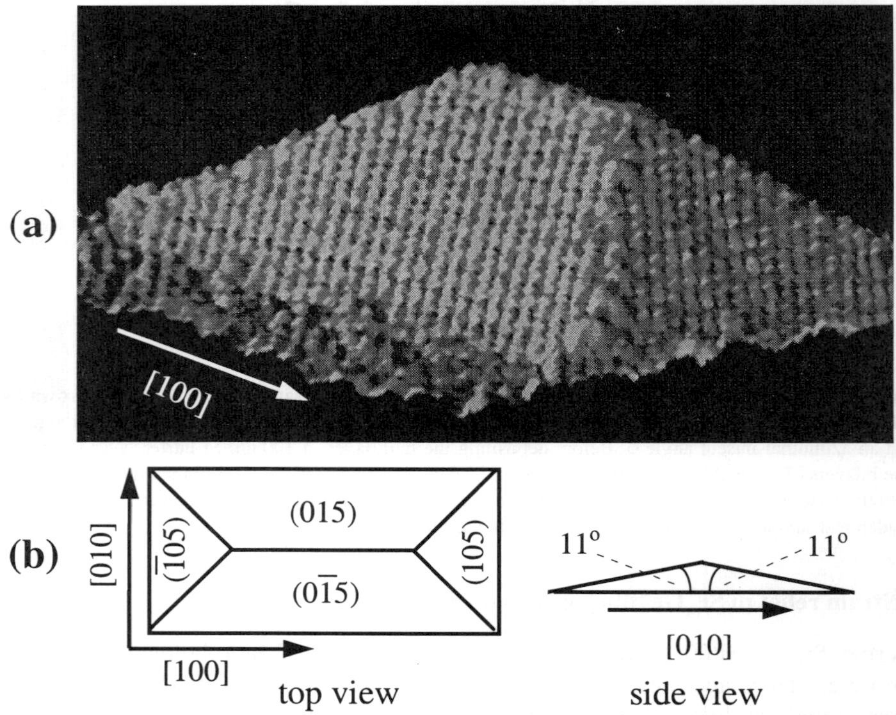

Figure 2: Three-dimensional scanning tunneling microscope image (a) and schematic top and side views (b) of a single prism-like island formed during the deposition of pure Ge on Si(001). The height of the island is ~2 nm while the base dimensions are of the order of 20 and 40 nm. The elongation of the rectangular base is either in [100] or [010] direction. All faces of the prism-like island are {105} facets. The vertical scale of the STM image is exaggerated: the islands actually are fairly flat with an 11° apex angle as it is shown in b). After Ref. 17.

image together with a schematic diagram of such an island formed during the deposition of pure Ge on Si(001) [17]. The sides of the islands have been identified as {105} facets, that are tilted by a shallow angle of 11.3°. The driving force of the formation of the {105} facets is an effective strain relief along that crystallographic direction that overcompensates the increase in surface energy [17,32]. The island bases are mostly rectangular with an elongation in either the [100] or the [010] direction rather than square. Thus the island shape is preferentially prism-like leading to their nickname "hut clusters". The island elongation, which may be related to the strain relaxation [33] causes the broad size and shape distribution in the island array. At a later stage 3D islands with steeper {113} and {518} facets arise allowing the islands to grow larger and dislocations begin to form [19]. The final stage of stress relief is coalescence of incoherent islands into a continuous layer, with the strain relieved by a "cross-hatched" pattern of dislocations [34,35]. At this stage the GeSi film is fully relaxed. This picture holds for a wide range of Ge concentrations, x. The film thickness at which a particular mechanism becomes active increases with lower x. For very low Ge concentration, the layer can grow quite thick, and eventually relax its strain with dislocations, without any island growth [36,37]. We have to notice that this picture is based on the morphological observations during the film growth, it does not involve the Ge segregation during growth [38] and the detailed strain distribution in the alloy film [39,40]. In this review we will focus on the stage of strain relief that causes the formation of {105} faceted islands. We will investigate the evolution of the island arrangement in the multilayer stack and develop a simple model that explains the observed self-organization of the islands.

4 Self-organization of {105} faceted SiGe islands in $Si_{1-x}Ge_x$/Si multilayer films

Figure 3 shows the surface morphology of a single 2.5 nm $Si_{0.25}Ge_{0.75}$ alloy film. The AFM image (Fig. 3a) exhibits a labyrinth-like array of interlocked rectangular and square based islands. Inspection of the image just by eye reveals a broad size and shape distribution of the islands. An analysis of the surface normals of the AFM image following a procedure proposed by Lutz et al. [19] is presented in Fig. 3b). It shows the polar plot of the orientations of the local surface normals that exhibits four clear peaks of equal probability at a polar angle of 11° ± 3° and azimuthal orientations along <100> corresponding to four {105} facets (exact polar angle 11.3°). Thus we can conclude first, that the rectangular and square based islands are the {105} faceted islands (see Fig. 2). Second, from the vanishing peak in the center of the polar plot (corresponding to the (001) orientation) we conclude that the surface is completely faceted at these particular growth conditions. With other words, the prism-like and pyramid-like islands touch each other, or at least approach within ~ 4 nm, the resolution owing to size and shape of the AFM tips used. The situation of a random island array with broad size distribution does not change to the better if the layer thickness is increased: Due to the formation of steeper facets the islands become larger and more compact but less defined in shape [19] (see also Fig. 8a).

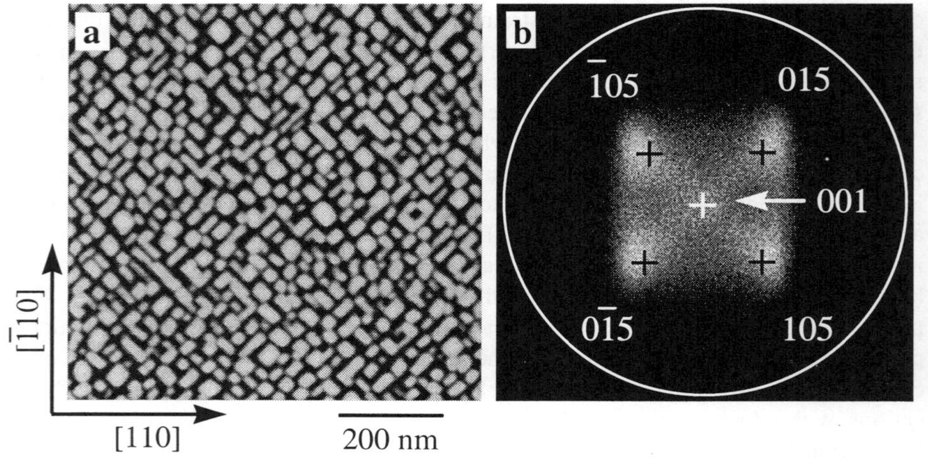

Figure 3: Surface morphology of a 2.5 nm $Si_{0.25}Ge_{0.75}$ single layer. (a) 0.8 μm x 0.8 μm AFM image. Gray-scale range is 5 nm. (b) Two-dimensional histogram of the orientations of local surface normals calculated from the image shown in (a). The histogram is displayed as a polar plot, with the white circle marking a polar angle of 30°. The gray-scale is given by the number of image points with the specified orientation. The [001] direction as well as the four relevant <105> directions are marked by crosses.

It arises now the question what would happen to the island morphology if we grow on top of these islands a silicon layer before continuing the alloy growth, and so forth. And indeed, upon growing a SiGe/Si multilayer stack the island arrangement changes dramatically. This is vividly illustrated by 3D AFM images. Figure 4 shows the surface of the 20th alloy layer of a $Si_{0.25}Ge_{0.75}$/Si multilayer (a) in comparison to the single layer (b). Although both alloy layers show the same kind of {105} faceted islands (as confirmed by the corresponding polar plots) there is a striking difference: In the multilayer stack the arbitrary arranged and shaped hut islands transform into uniformly sized and spaced pyramids. A look at the edges of the 3D images reveals that also for the 20th alloy layer the islands touch each other.

The transition from the random island arrangement to a periodic island array is demonstrated in Figure 5, that shows gray-scale AFM images of the surface of (a) a single layer of $Si_{0.25}Ge_{0.75}$ alloy film, (c) the 2nd alloy layer and (e) the 20th alloy layer of a $Si_{0.25}Ge_{0.75}$/Si multilayer film. With increasing bilayer number, two morphological changes occur. First, the islands become larger. Second and most surprising, the islands exhibit a degree of self-organization: they become more uniform in size, they adopt more or less square bases, and they arrange preferentially on chains along <100> or even on a square pattern (Fig. 5e). The self-assembly can be

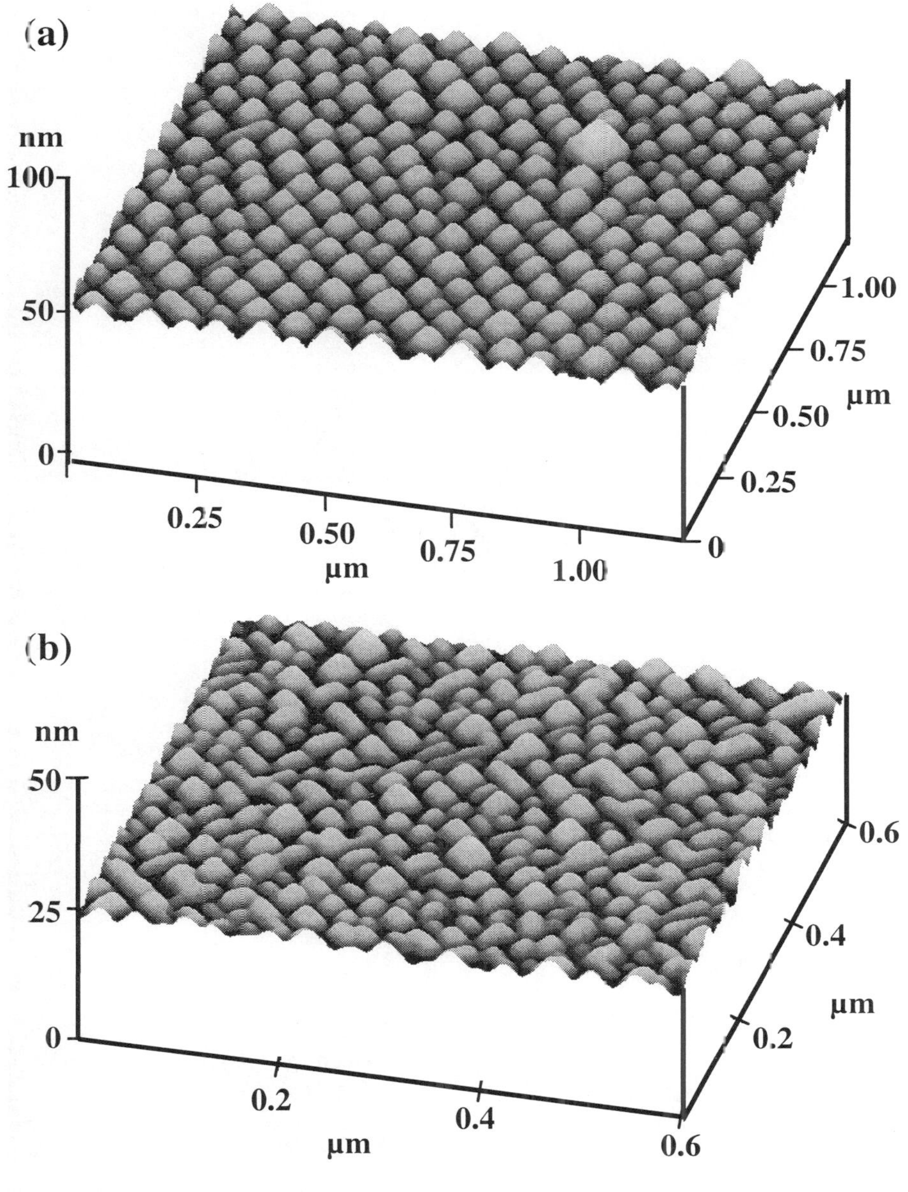

Figure 4: Three-dimensional AFM images of a single SiGe alloy layer (a) and the 20th alloy layer (b). The randomly arranged prism-like islands transform in the multilayer stack into a periodic array of larger and more uniform four-sided pyramids. Please, note the different scales.

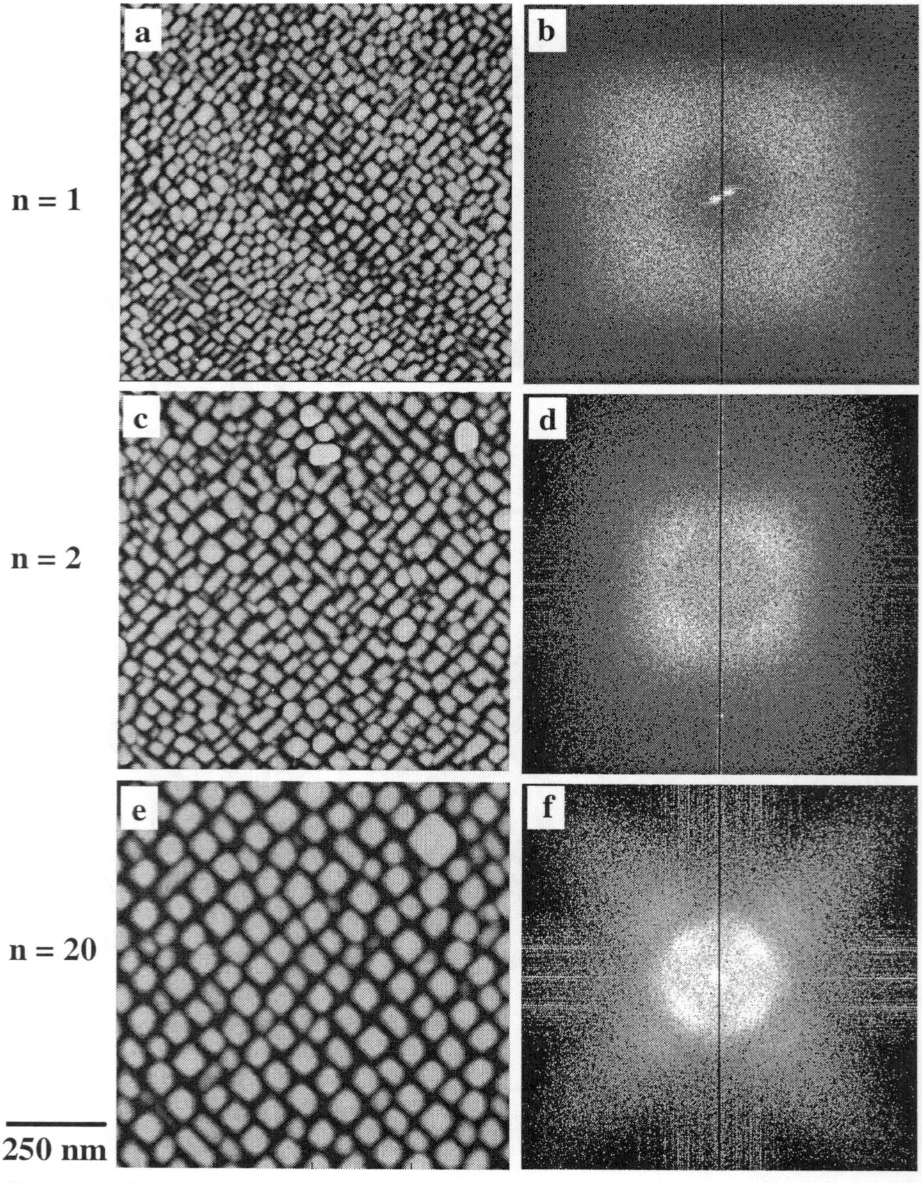

Figure 5: AFM images of SiGe alloy-terminated surfaces of SiGe/Si multilayer films and the corresponding 2D power spectra as a function of the number of alloy layers n. The horizontal direction is [110]. a), b) 2.5 nm $Si_{0.25}Ge_{0.75}$ single layer; c), d) 2.5 nm $Si_{0.25}Ge_{0.75}$/10 nm Si/2.5 nm $Si_{0.25}Ge_{0.75}$ sandwich film; e), f) 20 x (2.5 nm $Si_{0.25}Ge_{0.75}$/10 nm Si) multilayer film. The gray-scale range of the real space

images is (a) 5 nm, (b) 10 nm and (c) 20 nm, the image size is 1 μm x 1 μm. The 2D power spectra, ranging from -50 μm^{-1} to 50 μm^{-1}, have been calculated from 5 μm x 5 μm images. The faintly visible bright and dark bands extending from upper left to lower right in (a) are step-bunched ripples (see Ref. [4]) with a wavelength of ~600 nm. The split peak in the center of the power spectrum in (b) reflects the orientation (25°) and periodicity of these ripples.

quantified by analyzing the power spectral density of the surface roughness [41]. Along with each image in Fig. 5 we show the gray-scale presentation of the corresponding two-dimensional power spectrum. The spectra exhibit clear fourfold patterns, that reflect the well defined orientation of the island edges. For a single alloy layer, the pattern is broad and frame-like (Fig. 5b); with increasing bilayer number it converts into four relatively sharp peaks with corresponding higher orders (Fig. 5f). This transformation indicates an increase of uniformity in size and shape of the islands.

The predominant island spacing along <100> (and equivalently the base size, since the clusters touch each other) and the island size distribution can be obtained from the peak position and the peak width in a cut through the 2D power spectrum along that direction. Figure 6 shows power spectra for the first and the 20th alloy layer. The narrowing of the FWHM of the peak and the evolution of the second order peak for the 20th alloy layer indicate improved ordering.

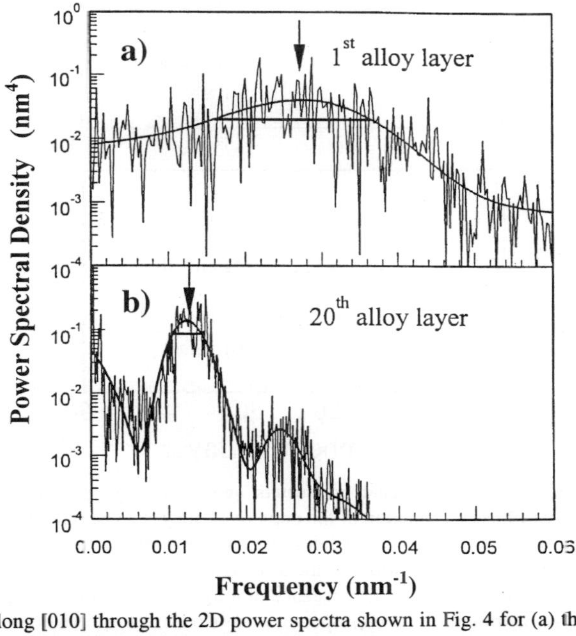

Figure 6: Cuts along [010] through the 2D power spectra shown in Fig. 4 for (a) the first alloy layer and (b) the 20th alloy layer along with spline fits. The location of the peaks (arrows) yields the mean island separation along [010]. The full width of half-maximum (horizontal lines) gives the width of the island size distribution.

Figure 7 summarizes the results for a series of bilayer numbers. It shows the evolution of (a) the mean island spacing <l> and the width of the size distribution Δl/<l>, (b) the average aspect ratio of the island bases, and (c) the rms-roughness of the multilayer surface. The mean length of the island base increases with increasing bilayer number from ~35 nm for a single layer and approaches a stationary value of ~100 nm. The width of the size distribution relative to the average island size decreases from 1.1 for the first layer to 0.3 for 20 bilayers. This narrowing of the size

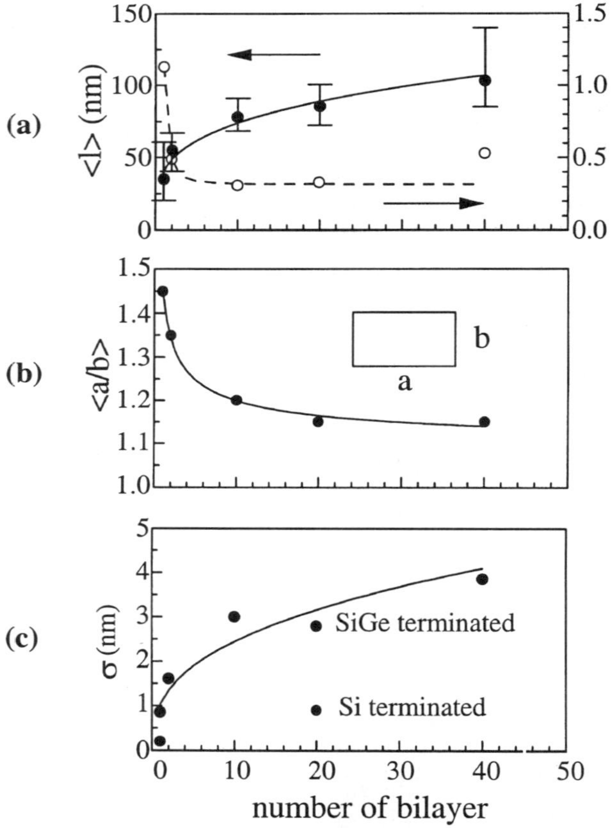

Figure 7: Evolution of characteristic island parameters and surface roughness of alloy terminated 2.5 nm $Si_{0.25}Ge_{0.75}$/10 nm Si films as a function of the number of bilayers. The lines are guides to the eye. a) Mean island spacing <l> along <100> (solid line) and relative width of the size distribution, Δl/<l> (dashed line). The error bars indicate the width of the distribution of island spacings. The deviating data point in Δl/<l> at 40 bilayers is caused by the existence of a few large, no longer coherent clusters [19] that also distorts the size distribution (see corresponding error bar in (a)). The onset of formation of these clusters can be avoided by modifying the alloy composition or alloy layer thickness. b) Average aspect ratio of the cluster bases, <a/b> (see inset). c) rms roughness calculated and averaged from 5 μm x 5 μm images. For one and 20 bilayers also the values of the Si terminated samples are shown.

distribution – mainly occurring within the first ten layers – is clear evidence for a self-organization process. The evolution of the average aspect ratio of the island bases (Fig. 7b), decreasing from 1.5 for a single alloy layer to 1.15 for 40 bilayers, indicates a change in crystal shape from prism-like hut to approximately a four-sided pyramid. The island height changes from ~3 nm to ~10 nm. Using the measured dimensions, we can estimate the volume of the islands. The mean volume of islands on a single layer is about 1300 nm^3, consisting of 3.2 x 10^4 atoms, but with a wide distribution. At the limit of 40 bilayers for this set of parameters, the volume of the predominant-size island is about 3.6 x 10^4 nm^3, containing about 9 x 10^5 atoms. With increasing island height of course also the vertical roughness of the alloy layers increases as it can be seen in Fig. 7c). Fig. 7c also presents some roughness values for the Si terminated multilayer films. It is known e.g. from cross-sectional transmission electron microscopy (TEM) that pure Si deposited on top of the alloy layer smoothes out the island morphology [42-44]. Here the 10 nm Si spacer layer was sufficient to smooth the growth front to nearly what it was for the substrate.

Because the Si spacer layers do not leave any morphological indications of the buried SiGe islands one might ask whether the Si spacers are needed at all for the self-organization, and – if they are necessary – how their thickness acts on the island arrangement in the alloy layers. Figure 8 shows the surface morphologies of two reference samples that should be compared with the island array on top of the

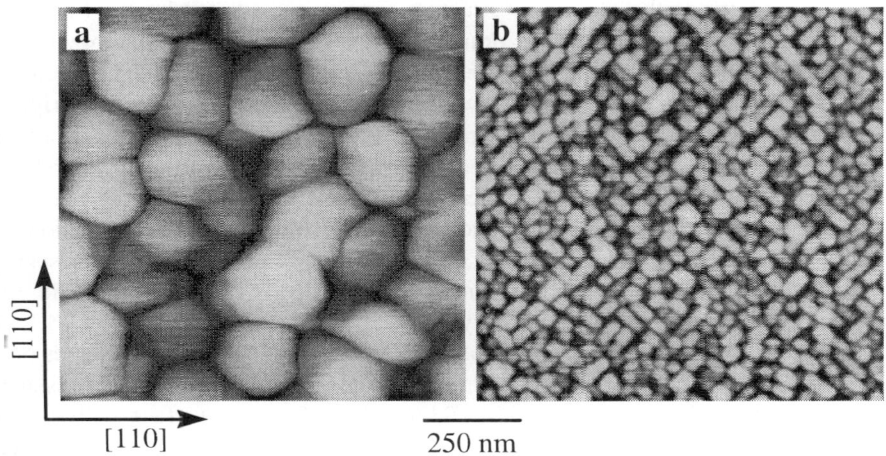

Figure 8: Influence of the Si spacer layers on the self-organization. a) AFM image of a 50 nm $Si_{0.25}Ge_{0.75}$ film, corresponding to 20 x 2.5 nm $Si_{0.25}Ge_{0.75}$ without the Si spacers. The islands with an average lateral extension of 300 nm are randomly arranged and shaped. They are showing {105}, {113}, and {518} facets and (001) terraces on top. These islands are no longer coherently strained. b) AFM image of a 20 x (2.5 nm $Si_{0.25}Ge_{0.75}$/30 nm Si) multilayer film exhibiting an average island size of 50 nm. The comparison with the 20 x (2.5 nm $Si_{0.25}Ge_{0.75}$/10 nm Si) multilayer film (see Fig. 5e) shows that the absence of the Si spacer layers does not yield any self-organization at all, whereas 30 nm Si spacers result in a poor self-organization effect. Image sizes: 1 µm x 1 µm, gray-scale range: (a) 25 nm, (b) 10 nm.

20 bilayer film (Fig. 5e) to judge the influence of the Si spacer layers on the observed self-organization of the islands. Without the interspersal of Si spacer layers, the self-organization does not occur at all (Fig. 8a): a 50 nm thick $Si_{0.25}Ge_{0.75}$ film (corresponding in thickness to 20 x 2.5 nm $Si_{0.25}Ge_{0.75}$ without the Si spacer layers) exhibits randomly arranged, irregularly shaped large islands of about 300 nm lateral size. These islands are still faceted. However, in addition to the {105} facets they are bounded by the steeper {113} and {518} facets [19] and have a (001) face on top. Consequently, the Si spacer layers are necessary for the self-organization of the islands. Fig. 8b) shows the surface of a multilayer film where the Si spacer thickness has been increased to 30 nm at conditions otherwise identical to those of the 20 bilayer film Here we observe a poorly organized array of 50 nm large islands that is very similar to that found for the second bilayer in the case of a 10 nm Si spacer layer (see. Fig. 5c). Thus the thick Si spacer layers result in a much less pronounced self-organization, in agreement with TEM results for multilayers of III-V semiconductors [45]. Hence the Si spacer layers act as a control on the self-organization.

5 Model of stress-induced self-organization of 3D islands in a multilayer film

How can we explain the observed self-organization? As we have seen above, it is clearly a property of the multilayer structure, i.e., it is mediated by the Si spacer layers. The spacer layer may act in principle either to copy the morphology of the surface on which it is deposited or to mediate the strain. The former can be ruled out because the deposition of a 10 nm Si layer smoothes the growth front to the original substrate roughness, as we have pointed out. However, x-ray diffraction measurements on the same multilayers reveal a partially vertically correlated roughness in the multilayer stack [22]. Therefore we have to assume that the Si spacer layer acts to create a strain distribution that is favorable for the formation of regularly spaced and sized islands. In the following we will examine the spatial arrangement of 3D islands with a simple and relatively generic model in the framework of continuum elasticity theory [21].

Consider a one-dimensional array of small coherently strained SiGe islands on a flat Si surface. For simplicity the islands are treated as small spheres. These islands are than buried under a thickness L of additional Si. The new surface will also be flat, but with local strain due to the buried islands. The strain is the only factor biasing nucleation at this surface. Following Maradudin and Wallis [46] the surface strain ε due to an island buried at a depth L and lateral position x = 0 is

$$\varepsilon(x) = C\,(x^2 + L^2)^{-3/2}\,[1 - 3L^2 / (x^2 + L^2)]. \tag{1}$$

The coefficient C is proportional to the volume of the buried island and the misfit, and also involves the elastic constants [46]. The resulting strain is shown in

Figure 9a,b). The favored nucleation position is at the minimum of ε, i.e. directly above the buried island. Nucleation is disfavored at lateral offsets larger than √2 L (as can be seen in Fig. 9b), with the strain dropping smoothly to zero at large distances. This preferential formation of a new island in the multilayer stack just above the buried one is sketched in Fig. 9c). Such piling up of islands has indeed been found by cross-sectional TEM measurements for the SiGe/Si system [43,44,47] as well as for III-V systems [48,49,45,50]. However, this behaviour alone would just result in copying the island morphology from alloy layer to alloy layer. Self-organization comes into play if we consider the lateral interaction of the strain fields of adjacent buried islands. This is demonstrated in Figure 10 for an illustrative arrangement of buried islands. The strain ε at the surface of the Si spacer layer is calculated by summing Eq. (1) over all the buried islands. In this deterministic model islands are assumed to nucleate at each minimum of ε, with volumes proportional to the areas of the corresponding Voronoï polygons (the region closer to that island than to any other). Besides replication of the lateral island positions (labeled A in Fig. 10) other mechanisms will occur. Extremely closely spaced buried islands cause only one minimum, resulting in a "thinning out" of closely spaced islands (labeled B). The shallow minimum of ε between widely spaced islands (see Fig. 9b), or random nucleation, could also cause filling in of gaps in the island arrangement (labeled C), although this is more speculative. As a result of processes B and C the nucleation positions in Fig. 10 are much more uniformly spaced than the initial island locations.

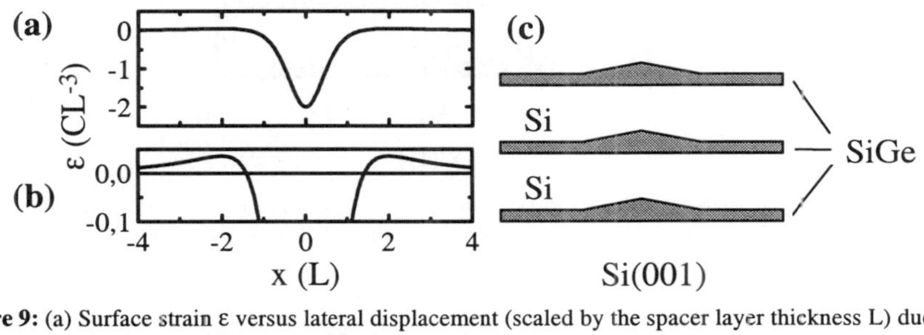

Figure 9: (a) Surface strain ε versus lateral displacement (scaled by the spacer layer thickness L) due to a strained island buried at a depth L and at lateral position x = 0. There is a deep minimum just on top of the buried island as well as an area with a positive strain (better visible in (b) at an enhanced ε scale). The minimum causes the preferential formation of an island in a subsequent alloy layer directly on top of a buried island as it is shown in the cross-sectional sketch (c).

In Figure 11 we present simulation results (for details see Ref. 21) for closely spaced islands. We start with an arbitrary arrangement of randomly sized islands in the first alloy layer. For successive layers, the island separation and size becomes progressively more uniform, because smaller and misaligned clusters are "thinned out" by the strain mediation through the Si spacer. Interestingly, the lateral island spacing approaches a finite value, that is about 3.5 L. Thus, the spacer layer acts as a

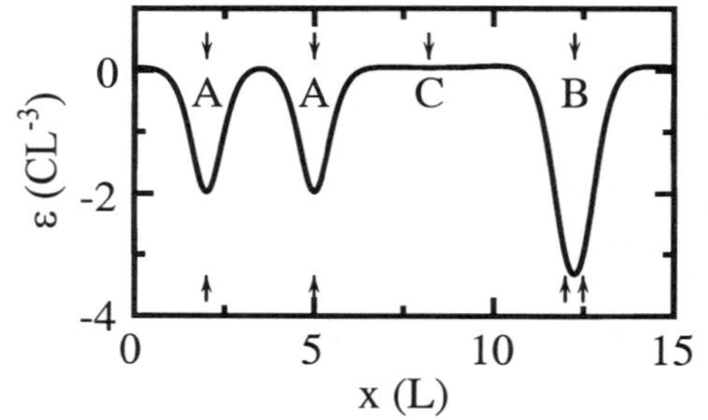

Figure 10: Surface strain ε versus lateral displacement (in units of the spacer layer thickness L) due to four strained islands. Arrows at bottom indicate lateral position of buried islands. Arrows on top indicate minima in ε, i.e., favored positions for subsequent island formation. Properly spaced islands replicate the lateral positions of the underlying ones (labeled A). In addition, there is a "thinning out" of closely spaced islands (B) and possibly even a filling in of gaps in the island arrangement (C).

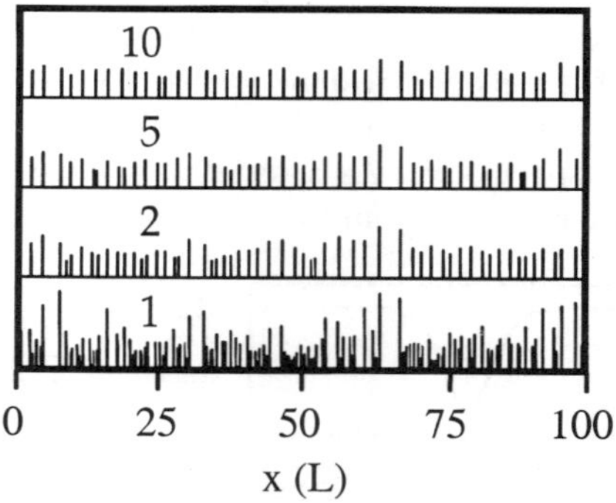

Figure 11: Calculated island positions and sizes in selected successive layers starting from a random arrangement of closely spaced islands. Layer numbers are indicated. Heights of vertical lines represent island volumes, relative to average for that layer.

band-pass filter for the morphology. Although it is very simple, the model qualitatively explains the main experimental results: the narrowing of the island size distribution and the increase of the average island size to a saturation value with increasing layer number. A more quantitative agreement is expected, if considering realistic island shapes.

It is also interesting to notice in Fig. 11, that the size variations among islands in the self-organized upper layers are not random. Adjacent islands tend to have especially similar sizes, with a characteristic length over which such correlations are lost. The band-pass filter analogy implies that short range variations are easier to filter out than long-period modulations. The AFM image of the 40 bilayer film surface presented in Figure 12 shows very similar fluctuations in cluster size. We observe a roughness modulation with a lateral correlation length of about 1 μm that is superimposed by the islands. Interestingly, the higher areas show a smaller island density but larger island size and vice versa. So, indeed larger islands and smaller islands are grouped together. However, in addition to the well ordered {105} faceted islands we observe for 40 bilayers already a considerable fraction of larger islands forming the steeper {113} facets. With other words, the next stage of strain relief is reached. The {113} facetted islands grow much larger in base size and height than the {105} faceted ones [19]. Due to this fact and their different base shape the formation of {113} faceted islands causes a reduction in size uniformity (see Fig. 8a). The randomly distributed larger islands are also assumed to cause the large scale roughness correlation visible in Fig. 12, because the 10 nm thick Si spacer layer is no longer able to smooth the surface of the alloy layer completely. A way to proceed the self-organization of the islands beyond 20 bilayers would be to reduce the Ge concentration with increasing bilayer number. By such kind of dynamic multilayer growth one should be able to delay the next stage of strain relief.

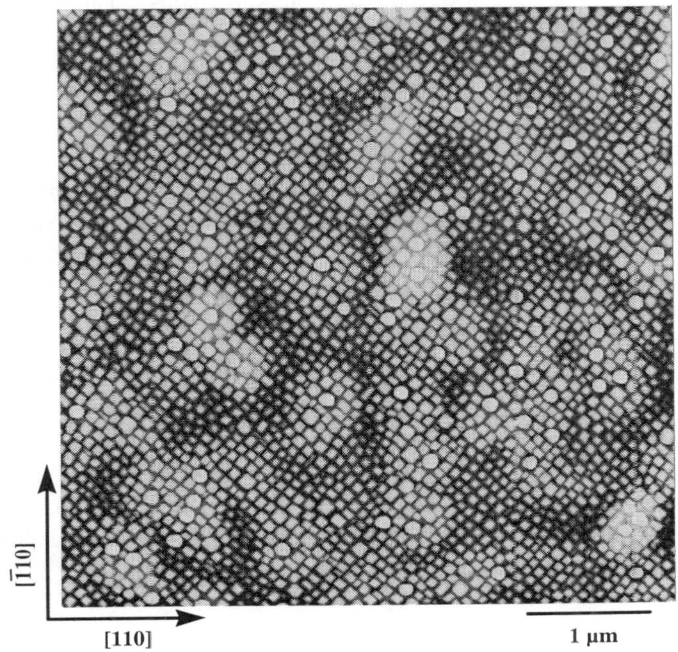

Figure 12: 5 μm x 5 μm AFM image of a 40 x (2.5 nm $Si_{0.25}Ge_{0.75}$/10 nm Si) multilayer film. Gray-scale range: 20 nm. The surface is still characterized by the self-organized pyramids. However, a few larger islands bounded by the steeper {113} and {518} facets have formed. Besides the islands, there is an additional surface roughness modulation with a lateral scale of ~1 μm visible. The island densities on the higher areas (bright) are smaller than those on the lower areas (dark).

6 Conclusion

In conclusion, we have demonstrated a way to create regular nanoscale structures in SiGe/Si multilayer films through the self-organization of regular, coherent, pyramid shaped 3D SiGe islands that form to relieve stress. This self-organization occurs through the strain mediation by the Si spacer layers and results finally in a 3D array of uniform islands. The self-organization, also present in similar multilayer systems, is understood in a simple model of interacting surface strain due to buried islands, where the spacer layer act as a morphological band pass filter. We expect that one can produce a variety of similar and other nanoscale structural patterns by exploiting knowledge of stress-induced self-organization mechanisms in single alloy and multilayer films. For instance, Pennycock et al. have shown that alloy growth at reduced temperatures followed by subsequent annealing results in square based SiGe islands with an improved size uniformity compared to the hut clusters [51]. Further, the self-organization can be influenced by substrate miscut direction and magnitude. In this case the interplay of stress-induced step-bunching and subsequent island formation leads, e.g., to alignment of the hut clusters in chains [4]. This concept also works for III-V semiconductor systems [52,53].

Finally we would like to point out that besides their beauty and besides revealing the underling physical principals these self-organized island arrays offer quite intriguing prospective applications. First, the uniformity of the islands might become sufficient to serve as quantum dots and thus opening the door to SiGe based optoelectronics. Second, the strain induced self-organization of faceted surface structures may become competitive to lithography based methods to produce large-scale arrays of periodic nanostructures as they might be used e.g. for ultra-high density storage devices.

Acknowledgements

This work was supported by NSF, Grant No. DMR 92-01856. We are grateful to L. J. Peticolas and J. C. Bean for growing the SiGe/Si films. We acknowledge helpful discussions with Y. H. Phang, D. E. Savage, and M. B. Webb, as well as technical assistance by K.-C. Liu and P.-F. Lam. We are grateful to G. Bauer for bringing a reference to our attention. C. T. acknowledges fellowship support from the German Academic Exchange Service.

References:

1. H. Haken, *Synergetics. An Introduction* (third edition) (Springer, Berlin 1983).
2. I. N. Stranski, L. Krastanov, Sitzungsber. Akad. Wien, Math. Nat. Kl. IIb **146**, 797 (1937).
3. M. Shinohara, N. Inoue, Appl. Phys. Lett. **66**, 1936 (1995).

4. C. Teichert, Y. H. Phang, L. J. Peticolas, J. C. Bean, M. G. Lagally in *Surface Diffusion: Atomistic and Collective Processes* ed. M. C. Tringides, (Plenum Press, New York 1997) p. 297.
5. J. Tersoff, Y. H. Phang, Z. Zhang, M. G. Lagally, Phys. Rev. Lett. **75**, 2730 (1995).
6. D. Leonard, M. Krishnamurthy, C. M. Reaves, S. P. Denbaars, P. M. Petroff, Appl. Phys. Lett. **63**, 3203 (1993).
7. R. Nötzel, T. Fukui, H. Hasegawa, J. Temmyo, T. Tamamura, Appl. Phys. Lett. **65**, 2854 (1994).
8. D. Bimberg, M. Grundmann, N. N. Ledentsov, S. S. Ruvimov, P. Werner, U. Richter, J. Heydenreich, V. M. Ustinov, P. S. Kop'ev, Zh. I. Alferov, Thin Solid Films **267**, 32 (1995)
9. J. M. Moison, L. Leprince, F. Barthe, F. Houzay, N. Lebouché, J. M. Gérard, J. Y. Marzin, Appl. Surf. Sci. **92**, 525 (1996).
10. S. Tanaka, S. Iwai, Y. Aoyagi, Appl. Phys. Lett. **69**, 4096 (1996).
11. S. H. Xin, P. D.Wang, A. Yin, C. Kim, M. Dobrowolska, J. L. Merz, J. K. Furdyna, Appl. Phys. Lett. **69**, 3884 (1996).
12. C. Priester, M. Lannoo, Phys. Rev. Lett. **64**, 1943 (1990).
13. V. A. Shchukin, N. N. Ledentsov, P. S. Kop'ev, D. Bimberg, Phys. Rev. Lett. **75**, 2968 (1995).
14. T. T. Ngo, P. M. Petroff, H. Sakaki, J. L. Merz, Phys. Rev. **B 53**, 9618 (1996).
15. H. T. Dobbs, D. D. Vvedensky, A. Zangwill, J. Johansson, N. Carlsson, W. Seifert, Phys. Rev. Lett. **79**, 897 (1997).
16. J. Eaglesham, M. Cerullo, Phys. Rev. Lett. **75**, 93 (1995).
17. Y.-W. Mo, D. E. Savage, B. S. Swartzentruber, M. G. Lagally, Phys. Rev. Lett. **65**, 1020 (1990).
18. A. J. Pidduck, D. J. Robbins, A. G. Cullis, W. Y. Leong, A. M. Pitt, Thin Solid Films **222**, 78 (1992).
19. M. A. Lutz, R. M. Feenstra, P. M. Mooney, J. Tersoff, J. O. Chu, Surf. Sci. **316**, L1075 (1994).
20. C. Teichert, M. G. Lagally, L. J. Peticolas, J. C. Bean, J. Tersoff, Phys. Rev. B **53**, 16334 (1996).
21. J. Tersoff, C. Teichert, M. G. Lagally, Phys. Rev. Lett. **76**, 1675 (1996).
22. Y. H. Phang, C. Teichert, M. G. Lagally, L. J. Peticolas, J. C. Bean, E. Kasper, Phys. Rev. **B 50**, 14435 (1994).
23. Y.-W. Mo, M. G. Lagally, J. Cryst. Growth **111**, 876 (1991).
24. U. Köhler, O. Jusko, B. Müller, M. Horn-von Hoegen, M. Pook, Ultramicroscopy **42/44**, 832 (1992).
25. R. Butz, S. Kampers, Appl. Phys. Lett. **61**, 1307 (1992).
26. X. Chen, F. Wu, Z. Zhang, M. G. Lagally, Phys. Rev. Lett. **73**, 850 (1994).
27. J. Tersoff, Phys. Rev. **B 45**, 8833 (1992).
28. F. Wu, X. Chen, Z. Zhang, M. G. Lagally, Phys. Rev. Lett. **74**, 574 (1995).

29. F. Wu, M. G. Lagally, Phys. Rev. Lett. **75**, 2534 (1995).
30. F. Liu, M. G. Lagally, Phys. Rev. Lett. **76**, 3156 (1996).
31. R. L. Headrick, J. M. Baribeau, J. Vac. Sci. Technol. **B 11**, 1514 (1993).
32. F. Tuinstra, P. M. L. O. Scholte, W. I. Rijnders, A. J. van den Berg, Surf. Sci. **317**, 58 (1994).
33. J. Tersoff, R. M. Tromp, Phys. Rev. Lett. **70**, 2782 (1993).
34. P. M. Mooney, F. K. LeGoues, J. Tersoff, J. O. Chu, J. Appl. Phys. **75**, 3968 (1994).
35. M. A. Lutz, R. M. Feenstra, F. K. LeGoues, P. M. Mooney, J. O. Chu, Appl. Phys. Lett. **66**, 724 (1995).
36. F. K. LeGoues, P. M. Mooney, J. Tersoff, Phys. Rev. Lett. **71**, 396 (1993).
37. J. Tersoff, F. K. LeGoues, Phys. Rev. Lett. **72**, 3570 (1994).
38. G. G. Jernigan, P. E. Thompson, C. L. Silvestre, Surf. Sci. **380**, 417 (1997).
39. S. Christiansen, M. Albrecht. H. P. Strunk, H. J. Maier, Appl. Phys. Lett. **64**, 3617 (1994).
40. A. J. Steinfort, P. M. L. O. Scholte, A. Ettema, F. Tuinstra, M. Nielsen, E. Landemark, D.-M. Smilgies, R. Feidenhans'l, G. Falfenberg, L. Seehofer, R. L. Johnson, Surf. Sci. **317**, 58 (1994).
41. See, e.g., J. M. Elson, H. E. Bennett, J. M. Bennett, Appl. Opt. Opt. Eng. **69**, 191 (1979); H.-N. Yang, G.-C. Wang, T.-M. Lu, *Diffraction from Rough Surfaces and Dynamic Growth Fronts* (World Scientific, Singapore 1993), p. 64; W. M. Tong, R. S. Williams, Ann. Rev. Phys. Chem. 45, **401** (1994).
42. H. Sunamura, Y. Shiraki, S. Fukatsu, Appl. Phys. Lett. **66**, 953 (1995).
43. T. S. Kuan, S. S. Iyer, Appl. Phys. Lett. **59**, 2242 (1991).
44. L. Vescan, W. Jäger, C. Dieker, K. Schmidt, A. Hartmann, H. Lüth, Mat. Res. Soc. Symp. Proc. **263**, 23 (1992).
45. Q. Xie, A. Madhukar, P. Chen, N. P. Kobayashi, Phys. Rev. Lett. **75**, 2542 (1995).
46. A. A. Maradudin, R. F. Wallis, Surf. Sci. **91**, 423, (1980).
47. P. Schittenhelm, G. Abstreiter, A. Darhuber, G. Bauer, P. Werner, A. Kosogov, Thin Solid Films **294**, 291 (1997).
48. L. Goldstein, F. Glas, J. Y. Marzin, M. N. Charasse, G. Le Roux, Appl. Phys. Lett. **47**, 1099 (1985).
49. J. Y. Yao, T. G. Andersson, G. L. Dunlop, J. Appl. Phys. **69**, 2224 (1991).
50. G. S. Solomon, J. A. Trezza, A. F. Marshal, J. S. Harris, Jr., J. Vac. Sci. Technol. **B 14**, 2208 (1996).
51. K. M. Chen, D. E. Jesson, S. J. Pennycook, T. Thundat, R. J. Warmack, J. Vac. Sci. Technol. **B 14**, 2199 (1996); S. Pennycook, D. E. Jesson, this book.
52. D. Leonard, K. Pond, P. M. Petroff, Phys. Rev. **B 50**, 11687 (1994); P. M. Petroff, this book.
53. M. Kitamura, M. Nishioka, J. Ostinovo, Y. Arakawa, Appl. Phys. Lett. **66**, 3663 (1995).

MORPHOLOGICAL EVOLUTION OF STRAINED SEMICONDUCTOR FILMS

David E. Jesson
Solid State Division, Oak Ridge National Laboratory, Oak Ridge, TN 37831, USA

1 Introduction

Strain induced roughening of semiconductor films is a major factor limiting the fabrication of coherently strained device structures. It is therefore important to understand the initial stages of the strain induced roughening transition to avoid or suppress three-dimensional (3D) growth. On the other hand, the transition is of benefit for the self assembly of quantum dots where the interest is to create regularly spaced 3D islands of uniform size[1-5]. In both cases it is clearly important to identify and understand the kinetic pathways to island formation. From a more basic perspective, the transition enables the interplay between elastic stresses and surface energy to be studied in a controlled environment. This should provide new insight into related phenomena which occur in other areas of materials science and physical metallurgy.

2 Surface Rippling

Often, the strain induced roughening of semiconductor films is manifested by the appearance of a continuous ripple morphology as shown in Fig 1. This is generally interpreted in terms of the Asaro-Tiller-Grinfeld (ATG) instability [6,7] illustrated in Fig. 2. The Si_xGe_{1-x} film in Fig. 1 is under compressive stress so that an undulation in the surface allows lattice planes, which are compressed in the bulk, to relax towards the ripple peaks (Fig. 2). This lowers the elastic energy stored in the film but also increases the surface energy relative to the planar layer. A balance between the elastic energy released and the associated cost in surface energy gives a condition for the minimum wavelength λ_c for which the undulation is stable

$$\lambda_c = \frac{2\mu\pi\gamma}{(1-\upsilon)\sigma^2} \qquad (1)$$

Figure 1: Crossectional TEM image showing a typical rippled surface of a Si_xGe_{1-x} film grown on Si (100).

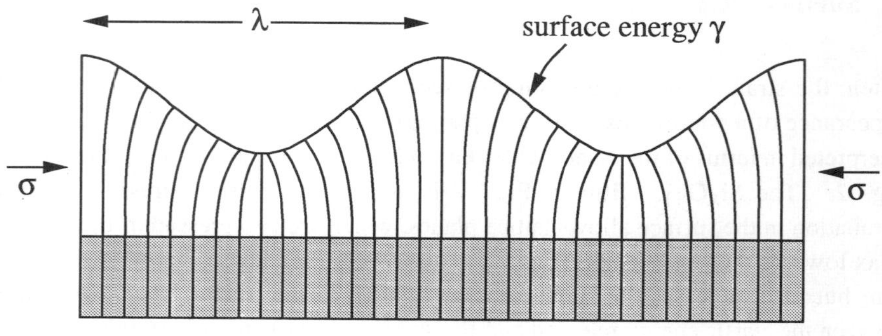

Figure 2: Crossectional schematic illustrating the physical basis of the ATG instability.

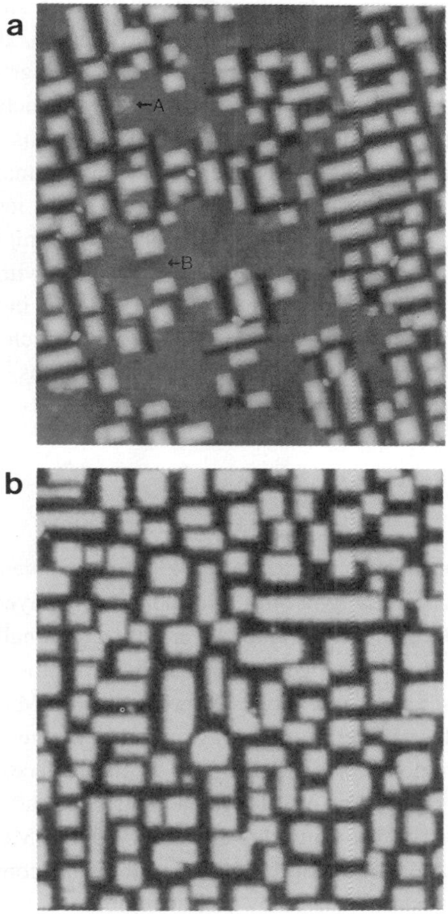

Figure 3: Atomic-force-microscopy image of a $Si_{0.5}Ge_{0.5}$ alloy layer grown on Si(001) and annealed for 5 min at 590°C. This image reveals cooperative nucleation of surface ripple domains where islands and pits nucleate adjacent to each other as indicated. Eventually the domains merge to form the continuous ripple array shown in (b). The scanned area in both images is $1{\times}1\ \mu m^2$, and the typical height (depth) of islands (pits) is 2.5 nm in (a) and 5.0 nm in (b). In both images, the islands and pits are bounded by {501} facets.

where γ is the surface energy density, σ is the misfit stress, μ is the shear modulus and ν Poisson's ratio of the layer. Beyond λ_c, the film will roughen naturally via surface diffusion. The ripple forms via gradients in the surface chemical potential such that atoms at the valley which experience a stress concentration will prefer to migrate to more relaxed regions at the peaks. In this model there is no energy barrier to roughening except for mass transport along the surface. The planar surface is therefore unstable to strain induced roughening and one would anticipate a continuous evolution in ripple amplitude across extended regions of the film. At first sight, this is in agreement with typical experimental observations such as those contained in Fig. 1. However, in the following section we demonstrate that rippling occurs by a cooperative nucleation mechanism [8]. This kinetic pathway for morphological evolution turns out to be useful for manipulating structures on the mesoscopic scale [9,10].

3 Cooperative Nucleation

To investigate the mechanism of surface ripple formation we have utilized a simple annealing experiment. Initially, a 5nm thick $Si_{0.5}Ge_{0.5}$ layer is grown on Si(100) at 400°C to produce a coherently strained layer with a nominally planar surface. The temperature of the film is then gently increased up to the anneal temperature of 590°C. Figure 3 is an Atomic Force Microscopy (AFM) image capturing the formation of the surface ripple morphology. In this image, the bright regions are islands and the dark regions correspond to pits in the surface. The islands and pits are bounded by {501} facets and are elongated along elastically soft <100> directions which is reminiscent of the hut clusters observed by Mo et al. [11]. Most importantly, the image shows surface ripple domains consisting of alternating islands and pits which are separated by planar regions of the strained film. This indicates that the ripple is forming by domain growth and the propagation of individual domains involves the sequential nucleation of islands and pits [8].

The physical origin of this cooperative nucleation mechanism reflects the fact that above a critical size, an isolated island or pit will lower the elastic energy of the strained layer. However, to nucleate an island or pit it is first necessary to overcome an energy barrier [12]. Since the elastic interaction between an island and a pit is negative (i.e. attractive) it is conceivable that cooperative effects may play an important role by lowering the activation barrier for domain growth. The essential physics of cooperative nucleation is captured in the following simple 2D model for surface ripple propagation. In Fig. 4 we represent a ripple domain in terms of rectangular islands and pits of fixed width L and height A. For the domain to propagate to the right as indicated by the arrow in Fig. 4(a), the shaded island of increasing height h must nucleate adjacent to the existing domain in the position shown in Fig. 4(b). The energy barrier to nucleate the island can be estimated by noting that the increase in surface energy per unit length is simply the strain energy per unit area γ_f associated with the vertical island facets multiplied by the height of the islands or depth of the pits. This is compensated by the elastic energy released per unit length which is found by summing over the elastic monopole interactions between vertical facets. The interaction energy between two facets a distance L apart is given by $C'_m \sigma\, h_1 h_2 \ln(L/a)$ where h_1 and h_2 are the facet heights, σ is the misfit stress and C'_m is an elastic constant. The microscopic cutoff length a is usually taken as a lattice constant. The interaction is positive (attractive) if the facets are the same sense (e.g., both down facets) or negative (repulsive) for opposite sense facets (e.g., up and down facets). The total energy is the sum of these two contributions and if we plot this as a function of h we can determine the energy barrier to form the island. The energy change associated with the nucleation of a single island (N=1) is shown, e.g., in Fig. 5.

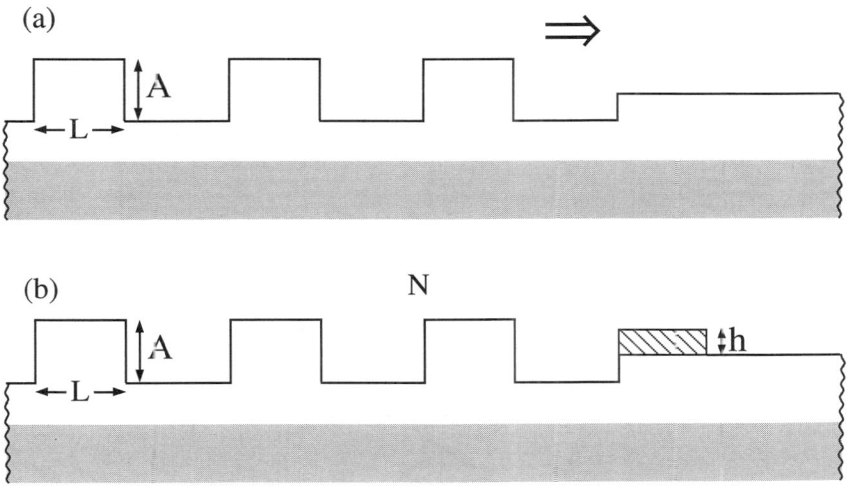

Figure 4: Atomic-force-microscopy image of a $Si_{0.5}Ge_{0.5}$ alloy layer grown on Si(001) and annealed for 5 min at 590°C. This image reveals cooperative nucleation of surface ripple domains where islands and pits nucleate adjacent to each other as indicated. Eventually the domains merge to form the continuous ripple array shown in (b). The scanned area in both images is 1×1 μm^2, and the typical height (depth) of islands (pits) is 2.5 nm in (a) and 5.0 nm in (b). In both images, the islands and pits are bounded by {501} facets.

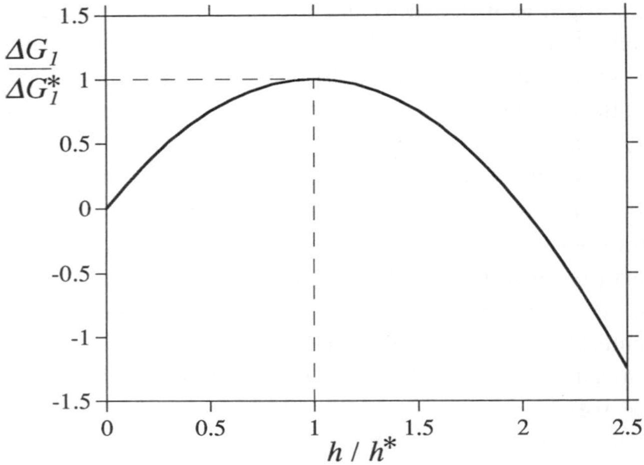

Figure 5 Energy ΔG_1 for the nucleation of a single island (N=1) as a function of height h. To attain a stable geometry the island must first overcome the energy barrier ΔG_1^* which occurs at height h^*.

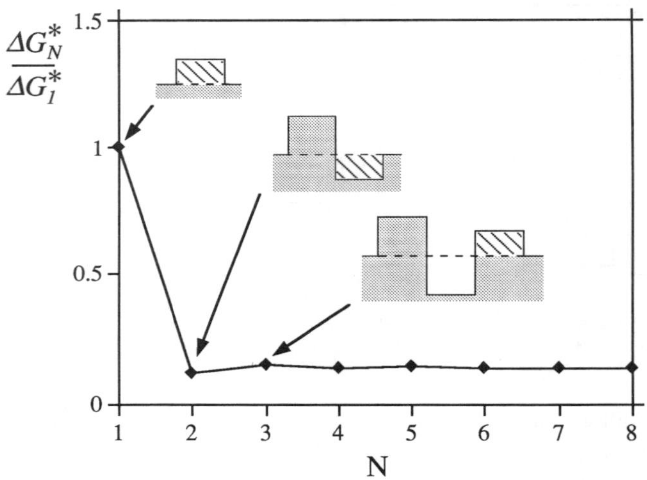

Figure 6. Energy barrier for the growth of a domain as a function of the total number of islands and pits N. We assume L=45 nm, A=2.5 nm, a=2 Å. γ_f=1.0 Jm^{-2}, with a shear modulus μ=40 GNm^{-2} and Poisson's ratio ν=0.3 in the elastic constant $C_m=(1-\nu)/\pi\mu$. The misfit ε=0.02.

The next event in the domain propagation is the nucleation of a pit adjacent to the freshly nucleated island followed by the nucleation of another island next to the pit and so on. We can evaluate the energy barriers for each nucleation event as a function of domain size N where N is the total number of islands and pits in the domain. This is plotted in Fig 6 where the first point (N=1) corresponds to the activation barrier to nucleate an isolated island. However, the energy barrier falls on nucleating a pit next to the island because of the elastic attraction. The barrier then remains approximately constant as further islands and pits are added to the domain. This simple model is therefore very suggestive that an isolated island or pit can first act as a seed for domain growth and the domain then grows rapidly through a cooperative nucleation mechanism involving the sequential nucleation of islands and pits. This is rather different from the conventional view of ripple formation as an instability and emphasizes the importance of nucleation phenomena in the evolution of strained layer morphology.

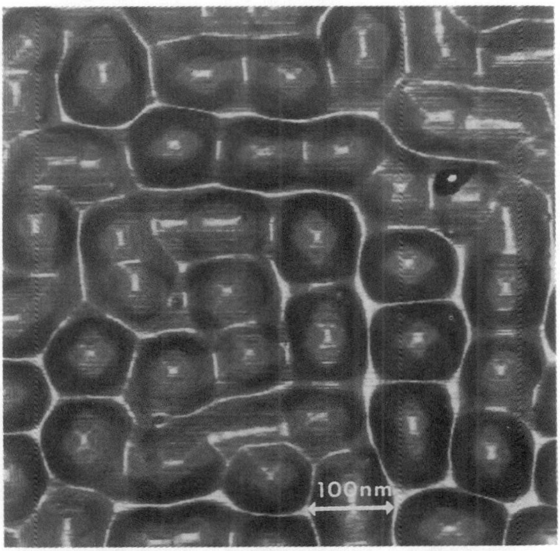

Figure 7. Atomic-force-microscopy image of ridge splitting during annealing. The elongated ridges are fissioning into smaller islands.

4 Fissioning of Elongated Ridges

Eventually, the domain structures merge leaving a continuous ripple array consisting of elongated islands and pits. At this stage, annealing the elongated ridges at a slightly greater temperature can give rise to island fissioning, as indicated in Fig. 7. The splitting process is reminiscent of a Rayleigh instability and can be explained by the destabilizing repulsive elastic interaction between the closely packed ridges. By forming separate islands, the system can relax along orthogonal [100] directions. This is optimized further by the increased aspect ratio of the islands which is achieved at the expense of low energy {501} facets. This might explain the 'dome-like' shape of the islands during the transition which corresponds to the appearance of higher order facets. Note that fissioning can be thought of as inverse Ostwald ripening since the mean particle size decreases with time.

The resulting morphology shown in Fig. 8 consists of a high density of self assembled islands with good size uniformity. They are also self-organized to some extent due to ridge fissioning, which tends to spatially align islands along [100] directions. In view of the interest in spontaneously generating quantum dots through self organization, cooperative nucleation phenomena coupled with island fissioning might provide a particularly attractive kinetic pathway for this purpose.

Figure 8. Bubble raft of quantum dots resulting from cooperative nucleation and island fissioning. The image scan is $5 \times 5 \mu m^2$.

5 Dislocations and Surface Morphology

A network of stress concentrations weaves in between the array of islands and eventually causes dislocations to nucleate and propagate in the film. In the AFM image in Fig. 9, a misfit dislocation has nucleated and propagated, regimenting the surface island above it. This is a useful feature because it readily facilitates the study of dislocation nucleation and propagation by AFM. A most interesting feature of the image is the observation that a misfit dislocation can reproduce itself through interaction with the surface morphology [13]. This can be understood from the Transmission Electron Microscope (TEM) images shown in Fig. 10. Large stress concentrations, manifested by the bright lines in the weak-beam image, develop between the islands which decorate the dislocation line. This is because the Burgers vector of the edge dislocation lies orthogonal to its line direction so that significant stress exists along the line. Such stresses offer preferred nucleation sites for dislocation nucleation [13] although at present the detailed nucleation mechanisms are unclear. Dislocations could in principle occur by a mechanism of direct 'crack-blunting' [13] or by a more gentle process of vacancy condensation at the stress concentration, followed by the reaction of dislocations to produce the observed edge dislocation [14]. In any event, it is clear from Figs 9 and 10 that the interaction between surface morphology and dislocation nucleation can certainly play an important role in dislocation multiplication. (See also [15]).

6 Dynamical Roughening of Strained Layers

Our discussion of surface ripple formation has been based entirely on nucleation theory. This reflects anisotropies in surface energy below the roughening temperature so that there is an energy barrier to nucleate an island or pit [12]. However, facetting can also be destroyed by dynamical roughening of the surface [16]. This will occur if the critical nucleus size for monolayer island nucleation is smaller than the step width or correlation length $\xi(T)$. This gives a condition for the critical supersaturation F^* required for the dynamical roughening of strained layers

$$F^* >> \frac{\gamma_S}{a\xi(T)} + \frac{\sigma^2}{2E} \qquad (2)$$

where γ_s is the step energy per unit length, a is the step height, σ is the misfit stress and E is Young's modulus for the layer. Relaxation effects have been neglected for simplicity. In the limit of very high supersaturation, the surface will be rough, even below the roughening temperature and we might therefore anticipate that the stress driven transition will attain some of the character of the ATG instability.

To investigate dynamical roughening, it is important to maintain the supersaturation while simultaneously monitoring the evolving surface profile. In particular, smoothing effects might be particularly important in the presence of stress enhanced migration during quench and look approaches. We therefore use a simple experimental approach in which we deposit thin Ge spikes at selected intervals during the growth of a $Si_{0.5}Ge_{0.5}$ sample [17,18]. We then image the sample in crossection using the Z-contrast imaging technique [19-20] to provide a map of the evolving surface morphology (Fig. 11). Initially the surface of the film grows with a planar morphology until a thickness of about 20nm where a wavy surface develops. This appears to evolve gradually and simultaneously across extended regions of the film at the growth temperature of 400°C. Note that the 5nm thick annealed $Si_{0.5}Ge_{0.5}$ film remained planar when held at this temperature so the roughening in Fig. 1 is related to growth and is characteristic of the ATG instability.

The ripple morphology in Fig. 11 rapidly develops into a sharp cusp-like geometry which is consistent with the non-linear behavior of the instability [17, 22-24]. A cusp forms from a sinusoidal profile because of gradients in the surface chemical potential. Adatoms prefer to migrate away from areas of large stress concentration at the valleys to more relaxed regions at the peaks (see Fig. 2). This deepens and sharpens the valley of the undulation which in turn increases the stress concentration and the rate of migration towards the peak. This positive feedback situation continues until a cusp is formed.

The veracity of the above scenario has been confirmed by approximate calculations [17] as shown in Fig. 11 as well as more rigorous computations [22-24]. A cusp is therefore a formal consequence of the ATG instability [25]. The fact that cusps are observed at high deposition rates suggests that facetting is indeed destroyed, or at least suppressed, and that the ATG instability may provide a useful semi-quantitative description of the transition under these conditions.

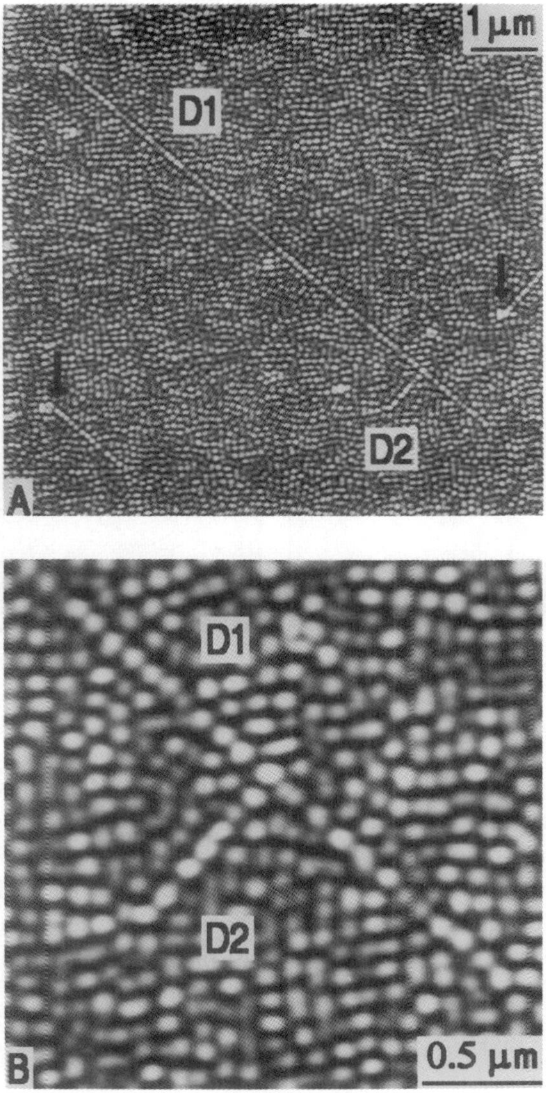

Figure 9. (a) Atomic-force-microscopy image showing surface islands that have aligned along dislocation lines D1 and D2. An enlarged view of the intersection region between the dislocations is contained in (b).

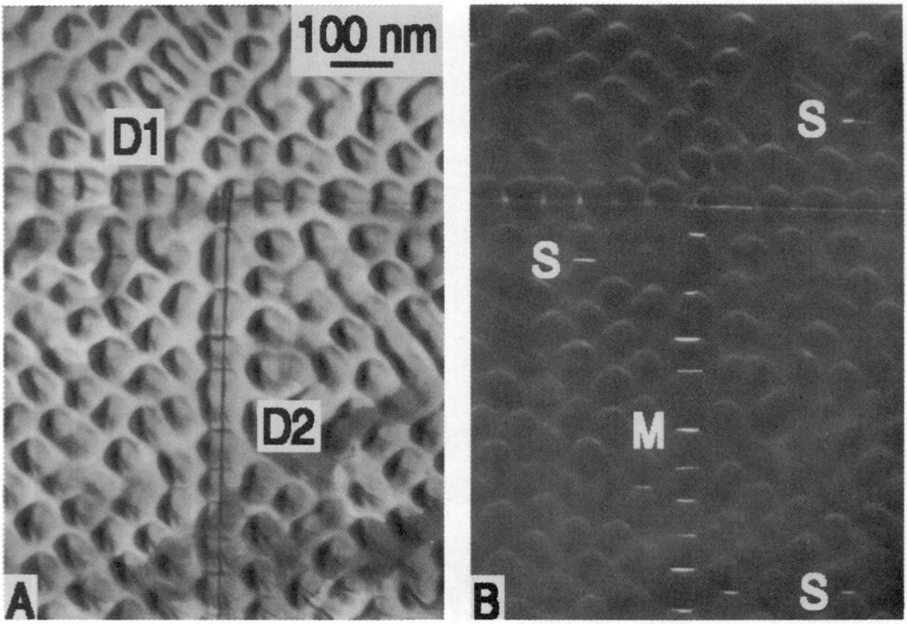

Figure 10. (a) Bright-field (220) two-beam and (b) $(2\bar{2}0)$ (g, 3g) weak beam TEM images of dislocation multiplication induced by morphological instability. High stress concentrations associated with the crack-like instabilities (S and M) are revealed as short bright segments in (b). The dislocation lines D1 and D2 define [110] and $[1\bar{1}0]$ directions, respectively. The nature of the diffraction conditions means that dislocation D2 is in strong contrast in (a), and D1 is in contrast in (b). Similarly, stress lines (M) associated with D2 are in strong contrast in (b).

7 Surface Cusps and Dislocation Nucleation

Unlike a slit crack, a surface cusp under compression has a large stress concentration at the tip [26]. This is because the material at the tip is not in contact. According to continuum theory, this large stress will continue to increase as the cusp forms from a sinusoid [17, 22-24]. In practice, we believe that such large stresses will eventually lead to the nucleation of misfit dislocations, even at the relatively low growth temperature of 400°C [17]. A schematic representation of how this might occur is given in Fig. 12. The total energy of the expanding loop as a function of its radius R_L can be written as

$$U(R_L) = R_L \frac{b^2\mu}{8} \frac{2-\nu}{1-\nu} \ln\left[\frac{\alpha R_L}{b}\right] - \frac{R_L \mu b^2}{4} \sin\beta + \frac{\pi R_L^2}{2}\gamma - C(R_L)\cos\frac{\phi}{2}\sin\phi\sin\beta \tag{3}$$

This expression is directly comparable with the treatment of cleavage crack stability by Rice and Thomson [27]. The first term specifies the energy cost of a dislocation of core parameter α and Burgers vector b in a medium of shear modulus μ and Poisson's ratio ν. The second term describes the energy gained by the removal of a surface step, β being the angle between $\boldsymbol{b}$ and the dislocation line. The energy cost per unit area γ of the stacking fault associated with partial dislocations is included in the third term. The fourth term describes the elastic energy released by the loop, where the angle ϕ is defined in Fig. 12, and $C(R_L)$ depends on the model used for the radial (R) dependence of the stress field away from the tip. For the Westergaard sharp crack solution [28], we obtain

$$C(R_L) = 2.47 d^{1/2} R_L^{3/2} \mu \frac{1+\nu}{1-\nu} \varepsilon b, \tag{4}$$

where ε is the applied uniaxial strain and d is the crack depth (Fig. 12). From Eq. (3), it is possible to determine the critical cusp depth at which the activation barrier is equal to the available energy for nucleation (~ 37 kT [29]). This leads to the simple condition for d^c

$$\sqrt{d^c}\,\varepsilon = A^{\beta}(T,\gamma), \tag{5}$$

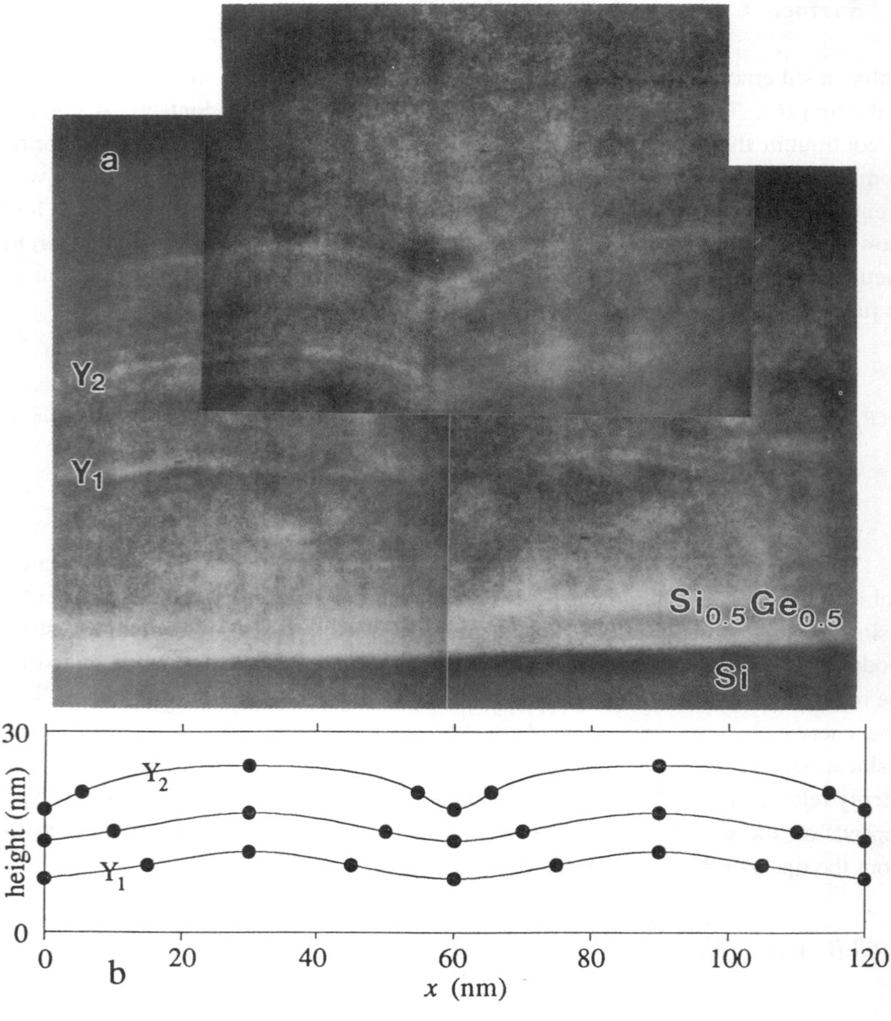

Figure 11. (a) [110] Z-contrast image of a $Si_{0.5}Ge_{0.5}$ alloy grown by MBE at 400°C and 2Ås^{-1}. The bright horizontal lines represent 2-monolayer-thick Ge marker layers. The surface profile simulations in (b) correspond to the period of cusp formation in (a) between vertical ordinates Y_1 and Y_2.

where the constant A^{β} is dependent on the nature of the dislocation (i.e., β = 30°, 60°, or 90°), stacking fault energy and temperature. For a given dislocation type, $A^{\beta}(T,\gamma)$ can be estimated from a single energy calculation using Eq. (3). The critical cusp geometry d^c is, therefore, simply related to misfit via Eq. (5), and the results are summarized in Fig. 13 using A^{30} = 0.402, A^{60} = 0.276, and A^{90} = 0.177. The calculation assumes a core parameter of 4 for all dislocation types.

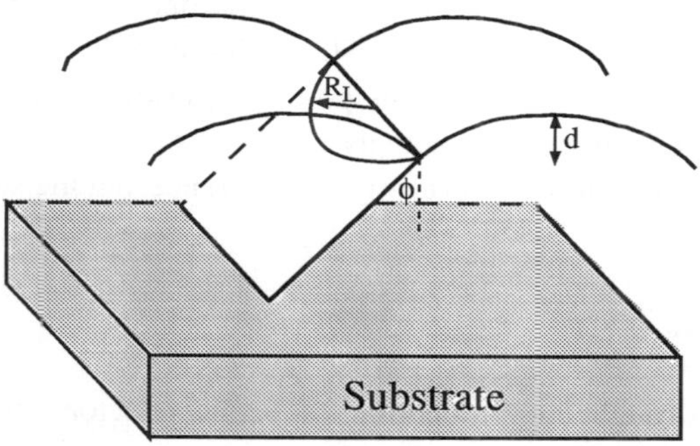

Figure 12. The geometry of dislocation half loop emission at the tip of a surface cusp (see text).

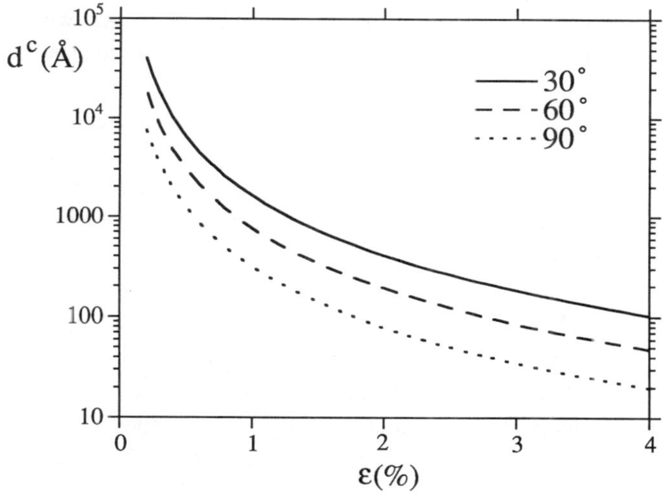

Figure 13. Critical cusp depth d^c required to nucleate 30°, 60° and 90° dislocations as a function of misfit ε assuming an atomically sharp cusp. The curves are evaluated using Eq. (3).

An interesting feature of Fig. 13 is that it is energetically more favorable to nucleate a complete 60° half loop from a cusp rather than a 30° partial. This is important because the glide plane geometry dictates that it is necessary to nucleate the 30° partial before the energetically favorable 90° partial. The nucleation of complete 60° dislocations is, therefore, favored in this case for atomically sharp cusps. It is interesting to compare half loop nucleation at cusps with half loop nucleation at flat surfaces where, assuming a step is removed, it is always favorable to inject a complete 60° dislocation before a 30° partial. However, unlike the cusp, the energy barrier for the nucleation of a 90° partial is only lower than the 60° half loop barrier down to 1% misfit. At this point, the critical radius is very large, and the stacking fault energy correspondingly high. Note that in the case of a critical cusp geometry the critical radius is always small, rendering the stacking fault energy less important.

8 Conclusions

The stress driven roughening transition of semiconductor films exhibits a rich variety of interesting phenomena on the mesoscopic scale. In the case of annealed films, nucleation effects dominate the nature of the surface ripple which forms by a cooperative mechanism. High deposition rates however appear to suppress facetting resulting in a transition with characteristics of the ATG instability including the non-linear evolution to surface cusps. In the general case we might expect intermediate behavior between nucleation dominated and diffusive roughening regimes.

A fascinating consequence of stress driven morphological evolution is the highly localized surface stresses which develop at features such as cusps Such stresses can play an important role both in dislocation nucleation and multiplication, revealing an intimate link between surface morphology and defect introduction during strain relaxation.

ACKNOWLEDGMENTS: This research was sponsored by the Division of Materials Science, US Department of Energy, under Contract No. DE-AC05-84OR21400 with Lockheed Martin Energy Systems, Inc. It is a pleasure to acknowledge the invaluable contributions of my colleagues K. M. Chen, S. J. Pennycook, T. Thundat and R. J. Warmack at Oak Ridge National Laboratory and J.-M. Baribeau and D. C. Houghton at the National Research Council, Canada.

References

[1] R. F. Service, *Science* **271**, 920 (1996).

[2] M. Krishnamurthy, J. S. Drucker, and J. A. Venables, *J. Appl. Phys.* **69**, 6461 (1991).

[3] D. Leonard, K. Pond, and P. M. Petroff, *Phys. Rev. B* **50**, 11687 (1994).

[4] G. S. Solomon, J. A. Trezza, and J. S. Harris, Jr., *Appl. Phys. Lett.* **66**, 991 (1995).

[5] J. M. Moison et al., *Appl. Phys. Lett.* **64**, 196 (1994).

[6] R. J. Asaro and W. A. Tiller, *Metall. Trans.* **3**, 1789 (1972).

[7] M. A. Grinfeld, *Sov. Phys. Dokl.* **31**, 831 (1986).

[8] D. E. Jesson, K. M. Chen, S.J . Pennycook, T. Thundat, and R. J. Warmack, *Phys. Rev. Lett.* **77**, 1330 (1996).

[9] D. E. Jesson, K. M. Chen, S. J. Pennycook, T. Thundat, and R. J. Warmack, *Proc.23rd Int. Conf. Physics of Semiconductors* **2**, 1253 (1996).

[10] D. E. Jesson, K. M. Chen and S. J. Pennycook, *MRS Bulletin* **21**, 31 (1996).

[11] Y-W. Mo, D. E. Savage, B. S. Swartzentruber and M.G. Lagally, *Phys. Rev. Lett.* **65,** 1020 (1990).

[12] J. Tersoff and F. K. LeGoues, *Phys. Rev. Lett.* **72**, 3570 (1994).

[13] D. E. Jesson, K. M. Chen, S. J. Pennycook, T. Thundat, and R. J. Warmack, *Science* **268**, 1161 (1995).

[14] A. G. Cullis, A. J. Pidduck, and M. T. Emeny, *Phys. Rev. Lett.* **75,** 2368 (1995).

[15] M. Albrecht, S. Christiansen, P. O Hansson, J. Michler, H. P. Strunk and E. Bauser, *Proc. Roy. Microsc.Soc., Oxford,* ed A. G. Cullis and J. L. Hutchinson (in press).

[16] P. Nozières and F. Gallet, *J. Physique*, **48**, 353 (1987).

[17] D. E. Jesson, S. J. Pennycook, J-M. Baribeau and D. C. Houghton, *Phys. Rev. Lett.* **71**, 1744 (1993).

[18] D. E. Jesson, S. J. Pennycook, J-M. Baribeau and D. C. Houghton, *Scanning Microcs.* **8**, 849 (1994).

[19] D. E. Jesson and S. J. Pennycook, *MRS Bulletin* **16**, 34 (1991).

[20] S. J. Pennycook and D. E. Jesson, *Phys. Rev. Lett.* **64**, 1943 (1990).

[21] D. E. Jesson, S. J. Pennycook, J-M. Baribeau and D. C. Houghton, *MRS Symp. Proc.* **312**, 47 (1993).

[22] W. H. Yang and D. J. Srolovitz, *Phys. Rev. Lett.* **71**, 1593 (1993).

[23] B. J. Spencer and D. I. Meiron, *Acta. Met. Mater.* **42**, 3629 (1994).

[24] C-H. Chiu and H. Gao, *Mat. Res. Soc. Symp. Proc.***317**, 369 (1994).

[25] H. Gao, *J. Mech. Phys. Solids* **39**, 443 (1991).
[26] C-H. Chiu and H. Gao, *Int. J. Solids Structures* **30**, 2983 (1993).
[27] J. R. Rice and R. Thomson, *Philos. Mag.* **29**, 73 (1974).
[28] See, for example, J. F. Knott, *Fundamentals of Fracture Mechanics*, (John Wiley and Sons, New York, Toronto), 57 (1973).
[29] S. V. Kamat and J. P. Hirth, *J. Appl. Phys.* **67**, 6844 (1990).

LARGE SCALE SURFACE EVOLUTION DURING MBE GROWTH: MOUNDING

C. ORME, M. D. JOHNSON, B. G. ORR
The Harrison M. Randall Laboratory, The University of Michigan, Ann Arbor, MI 48109-1120, USA

We consider the growth of films by molecular beam epitaxy in the presence of step-edge barriers. We show that the growth on a singular surface is unstable. For these surfaces the instability gives rise to the formation of large mounded structures.

1 Introduction

Evolution of growing interfaces is a ubiquitous and important phenomena. In the study of film deposition and growth the interface often influences or controls the microstructure of the resulting layer thus determining the film properties and quality. From a purely scientific standpoint, this phenomena allows the study of far from equilibrium processes under extremely controlled conditions. The goal of such studies is often to find the essential features or themes common to all systems undergoing such an evolution. In this chapter we will describe several homoepitaxial systems grown by molecular beam epitaxy (MBE). For almost 15 years MBE surface evolution has been examined using reflection high energy electron diffraction (RHEED)[1,2]. In this discussion we will be concerned about stuctures which are formed on large length scales, scales over which RHEED does not sample.

2 Mounding

In this chapter we will examine the phenominal of surface mounding during growth. There are many models of film growth which predict a self affine or scaling surface during growth [3,4,5] However, it is found [6,7,8] that frequently surfaces are not self-affine but rather that they are composed of reasonably uniformly sized mounds. Or, said another way, these surfaces have a peak in the power spectrum rather than having many wavelengths with a $1/k^p$ dependence. In this chapter we show that the mounds come about due to a step-edge barrier. This barrier causes a growth instability which results in mounding on singular surfaces and step-flow growth on miscut surfaces. Mounds have been imaged in GaAs by Smith et al. [9] who attribute the mounds to a kinetic effect and Heller and Lagally [10] who attribute the mounds to step-edge barriers. Johnson et al. [6,8] couples the observations with Monte Carlo simulation and continuum modeling putting together a self-consistent picture for the formation and evolution of these surface structures. They also emphasize that mounding can not be analyzed within the scaling hypothesis and that mound formation provides a new paradigm for film growth. This effect has been seen in many materials, both metals[11,12,13] and semiconductors[7,9,10,14]. It is, in fact, expected that most materials have step-edge barriers and are thus prone to the unstable growth mode which produces mounding.

Figure 1 illustrates the progression from the small scale to larger scale features. At the smallest scale shown, (260 Å × 260 Å), the corrugations due to the surface reconstruction can be seen. While there is not enough resolution to see the individual arsenic atoms in the unit cell, it is known from RHEED data that this a 2x4 reconstruction. The dark lines are the missing dimer rows that run along the [$\bar{1}$10] direction.

As we move up in scale (figures 1b-d) the reconstruction becomes less clear and the distribution of islands, and then terraces become the dominant features of the images. The islands are elongated along the [$\bar{1}$10] direction. This is thought to be due to anisotropic surface diffusion rates [10] (faster along [$\bar{1}$10] than along [110]) but could also be caused by anisotropic sticking coefficients[10].

This surface is currently growing by nucleation and islands can be seen on the terraces. However, if the slope becomes steeper then the distance between nucleating islands, σ, will become comparable with the terrace width and the growth will proceed by step-flow.

Continuing to increase the scan range, we see in figure 1e a mound with fifteen terraces. This image was taken from the same sample but in a slightly different region. This is rather different than the images in figure 1a-d not only because the scale is large enough to tell that there are both ascending and descending staircases but also because the terrace width are narrower. These steps have widths of 200Å which from the geometry (200Å wide 2.8Å tall) corresponds to a slope of almost 1°. This is to be compared to the overall sample miscut which has a slope of 0.2-0.3°. From this image it is clear that the step structure is not simply due to the miscut of the substrate.

Figure 1f shows the same surface with an approximately 10-fold increase in scale. To accommodate the larger scan size it was necessary to take the sample out of the UHV chamber to use an ambient environment force microscope. With this larger range the atomic step heights can no longer be resolved but it is clear that the mound in figure 1e is not an isolated event but rather that these mounding structures cover the surface. This type of mounding behavior was not anticipated from the smaller scale STM images and cannot be understood within the scaling hypothesis.

Figure 1: Homoepitaxial GaAs grown on a singular (001) substrate. This progression represents a 150-fold increase in scale from 260Å to 4 μm. a-e are STM images with scan sizes: a) 260Å × 260Å, b) 700Å × 700Å, c) 1300Å × 1300Å, d) 2700Å × 2700Å, e) 5000Å × 5000Å. Image e) is a force microscope image of a 4μm×4μm area taken on the same surface.

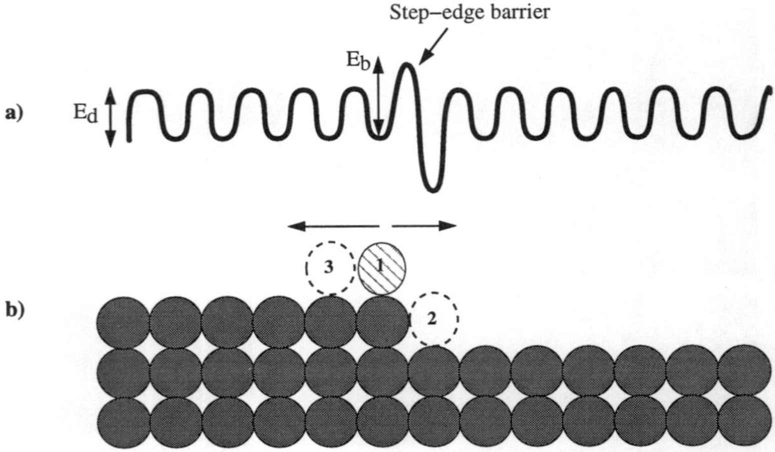

Figure 2: a.) The potential energy landscape for the surface shown in cross section in b. E_d is the activation energy needed for diffusing within a terrace–moving from 1 to 3 for example. E_{se} is the activation energy needed to cross the step (1 to 2). When $E_{se} > E_d$ more adatoms will be reflected away from the step edge than will be transmitted through it.

3 Step-edge Barriers

Mounding can be explained by the presence of a potential which impedes adatoms on the tops of steps from hopping down to the lower terraces – step-edge barriers. As the adatom diffuses it interacts with the surface through an interaction potential which is shown schematically in figure 2a. The minima of the potential are at the crystalline positions and the deepest minima is at the step where the adatom would find a well coordinated site (many nearest neighbors). Within the framework of an activated hopping model, the adatom must surmount a diffusion barrier (E_d) to move from site to site. If the material has step-edge barriers ($E_{se} > E_d$) then an adatom at the top of the step in position one of figure 2b, will need a higher activation energy to hop to position two on the lower terrace than it would to move away from the step edge to position three. This reduces interlayer transport by biasing the adatom motion away from the step-edge. This effect is sometimes termed *diffusion bias*[15] for this reason. The step-edge barriers are also called Ehrlich barriers[16] or Schwoebel barriers[17] after the experimentalist and theorist who discussed them in the 1960's.

The fact that flow is not symmetric across the step-edge leads to an instability in the growth. If a fluctuation causes an island to form then adatoms

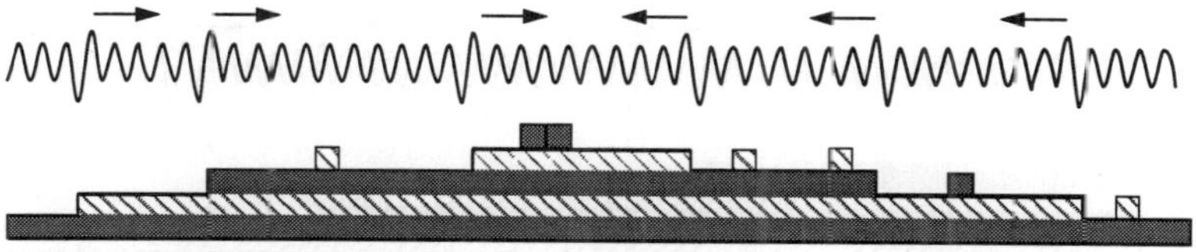

Figure 3: The effect of step-edge barriers on a singular surface. The adatoms have the color of the terrace they are more likely to attach to.

which land on top of the island will have a tendency to be reflected away from the island edges and will eventually nucleate new islands on the upper terrace consequently producing multi-level structures. This type of scenario is illustrated in figure 3. On the highest terrace, adatoms nucleate new islands and on the stepped regions adatoms have a greater probability of attaching to the upper terrace than to the lower terrace. This is an unstable process; a perturbation to the flat surface, i.e. a small mound, will tend to grow rather than to decay back to the initial plane. Although the result is three dimensional growth, the current is not literally flowing "uphill" in the sense that the current has a non-zero z-component. The adatoms do not hop up steps more than they hop down steps – rather, adatoms which land on a terrace are more likely to attach to an island or step within that layer rather than crossing the step-edge barrier and attaching to material in the lower level. Adatoms converge towards the initial perturbation they don't climb up it.

It should be emphasized that this is inherently a non-equilibrium effect. In equilibrium, by definition, the flow of adatoms across the step is symmetric, i.e. the net flow is zero. Using figure 2 as a guide, as many atoms move from position one to position two (N_{12}) as move from position two to position one (N_{21}). This is a statement of detailed balance. When the system is not in equilibrium, as is the case during growth, the occupancy of a site is not given by the energy of the potential well and detailed balance breaks down; during growth a diffusion bias can exist. In a system without step-edge barriers ($E_{se} = E_d$) the current across the step is symmetric independent of whether the system is in equilibrium or not. This can be seen by noting that when $E_{se} = E_d$ the probability of moving to the right or to the left is equal everywhere whether the occupancy is given by a Boltzmann factor or not.

3.1 Step Regularization

It has been shown that step-edge barriers induce mounding if growth fluctuations can produce the islands which initiate the mounding instability. When a

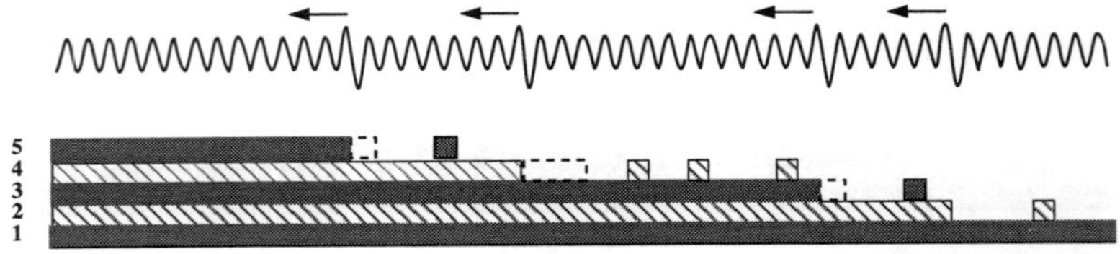

Figure 4: Step-edge barriers tend to regularize the terrace widths on vicinal surfaces. The adatoms have the color of the terrace they are more likely to attach to. For the purposes of demonstration the dashed line shows the new position of the step edge in the case where all of the adatoms attach to the upper step (i.e. an infinite barrier or 100% reflection). In reality the barriers are relatively small but the tendency is the same.

surface is growing in step-flow mode no islands are nucleated and a different effect is present. On vicinal surfaces the step-edge barrier acts to regularize the terrace spacings. The number of adatoms which land on a terrace is proportional to its size thus larger terraces tend to collect more adatoms than smaller terraces. With a step-edge barrier these adatoms tend to attach to the upper terrace more often than the lower terrace – effectively causing the largest terrace widths to become smaller. Thus, there is a feedback mechanism to keep the terrace width approximately constant. This is illustrated in figure 4 where more material is added to the fourth terrace than to the other terraces - the terrace width thus decreases in relation to the others. Notice that in the case where the initial starting substrate is singular, after mounds form, their sides will look like vicinal surfaces and the terrace spacings on the mound sides should regularize. This can be seen in image 1e where the terraces are seen to have approximately the same widths. Without a step-edge barrier, although large terraces still collect more adatoms these adatoms are as likely to attach to the lower terrace as the upper terrace and the terrace width distribution does not change as the surface grow. Equally spaced steps and a sharpening of the terrace distribution can be taken as evidence for the presence of a step-edge barrier.

3.2 Microscopic origin of step-edge barriers

The origin of barriers depends on the details of the material as well as the local configuration. However, there are two generic ideas which give an intuitive feeling for why step-edges are special and thus why one might anticipate an activation barrier at steps which is different from the barrier for diffusion within a terrace.

One such idea is presented in figure 5 where the adatom moves through a region with fewer neighbors while traversing the step. Being less bound to

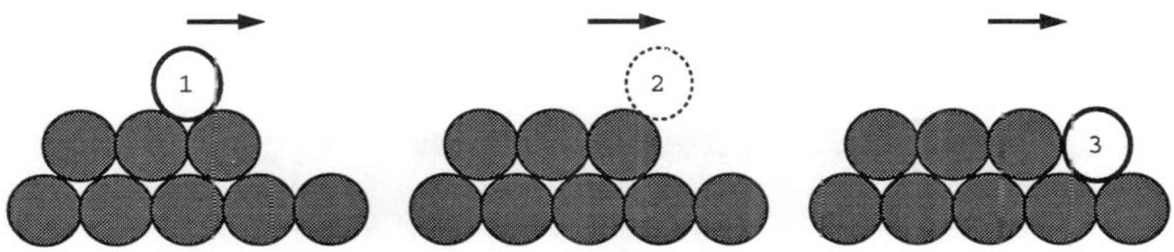

Figure 5: As the adatom moves from the top of the step (1) to the lower terrace (3) it moves through a region where it has few neighbors (2)

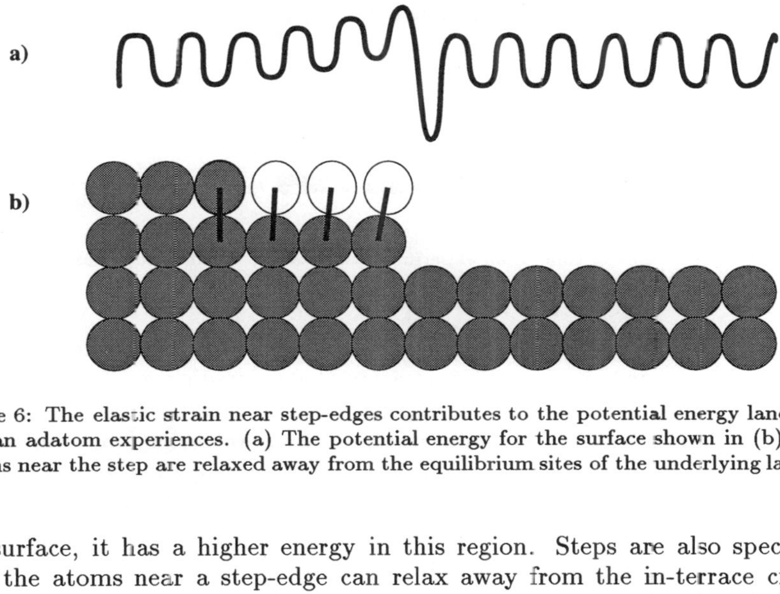

Figure 6: The elastic strain near step-edges contributes to the potential energy landscape that an adatom experiences. (a) The potential energy for the surface shown in (b). The atoms near the step are relaxed away from the equilibrium sites of the underlying lattice.

the surface, it has a higher energy in this region. Steps are also special in that the atoms near a step-edge can relax away from the in-terrace crystal positions. In a simplified picture, the potential energy can be thought of as the superposition of the energy due to the unstrained interaction potential and a harmonic contributions due to the strain ($U = k\Delta x^2$). This is depicted in figure 6. In this scenario the barrier has a longer range near the step edge rather than being a purely local effect.

Although it seems intuitive to expect the presence of a step-edge barrier the magnitude and sign of ΔE_{se} (i.e. whether E_{se} is larger or smaller than E_d) is difficult both to measure and to calculate. A better description of the origin of the step-edge barrier requires looking at realistic, material specific models.

The quantitative measurements of step barriers have only been made on metal systems. The earliest measurements used a field ion microscope[16] which limited its application to transition metals because these are stable at fields high enough to ionize He gas. Recently, Kern's group has used a variable

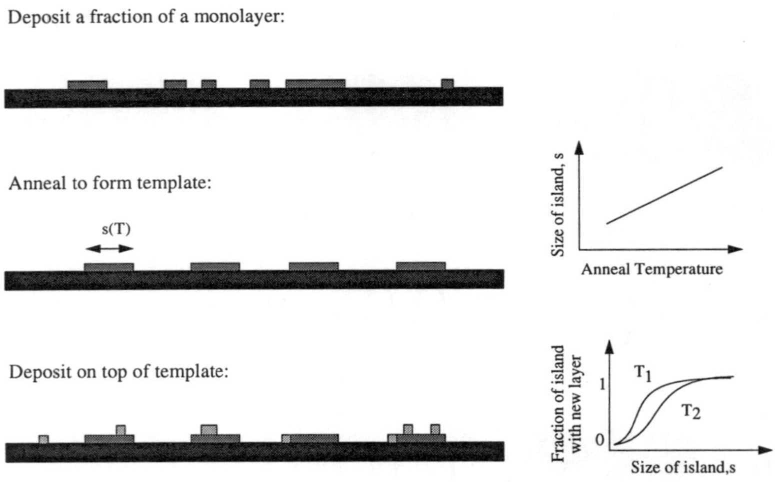

Figure 7: The outline of an experiment to measure the step-edge barrier.

temperature STM to obtain island distributions as a function of temperature. By fitting these distributions to nucleation theory predictions [18] they can extract a step-edge barrier. Another approach which provides an estimate of the step-edge barrier has been to quantitatively compare various distributions or signatures of experimental data with that obtained using an activated-hopping Monte Carlo simulation [19,20,21]. In this approach measured values are used for most of the parameters of the simulation and the barrier energy is varied until agreement is found with experiment.

To obtain hopping probabilities from STM images (which provide snapshots rather than trajectories) the surface morphology must be fit to theoretical predictions. Nucleation theory, which uses rate-equations to describe the formation and evolution of islands, provides the necessary framework for this analysis. The parameters which describe intralayer diffusion (ν_d and E_d) can be obtained by fitting the number of islands (in the submonolayer regime just before the islands begin to coalesce) to an Arrhenius distribution[22,23,24]. The application of nucleation theory requires knowledge of the critical island size (i). Islands with i atoms dissociate faster than material is added to them and are not stable. Thus, if dimers are stable then monomers are the critical islands and $i = 1$. When i is greater than one then the binding energy associated with the critical islands is another parameter which must be known to compare with theory.

In a systematic set of experiments of Ag grown on Ag(111) Kern's group has measured activation barriers on terraces and across steps[25,26]. What makes this possible is a variable temperature STM. Part of the success of these experiments lies in the ability to tune to a temperature where the critical island size is a monomer; this allows the simplest application of nucleation theory. The experiment is shown schematically in figure 7. First a fraction of a monolayer is grown on a smooth substrate. Then the surface is annealed to produce a surface with a reasonably narrow distribution of island sizes. By changing the anneal temperature the island size can be tuned. More material is then grown on top of this template to monitor the number of islands which acquire a third layer of growth. Small islands will have very little three dimensional growth whereas large islands have a high probability of having an island nucleate on top of them. The probability depends on the step-edge barrier. Large step-edge barriers will increase the likelihood that an island will nucleate on the upper surface. The fraction of islands with additional growth on their upper surface as a function of the size of the initial island can be fit using nucleation theory[18] provided E_d, ν_d, and i are known. Again it is necessary to do this at several temperatures to get a unique fit for both the attempt frequency (ν_{se}) and the activation barrier(E_{se}).

Neither the barrier obtained by fitting to nucleation theory nor the barrier obtained by comparing to Monte Carlo simulations distinguish between different step-edge morphologies or different mechanisms for diffusing to a lower terrace. The barrier that is measured, is an effective barrier which incorporates the different means of moving between layers.

4 Coarsening

Once we have established that mounds develop when the growth mode is nucleating islands and that the existence of mounds can be understood by the presence of a step-edge barrier. It is natural to address the question of how the mounds coarsen as growth proceeds. The pertinent measurements here are the time evolution of the mound-mound spacings and the mound heights. In particular the ratio of these two quantities (the slope) is of interest. One issue is whether the slope is constant in time. A related issue is what determines the slope.

Figure 8 demonstrates this evolution for a GaAs growing from a commercial GaAs substrate[7]. The starting surface is quite rough due to the desorption of the oxide layer. Initially the growth smooths the pitted surface. Eventually, (somewhere between 100 and 300 monolayers for these growth conditions), the long wavelength oscillations of the mounds emerge. With further growth both

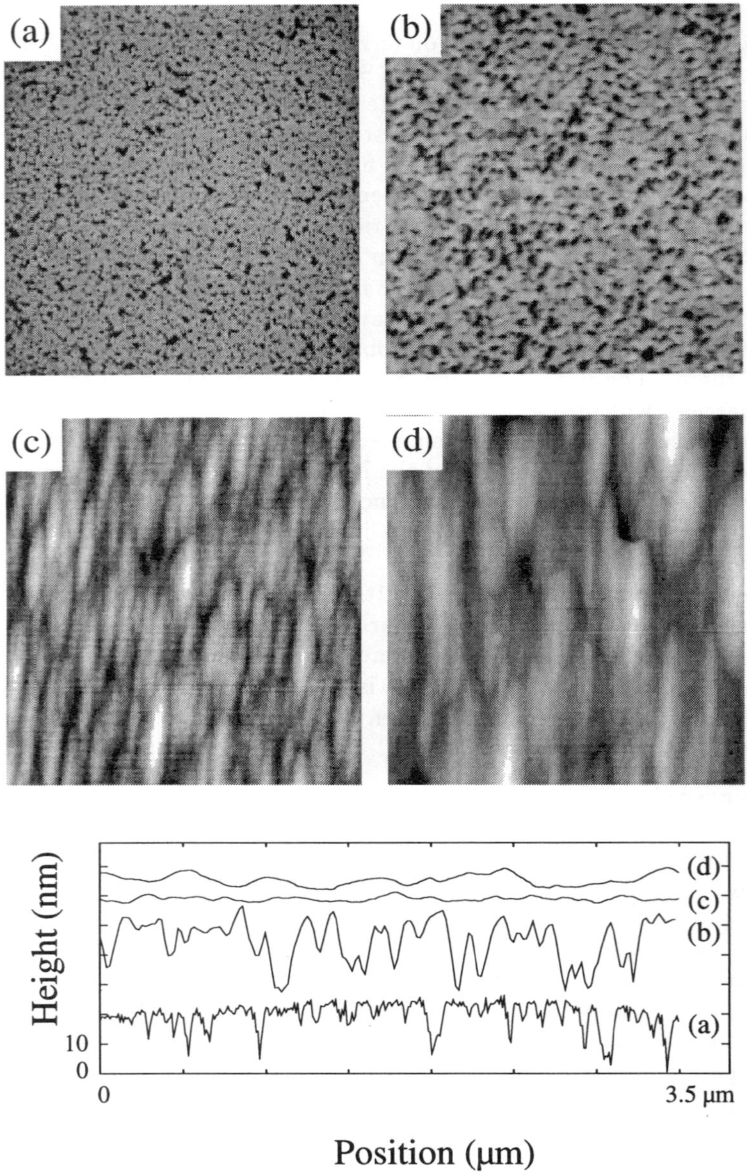

Figure 8: AFM images of a 3.5 μm × 3.5 μm area of homoepitaxial GaAs for various film thicknesses. (a) Substrate after oxide desorption. (b) 90 bilayers. (c) 270 bilayers. (d) 540 bilayers. (e) Cross-sectional view of images (a)-(d) with arbitrary offset between each. The scale on the y-axis is 10nm per division.

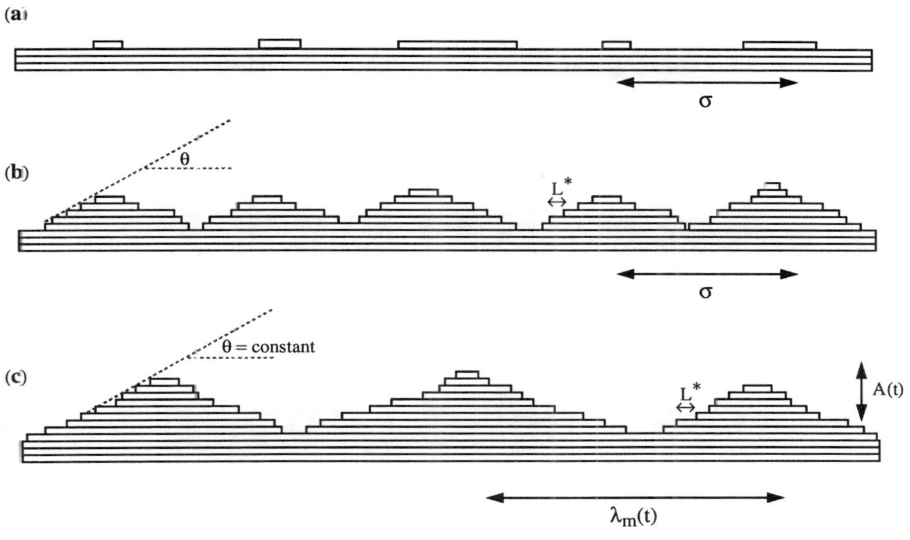

Figure 9: The mounds become larger as more epilayers are grown. An average slope can be obtained from the height and lateral extent of the mound, $m = 2\mathrm{A}/\lambda_\mathrm{m}$ or a local slope can be obtained from the atomic step height and the terrace width, $m = \mathrm{a_z}/\mathrm{L}^*$. The angle is given by $\theta = \arctan(\mathrm{m})$.

the wavelength and the amplitude of these corrugations increases, however their ratio, the slope, remains approximately constant. The angle on the sides of the mounds is quite shallow (1°) and is consistent with a terrace width that is set by σ (the distance between nucleating islands on a flat surace). In other words, the sides have achieved a slope which allows them to grow by step-flow.

In the case where the initial substrate is smooth, one scenario for coarsening is a rapid evolution of the surface to mounded structures with a fixed slope. As the film grows thicker both the feature separation (λ_m) and the height of the mounds (A) grow but the slope ($\overline{m} = 2A/\lambda_m$) remains unchanged.

This scenario is perhaps best brought out in the work of Stroscio et al. [12] where they study iron grown on iron whiskers. Part of what makes this work so clean is the incredible flatness of the initial substrates which are monatomically flat over microns. They show that the initial nucleation process in the submonolayer regime creates a template from which the mounds grow. During the first several monolayers the island-island (or mound-mound) spacing remains constant but the slopes quickly increase and then eventually saturate

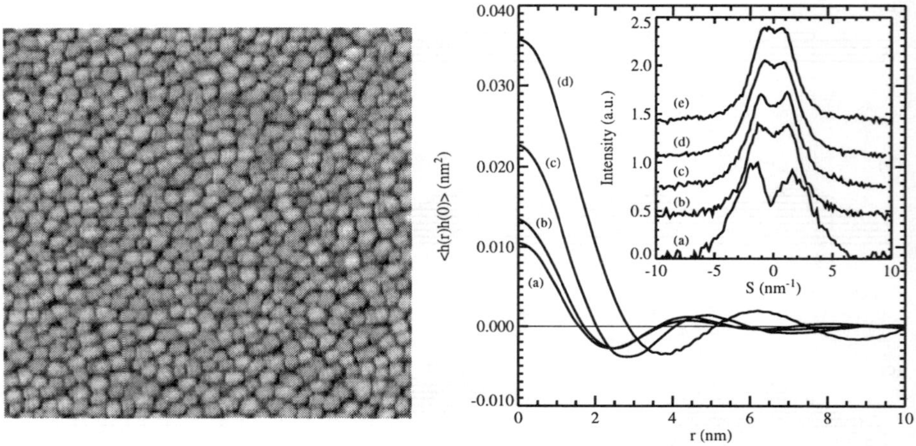

Figure 10: Mound formation during homoepitaxial growth on an Fe whisker. Courtesy of J. Stroscio. [12]

at some value. At this point, the mounds begin to coarsen leaving the slopes constant as the lateral width and amplitude both increase. This is illustrated schematically in figure 9 where σ is the initial nucleation distance between islands. It is not known what sets the angle, θ, or equivalently the terrace spacing L^*.

Figure 10 shows the mounded Fe surface after ten monolayers have been deposited. The mounds are fairly uniform both in height and in lateral extent. A statistical measure of these features can be obtained from the root mean square width and the height-height correlation function. The height-height correlation function is plotted in figure 10 and gives a measure of the wavelength. The oscillations in the height-height correlation function show the presence of a periodic structure. If the mounds were not uniform but had a broad distribution of sizes then the height-height correlation function would not oscillate and the system could possibly be modeled with scaling concepts. The time evolution of mound size $(\lambda_m(t))$ can be seen in the shift of the peak-to-peak distance at different coverages.

The slopes on the sides of the mound are most accurately determined from STM or force microscopy images but it is difficult to gather statistics if each measurement has to be made by hand. A more convenient measure of the time evolution of the slope where the entire image can be processed at once is given

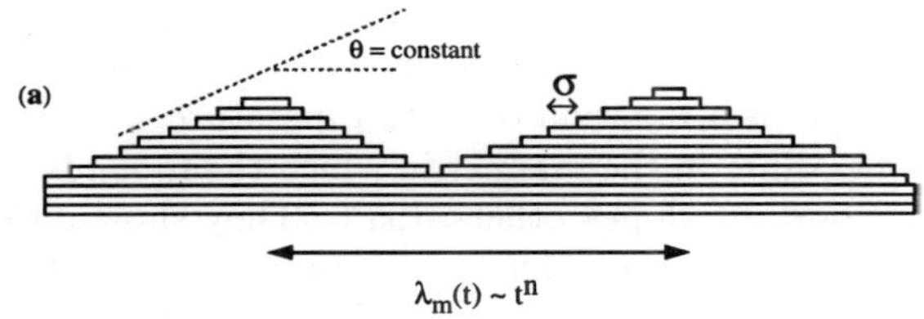

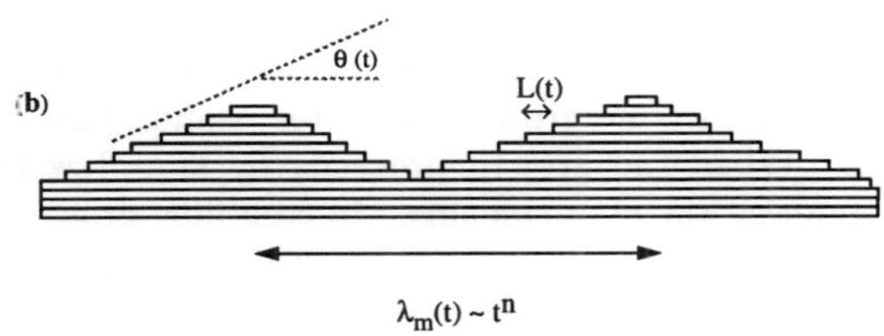

Figure 11: The mounds can coarsen in ways different from that illustrated in figure 9. (a) Mounds seen in GaAs and Rh grown above 600°K have approximately constant slopes during growth. The slope is consistent with $m = a_z/\sigma$. (b) In germanium the slope varies during growth.

by $m \sim w_{rms}/\lambda_m$. It is convenient to take the measure of λ_m from the first zero crossing of the height-height correlation function. This measure averages over all slopes, the tops and bottoms as well as the sides of the mounds and will therefore under estimate the value of the slopes on the sides but the time-evolution of this measure should mimic the behavior of the slope provided that the shapes of the mounds are not changing too much in time.

Interestingly, Rh/Rh(111) shows two mounding regimes[13]. At low temperature the behavior is very similar to iron (which was also grown at low temperature) and is thus like that depicted in figure 9. However at higher temperature the templating is less clear because the layers fill in more readily. In this regime, it is the terrace width rather than the original mound template that is consistent with σ, the distance between nucleating islands in the submonolayer regime. This is depicted in figure 11a. Whether the terrace spacing is exactly sigma or whether it is simply related in some way because sigma embodies the growth parameters is still an open question. This higher temperature regime is more akin to what has been seen in GaAs [7,9]. Yet another situation seems to exist in germanium where the slope increases slowly in time[14].

The experimental results from GaAs, Ge, Fe, and Rh agree that the instability produces surfaces that have almost equally sized mounds. These mounds grow both laterally and vertically as the epilayer thickness is increased. The slope that they achieve is a function of the temperature.

The experiments differ in the manner in which the mounds coarsen – some experiments show constant slopes others don't. They show differences in the numerical value of the exponents. That is, assuming the wavelength coarsens according to $\lambda_m \sim t^n$, n varies from material to material($n = 1/6$ in Fe, $n = 1/3$ in Rh, and $n = 0.4$ in Ge). In Rh while the low temperature and high temperature growth have different slopes they coarsen according to the same power law.

To summarize, materials which have a step-edge barrier are unstable to mound formation. When the growth nucleates islands then the mounds will eventually form. When the growth proceeds by step-flow the effect of the barrier is to regularize the step spacings. In step-flow growth mounds do not form. The mounding instability is an inherently non-equilibrium process. It only occurs when the evolution of the surface is dominated by kinetics rather than by energetics. When growth is turned off and the surface allowed to relax to its equilibrium state by annealing, the mounds dissolve.

5 Simulating growth

One of the standard ways people try to understand growing interfaces is to simulate their growth on the computer. This has the advantage of complete control over the dynamics and the hazard that the chosen dynamics may not represent reality well. Not surprisingly, detail is in direct conflict with speed.

While none of the molecular dynamics methods are fast enough to model growth of a surface, the diffusion barriers that they calculate are essential for the Monte Carlo simulations which are used to "grow surfaces".

Monte Carlo methods provide an extremely useful complement to experiment when trying to pinpoint the cause of particular morphologies. In experiments, the surface evolves according to the full dynamics. This brings along with it not only the complexities of the material but also non-ideal effects such as impurities, temperature gradients, etc. It is difficult to seperate out the effects due to material and growth parameters from those due to experimental inconsistencies. Monte Carlo simulations allow complete control of both the environment and also the dynamics. While this means that the dynamics are always simplified, it provides a way to link cause with effect. Both techniques bear criticism: experiment usually due to issues of contamination during growth and simulation due to over-simplification of the dynamics. Together however, they can provide a self-consistent picture where macroscopic morphologies can be attributed to microscopic processes.

5.1 Activated hopping models

In a realm of complexity between first principle calculations and single-step models are activated hopping Monte Carlo simulations, often called kinetic Monte Carlo (KMC). Adatoms are randomly deposited onto a grid and allowed to diffuse. The diffusion is modeled as an Arrhenius hopping probability with activation energies that depend on the local environment. The hop rate is $h = \nu e^{-E_{act}/k_BT}$ where ν is the attempt frequency and E_{act} is the activation barrier. There are two flavors of this type of simulation which differ in how the activation energies are calculated.

A simple, yet surprisingly successful, model uses a bond counting scheme to determine the activation energies[27,28,29]. This model was discussed earlier but here we include a term to mock up the effect of a step-edge barrier. The activation energy is simply computed by counting the number of nearest neighbors in the initial state, n_{nn}, and the change in the number of next-nearest neighbor bonds between the initial and final states. The activation energies are given by,

$$E_{act} = E_0 + n_{nn}E_{nn} + \Delta_{nnn}E_{nnn}. \tag{1}$$

where,

$$\Delta_{nnn} = \begin{cases} 0 & n_i < n_f \\ n_i - n_f & n_i > n_f \end{cases} . \tag{2}$$

and n_i and n_f are the number of next-nearest neighbors in the initial and final states. The energy, E_0, represents the binding energy due to the underlying atom. Because E_0 appears in all hopping probabilities independent of configuration it can be thought of as setting the hopping rate relative to the flux rate, $h = (\nu e^{-E_0/k_BT})h(n_{nn}, n_{nnn})$. The next contribution can be thought of as the energy required to break the in-plane bonds before the adatom is free to move to a new location. This heuristic picture with $E_{act} \propto n_{nn}$, has been verified using molecular dynamics for some metal systems. The next-nearest neighbors are included to provide a convenient way of creating a barrier near step edges. It is not thought that this is in fact how step-edge barriers are produced in materials, rather equation 2 is a computationally reasonable way to create a barrier. An adatom approaching an edge goes from a position where it has four next-nearest neighbor bonds to one where it has three next-nearest neighbor bonds ($\Delta_{nnn} = +1$); this produces a step-edge barrier of E_{nnn}. The energies: E_0, E_{nn}, E_{nnn}, and kT as well as the flux rate are the input parameters for this type of simulation. Values for the energy parameters can be obtained by fitting experimental RHEED data with simulated RHEED data[30]. For convenience, the attempt frequency is usually assumed to be constant independent of configuration.

Figure 12 shows the potential energy landscape generated by this algorithm for two sample trajectories. It should be noted that the sine wave interpolation between the well energies and the activation energies is purely for viewing convenience.

Another Monte Carlo approach marries the utility of the activated hopping algorithm with the more detailed energy barrier calculations. Rather than counting local bonds, this method uses activation barriers calculated from some other method. A table is created which lists the local configurations and their corresponding activation energy for diffusion. Initially Voter used classical potentials [31] to calculate the energies and more recently this approach has been used with effective medium theory calculations [32]. This is in principle more accurate than the bond-counting method but it is only as good as the approximation used to calculate the energy barriers. For the effective medium theory calculations, it is found that these values need to be scaled up to reproduce the surface morphologies observed with scanning tunneling microscopy. The use of tabulated barriers offers more flexibility to test how specific local geometries effect the morphology. For example Jacobsen et al.[32], guided by their molecular dynamics calculations, make a distinction between motion over a straight step and motion over a step with a kink in it (see figure 12b). The bond counting algorithm makes no such distinction.

One of the questions which can be tested with simulation is how the mounding behavior depends on growth mode. This is shown in figure 13 using an activated hopping bond-counting algorithm[6,33,34]. The first column shows the evolution of a singular (flat) surface during growth; it is seen that mounds eventually form. The second column shows the evolution of a vicinal (stepped) surface under growth conditions which allow most adatoms to attach to step-edges (i.e. the distance between nucleating islands is larger than the distance between steps). This surface grows by step-flow and mounds do not form. Here we see how the growth mode effects the formation of mounds. This has been tested by experiment and the agreement is good.

In a similar manner KMC can be used to determine the mound size as a function of the step-edge barrier, sigma (the distance between nucleating islands), and time. At present much research is directed at determining the magnitude of the slope as a function of time. Studies along these lines are currently being carried out by several research groups [35,36,37,38,39].

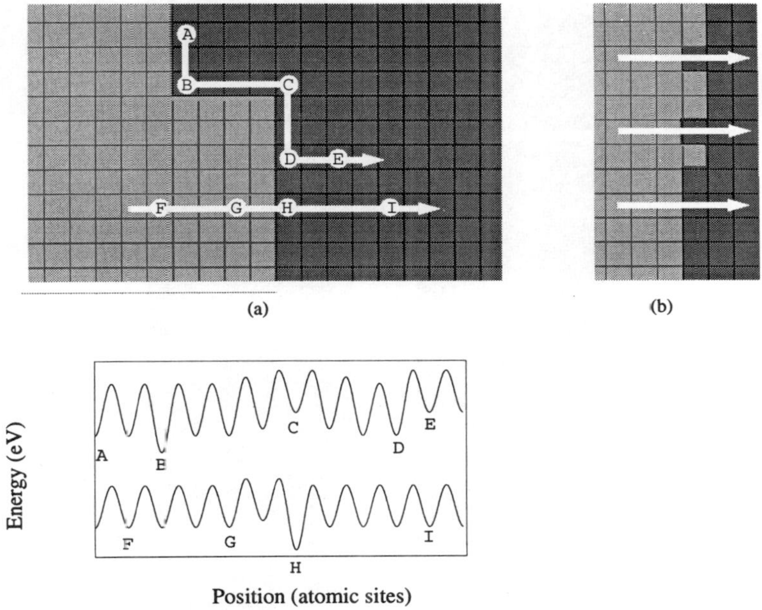

Figure 12: A sample energy landscape using a bond counting algorithm. The lighter terrace is one step higher than the darker terrace. (a) The potential energy surface is shown for the two trajectories marked by the white arrows. The parameters are $E_0 = 1.3$eV, E_{nn}=0.3eV, and E_{nnn}=0.2eV. The nearest neighbor and next-nearest neighbor energies were made higher than they are normally taken in simulating GaAs so that the line shape would be more apparent. In the trajectory along the step edge it is seen that there is a barrier to moving around the corner (site C). Adatoms which are attached to the step edge are more likely to diffuse along the step than to move away from the step (site D). In the trajectory which descends the step, an adatom at position G sees an additional barrier $\Delta E = E_{nnn}$ hindering motion towards the step. In (b) the trajectory over the step is shown with different local configurations. In the bond-counting algorithm these all have identical energy landscapes (same as F-I). The energies calculated by molecular dynamics methods make a distinction between these pathways. See, for example,[32].

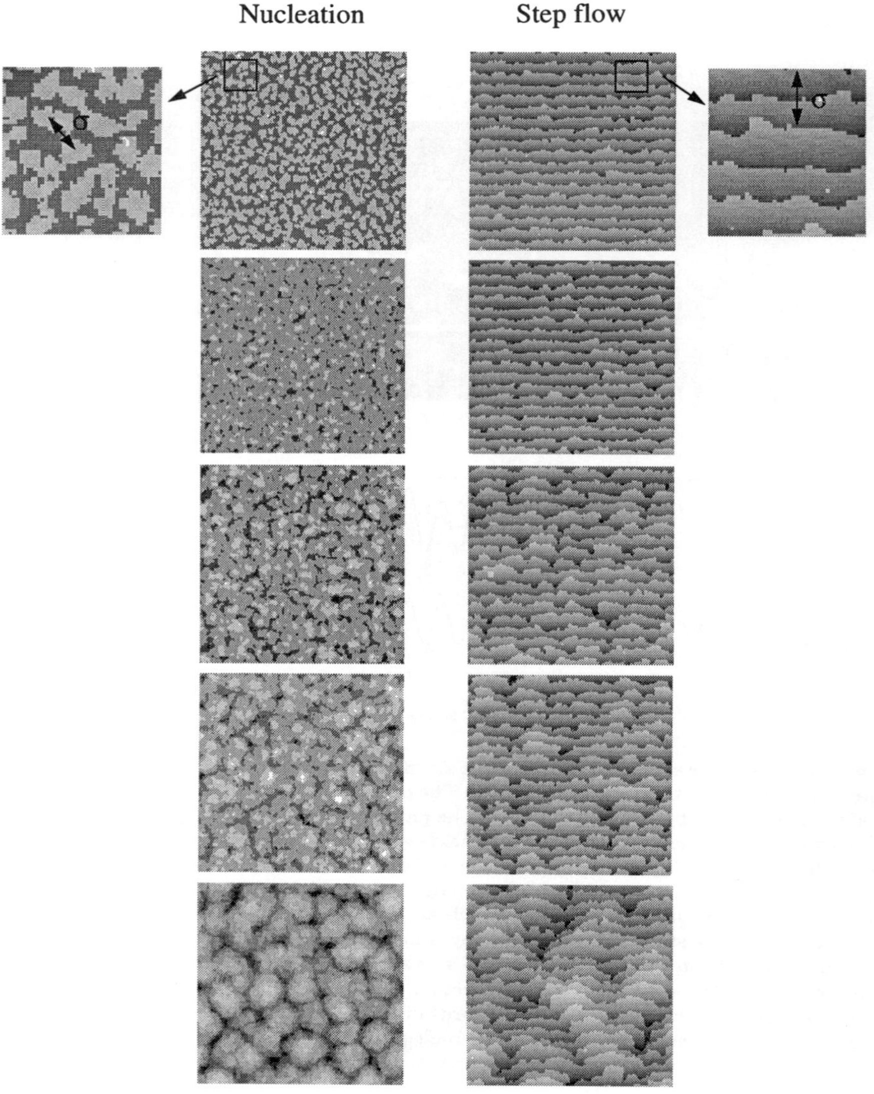

Figure 13: The effect of growth mode on mound formation. Runs using a bond counting activated hopping simulation. From top to bottom the images have 0.5, 1.0, 5.0, 10, and 50 monolayers of growth. The run parameters are: E_0 = 1.5 eV, E_{nn} = 0.24 eV, E_{nnn} = 0.036 eV, and kT = 0.07 eV. This produces a step-edge barrier which reflects 63% of the adatoms (where 50% reflection indicates that there is no barrier). The distance between nucleating islands, σ, is indicated in the blown up sections. The lattices are 200 x 200.

5.2 Continuum equations

Continuum modeling of surface growth tries to emulate the processes of random deposition, surface diffusion, and desorption [40]. In MBE growth desorption is usually negligible (that is, desorption is an unlikely event in the time it takes to deposit one monolayer). This coupled with the fact that voids and overhangs are not typically formed during growth means that the average height depends linearly on the time according to $\overline{h} = v_0 \overline{F} t$, where $\overline{F}$ represents the average flux (average number of atoms deposited/(unit area$\times$ time)), v_0, the unit volume, and t, the time of growth. Continuum equations are normally written in a frame moving with the average height and thus describe the fluctuations in height about the average value. The time evolution of the interface is described by two effects: the noise in the flux, $\eta(x,t)$, which tends to roughen the surface and the surface diffusion which, during growth, can either roughen or smooth the surface. The diffusing adatoms can be thought of as a surface current which rearranges the height profile without changing its average value. In the co-moving frame, the lack of desorption implies that mass is conserved on the surface. When void formation is also negligible then this mass conservation leads to volume conservation and the time evolution of the height can be represented as a continuity equation which reflects this conserved volume[41,42].

$$\frac{\partial h(x,t)}{\partial t} = -\nabla \cdot \mathbf{j}(\mathbf{x},t) + \eta(\mathbf{x},t) \tag{3}$$

where $h(\mathbf{x},t)$ is the height of the surface relative to the average height at position $\mathbf{x}$ and time t, $\mathbf{j}(\mathbf{x},t)$ is the conserved current parallel to the average surface, and $\eta(\mathbf{x},t)$ is the noise in the flux. The average flux determines the average height whereas $\eta(\mathbf{x},t)$ represents the fluctuations in the beam.

To model mounding using continuum equations the effect of the step-edge barrier must be added[6,8,15,35,43,44]. Models which do this have several common features:

1. A conserved current - this is a reflection of the MBE process
2. A current ($\mathbf{j}$) which is a function of the local slope (m) of the surface (not the global slope or miscut).
3. A small slope region where the current increases as the slope increases, $\partial\mathbf{j}/\partial m > 0$, and a turnover to $\partial\mathbf{j}/\partial m < 0$ at some higher slope.
4. The terms responsible for the mounding have a flux dependent coefficient whereas the flux independent terms smooth the surface. A flat surface is obtained from a rough surface by turning off the flux.

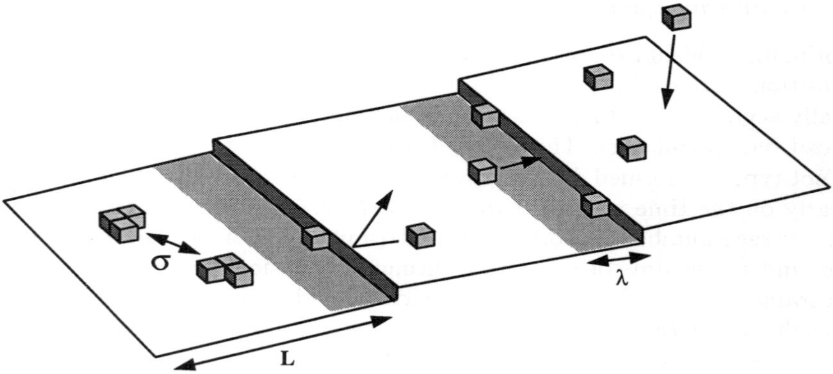

Figure 14: A miscut surface. When the barrier is 100% reflecting, it is only the adatoms which land within the shaded band which participate in the current.

5. The interface width grows exponential in the small slope regime.
6. The growth process makes certain wavelengths grow faster than others leading to wavelength selection.

5.3 The current

Villain[15] gave a simple one-dimensional argument for the form of the current in the presence of step-edge barriers. The current depends on the density of adatoms on the terrace, their average velocity, and the fraction of them which participate in the motion of the steps. Using figure 14 as a guide,

$$\begin{array}{ccccc} \text{Current density} & = & (\text{density}) & \overline{(\text{velocity})} & (\text{fraction}) \\ \mathbf{j} & = & (\overline{F}\tau) & \frac{1}{2}(\lambda/\tau) & (\lambda/L) \\ & = & \frac{1}{2}\overline{F}\lambda^2\nabla h & & \end{array}$$

where $\overline{F}$ is the average flux, τ is the typical time it takes to deposit a monolayer, λ is the typical distance an adatom travels in this time, and $\nabla\mathbf{h}$ has been substituted for $1/L$ where L is the terrace width. In the extreme case where the step-edge barrier is taken to be 100% reflecting, only the adatoms within some distance, λ, (determined by their mobility) of the upper step add to the current. This is the case shown in figure 14. More generally, the net current is the difference between the current of adatoms which attach to the upper step ($\mathbf{j}_+$) and the current of adatoms which attach to the lower step ($\mathbf{j}_-$). Only a fraction of the adatoms which have enough mobility to reach the lower step-edge will be transmitted through the barrier.

$$\begin{aligned} \mathbf{j} &= \mathbf{j}_{+} - \mathbf{j}_{-} \\ &= \frac{1}{2}\overline{F}\lambda^2 \nabla h(1-2T) \end{aligned}$$

where T is the transmission across the step-edge and $S = (1-2T)$ is a parameter which ranges from -1 to 1 and gives the strength of the step-edge barrier. When S=0 (no barrier), half of the adatoms are reflected and half are transmitted hence no net current flows across the step-edge. When S=1 (infinite barrier), 100% of the adatoms are reflection at the step-edge.

The two length scales in the problem are the terrace width, L, and the distance between nucleating islands, σ. The current can be looked at for different extremes of the ratio σ/L. In the regime where $\sigma/L \ll 1$ (either small slopes or low diffusion) islands can form on the terrace. These islands don't contribute to the net current across the steps - it is only the adatoms which are within a length σ of the step which contribute. In this case $\lambda = \sigma$ (see figure 14) and the current is proportional to ∇h. In the other extreme $\sigma/L \gg 1$ (large slopes or high diffusion) the growth proceeds by step flow and all of the adatoms participate in the current. Thus $\lambda = L = 1/\nabla h$ and the current is proportional to $1/\nabla h$. The simplest interpolation between these two regimes produces the current[6,8]

$$\mathbf{j} = \frac{1}{2}\overline{\mathrm{F}}\mathrm{S}\frac{(\sigma^2 \nabla \mathrm{h})}{(1+\sigma^2 \nabla \mathrm{h}^2)} \tag{4}$$

which is the solid line in figure 15. We were also able to obtain this form of the current by measuring the adatom current as a function of miscut in the activated hopping Monte Carlo simulation[6,8].

The current 4 depends linearly on flux. The total current also needs a flux independent Mullins[45] term which describes how the surface moves back towards its equilibrium surface,

$$\mathbf{j} = \frac{1}{2}\overline{\mathrm{F}}\mathrm{S}\frac{(\sigma^2 \nabla \mathrm{h})}{(1+\sigma^2 \nabla \mathrm{h}^2)} + \mathrm{K}\nabla^3 \mathrm{h}. \tag{5}$$

This form of the current contains the generic features discussed earlier on. A fluctuation in the flux which produces a local slope will generate a current proportional to the slope which will in turn cause the perturbation to grow; in other words, the surface is unstable to fluctuations. The time evolution of the surface which has a current of the form 5 and equation of motion 3, is shown if figure 16. Mounds form, their lateral and vertical dimensions grow in time, they merge together. A Fourier transform of the surface reveals a

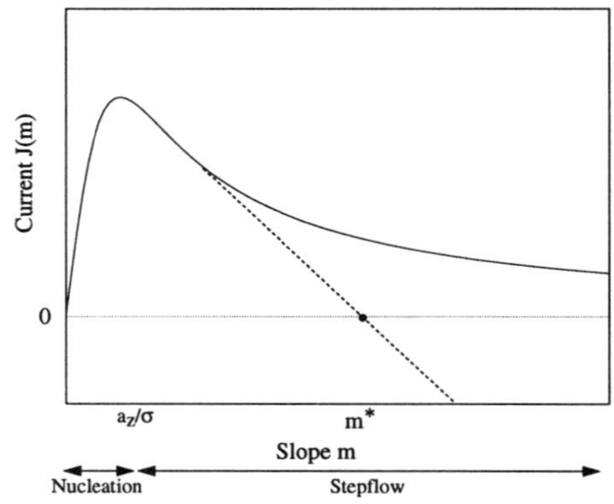

Figure 15: The current (using equation 4) as a function of local slope. For small slopes the adatoms nucleate islands on the terraces which don't contribute to the current. The peak is at a slope of a_z/σ where a_z is the atomic step height and σ is the distance between nucleating islands. This marks the boundary between nucleation and step-flow.

dominant wavelength which becomes larger as a function of time according to $\lambda_m \sim t^{1/4}$. [8] The presence of a dominant wavelength rather than a power spectrum which has all wavelengths up to the correlation length means that the dynamic scaling hypothesis which has been useful for analyzing self-affine surfaces can not be applied to this surface.

The current, equation 5, however, has the problem that it predicts that the slope along the sides of the mounds which form will increase without bound (albeit very slowly). Although some of the semiconductor data shows slowly increasing slopes[14], many of the experiments indicate that there exists a selected slope, m^*. This point is made very clearly with some of the recent metal data[12,13] where, during growth, the slopes on the sides of the mounds increase rapidly to m^* then remain approximately constant for the duration of growth. To achieve such a result in the continuum equation, the current must go through zero at some point other than at the singular ($m = 0$) surface[35,38,44]. The dashed line in figure 15 schematically shows such a current. Slopes smaller than the selected slope produce a current which causes the slope to increase and slopes larger than the selected slope produce a negative current which causes the slope to decrease. The slope is therefore stabilized at a particular value, the selected slope or magic slope.

Some of the data seem to indicate that this stable slope is achieved at an equilibrium facet[11] but other data seem to indicate that the slope is determined by the kinetics and is not related to a crystal facet[6,7,9,13,14]. Experimentally the

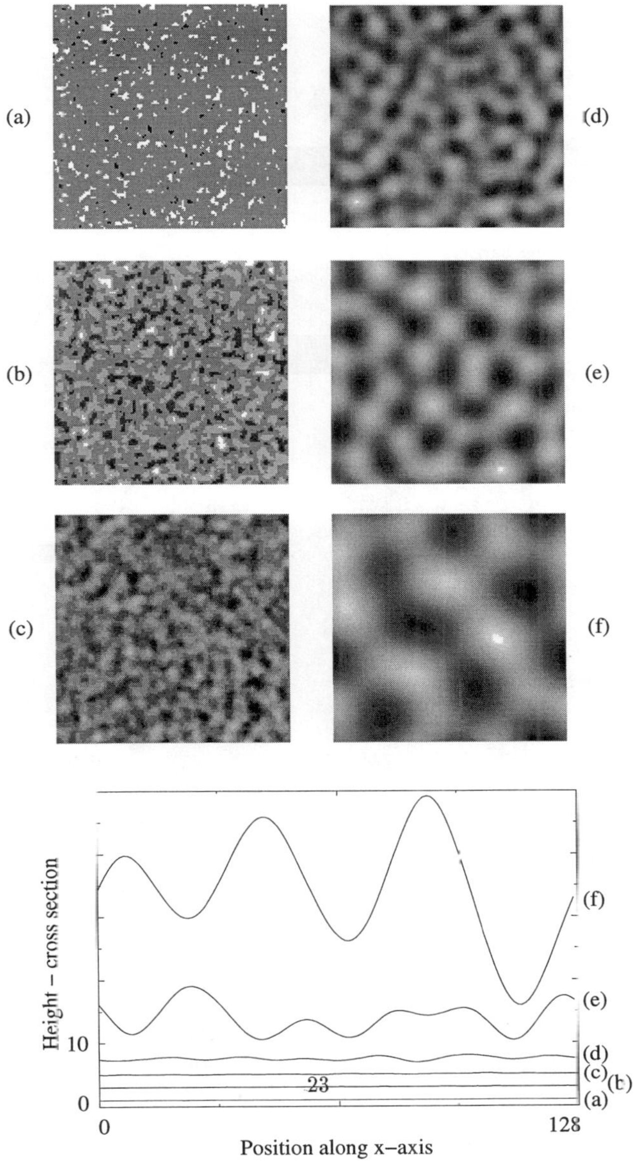

Figure 16: The time evolution of the surface generated from a current of the form 5.

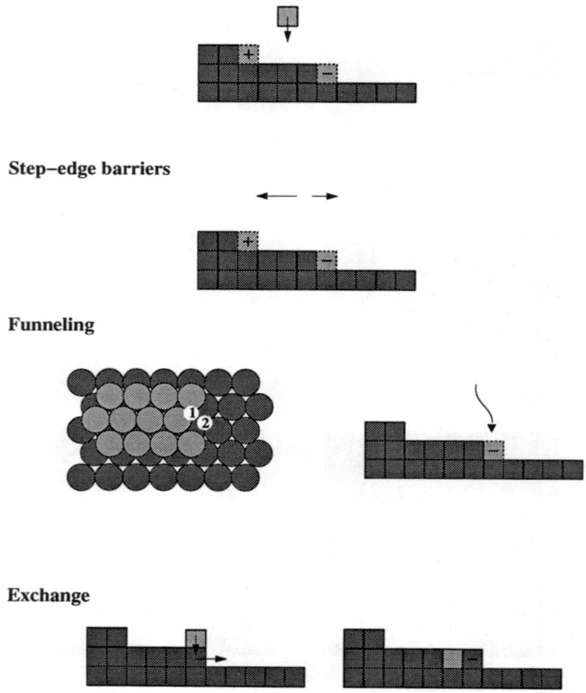

Figure 17: Mechanisms for producing contributions to the current. Positive contributions to the current preferentially join a step at the same height. This is the case for step-edge barriers. While adatoms join both in-plane steps (+) as well as steps on terraces below (-), the net current is positive. Funneling is a mechanism which occurs due to the geometry of an fcc or hcp lattice and doesn't occur for solid-on-solid models. Adatoms which land in position 1 cannot stay there because the underlying layer which defines the three-fold hollow is not complete. Instead the adatom will be deflected to position 2. This produces a negative current; more adatoms "land" on the lower terrace than land on the upper terrace (per number of sites). Exchange processes involve the motion of two-atoms. An adatom which lands near an edge site can push away the underlying atom and takes its place. This has the same effect as jumping over a step-edge (although with a lower activation energy) and produces a negative contribution to the current.

cases should be easy to distinguish - a kinetically determined slope will decay away after growth is stopped whereas an equilibrium facet will not. A theory to discuss mounding will have to accommodate these two different scenarios.

By definition, the current goes through zero at an equilibrium facet but it is less clear why the current should go through zero in the case where the slope is determined kinetically. A process is needed which becomes more likely when the terraces become smaller and which has the net result of pushing adatoms downhill i.e. a negative current.

5.4 Mechanisms for a negative current

Several possible mechanisms for producing a downhill current have been suggested and these are summarized in figure 17. One such candidate is transient mobility[46] or "hot adatoms"[47]. Incoming adatoms initially have extra thermal energy which comes from condensing out of the vapor. Having an energy greater than the potential energy corrugations of the substrate they may easily diffuse over step-edge barriers until they have dissipated their extra energy and come into equilibrium with the substrate. However, a hybrid molecular dynamics and Monte Carlo simulations using corrected effective medium theory on metal (100) faces suggest that thermalization occurs extremely quickly and does not produce additional diffusion[48,49].

Another form of transient mobility suggested to produce a negative current is a knockout process where a freshly landed adatom on the upper terraces pushes the underlying atom away from the step and takes its place. This has the same effect as stepping over the terrace but the energetics for the process may be less. Again, the same molecular dynamics studies show that knockout is not a typical process[50].

Another possibility comes from considering the geometry of the crystal[38,51]. In fcc, bcc, and hcp crystals adatoms land in the "hollows" created by the underlying layer. It has been proposed that adatoms which land in a hollow that does not have a complete underlayer will be deflected to a lower terrace where the layer is complete. This is demonstrated for an fcc (111) face in figure 17. These effects become more important as the proportion of edge atoms becomes larger. This effect is born out by molecular dynamics simulations[48,50]. It has also been shown that mounds with fixed slopes develop from Monte Carlo simulations on fcc lattices which add this funneling effect [38,52].

6 Conclusion

In this paper we have attempted to put together, in a pedagogical way, a base of concepts needed to study surface evolution and roughening. At the largest length scales mounding structures are observed experimently and they could be successfully modeled using both a Monte Carlo simulation and a continuum equation for the height evolution. The ideas of scaling proved not to capture the dynamics leading to the unstable mounds. Once the root cause of the mounding instability was understood it was clear that it should be present in a great number of different system. This has proven true and the phenomena has been shown to be very common in systems where kinetics dominate the surface morphological evolution.

Acknowledgments

We would especially like to thank L. M. Sander for continuous theoretical support and advice. In addition, we had many stimulating discussions with D. Williams, D. Vvedensky, and A. Zangwill. A. Hunt and C. Gordon assisted with the numerical simulations. This work was supported in part by ONR grant N00014-89-J-1519.

References

1. J. H. Neave, B. A. Joyce, P. J. Dodson and N. Norton, Appl. Phys. **A31**, 1 (1983).
2. J. M. Van Hove, C. S. Lent, P. R. Pukite and P. I. Cohen, J. Vac. Sci & Technol. **B2**, 741 (1983).
3. F. Family, J. Phys. A **19**, L441 (1986).
4. D. E. Wolf and J. Villain, Europhys. Lett. **13**, 389 (1990).
5. Several other single-step models with slightly different rules were examined at about the same time. S. Das Sarma and P. Tamborenea, Phys. Rev. Lett. **66**, 325 (1991); Z. -W. Lai and S. Das Sarma, Phys. Rev. Lett. **66**, 2348 (1991); D. A. Kessler, H. Levine, and L. M. Sander, Phys. Rev. Lett. **69**, 100 (1992).
6. M. D. Johnson, C. Orme, A. W. Hunt, D. Graff, J. Sudijono, L. M. Sander, and B. G. Orr, Phys. Rev. Lett. **72**, 116 (1994).
7. C. Orme, M. D. Johnson, J. Sudijono, K. T. Leung, B. G. Orr, Appl. Phys. Lett. **64**, 860 (1994); C. Orme, M. D. Johnson, K. T. Leung and B. G. Orr, in *Compound Semiconductor Epitaxy* ed. C. W. Tu, L. A. Kolodziejski, and V. R. McCray (Materials Research Society, Pittsburgh,

1994); C. Orme, M. D. Johnson, K. T. Leung, and B. G. Orr, Mat. Sci. Engin. B **30**, 143 (1995).

8. A. W. Hunt, C. Orme, D. R. M. Williams, B. G. Orr, and L. M. Sander, Europhys. Lett. **27**, 611 (1994); A. W. Hunt, C. Orme, D. R. M. Williams, B. G. Orr, and L. M. Sander, *Patterns and Scaling in the Growth of Thin Films* from Proceedings of NATO ASI, Cambridge UK 26-30 June, 1994.
9. G. W. Smith, A. J. Pidduck, C. R. Whitehouse, J. L. Glasper, A. M. Keir, and C. Pickering, Appl. Phys. Lett. **59**, 3282 (1991); G. W. Smith, A. J. Pidduck, C. R. Whitehouse, J. L. Glasper, and J. Spowart, J. Crystal Growth **127** (1993); A. J. Pidduck, G. W. Smith, A. M. Keir, and C. R. Whitehouse, in *Mechanisms of Thin Film Evoultion* ed. S. M. Yalisove, C. V. Thompson, and D. J. Eaglesham (Materials Research Society, Pittsburgh 1994).
10. E. J. Heller and M. G. Lagally, Appl. Phys. Lett. **60**, 2675 (1992).
11. H. -J. Ernst, F. Fabre, R. Folkerts, and J. Lapujoulade, Phys. Rev. Lett. **72**, 112, (1994).
12. J. A. Stroscio, D. T. Pierce, M. Stiles, A. Zangwill, and L. M. Sander, Phys. Rev. Lett. **75** 4246 (1995).
13. F. Tsui, J. Wellman, C. Uher, and Roy Clarke, Phys. Rev. Lett. **76** 3164 (1996).
14. J. E. Van Nostrand, S. J. Chey, M. -A. Hasan, D. G. Cahill, and J. E. Greene, Phys. Rev. Lett. **74** 1127 (1995).
15. J. Villain, J. Phys. I. (France) **1**, 19 (1991).
16. G. Ehrlich and F. G. Hudda, J. Chem. Phys. **44**, 1039 (1966); S. C. Wang and G. Ehrlich, Phys. Rev. Lett. **70**, 41 (1993).
17. R. L. Schwoebel, J. Appl. Phys. **40**, 614 (1969); R. L. Schwoebel and E. J. Shipsey, J. Appl. Phys. **37**, 3682 (1966).
18. J. Tersoff, A. W. Denier van der Gon, and R. M Tromp, Phys. Rev. Lett. **72**, 266 (1994).
19. E. Z. Luo, J. Wollschlager, F. Wegner, and M. Henzler Appl. Phys. A **60**, 19 (1995).
20. J. Amar and F. Family, Phys. Rev. Lett. **74**, 2066 (1995). SEB for Fe (1995).
21. M. C. Bartelt and J. W. Evans, Phys. Rev. Lett. **75**, 4250 (1995).
22. J. A. Venables, G. D. T. Spiller, and M. Hunbücken, Rep. Prog. Phys. **47**, 399 (1984).
23. M. C. Bartelt and J. W. Evans, Phys. Rev. B. **46**, 12675 (1992); Surf. Sci. **298**, 421 (1993); J. W. Evans and M. C. Bartelt, J. Vac. Sci. Technol. A**12**, 1800 (1994).

24. G. S. Bales and D. C. Chrzan, Phys. Rev. B. **50**, 6057 (1994).
25. H. Brune, H. Roder, C. Boragno, and K. Kern, Phys. Rev. Lett. **73**, 1955 (1994).
26. K. Bromann, H. Brune, H. Roder, and K. Kern, Phys. Rev. Lett. **75**, 677 (1995).
27. J. D. Weeks and G. H. Gilmer, Adv. Chem. Phys. **40**, 157 (1979); J. D. Weeks, G. H. Gilmer, and K. A. Jackson, J. Chem. Phys. **65**, 712 (1976).
28. S. Clarke and D. D. Vvedensky, Phys. Rev. Lett. **58**, 2235 (1987).
29. P. Šmilauer, M. R. Wilby, and D. D. Vvedensky, Phys. Rev. B **47**, 4119 (1993).
30. T. Shitara, D. D. Vvedensky, M. R. Wilby, J. Zhang, J. H. Neave, and B. A. Joyce, Appl. Phys. Lett. **60**, 1504 (1992).
31. A. F. Voter, Phys. Rev. B **34**, 6819 (1986).
32. J. Jacobsen, K. W. Jacobsen, P. Stoltze, and J. K. Nørskov, Phys. Rev. Lett. **74**, 2295 (1995).
33. M. D. Johnson, Ph.D. Thesis, unpublished The University of Michigan (1994).
34. A. Hunt, Thesis (unpublished), The University of Michigan (1994).
35. M. Siegert and M. Plischke, Phys. Rev. Lett. **73**, 1517 (1994).
36. P. Šmilauer and D. D. Vvedensky, Phys. Rev. B, **52**, 14263 (1995).
37. K. Thürmer, R. Kock, M. Weber, and K. H. Rieder, Phys. Rev. Lett. **75**, 1767 (1995).
38. M. C. Bartelt and J. W. Evans, Phys. Rev. Lett. **75**, 4250 (1995).
39. E. Somfai and L. M. Sander, MRS proceedings to be published.
40. For a pedagogical discussion see A. -L. Barabasi and H. E. Stanley, *Fractal Concepts In Surface Growth* Cambridge University Press, Cambridge, (1995).
41. Notably, the constraint that the height must obey a continuity equation rules out the lowest order non-linear term $(\nabla h)^2$, (which appears in the KPZ equation[42]) because there is no current, $\mathbf{j}(\mathbf{x},t)$, such that $-\nabla \cdot \mathbf{j}(\mathbf{x},t) = (\nabla h)^2$.
42. M. Kardar, G. Parisi, and Y. C. Zhang, Phys. Rev. Lett. **56**, 889 (1986).
43. J. Krug, M. Plischke, and M. Siegert, Phys. Rev. Lett. **70**, 3271 (1993).
44. M. Siegert and M. Plischke, Phys. Rev. Lett. **73**, 1517 (1994).
45. W. W. Mullins, J. Appl. Phys. **28**, 333 (1957).
46. W. F. Egelhoff, Jr. and I. Jacob, Phys. Rev. Lett. **62**, 921 (1989).
47. D. D. Vvedensky, A. Zangwill, C. N. Luse, and M. R. Wilby, Phys. Rev. E **48**, 852 (1993).
48. P. Stoltze and J. K. Nørskov, Phys. Rev. B **48**, 5607 (1993).

49. D. E. Sanders and A. E. DePristo, Surf. Sci. **254**, 341 (1991).
50. J. W. Evans, D. E. Sanders, P. A. Thiel, and Andrew E. DePristo, Phys. Rev. B **41**, 5410 (1990); D. E. Sanders, D. M. Halstead, and A. E. DePristo, J. Vac. Sci. Technol. A **10**, 1986 (1992); D. M. Halstead and A. E. DePristo, Surf. Sci. **286**, 275 (1993).
51. H. C. Kang and J. W. Evans, Surf. Sci. **271**, 321 (1992); J. W. Evans, Phys. Rev. B, **43**, 3897 (1991).
52. J. Amar and F. Family, Phys. Rev. B. **54**, 14742 (1996); J. Amar and F. Family, Phys. Rev. B. **54**, 14071 (1996).

GROWTH STRUCTURES OF SILICON HOMOEPITAXY BY CHEMICAL VAPOR DEPOSITION

H. RAUSCHER and R. J. BEHM
Abteilung Oberflächenchemie und Katalyse, Universität Ulm, 89069 Ulm, Germany

Recent progress in understanding Si homoepitaxial layer formation during ultrahigh vacuum chemical vapor deposition (UHV-CVD) from SiH_4 and SiH_2Cl_2 on Si(100)2 $\times$ 1 and Si(111)7 $\times$ 7 is reported. Structures observed by scanning tunneling microscopy (STM) under various growth conditions (gas flux and surface temperature) and the mechanisms relevant for the formation of these structures are discussed. The topographic information is correlated with the kinetics of precursor decomposition and island formation and additional kinetic model calculations. The results are also compared to Si layer formation under molecular beam epitaxy (MBE) conditions.

1 Introduction

Thin silicon films for industrial applications such as integrated circuits or discrete devices are mainly grown by chemical vapor deposition (CVD) techniques. With the continuous miniaturisation of these devices knowledge and understanding of the processes relevant for film growth at the atomic level is essential since these basic steps ultimately determine the crystalline quality of the layers and hence the functionality of subsequently fabricated devices. A number of gaseous precursors are currently in use for Si CVD, including monosilane (SiH_4) and dichlorosilane (SiH_2Cl_2). Elementary (atomic level) processes in CVD growth include precursor adsorption, which is in most cases followed by several dissociation steps via intermediate species, diffusion of the dissociation products, epitaxial island nucleation and growth of these islands by further incorporation of diffusing species. All these processes may be dependent on each other and hence influence each other mutually.

In this chapter we give a short account of the recent progress in understanding the basic mechanisms relevant for growth of ultrathin silicon films on silicon substrates by ultrahigh vacuum chemical vapor deposition (UHV-CVD), in the precursor pressure region 10^{-8} - 10^{-4} mbar. We will confine ourselves to the precursors SiH_4 and SiH_2Cl_2 and concentrate mainly on structures formed in the initial stages of chemical vapor deposition, on nucleation and growth of monolayer silicon islands, and less on atomic reaction mechanisms. Excellent reviews of the surface chemistry of silicon in general can be found in refs. [1]

and [2], and a comprehensive general overview of STM studies on semiconductors is given in [3], while CVD of electronic materials is, e.g., discussed in [4]. Because of the limited space available here this article is based mainly on recent results from our own laboratory.

This chapter is organized as follows: (1) First the initial reaction step of the gas phase precursor on the surface, a prerequisite for CVD, is briefly introduced using the dissociative adsorption of SiH_2Cl_2 on Si(111)7 × 7 as an example. For further examples, the reader is referred to refs. [1] and [2]. (2) The next section deals with the structures formed on Si(100)2 × 1 upon interaction with SiH_4 at various temperatures, leading to island formation. Particular attention is given to the correlation between the precursor dissociation kinetics and the surface topography. Similarities and differences to Si MBE will be discussed. (3) In the final section, bilayer formation on Si(111)7 × 7 upon interaction with SiH_4 and with SiH_2Cl_2 is discussed.

2 Primary processes: SiH_2Cl_2 dissociation on Si(111)7 × 7

The primary processes are the fundamental reaction steps which lead from precursor adsorption over the first step of dissociation and possibly via several subsequent dissociation steps to the elementary species which nucleate the growth structures on the surface or which are incorporated into existing structures, such as islands. One example is reported in this section, namely dissociative adsorption of SiH_2Cl_2 on Si(111)7 × 7. This species has been identified as suitable precursor for CVD and also for atomic layer epitaxy (ALE) [5].

SiH_2Cl_2 decomposes on Si(111)7 × 7 and on Si(100) at room temperature by first splitting a Si-Cl bond, leading to adsorbed SiH_2Cl, which continues to dissociate at higher temperatures [6,7]. In constant current STM images recorded after room temperature exposure of Si(111)7 × 7 to SiH_2Cl_2 (Fig. 1), part of the Si adatoms are darkened compared to the initial surface. This is explained by a reduction of the local density of states (LDOS) close to the Fermi niveau due to saturation of the adatom dangling bond by reaction with Cl atoms, which evolve during dissociation of SiH_2Cl_2 molecules. An analogous darkening of the adatoms is observed if the dangling bonds are saturated by hydrogen, which is known to cause a similar reduction of the LDOS [8-11]. The STM tip has to move closer to the surface in order to maintain a constant current while imaging the surface, and consequently these adatoms appear darker in the STM images in both filled and empty states, as discussed in refs. [8] and [12], and can clearly be distinguished from bare adatoms on the surface [8,13].

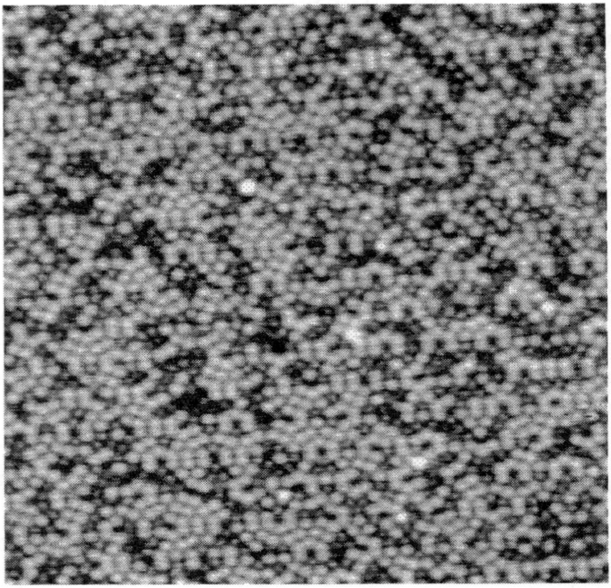

Figure 1: Si(111)7 × 7 after exposure to 3.6 L SiH_2Cl_2 at 300 K. Image size 290 Å × 290 Å. Sample voltage +1.7 V. [14]

The other dissociation product of SiH_2Cl_2 molecule, namely the SiH_2Cl group, remains virtually invisible at the tunnel parameters employed in Fig. 1. It is supposed to be adsorbed on the rest atoms. Further information on the nature of the adsorption process and the adsorption mechanism was obtained from the adsorption kinetics which, utilizing Auger electron spectroscopy (AES) measurements, was determined to be second order, from the ratio of reacted vs. unreacted adatoms after saturation (60:40) and from the absolute room temperature saturation coverage of Cl (0.33 ML) after reaction with SiH_2Cl_2 [14]. While half of the Cl is adsorbed as Cl atoms on the Si adatoms and can thus be identified in the STM images (e.g., in Fig. 1) due to their darkening of the adatoms, the other half is contained in the SiH_2Cl groups which are adsorbed on the rest atoms. This reaction is self-limiting at room temperature as soon as all rest atom dangling bonds, and most likely also the corner hole dangling bonds, are terminated by SiH_2Cl_2 dissociation products, i.e., by SiH_2Cl. This leaves 40 % of the adatom dangling bonds unreacted, even at saturation coverage. Based on these data it was proposed that the first reaction step of SiH_2Cl_2 on Si(111)7 × 7 at 300 K is a concerted four-center-mechanism where the split-off Cl atoms react with an adatom dangling bond while the residual SiH_2Cl groups adsorb on a rest atom or in a corner hole next to the Cl terminated adatom. These reaction and adsorption processes and the resulting model are discussed in more detail in ref. [14]. In subsequent processes, which proceed at higher temperatures, the SiH_2Cl_2 molecules dissociate completely, leading eventually to Si layer growth (see section 4).

3 CVD growth structures on Si(100)2×1

Monosilane (SiH_4) decomposes on Si surfaces by sequential scission of its Si-H bonds, and it depends on the surface temperature if the molecule dissociates completely upon adsorption or if stable SiH_x species remain on the surface. Details of the dissociation process depend also on the surface hydrogen coverage, i.e., the number of dangling bonds in the vicinity of the dissociating SiH_x species [15]. As it is common to all Si CVD precursors, growth of Si from SiH_4 on Si(100) is either self-limiting or continuous, depending on the surface temperature during exposition: at temperatures below the hydrogen desorption regime ($\approx$600 K) the surface becomes saturated with hydrides released by SiH_4 which leads to passivation, while at higher temperatures hydrogen desorbs continuously and the Si growth rate is either hydrogen desorption limited or, at even higher temperatures, SiH_4 adsorption limited [16]. The temperature of the transition from hydrogen desorption to SiH_4 adsorption limited growth depends also on the SiH_4 flux onto the surface. The structures which are formed on the surface during layer formation depend strongly on the growth region in which SiH_4 exposure is carried out. The correlation between the structures and the SiH_4 decomposition kinetics is illustrated in the following.

The Si surface structures formed during UHV-CVD on the well-known Si(100)2 × 1 surface are different from those known from MBE. The latter have been subject of several investigations [17-20]. At not too high temperatures, where island formation and growth mainly proceeds by diffusion and incorporation of Si monomers, and not by migration of Si oligomers or re-evaporation of island atoms, the shape of the MBE islands is kinetically determined. It is highly anisotropic due to preferential growth perpendicular to the dimer row direction of the underlying layer. This is due to a highly anisotropic sticking coefficient for diffusing Si atoms, which is much larger at the end of the dimer rows than at their sides [21]. Epitaxial islands are also formed during interaction of SiH_4 with Si(100)2 × 1, provided the mobility of the surface species is high enough for island nucleation.

SiH_4 impinging on the surface dissociates in a sequence of steps, leading through SiH_x intermediates [15,22]: higher hydrides decompose below 670 K, while the monohydride is stable up to the onset of hydrogen desorption from this state [8,23]. Depending on the temperature, the stable species remain either isolated on the surface or form islands. The latter is certainly the case if the surface temperature is at the onset of hydrogen desorption from the monohydride state, where no more SiH_x species are stable (≈700 K). Structures typical for Si growth from SiH_4 in this temperature region are reproduced in Fig. 2, which shows an STM image of the Si(100) surface after interaction with SiH_4 at 690 K [13]. This surface still exhibits the (2×1) dimer row structure where the dimer rows are rotated by 90° relative to each other on consecutive terraces.

Upon interaction of SiH_4 with Si(100)2 × 1 under these conditions, Si and hydrogen atoms are released onto the surface. Hydrogen terminated adatoms appear darker in the constant current STM images than unreacted adatoms, as explained in section 2. At 690 K always both atoms of a Si dimer are hydrogen terminated, similar to the H/Si(100) system [8]. The deposited Si atoms initially form small, anisotropic islands with the same dimer row structure as the terraces, the orientation of the dimer rows being perpendicular to the dimer row direction of the substrate terrace. These CVD-grown islands are qualitatively similar to MBE-grown Si islands as far as the dimer row structure and the orientation with respect to the substrate are concerned. Quantitatively, there are considerable differences between MBE-grown and CVD-grown Si islands. After MBE growth the islands are mostly formed by individual dimer strings with an aspect ratio of up to 30:1 at 580 K and still up to 20:1 at 690 K [19], while for CVD growth the anisotropy is considerably less pronounced under similar thermal conditions. Many of the islands have a length of only a few dimers, resulting in an average aspect ratio of 2:1 (Fig. 2a). For MBE deposition the anisotropic shape of the islands was attributed to a combined effect of a strongly anisotropic Si adatom mobility on bare Si(100)2 × 1, which is much faster along the dimer rows than in the perpendicular direction [20], and a strongly anisotropic sticking coefficient of diffusing Si adspecies at the island edges. The smaller aspect ratio of CVD islands can then be assigned to the influence of hydrogen adatoms, which hinder the diffusion of Si containing dissociation products along the dimer rows [13]. Interestingly, the newly grown Si islands are almost free of hydrogen atoms; these tend to aggregate on the surface, forming hydrogen islands [24].

At higher coverages (Fig. 2b) the islands gradually coalesce and antiphase domain boundaries can be formed, either along (type AP1) or perpendicular (AP2) to the dimer rows, depending on the phase shift and the orientation of the two islands relative to each other [19]. Second and higher layer nucleation and growth starts preferentially at AP2 boundaries, long before the underlying

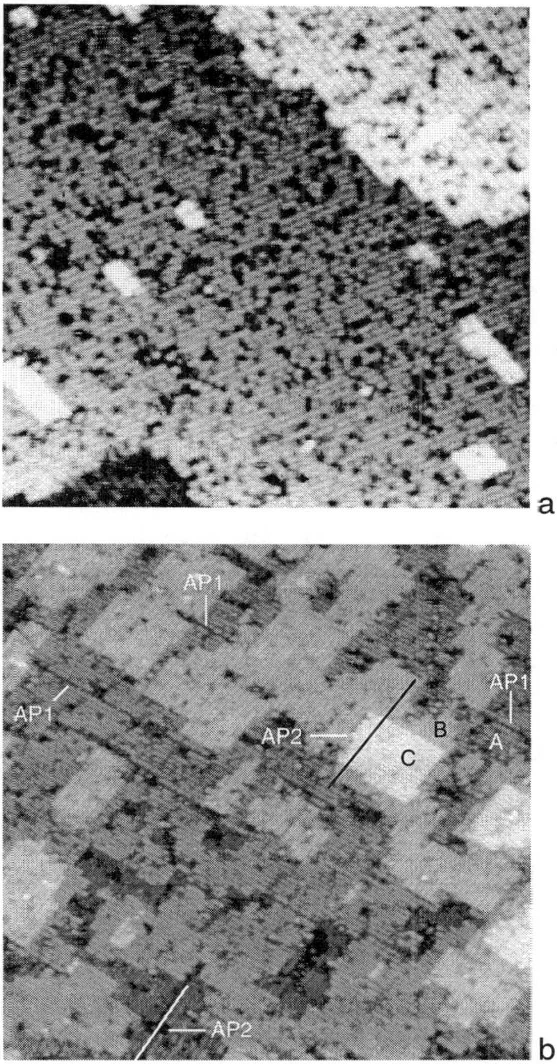

Figure 2: STM images of $Si(100)2 \times 1$ after interaction with SiH_4 ((a) 170 L; (b) 4100 L) at 690 K ($p(SiH_4)=2\times10^{-6}$ mbar). Deposited Si: (a) 0.04 ML, (b) 1.2 ML. Image sizes: (a) 650 Å × 650 Å, (b) 780 Å × 780 Å. [13]

layer is complete [25], which was equally observed for MBE [19] or Si_2H_6-CVD growth [26]. Multilayer growth starts on a percolation network of deposited Si islands (Fig. 2b). Apparently the size and the shape of the islands are not determined by underlying type AP1 domain boundaries, which are simply overgrown, while nucleation at AP2 boundaries is always strongly preferred. To illustrate this behavior some AP1 and AP2 boundaries and also three levels of newly deposited Si (A, B, and C) are marked in Fig. 2b, where the grown Si amounts to 1.2 ML in total.

The structures observed after SiH_4 CVD at different growth temperatures differ considerably from each other. In the following it is shown that the decomposition mechanism of SiH_4 on Si(100)2 × 1 and its dissociation kinetics can be used in helping to understand the various structures seen during UHV-CVD. Following refs. [15,24] SiH_4 decomposition on Si(100)2 × 1 proceeds in a sequence of reaction steps, via several intermediate species:

$$2\mathrm{SiH_4(g)} + 4\mathrm{db} \xrightarrow{k_1} 2\underline{\mathrm{H}} + 2\underline{\mathrm{Si}}\mathrm{H_3} \quad (1)$$

$$2\underline{\mathrm{Si}}\mathrm{H_3} + 2\mathrm{db} \xrightarrow{k_2} 2\underline{\mathrm{H}} + 2\underline{\underline{\mathrm{Si}}}\mathrm{H_2} \quad (2)$$

$$2\underline{\underline{\mathrm{Si}}}\mathrm{H_2} \xrightarrow{k_3} \mathrm{H_2(g)} + \mathrm{H} - \underline{\underline{\mathrm{Si}}} - \underline{\underline{\mathrm{Si}}} - \mathrm{H} \quad (3)$$

and/or

$$2\underline{\underline{\mathrm{Si}}}\mathrm{H_2} + 2\mathrm{db} \xrightarrow{k_{3a}} 2\underline{\mathrm{H}} + \mathrm{H} - \underline{\underline{\mathrm{Si}}} - \underline{\underline{\mathrm{Si}}} - \mathrm{H} \quad (4)$$

$$\mathrm{H} - \underline{\underline{\mathrm{Si}}} - \underline{\underline{\mathrm{Si}}} - \mathrm{H} \xrightarrow{k_4} \mathrm{H_2(g)} + 2\mathrm{db} + \underline{\underline{\mathrm{Si}}} = \underline{\underline{\mathrm{Si}}} \quad (5)$$

$$4\underline{\mathrm{H}} \xrightarrow{k_4} 2\mathrm{H_2(g)} + 4\mathrm{db} \quad (6)$$

The number of surface bonds of the individual species is indicated by single or double underlining, respectively, and the symbol "db" is used for the dangling bonds. Using the kinetic parameters for the individual reaction steps reported in the literature [13,15,22,27,28], the evolving coverage of the different surface species for any combination of growth parameters (surface temperature and SiH_4 flux) can be calculated from this dissociation cascade [29]. For example, in Fig. 3a the temporal evolution of the various SiH_4 decomposition species at 655 K and a SiH_4 pressure of 2×10^{-6} mbar is shown as a function of SiH_4 exposure (SiH_3 and SiH_2 dissociate very rapidly under these conditions). Even at this relatively low temperature, continuous growth of Si is possible, albeit with an extremely small growth rate of 0.29 ML/h. The steady state growth regime is reached at ≈5000 L where the dangling bond concentration is ≈0.49 ML and the hydrogen coverage ≈0.54 ML. As a second example we

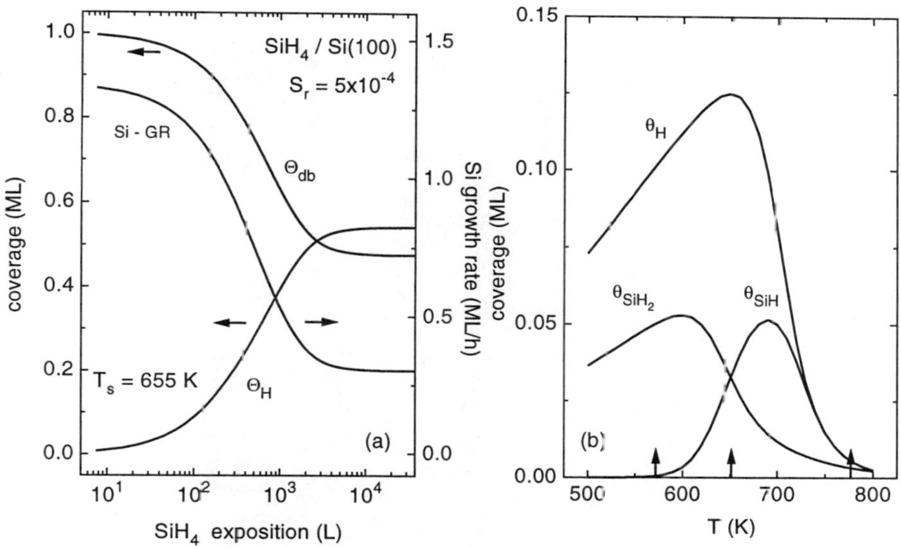

Figure 3: (a) Calculated temporal evolution of the hydrogen and dangling bond concentration and the Si growth rate upon exposure of Si(100)2 $\times$ 1 to SiH_4 (655 K, $p(SiH_4)=2\times10^{-6}$ mbar); (b) Calculated coverages of the SiH_4 decomposition products SiH_2, SiH, and H after growth of 0.1 ML Si, for exposure at $p(SiH_4)=2\times10^{-6}$ mbar at various temperatures. [29]

show in Fig. 3b the population of the various surface species which is found after growth of 0.1 ML of Si as a function of the deposition temperature, but at constant SiH_4 pressure. SiH_3, which is formed in the first decomposition step, decomposes so rapidly that its coverage is negligible in the temperature and pressure regime considered here. Growth at lower temperatures (<650 K) leads to accumulation of atomic H and SiH_2 on the surface, while SiH, which is formed from SiH_2 decomposition, is found for temperatures larger than ≈600 K. After reaching a maximum at medium temperatures the concentration of these species decreases again because of rapid dissociation of the SiH_2 species and desorption of hydrogen, and at 800 K only pure Si islands (or Si step flow, depending on the terrace width) will be present.

The structures actually found in the STM topographies can now be related and compared to the kinetic calculations introduced above. This is illustrated with the STM images in Fig. 4a-c. These STM images, recorded after deposition at $p(SiH_4)=2\times10^{-6}$ mbar at 570 K, 655 K and 775 K and immediate quenching to room temperature after the end of the exposure, correspond to

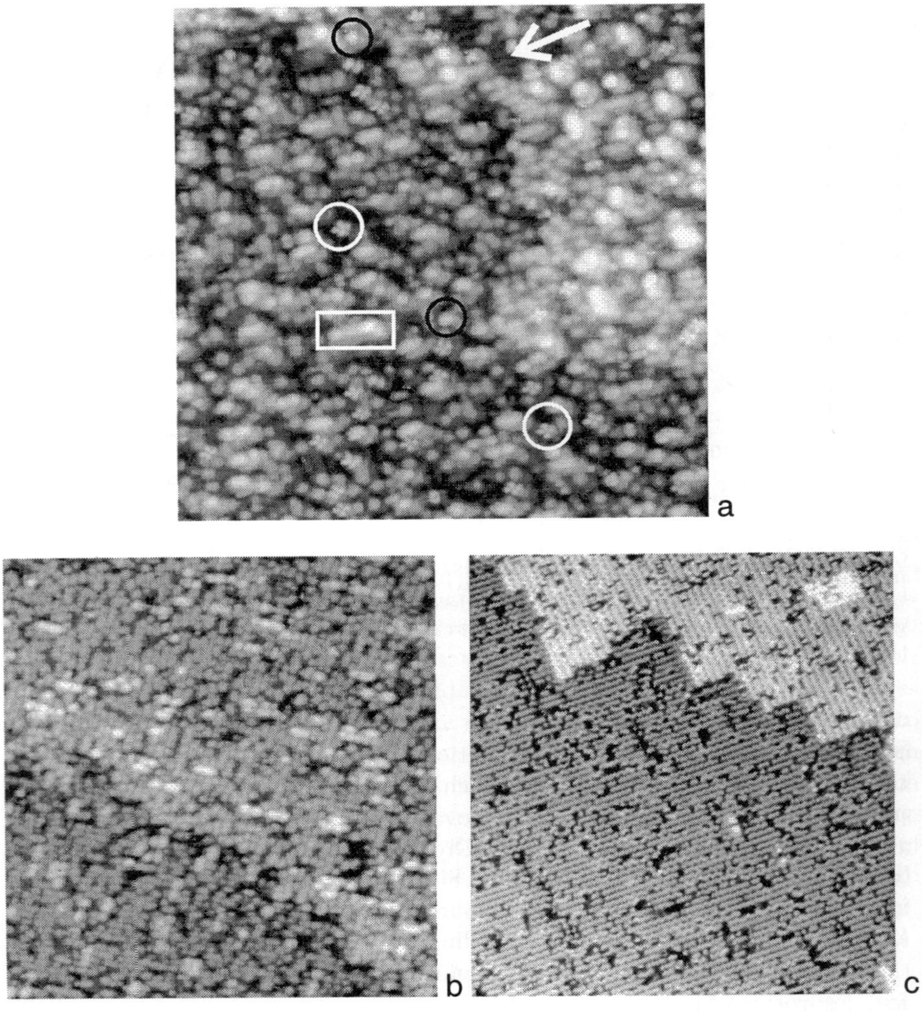

Figure 4: STM images of the Si(100) surface after growth of 0.1 ML Si at $p(SiH_4)=2\times10^{-6}$ mbar. Deposition temperatures: (a) 570 K, (b) 655 K, (c) 775 K. Image sizes (a) 470 Å × 470 Å, (b) 560 Å × 560 Å, (c) 620 Å × 620 Å. [29]

the deposition conditions indicated by the three arrows in Fig. 3b. A variety of different structures can be found after growth at 570 K (Fig. 4a), where SiH_4 adsorption is self-limiting and the hydrogen saturation coverage does not exceed one ML. Most of the structures in Fig. 4a can be unambiguously identified. Epitaxial Si islands are formed even at this low temerature. They are one dimer in width and mainly consist of 2-4 dimer units (e.g., the white rectangle), the majority is not covered by adsorbed hydrogen. In some cases epitaxial, dimerized monohydrides are also found. Furthermore, there are many round or oval protrusions (black circles), which most probably can be assigned to rotated and/or non-rotated dimers, respectively. Additionally, monohydride islands are detected on the original substrate surface (e.g., white arrow), similar to observations in the Si_2H_6/Si(100) system [12]. Most remarkable in Fig. 4a, however, are cross shaped structures (indicated by white circles), which were previously assigned to tetramers consisting of four SiH_2 groups [24,30]. These are metastable clusters which are formed under conditions where the concentration (see Fig. 3b) and the mobility of SiH_2 groups is sufficiently large, but their decomposition rate is still small enough so that tetramers can be formed.

The topography is different if the surface is exposed to SiH_4 at higher temperatures. This is illustrated in Fig. 4b, after deposition at 655 K. The silylen tetramer units are not observed anymore and longer epitaxial islands, one or two dimer units in width, have grown. Also, islands of aggregated hydrogen can be observed. This situation corresponds to a higher dissociation rate of the SiH_2 groups and a higher mobility of the SiH_4 dissociation fragments.

At even higher temperatures nucleation of Si or H_{ad} islands drops and Si adatoms condense mainly at step edges (Fig. 4c). It has been established that Si atoms are most efficiently incorporated at S_B type step segments [17], where the dimer rows are perpendicular to the step direction. As a consequence, large portions of the terraces are essentially free from individual Si islands and the remaining islands are considerably less anisotropic than at lower temperatures. The steady state Si hydrogen (monohydride) coverage under these conditions is $\approx$0.02 ML, that of higher hydrides even less, only a few tenths of a percent. Under these conditions the steady state growth rate is 1.84 ML/h.

A similar comparison between the kinetic growth model and STM images for deposition at various gas fluxes but constant surface temperature can be found in ref. [29].

Interesting similarities but also differences between the MBE Si/Si(100)2×1 system without [20] and with [31] coadsorbed hydrogen on the one hand and the SiH_4/Si(100)2×1 CVD system on the other hand can be found when analyzing the Si island density as a function of deposition temperature (Fig. 5). We here discuss the island densities in

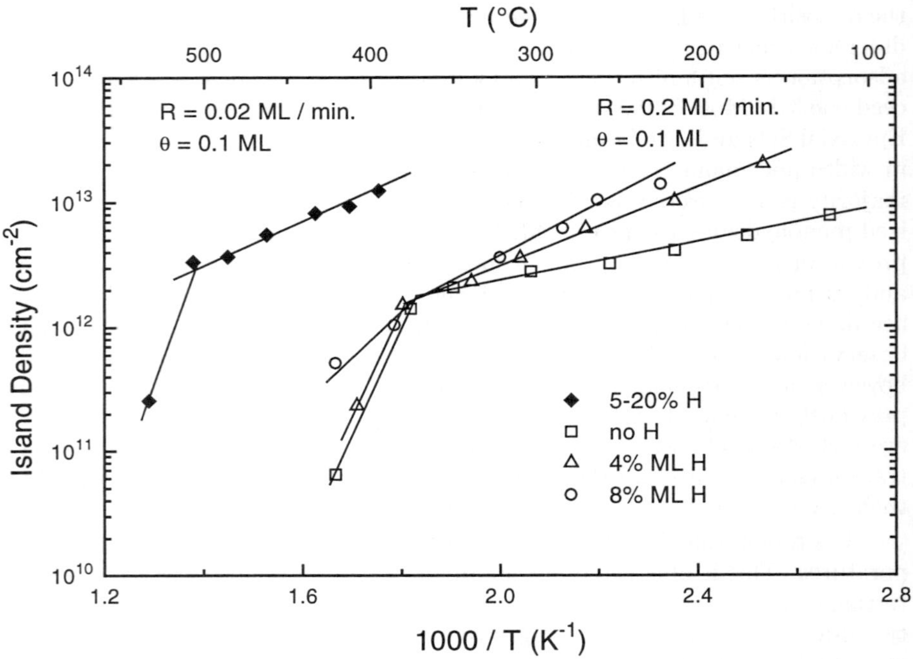

Figure 5: Arrhenius plots of the Si island densities in the MBE Si/Si(100)2×1 system with and without preadsorbed hydrogen (open symbols) [31] and in the CVD SiH_4/Si(100)2×1 system (solid symbols).

these systems only qualitatively; a more rigorous evaluation would require detailed knowledge of all atomic processes, which is not (yet) available. The quantification of these processes in terms of activation energies and prefactors still remains a tremendous task, since, e.g., a description of Si adatom motion on Si(100)2×1 alone requires consideration of at least six different hopping channels [32]. In the CVD system SiH_x dissociation and diffusion of H and SiH_x fragments render their microscopic description even more complex.

When comparing the Arrhenius plots of the Si island densities in Fig. 5 with each other, one finds that in each system the slope of the graphs changes at a certain temperature: from a larger slope at the higher temperature side to a smaller one at the lower temperature side. It can also be seen that in the MBE system the slope of the graph changes with H precoverage, which is most obvious at the lower T side of the curves: the higher the H precoverage, the larger the slope. This has been assigned to an increase of the *effective* diffusion barrier for Si atoms [20] by coadsorbed H from 0.68 meV for zero H coverage [20] to 1.24 eV for a H precoverage of 8 % of a ML, assuming that Si diffusion on Si(100) is essentially one-dimensional (although this might not be self-evident with coadsorbed H). The physical origin for the increased effective diffusion barrier, however, is not yet clear. Notably, the temperature where the slope changes does not significantly depend on the hydrogen precoverage in the MBE system. The change in the slope at higher temperature in the MBE system was attributed to an increased mobility of the Si atoms and possibly also Si dimer mobility [33]. On the one hand this leads to an increased incorporation rate of diffusing adatoms at the steps so that these adatoms cannot contribute to island nucleation, which decreases the effective island density found after deposition on the terraces [20]. On the other hand the Ostwald ripening of islands will also be accelerated, leading to faster coarsening and thus also to a decrease in island concentration.

In the CVD system, the low temperature slope is very similar to the MBE system with a H precoverage of 4 % of a ML, although it is clear that due to the presence of various SiH_x intermediates in the CVD system the surface composition must be very different from the MBE system, where only Si and H atoms are present on the surface. The effective surface diffusion barrier of 1.42 eV, which could be extracted from the low temperature section of the graph for the CVD system, is then the result of the interplay of all the different processes which contribute to Si growth (and hence also surface diffusion) in the SiH_4/Si(100) system and cannot be assigned to a specific process alone.

There is a remarkable difference between the SiH_4/Si(100) and the Si/Si(100) MBE system [20] concerning the temperature above which the island density begins to decrease rapidly. In the MBE system, the fast decrease of the island density begins at ≈550 K while in the CVD system this decrease starts only at ≈770 K, at a comparable terrace width and Si growth rate. The kinetics of SiH_4 decomposition and CVD layer growth discussed above and the mechanistics of epitaxial Si dimer formation from Si hydride species may be used to rationalize this difference. It has been proposed earlier that SiH_2 groups which are formed after dissociation of Si_2H_6 on Si(100)2 × 1 decompose similar as in the SiH_4/Si(100)2 × 1 system, i.e., via reaction 3, forming non-rotated monohydride dimers [12]. These non-rotated monohydride dimers then loose the hydrogen atoms and rotate to form elementary epitaxial Si islands or they may dissociate further leading to diffusing Si atoms. If reaction 3 is responsible for the initial formation of (monohydride) dimers, then this could also be the rate limiting step for island nucleation since it has the highest activation energy in the entire Si hydride decomposition cascade of SiH_4

(reactions 1 - 5).

4 CVD growth structures on Si(111)7×7

Deposition of silicon by SiH_4 UHV-CVD on Si(111)7 × 7 can be divided into the same three growth regions, depending on the surface temperature. At surface temperatures below ≈700 K Si deposition from SiH_4 is self-limiting; the surface becomes passivated by hydrides formed in the course of SiH_4 dissociation. Continuous film growth can be observed at temperatures above ≈700 K, which coincides with the onset of the β_1 hydrogen desorption peak [23]. In this temperature region reactivation of surface dangling bonds by hydrogen desorption is the Si growth limiting step, while for even higher temperature, in the third region, dissociative SiH_4 adsorption is rate limiting. The transition temperature between these latter two growth regimes depends on the gas flux; under UHV- and very low pressure- (VLP)-CVD conditions it seems to lie at ≈810 K [23]. Typical growth structures which are found on the surface in UHV-CVD in each of these three regions are discussed below.

In principle, the separation into these three regions also holds for Si deposition from SiH_2Cl_2: at low temperature growth is limited by surface desorption, while at higher temperatures the SiH_2Cl_2 flux is limiting. Using this precursor, the transition between desorption and adsorption limited growth is around 1150 K under UHV-CVD conditions, with the exact transition temperature depending on the SiH_2Cl_2 flux [34]. This higher transition temperature is mainly due to adsorbed Cl from SiH_2Cl_2 dissociation: in thermal desorption after SiH_2Cl_2 adsorption desorption of HCl (around 850 K) and $SiCl_2$ (around 950 K) is found [34], in addition to a hydrogen peak similar to the SiH_4 β_1 peak. Thus, the surface is passivated by chlorides and hydrides at temperatures below the onset of HCl desorption, while it is passivated by monochlorides only below the onset of $SiCl_2$ desorption (≈850 K). An example is also shown below.

Not only the growth behavior, but even more the topography of the Si structures formed during CVD-deposition differs significantly between these growth regimes which will be illustrated with a few examples.

The lowest possible temperature for continuous Si growth from SiH_4 on Si(111)7 × 7 is around 700 K, at the onset of hydrogen desorption from the monohydride state. At this temperature the original (7×7) surface is severely restructured (Fig. 6): the adatoms are removed from their original sites by hydrogen released from decomposing SiH_4 molecules, which can bond to adatoms and simultaneously break their backbonds to the substrate by insertion of H atoms. Here, center adatoms are etched preferentially as compared to corner

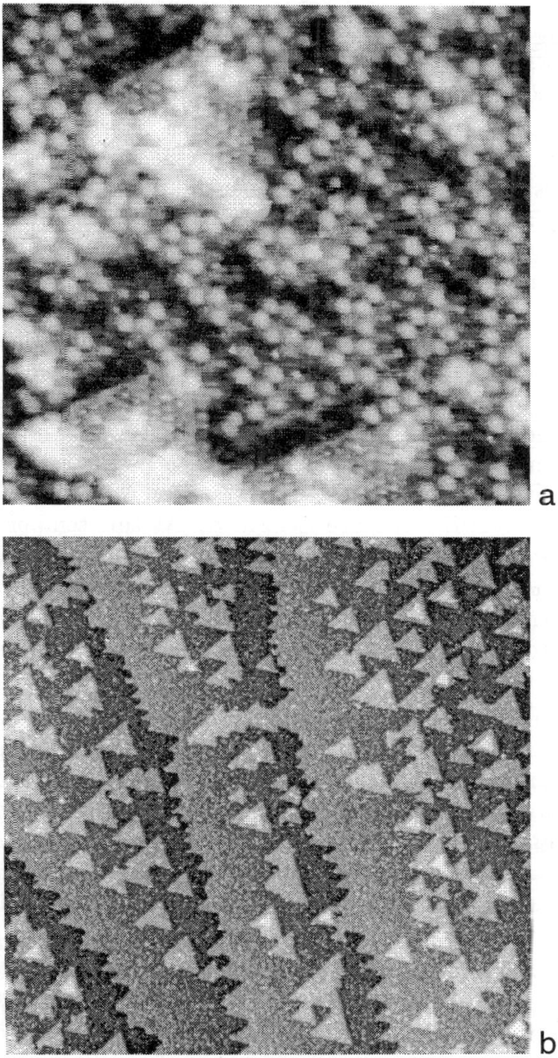

Figure 6: Si(111)7 × 7 after interaction with SiH_4 at 690 K ($p(SiH_4)$=2×10^{-6} mbar). (a) 4800 L, image size 180 Å×180 Å [13]; (b) : 21600 L, image size 2700 Å×2700 Å. [35]

adatoms [35]. With increasing coverage also the rest atoms are involved in the restructuring process. The stacking fault is lifted and gradually (1×1):H regions develop (Fig. 6a) until finally the former DAS structure has been completely removed and a monohydride terminated Si(111)-(1×1):H surface is formed [35], in a process analogous to the (7×7)→(1×1) transformation upon interaction with atomic hydrogen [9] or chlorine [36] at 650-670 K. The newly evolving islands which include former Si adatoms and Si supplied from the gas phase by SiH_4 also adopt a bulk-like monohydride terminated (1×1):H structure. In contrast to what is observed after interaction with atomic H [9] or Cl [36] these new monohydride islands are strictly triangular, and nearly all of them have the same azimuthal orientation (Fig. 6b). In those exceptional cases, where the islands appear to be rotated by 60° stacking faults have grown, while the vast majority of islands grow without a stacking fault. This suppression of stacking fault growth during interaction with SiH_4 can be assigned to the additional supply of Si from the gas phase during CVD which reduces the Si mass transport on the surface necessary to lift the stacking faults [35]. Additionally, Si is also transported to the steps, where the same quasi-triangular (1×1):H structure is formed. At this temperature of 690 K the surface is largely passivated, and further growth is probably limited by hydrogen desorption from the edges of the islands or the steps. Models for the growth of such triangular Si islands, which include the lifting of the stacking faults and the growth shapes observed at the step edges, are discussed in refs. [35,37,38].

If the surface is exposed to SiH_4 at temperatures in the region of the β_1 H_2 thermal desorption peak at not too high pressures, a different growth process can be observed, resulting in a change of the surface structures. For SiH_4 exposure at 770 K, at a pressure of 2×10^{-6} mbar, no island nucleation is found any more in the middle of the terraces and all islands exclusively nucleate at translational domain boundaries of the original (7×7) surface. Consequently, strings of islands are found on the surface (Fig. 7a). These islands still have a triangular shape, but their atomic structure is very different from the monohydride terminated (1×1) at lower temperature. Because of continuously desorbing hydrogen the a (1×1):H surface is not stable in this temperature region. Instead, in the islands a relatively disordered arrangement of adatoms with (7×7) and even (5×5) regions in between is observed (Fig.7b).

The original substrate surface shows a defective (7×7) structure, where the missing adatoms are etched by hydrogen and transported to islands or to the step edges. Growth at steps can be seen from the near-triangular structures at the edges as compared to the originally straight edges.

At even higher temperatures growth proceeds via a step flow process, and

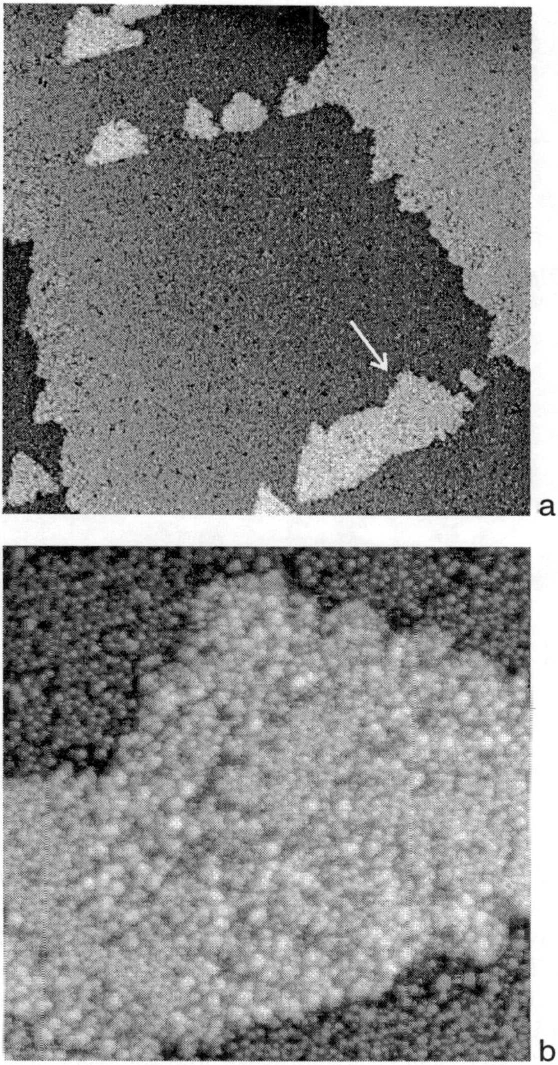

Figure 7: $SiH_4/Si(111)7 \times 7$ after exposure to 5400 L SiH_4 at 770 K ($p(SiH_4)=2\times10^{-6}$ mbar). Image sizes: (a) 2810 Å × 2810 Å, (b) 570 Å × 570 Å. The arrow in (a) points to the region corresponding to (b). [35]

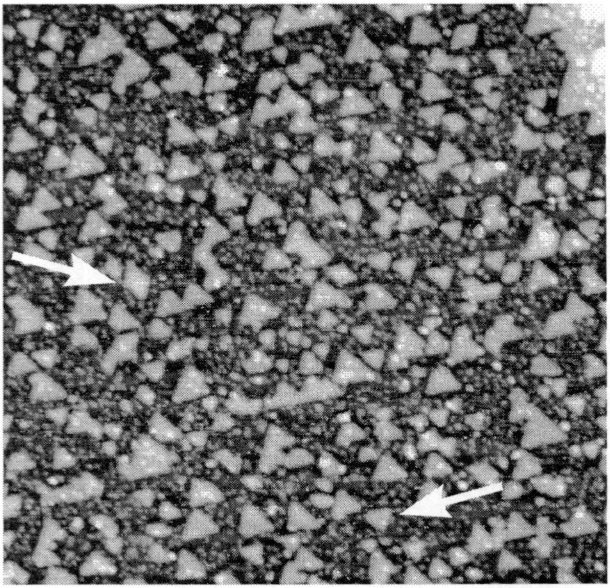

Figure 8: Si(111)7 × 7 after exposure to 9×10^4 L SiH_2Cl_2 at 850 K and $p(SiH_2Cl_2)=1\times10^{-3}$ mbar. Some grown islands with stacking faults are marked by arrows. Image size 1320 Å × 1320 Å. [39]

the formation of preferential step orientations can be observed at least up to 850 K [35].

An island structure similar to the Si(111)-(1×1):H surface evolves also with the SiH_2Cl_2 precursor, but with some important differences (Fig. 8). As mentioned above, exposure to SiH_2Cl_2 at 850 K leads to a hydrogen free surface, since hydrogen either desorbs as H_2 or it reacts with a Cl atom to desorb as HCl. The surface then finally becomes monochloride terminated and a Si(111)-(1×1):Cl structure is found, with a topography similar to the (1×1):H surface formed in the course of SiH_4 exposure at around 700 K. In this case also triangular (1×1) islands are formed, however, they are Cl terminated, as evidenced by XPS [7] and AES [39] results. Homogeneous nucleation in the middle of the terraces and heterogeneous nucleation at step edges and at translational domain boundaries of the substrate is also observed. Compared to the SiH_4-CVD (1×1):H surface growth involving stacking faults is found more frequently, as evidenced by triangular islands which are rotated by 60° (some are marked with arrows in Fig. 8) with respect to the majority of the islands. At higher temperatures, the growth also transforms into a step flow mode with the SiH_2Cl_2 precursor.

5 Summary

The various surface reaction steps which involve precursor species and the original substrate are closely associated with the growth structures observed in Si epitaxy. These growth structures can vary drastically, depending on the growth conditions. We have discussed here selected systems of Si homoepitaxy that reveal either novel chemical effects or strong connections between fundamental surface kinetics and chemistry and CVD growth. It is therefore obvious that the understanding of such surface structures and the processes by which they are formed is essential for the ultimate control of CVD processes.

Acknowledgment

We would like to thank M. Fehrenbacher, J. Spitzmüller, M. Pitter, and F. Haug for their excellent experimental work. We would also like to thank the Stiftung Volkswagenwerk for generous financial support of our work.

References

1. H. Neergard Waltenburg and J. T. Yates, jr., Chem. Rev. 95 (1995) 1589.
2. R. J. Hamers and Y. Wang, Chem. Rev. 96 (1996) 1261.
3. H. Neddermeyer, Rep. Prog. Phys. 59 (1996) 701.
4. S. M. Gates, Chem. Rev. 96 (1996) 1519.
5. S. M. Gates, J. Phys. Chem. 96 (1992) 10439.
6. P. A. Coon, P. Gupta, M. L. Wisa, and S. M. George, J. Vac. Sci. Technol. A 10 (1992) 324.
7. J. A. Yarmoff, D. K. Shuh, T. D. Durbin, C. W. Lo, D. A. Lapiano-Smith, F. R. McFeely, and F. J. Himpsel, J. Vac. Sci. Technol. A 10 (1992) 2303.
8. J. J. Boland, Adv. Phys. 42 (1993) 129.
9. F. Owman and P. Mårtensson, Surf. Sci. 303 (1994) L367; F. Owman and P. Mårtensson, Surf. Sci. 324 (1995) 211.
10. U. Memmert, A. Berkó, and R. J. Behm, Surf. Sci. 325 (1995) L441.
11. L. Bolotov, H. Rauscher, and R. J. Behm, Chem. Phys. Lett. 243 (1995) 445.

12. Y. Wang, M. J. Bronikowski, and R. J. Hamers, Surf. Sci. 311 (1994) 64.
13. M. Fehrenbacher, J. Spitzmüller, U. Memmert, H. Rauscher, and R. J. Behm, J. Vac. Sci. Technol. A 14 (1996) 1499.
14. M. Fehrenbacher, H. Rauscher, and R. J. Behm, Z. Phys. Chem. 198 (1997) 205.
15. S. M. Gates and S. K. Kulkarni, Appl. Phys. Lett. 58 (1991) 2963.
16. M. Liehr, C. M. Gates, S. R. Kasi, and M. Offenberg, Appl. Phys. Lett. 56 (1990) 629.
17. Y.-W. Mo, B. Swartzentruber, R. Kariotis, M. Webb, and M. G. Lagally, Phys. Rev. Lett. 63 (1989) 2393.
18. R. J. Hamers, U. K. Köhler, and J. E. Demuth, Ultramicroscopy 31 (1989) 10.
19. R. J. Hamers, U. K. Köhler, and J. E. Demuth, J. Vac. Sci. Technol. A 8 (1990) 195.
20. Y.-W. Mo, J. Kleiner, M. B. Webb, and M. G. Lagally, Surf. Sci. 268 (1992) 275.
21. C. Pearson, M. Krueger, and E. Ganz, Phys. Rev. Lett. 76 (1996) 2306.
22. S. M. Gates, C. M. Greenlief, and D. A. Beach, J. Chem. Phys. 93 (1990) 7493.
23. S. M. Gates, C. M. Greenlief, D. B. Beach, and P. A. Holbert, J. Chem. Phys. 92 (1990) 3144.
24. J. Spitzmüller, M. Fehrenbacher, M. Pitter, H. Rauscher, and R. J. Behm, Phys. Rev. B 55 (1997) 4659.
25. J. Spitzmüller, M. Fehrenbacher, F. Haug, H. Rauscher, and R. J. Behm, Appl. Phys. A, *submitted.*
26. M. J. Bronikowski, Y. Wang, and R. J. Hamers, Phys. Rev. B 48 (1993) 12361.
27. P. Gupta, V. L. Colvin, and S. M. George, Phys. Rev. B 37 (1988) 8234.
28. K. Sinniah, M. G. Sherman, L. B. Lewis, W. H. Weinberg, J. T. Yates, Jr., and K. C. Janda, Phys. Rev. Lett. 62 (1989) 567.
29. M. Fehrenbacher, J. Spitzmüller, M. Pitter, H. Rauscher, and R. J. Behm, Jpn. J. Appl. Phys. 36 (1997) 3804.
30. J. Spitzmüller, M. Fehrenbacher, H. Rauscher, and R. J. Behm, Surf. Sci. 377-379 (1997) 1001.
31. J. E. Vasek, Z. Zhang, C. T. Salling, and M. G. Lagally, Phys. Rev. B 51 (1995) 17207.
32. A. P. Smith and H. Jónsson, Phys. Rev. Lett. 77 (1996) 1326.
33. B. S. Swartzentruber, Phys. Rev. Lett. 76 (1996) 459.

34. P. A. Coon, M. L. Wise, and S. M. George, J. Cryst. Growth 130 (1993) 162.
35. M. Fehrenbacher, H. Rauscher, U. Memmert, and R. J. Behm, Surf. Sci. (1997) *in press.*
36. J. J. Boland and J. S. Villarrubia, Phys. Rev. Lett. 41 (1990) 9865.
37. H. Tochihara and W. Shimada, Surf. Sci. 296 (1993) 186.
38. W. Shimada and H. Tochihara, Surf. Sci. 311 (1994) 107.
39. M. Fehrenbacher, H. Rauscher, and R. J. Behm, *to be published.*

METAL-ON-METAL GROWTH

TWO-DIMENSIONAL ISLAND SHAPES

THOMAS MICHELY
Institut für Grenzflächenforschung und Vakuumphysik, Forschungszentrum Jülich,
52425 Jülich, Germany
E-mail: michely@igv016.igv.kfa-juelich.de

GEORGE COMSA
Institut für Physikalische und Theoretische Chemie, Wegelerstr. 12,
53115 Bonn, Germany
E-mail: comsa@uni-bonn.de

1 Introduction

"A snowflake is a letter to us from the sky" [1]. This sentence is probably the best introduction to this chapter. When receiving a letter, the first step is to be able to read it. The scanning tunneling microscope (STM) gave us this opportunity by revealing the shape of two-dimensional (2D) crystals down to atomic dimensions. The STM images in fig. 1 give a first impression of the variety and beauty of monatomic high 2D-islands obtained by molecular beam epitaxy (MBE) of Pt on a Pt(111)-surface at different substrate temperatures.

Intellectually more demanding, like in the case of letters, is to properly understand the message. Certainly, the island shapes reflect the growth conditions. But what is the relation between growth conditions and actual growth shapes, and how are they related to individual atomic processes? A prerequisite for answering these questions is a broad, detailed and faithful basis of experimental data, obtained under various and especially clear defined experimental conditions. STM-images, being in many cases persuasive per se, tempt us often to draw "obvious" conclusions. Internal consistency considerations and a tight interaction with kinetic Monte-Carlo (kMC) simulations may avoid misleading conclusions. The kMC simulations have, indeed, proved to be very useful, when reasonable assumptions and realistic input parameters where used. These parameters are obtained from available experimental data and from theoretical calculations, which range from empirical to ab initio ones. As will become apparent below, in some cases, where only a few processes and energy parameters, leading to certain island shapes, are involved, a rather unique interpretation could be obtained. In others, due to a large number of simultaneously active processes, often controversial interpretations, again like in the case of letters, in particular of those from the sky, are offered.

Islands like those shown in figs. $1e_1$ and e_2 are obtained by deposition at

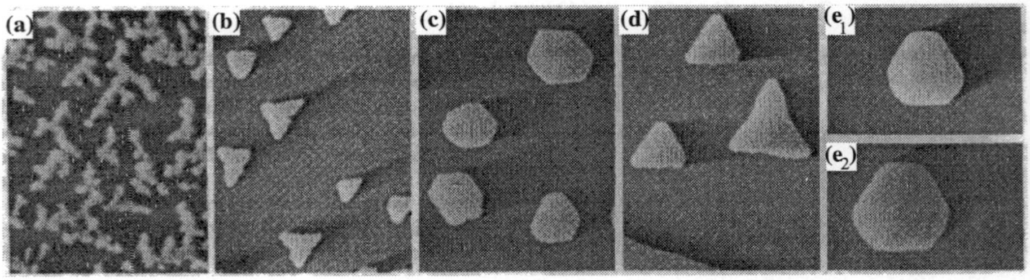

Figure 1: Shapes of monatomic high Pt-islands after MBE of Pt atoms on Pt(111) with a typical rate of 10^{-2} ML/s at different substrate temperatures T. (a) T = 200 K, Θ = 0.2 ML, scale: 280 Å × 400 Å; (b) T = 400 K, Θ = 0.08 ML, scale: 1300 Å × 1900 Å; (c) T = 455 K, Θ = 0.14 ML, scale: 770 Å × 1100 Å; (d) T = 640 K, Θ = 0.15 ML, scale: 2300 Å × 3300 Å; (e_1) T = 700 K, Θ = 0.08 ML, scale: 1540 Å × 1100 Å; (e_2) T = 450 K, Θ = 0.08 ML and subsequent 60 s annealing at 710 K, scale: 1540 Å × 1100 Å.

$T \geq 700\ K$ or by deposition at lower temperature and subsequent annealing at $\geq 700\ K$. They result from a close-to-equilibrium situation. Their characteristic shape depends neither on history (including time), nor on their size. It is the equilibrium shape. The relative values of the step edge free energies, which unequivocally determine the shape, can be obtained from the analysis. All other island shapes (e.g. figs. 1a-d), which are the consequence of a far-from-equilibrium situation, depend mainly on the kinetics of the growth process rather than on thermodynamics. Their shapes are called growth shapes. They depend sensibly on the deposition parameters, of which the most important are deposition temperature and rate. These deposition parameters determine which of the many potentially possible diffusion processes are actually effective during growth as well as their relative contribution to it. From the successful analysis one obtaines activation energies (barrier heights) and frequency factors of diffusion processes. The following sections are devoted correspondingly to the equilibrium and the growth shapes, respectively.

The study of 2D-island shapes is primarely motivated by their surprising beauty and by the endeavor to understand their origin and diversity. The information obtained is certainly useful for the modern field of nanotechnology, in general, and for thin film growth, in particular. Indeed, the characteristic shapes of the islands, even if confined - as here - to the submonolayer growth of a particular homoepitaxial system, appear to have in many cases a decisive importance for the nature of the growth far from equilibrium.

2 The equilibrium shape

In three dimensions (3D) the crystal shape corresponds to the minimum of the total surface free energy

$$\int_S \Gamma(\theta, \phi) ds = min! \tag{1}$$

with the number of crystal particles kept constant [2,3,4]. Under equilibrium conditions, the shape corresponding to Eq. (1) is uniquely determined by the state variables (p, T) and the material parameter $\Gamma(\theta, \phi)$, the specific surface free energy, and is independent of history and island size. In order to obtain in a reasonable time crystals of equilibrium shape, the material transport has to be strong enough in relation with the number of crystal particles. Accordingly, the few 3D equilibrium shapes achieved so far have been obtained for small crystallites and only near the melting point [5,6,7,8,9,10,11,12]. An interesting consequence of the uniqueness of Eq. (1) is the identity of the equilibrium shape of a crystal to that of a hollow volume in the same material.

The two dimensional 2D-crystal shapes, we are here dealing with, correspond in analogy with Eq. (1) to the minimum of the total step free energy. Obviously, also in this case, on a given face of a crystal the equilibrium shape of adatom islands is identical to that of vacancy islands. The first attempts to investigate 2D-equilibrium shapes were performed on monatomic high Si-islands on Si(001) by means of STM and by Low Energy Electron Microscopy (LEEM) [13,14]. In both cases the island shapes were not fully equilibrated, and accordingly, only a rough estimate of the ratio of the step energies could be obtained. Meanwhile, the correct equilibrium shape of Si-islands on Si(001) has been determined by LEEM [15]. The first accurate determination of the equilibrium shape of 2D-adatom and -vacancy islands has been performed on Pt(111) [16,17].

The proof that a given 2D-island has reached its equilibrium shape is not straightforward. Indeed, the perfect equilibrium shape of a 2D-island can be obtained in principle only if the island is located on a crystal which in turn is fully equilibrated. (Note that in this case at most one 2D-island can be present on a given crystal face, and this only on the most favorable type of face.) Such a state is hardly achievable in practice. In order to obtain island shapes at least very close to the equilibrium ones, one has to create at and around the islands conditions of local equilibrium. Thus one has to analyse very carefully the island creation conditions and in particular whether the properties and the behavior of the islands correspond to those expected for ideally equilibrated ones. This has been done for the monatomic deep vacancy islands and monatomic high Pt-adatom islands on Pt(111) shown in figs. 2b, c, e and f . The proof that they possess very close to equilibrium shapes is based on the following observations:

1. Irrespective of the generation procedure the islands exhibit within measurement error identical shapes after a few minutes of annealing above

700 K. This is illustrated in figs. 2a, b and c for vacancy islands created by ion bombardment and in figs. 2d and e for adatom islands obtained by MBE. See also figs. 1e_1 and e_2.

2. Irrespective of their size (> 100 Å in diameter) the island shapes are identical.

3. The island shape does not change during long time annealing.

4. The shapes of adatom and vacancy islands are identical. This is obvious from the comparison of figs. 2b and c with fig. 2e and in particular from fig. 2f, where both types of islands are present. The adatom and vacancy islands are mutually rotated by 180°. This is due to the fact that step edges of the same structure are interchanged in adatom and vacancy islands, respectively (see fig. 3).

5. The shape of the islands is nearly hexagonal, with the sides oriented along the $< 110 >$ directions. Deviations from the exact hexagonal shape are due to the slightly rounded corners.

From a large number of equilibrated adatom and vacancy islands obtained from STM-images like those shown in fig. 2, the ratio of the distances to the hexagon center of the long to that of the short hexagon side for Pt-islands on Pt(111) has been determined to 0.87 ± 0.02. As shown further below this is also the ratio of the specific step free energies, γ_A/γ_B, of the two sides. (The short (long) ones are made up of {100} ({111}) microfacets and are usually named A (B) steps - see fig. 3.)

Two comments:

The first concerns the equilibration mechanism, which is per definition irrelevant only for perfectly equilibrated systems. However, as discussed above, the close to equilibrium shapes, which are actually obtained in practice, result under local equilibrium conditions. In this case the equilibration mechanism is certainly relevant. The equilibration of 2D-islands proceeds by material transport mainly through adatom 2D-evaporation from and attachment to step edges. This equilibration mechanism, which has been proved in detail in the case of vacancy islands to be the determining one [18], probably also holds for adatom islands. On this basis we can infer the conditions that are favorable for the generation of adatom islands with close to equilibrium shapes. The optimal situation appears to be the presence of a large field of adatom islands of similar size, remote from preexistent steps. This leads to a homogeneous adatom 2D-gas concentration $\rho(r)$ between the islands, determined by the average radius of the islands. Thus the adatom evaporation and reattachment fluxes are identical, a necessary condition for equilibrium. In contrast, if the island is located in the vicinity of a preexistent, mostly linear step, the adatom

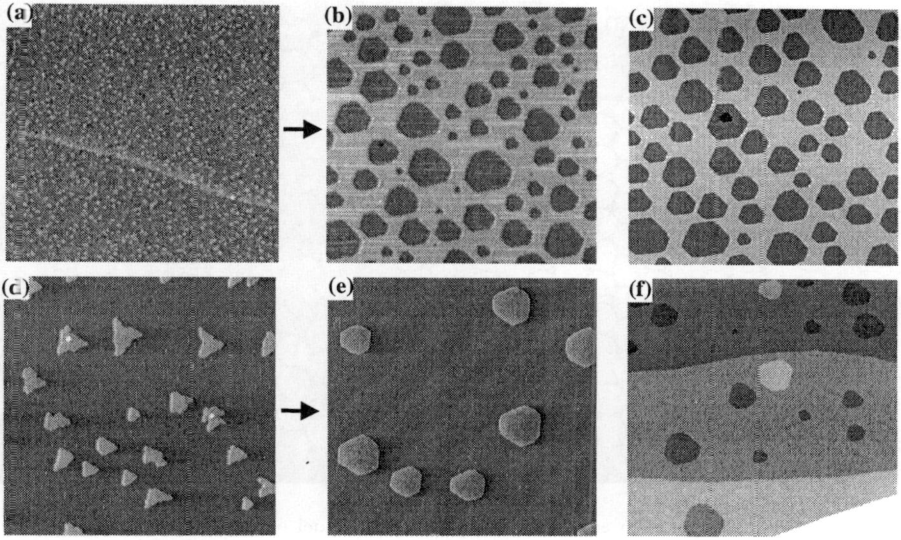

Figure 2: Vacancy and adatom islands generated by ion bombardment (600 eV Ar^+, 2.8×10^{16} $ions/m^2s$) and by MBE of Pt-atoms on Pt(111). (a) 0.45 ML removed during 120 s bombardment at T = 245 K. (b) Same as (a), but with subsequent 120 s annealing at T = 750 K. (c) 0.45 ML removed during 120 s bombardment at T = 760 K. (d) MBE of 0.1 ML at T = 425 K. (e) Same as (d), but with subsequent 60 s annealing at T = 700 K. (f) 0.45 ML removed during 120 s bombardment at 625 K and subsequent 120 s annealing at 850 K. Scales: (a), (b) and (c) : 1650 Å × 1650 Å; (d) and (e) :2000 Å × 2000 Å; (f): 3300 Å × 3300 Å.

gas density is highly inhomogeneous due to the different vapor pressure of a linear step and an island edge with a relatively small radius of curvature. In this case the local equilibrium situation is not established and thus the equilibrium shape can hardly develop.

Concerning the shape equilibration of vacancy islands, we have to consider two further aspects. The first one is the "negative" material transport ensuing from the thermal generation of surface monovacancies. At least in the case of the Pt(111)-surface, the activation energy for the creation of a monovacancy is larger than that of an adatom so that the vacancy fluxes can be neglected. The second particularity of vacancy islands results from the possible presence of a significant Ehrlich-Schwoebel barrier [19,20] at the step edge. This barrier, with a height of several tenths of an eV for the close packed steps on Pt(111) [21], hampers the adatoms from crossing the step edge. At not too high temperatures, these two aspects isolate the interior of the vacancy islands, so that their equilibration proceeds by "internal" adatom exchange, almost independently

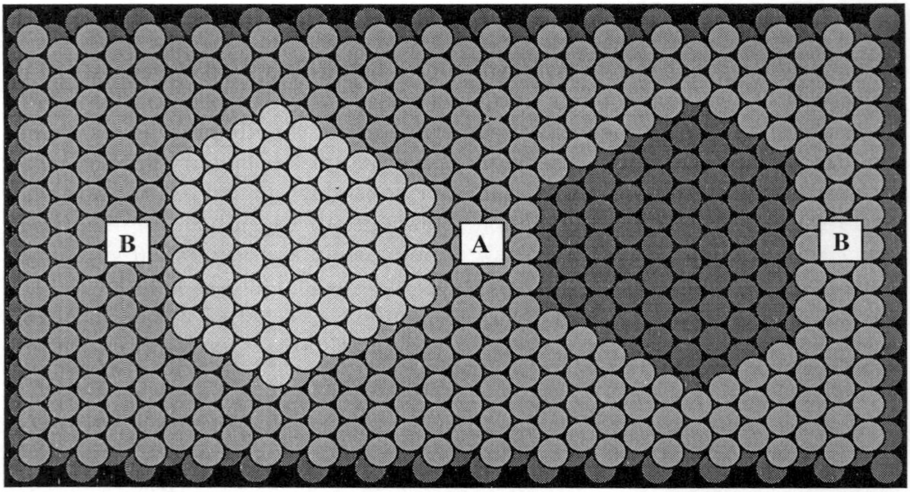

Figure 3: Definition of step edges A and B in a ball model of an adatom and a vacancy island with idealized equilibrium shape. The model evidences the cause for the mutually 180° rotated orientation of the adatom and vacancy island.

from the state of the surrounding adatom gas.
Although the formation of a significant adatom 2D-gas density (above 700 K for Pt(111)) is the decisive and rapid equilibration mechanism, the diffusion along step edges can lead at lower temperatures and for smaller islands to shape equilibration on a longer time scale. This has been, indeed, observed on Pt(111) for small islands after annealing between 620 K and 700 K. At lower annealing temperatures equilibrated island shapes have not been observed.

The second comment concerns the accuracy of the island equilibrium shapes generated experimentally. The islands observed in a large number of deposition and sputtering experiments performed between 700 K and 900 K over the years exhibited in average the features indicated above. However, slight deviations were present both between islands produced in the same (see, e.g. fig. 2), as well as (more systematic ones) between islands generated in different experiments. The shape fluctuations within the vacancy island ensemble shown in fig. 2c can be explained by the natural fluctuation of the island shape at the equilibration temperature and by the mutual influence of neighbouring islands (i.e., by inhomogeneities in the local 2D-gas density, due to island size differences), which cause inhomogeneous island growth and shrinking. The slight systematic deviations between different experiments can be due to differences in average island sizes or densities, to the presence of impurities or to alterations during cooling to the imaging temperature. Significant island shape modifications during cooling from equilibration temperatures only slightly larger than 700 K are improbable for Pt(111), because the dominant and most rapid material transport mechanism "freezes" below 700 K. In contrast, for higher equilibration temperatures the corresponding equilibrium shape may undergo significant alterations during the initial cooling period (until crossing the 700 K mark), because of the effectiveness of the material transport in this temperature range. In spite of these uncertainties one may infer two trends from experiments performed at various equilibration temperatures: i) with increasing temperature the hexagon sides become less pronounced; instead of the A-steps, one observes rounded edges and even the B-steps exhibit a slight curvature, which increases with temperature, and ii) above 900 K the difference between the distances of sides of different structure to the island center decreases. More reliable information on the temperature dependence of the island equilibrium shape can only be obtained from more appropriate experiments. STM-measurements at the equilibration temperature would be a decisive progress. By such "in situ" experiments one might also monitor temperature dependent shape changes.

The experimental determination of the equilibrium shape of 2D-islands is worthwhile because, similar to the case of 3D-crystals, it supplies basic energetic parameters via the Wulff construction [22]. In two dimensions the Wulff construction establishes the relation between the γ-plot, i.e. the angular dependence of the specific step free energy $\gamma(\theta)$, and the equilibrium shape[23,4].[a] This relation is unequivocal, if the equilibrium shape has no sharp corners. From the knowledge of the equilibrium shape one can determine the relative value of the step free energy in any direction. Furthermore, from the variation of the step free energy with the step orientation around that of a kink free, close packed step one may get information on the kink generation energy. The construction of the γ-plot from the measured shape of an island equilibrated at 850 K is illustrated in fig. 4. The γ-plot evidences that the close packed steps A and B correspond to orientations with minimum step free energy. While the minimum of the A-step is relative, that of the B-step is an absolute one. These are the only minima present in the γ-plot. This is obviously due to fact that these steps correspond to the only step orientations which do not contain kinks. Between 700 K and 900 K, the average ratio of their step energies is $\gamma_B/\gamma_A = 0.87 \pm 0.02$ [17]. The difference in step free energy is due to structural difference of the two types of close packed steps mentioned above and obvious in fig. 3.

Meanwhile, the step energy ratio has been determined experimentally also for the fcc(111) surfaces of Ir (0.93 ± 0.01 at about 1000 K [24]), Ag (1.00 ± 0.02

[a]The Wulff construction is as follows: One draws at each point of the closed contour of the γ-plot a line perpendicular to the radius vector of that point; the internal envelope of this line family is geometrically similar to the equilibrium shape.

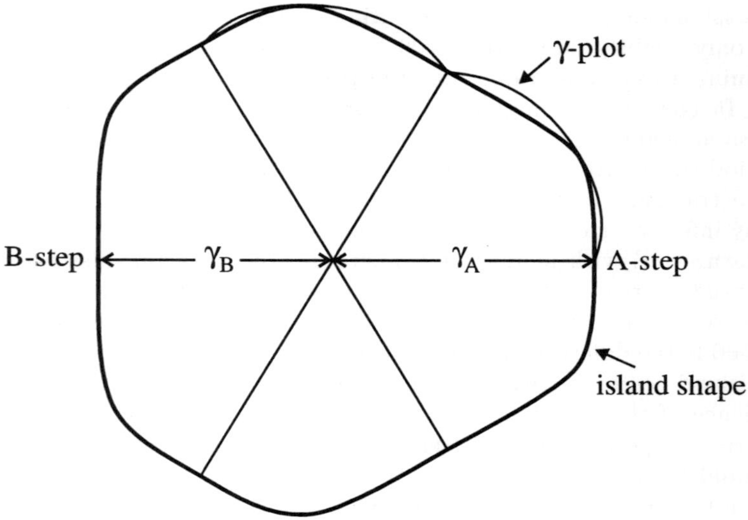

Figure 4: Schematics of the γ-plot (thin-line) of one of the 120°-segments of an island equilibrated at 850 K (thick line). $\gamma_B/\gamma_A = 0.88$

at about 300 K [25]) and Pb (0.91 ± 0.02 at 520 K [26]). Step energies and their ratio have also been calculated for Pt(111) [27,28,29,30], Ag(111) and Al(111) [27,28,31] using various approaches. For Pt(111) the theoretically inferred values show the same hierarchy as observed experimentally: a lower energy for B- than for A-steps. However, for most computations the step energy anisotropy calculated is generally significantly lower than the experimental one: 0.97-0.99 compared to 0.87. A remarkable exception is a recent ab initio calculation, which gives a ratio of 0.86 ± 0.07 [30] in good agreement with the experiment. The calculated values of the step edge energy per atom vary depending on the approach. The embedded atom method (EAM) yields the lowest values (0.34 eV [27]) while the equivalent crystal theory (ECT) the high ones ((0.76 eV [28]). The ab initio values ranging from 0.38 to 0.46 eV (B-step) and 0.43 to 0.47 eV (A-step) appear to be the most realistic ones [29,30].

The effect of temperature on the island equilibrium shape is illustrated in fig. 5. In view of the uncertainties related to the sample cooling from higher equilibration temperatures, which have been mentioned above, the following discussion is to a certain degree speculative and should be considered with caution. Three trends are observed with increasing equilibration temperature:

1. A rounding of the islands as obvious in fig. 5.

2. The ratio of the specific step free energies γ_B/γ_A appears to increase from

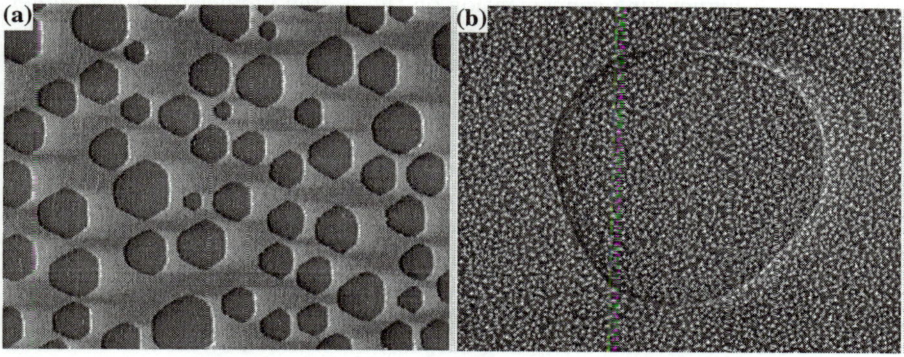

Figure 5: Equilibrium shape (a) after removal of about 0.4 ML at T =760 K and (b) after equilibration at 1270K. (The small adatom islands in (b) are due to the MBE of a small amount of Pt-adatoms which, however, being performed at low temperature did not influence the shape of the large vacancy island.) Scale: 3300 Å × 3300 Å.

0.87 at 760 K to 0.89 at 1270 K. Although the difference is within experimental error, the decrease of the anisotropy with increasing temperature seems to be in line with other conclusions inferred below.

3. A more detailed examination of fig. 5b shows that the most remote points of the island contour from the island center correspond to the A-steps. This means that the A-step free energy, which after equilibration below 900 K represents a relative minimum (see e.g. fig. 4), becomes an absolute maximum when the equilibration temperature is around 1300 K.

The three effects may be traced back to the step orientation dependence of the entropy. Indeed, the number of possible step configurations is essentially determined by the kink concentration[23]. While per definition open steps contain kinks even at 0 K, on the close packed A- and B-steps the kinks are created by thermal excitation. Accordingly, the configuration entropy of open steps is substantially larger than that of close packed ones, and thus the specific step free energy $\gamma = \gamma_0 - Ts$ (with s the specific step entropy) of open steps decreases much more rapidly with temperature. This leads to the observed rounding of the island shape with increasing equilibration temperature. It explains also why, with increasing equilibration temperature, the free energy of the open steps may become lower than that of the close packed ones. This happens first for the A-step, because the kink energy ϵ_A at the A-step is substantially smaller than the kink energy ϵ_B at the B-step[21] and thus the minimum of the A-step energy (0 K) is shallow compared to that of the B-step. This relation between the kink energies also implies that kinks are more easily generated on A-steps, and thus that their concentration is higher on the A-steps than on the B-ones. A higher kink concentration on the A-step implies in turn a higher configuration entropy and thus a stronger decrease of the free energy of the A-steps compared to that of the B ones, leading to the observed decrease of the anisotropy of the step free energies ($\gamma_B/\gamma_A \to 1$).

3 The growth shapes

The fascinating variety of 2D-island shapes observed during growth far from equilibrium (the "growth" shapes in figs. 1a-d) is in obvious contrast to the rather monotonous equilibrium shapes (figs. 1e_1 and e_2). As discussed in the preceding section, the equilibrium shape is per definition independent of (generation) history, time, size, etc. being uniquely determined by the minimization of the total step free energy (the 2D analogon of Eq. (1)). Accordingly, shape variations may only result from the anyhow faint variation of the step free energy with T and p. In contrast, the growth shape depends on the kinetics of the growth process and thus obviously on history, time, island size and many other parameters. Accordingly, the variety of growth shapes is practically unlimited. Because time is non-existent in equilibrium but determining during growth, one may compare the monotony of the equilibrium shapes with that of death, the variety of the growth shapes with the fascination of life.

In order to rationalize the genesis of the various growth shapes, one has to focus on the numerous diffusion processes, which play the major part during growth. Examples of such diffusion processes are: adatom diffusion on terraces, along and over step edges, as well as around corners and over kinks, dimer diffusion and rotation, etc.. All diffusion processes are frozen at low enough temperature. With increasing temperature, diffusion processes are, depending on their activation barriers, successively switched-on, leading to always new island shapes. On the (111)-faces of the fcc metals strongly ramified shapes grow below a certain temperature (fig. 1a). The careful analysis of these shapes at atomic level in connection with kinetic Monte Carlo (kMC) simulations supplies quantitative information on the diffusion processes. An example of such an analysis will be given in paragraph 3.1 below. When, at higher temperatures, many diffusion processes at step edges are active, the island shapes become more compact (fig. 1b-d). The various apparences of these shapes have also been well reproduced by kMC simulations[32,33]. However, due to the large number of processes (and thus of free parameters) involved, the determination of the individual contributions is less unequivocal. The compact shapes will be discussed in paragraph 3.2. Finally, in paragraph 3.3 we will present the effect of instabilities and of kinetic anisotropies on the growth shapes. For understanding these effects, a continuum approach, in contrast to the atomistic one used in the other two paragraphs, seems more appropriate.

A final remark: the successive switching-on of diffusion processes and their differing contribution to the overall growth process under far-from-equilibrium conditions are a rich source of information about these processes themselves. In contrast, at equilibrium, where, due to the absence of the parameter time, all diffusion processes independent of their activation energy are principially active no information on diffusion can be obtained.

3.1 The ramified island shapes at low temperatures

One of the most exciting results of the STM-investigations of epitaxial growth has been the observation of strongly ramified 2D-island shapes on metallic systems. Such ramified shapes have been observed for various systems: Au on Ru(0001) [34], Pt on Pt(111) [35,36], as well as Ag on Pt(111) and on Ag(111) [37,38,39]. These shapes are similar to those obtained earlier by MC-simulation in the frame of the diffusion limited aggregation (DLA) model [40,41]. Accordingly, they were interpreted to result from the reduced adatom mobility along step edges compared to that on the surrounding terraces [34,35,36,42,43,44]. The discussion about the instabilities during island growth in paragraph 3.3 will show that this interpretation is not always necessarily justified.

In this paragraph, however, we will analyze the nature of the limitations of the diffusion along steps which lead to the ramified island shapes observed for Pt/Pt(111) [45]. The essence of the analysis procedure consists in a qualitative (island shapes) and a quantitative (average coordination number of the island atoms) comparison between the experimental results at various growth temperatures and the output of kMC simulations. This comparison turns out to be a sensitive test of the assumptions concerning the relevant growth processes at atomic level.

Fig. 6 shows that adatom islands generated by MBE of 0.094 ML Pt on Pt(111) in the temperature range 150 K - 265 K are ramified. It appears that the main and the side branches have similar thicknesses. The average apparent branch thickness lies between 13 Å and 20 Å after deposition at 150 K (fig. 6a) and 265K (fig. 6b), respectively. The envelope of the ramified islands and the orientation of their branches are anisotropic. The branches grow preferentially in the $< \bar{1}\bar{1}2 >$ directions. In particular at deposition temperatures above 180 K the island envelopes have a clear tendency to become triangular, with their corners pointing into the $< \bar{1}\bar{1}2 >$ growth direction of the main branches. These ramified islands with preferential growth directions will be called dendritic, in contrast to those with no preferential growth direction which are usually called fractal. The whole class of ramified islands will be named dendritic-fractal.

In order to get a quantitative measure of the island structure suitable for comparison with kMC simulations, the average *apparent* branch thickness d_{app}

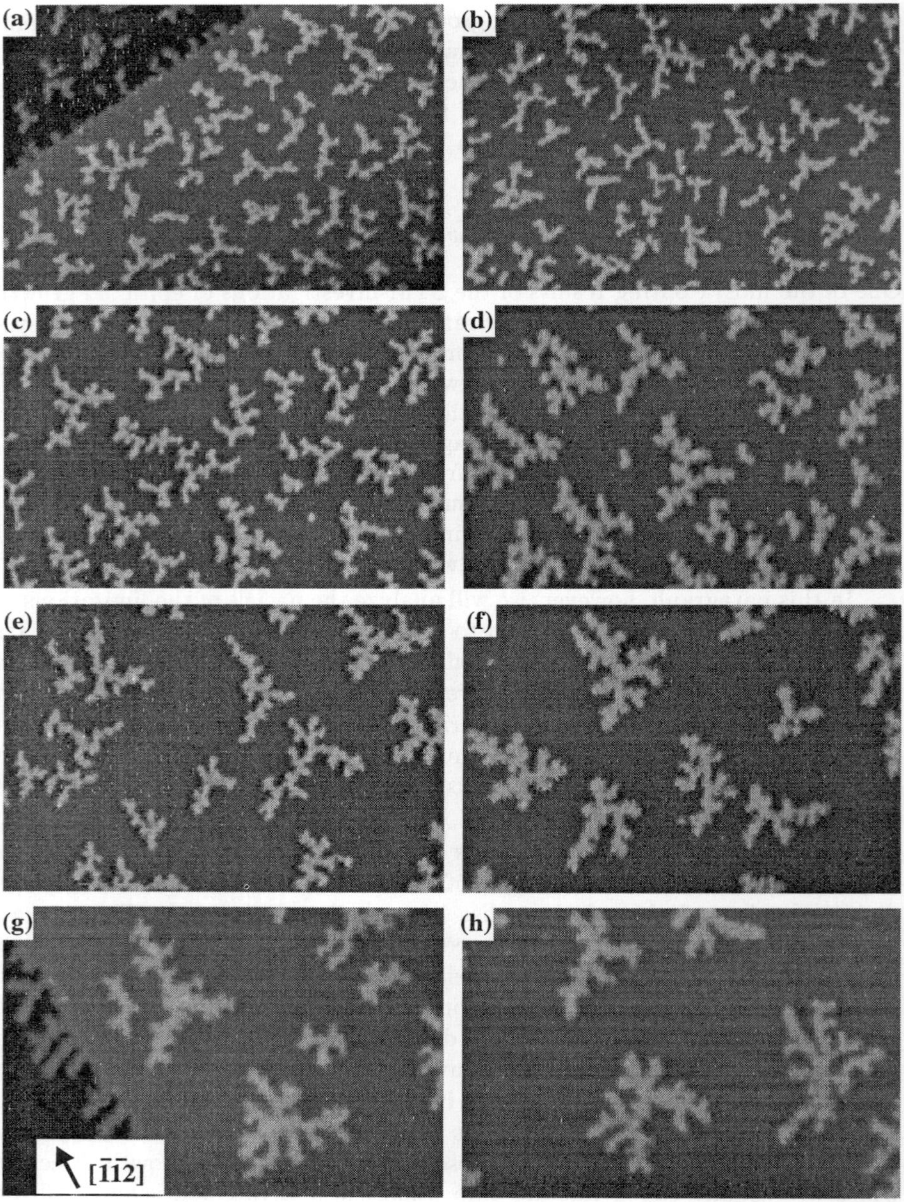

Figure 6: Islands after deposition of 0.094 ML Pt on Pt(111) at 150 K (a), 165 K (b), 180 K (c), 195 K (d), 210 K (e), 225 K (f), 245 K (g) and 265 K (h). Deposition rate: 6.7×10^{-4} ML/s. Scale: 800 Å × 500 Å.

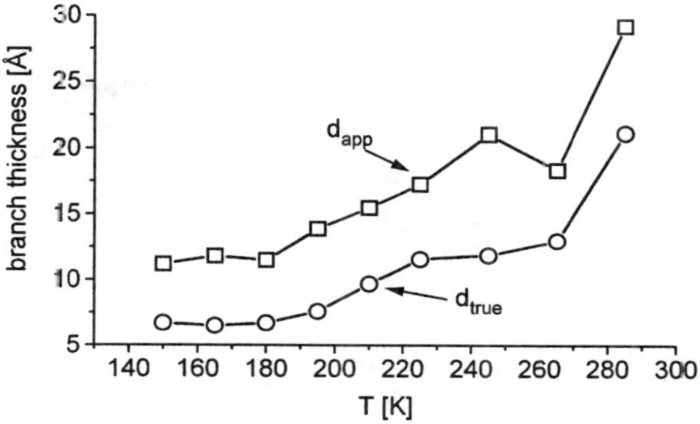

Figure 7: Temperature dependence of the apparent d_{app} and true d_{true} average branch thickness.

is transformed into an average *in plane*[b] coordination number of the atoms in adatom islands. Grey scale topographs like those in fig. 6 are first transformed into binary black-and-white images by setting a threshold value. By averaging the measured thicknesses of all branches in a given image, a value for $d_{app}(T)$ at the corresponding temperature is obtained. From a set of topographs like that shown in fig. 6 one gets the temperature dependence of d_{app} plotted in fig. 7.

Each measured value d_{app} results from a widening w of the *true* branch thickness d_{true} by the STM-tip: $d_{app} = d_{true} + 2w$. Since the STM-tip dependent widening w cannot be directly obtained, one proceeds as follows: The total *apparent* coverage Θ_{app} is determined for each image. The Θ_{app} values obtained from the images in fig. 6 range between 0.13 and 0.17 ML while the actual coverage Θ_{true} is 0.094 ± 0.005 ML.[c]. Since the branch thickness of the dendrites investigated here is substantially smaller than the branch length

$$d_{true} \approx d_{app} \frac{\Theta_{true}}{\Theta_{app}}. \tag{2}$$

The widening

[b]In the following we will deal exclusively with *in plane* coordination. Therefore we will call it simply "coordination".

[c]The actual coverage Θ_{true} has been determined in a separate experiment, in which under identical conditions, but at 400 K, the same Pt-amount has been deposited. For the large and compact islands obtained at this temperature, the widening w is negligible.

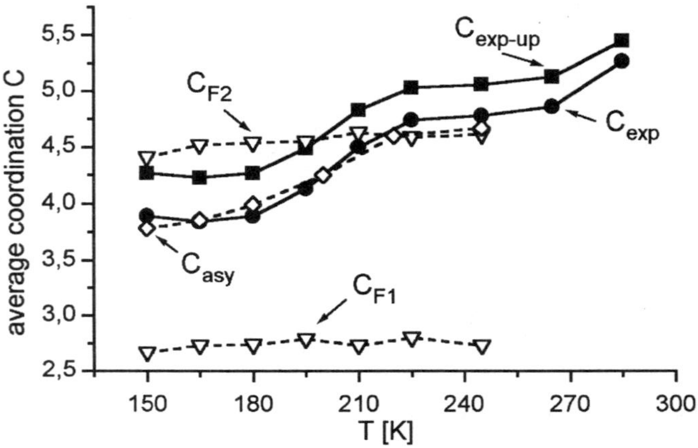

Figure 8: Temperature dependence of the average coordination of the island atoms inferred from (i) experiment via the "true" branch thickness d_{true} assuming different branch edge roughnesses (full lines) and from (ii) kMC simulations with various assumptions concerning the jump probability of branch edge adatoms (dashed lines). C_{exp-up}: smooth branch edges; C_{exp}: every fifth edge atom is only twofold coordinated; C_{F1}: all edge adatoms are immobile (hit-and-stick); C_{F2}: onefold coordinated adatoms jump to twofold coordinated sites, where they are immobile; C_{asy}: onefold coordinated adatoms jump with a path dependent probability to twofold coordinated sites, where they are immobile (see text).

$$w \approx \frac{1}{2} d_{app} (1 - \frac{\Theta_{true}}{\Theta_{app}}) \tag{3}$$

caused by the STM tip, ranges in the present experiments from 2 to 5 Å, consistent with values found previously[46]. The values of d_{true} obtained using Eq. (2) are plotted in fig. 7. The true branch thickness appears to stay nearly constant up to 180 K , then increases, reaching a small plateau of about 12 Å between 225 K and 265 K before a final rapid increase sets in.

In order to obtain the average coordination number of the island atoms one has to make an assumption concerning the roughness of the branch edges. An upper bound for the experimental coordination number is obtained, assuming that the branch edges are perfectly smooth. The corresponding values C_{exp-up} are plotted in fig. 8. The branch edges being actually rough, a better approximation for the experimental coordination number is obtained assuming that every fifth edge atom contributes to the branch roughness (i.e., every fifth branch edge atom is only twofold coordinated). This leads to the C_{exp} values in fig. 8.

In the following, kMC simulations are used to test simple assumptions

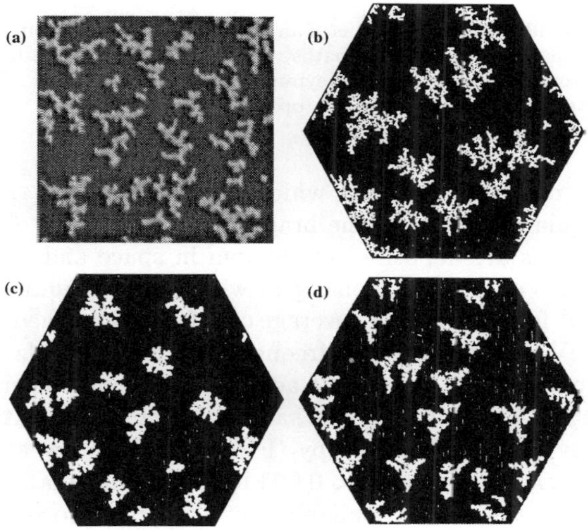

Figure 9: (a) Island shapes obtained after deposition of 0.094 ML Pt at 180 K and a rate of 6.7×10^{-4} ML/s. Scale: 570 Å × 570 Å. (b), (c), (d) Simulated island shapes assuming that onefold coordinated atoms are (b) immobile, (c) mobile, jumping to twofold coordinated sites, where they become immobile, and (d) mobile, jumping with a path dependent probability to twofold coordinated sites, where they become immobile (see text).

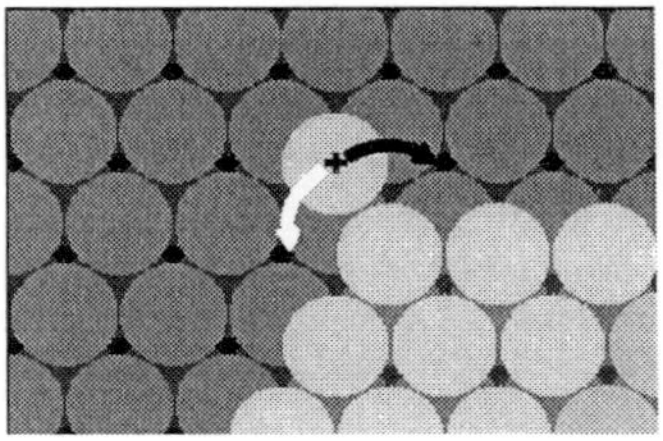

Figure 10: Ball model illustrating the paths of an onefold coordinated atom towards a twofold coordinated site. The only two possible paths, which do not break the bond to the island are indicated by arrows. They differ by the symmetry of the visited sites. White arrow: over bridge- and hcp-sites; black arrows: over on-top site.

with a minimum of free parameters, which reproduce the $C_{exp(T)}$ curve in fig. 8 and the triangular envelope of the branched islands in fig. 6. In the kMC-code used, Pt-atoms are deposited at random in space and time with a rate of 6.7×10^{-4} ML/s onto a hexagonal grid with periodic boundary conditions containing 3×10^4 fcc sites until a coverage of 0.1 ML is reached. An activation energy of 0.26 eV and an attempt frequency of 5×10^{12} Hz determined in independent experiments [47] for the Pt-adatom migration on terraces are used as fixed parameters. (Within error the same value of the activation energy has been obtained by field ion microscopy (FIM) [48]). The few atoms deposited directly onto the growing islands (< 0.003 ML) are immediately incorporated into the step edge at the nearest available site. The other processes and energies used in the simulations, which will be introduced below, are chosen to be as simple as possible, while being still consistent with available FIM observations.

The simplest assumption for the growth of ramified Pt-islands at low temperatures is that the only thermally activated diffusion mechanism allowed is adatom migration on terraces. An adatom arriving at the edge of an island sticks and stays immobile at the impact site, even if it is only onefold coordinated to the island (the lightly shaded atom in fig. 10). This assumption corresponds to the classical DLA-simulations (the hit-and-stick mechanism) [40,41] and the resulting island shape has been classified as fractal type-I [42]. The islands simulated with this assumption at 180 K and shown in fig. 9b are heavily branched and show no preferential orientation (isotropic). The simulated shape is clearly different from that observed experimentally at 180K (fig. 9a). Moreover, the resulting average coordination number C_{F1} is by far too low and practically temperature independent (fig. 8).

In order to increase the coordination number one may assume that onefold coordinated atoms are mobile, and jump to twofold coordinated island edge sites before another arriving adatom might stabilize them at the initial position. The twofold coordinated atoms are still assumed to be immobile. The island shape simulated with this assumption (fractal type-II [42]) at 180 K (fig. 9c) is much more compact. The shape is still isotropic and, compared with the experimental shape, too compact. Indeed, below 200 K the resulting coordination numbers C_{F2} are larger than the upper bound of the experimental values C_{exp-up} (fig. 8). Thus also assumption has to be rejected too. The result also implies that step edge diffusion of twofold coordinated atoms along dense packed steps, which would further increase the coordination number, must be absent below 200 K.

In order to get a physical basis for a slightly more involved assumption to be used in the kMC simulations, a closer inspection of fig. 10 is helpful. The two possible paths of a onefold coordinated atom to adjacent twofold coordinated sites, without loosing in-plane coordination are different: the atom moving along the white arrow is able to remain in a higher coordination with respect to the substrate (moving via bridge- and hcp-sites) than the atom moving along the black arrow (via an on-top site). It appears thus reasonable to assume that jumps from onefold to higher coordinated sites via bridge- and hcp-sites have a lower onset temperature than via on-top ones.[d] Defining the onset temperature as the temperature at which the jump frequency is $\nu = 1$ Hz, we choose for the onset temperatures of the two types of jumps 130 K (corresponding to $\nu_0 = 5 \times 10^{12}$ Hz and $E_d = 0.33$ eV) and 200 K (correponding to $\nu_0 = 5 \times 10^{12}$ Hz and $E_d = 0.51$ eV)), respectively. The coordination numbers C_{asy} simulated with this asymmetric jump probability assumption reproduce quantitatively the experimental average coordination numbers C_{exp} and their temperature dependence (fig. 8). Note also that the islands simulated with this assumption are at 180 K no more symmetric, but show (fig. 9d) a similar tendency towards anisotropy and a similar branch structure as the islands observed experimentally at the same temperature (fig. 9a). The correctness of the asymmetry assumption is further confirmed by the striking similarity of the island shapes observed at 245 K in experiment and simulation (fig. 11). In both cases, islands with triangular envelopes and preferential branch growth in the $<\bar{1}\bar{1}2>$ directions are visible, as well as less regularly shaped islands.

The increase in branch thickness, which is observed to resume beyond 265 K (fig. 8) has to be attributed to the onset of an additional diffusion mechanism which involves the motion of atoms with higher coordination. This is probably of the diffusion of twofold coordinated atoms along close-packed steps.[e]

---

[d] This assumption is consistent with the different frequencies observed for intra- and intercell jumps for Ir-dimers on Ir(111) [49].

[e] This is in accordance with FIM observations: the onset temperature of the movement of twofold coordinated atoms along the channels of the Pt(113) and Pt(331) surfaces, which

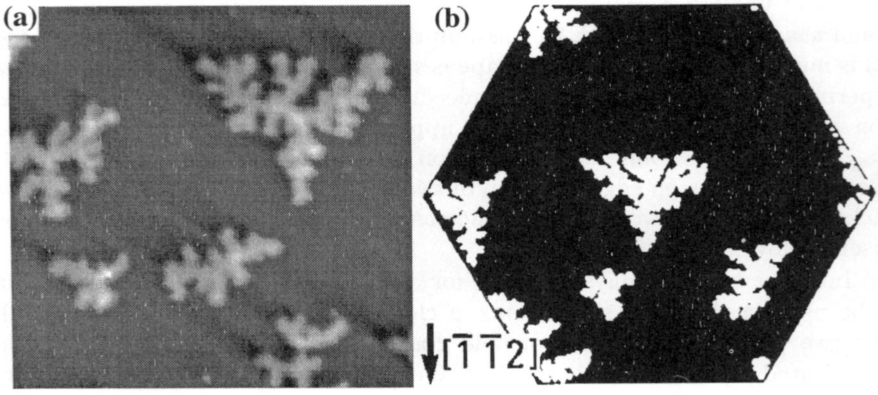

Figure 11: Island shapes upon (a) deposition at 245 K with a rate of 6.7×10^{-4} ML/s; scale 660 Å × 660 Å; (b) simulation with identical parameters but assuming asymmetric jump probabilities. A branch widening w = 2 Å has been included in the simulation in order to facilitate the comparison to experiment.

Further support for the assumption of the asymmetric jump probability of onefold coordinated atoms is provided by the nature of the aggregation of Pt atoms at the two types of close-packed < 110 > steps on Pt(111). The structures of the two types of steps are shown in the ball model in fig. 12a. The STM-image in fig. 12b shows that while the A-step grows very rough, the B-step remains relatively smooth. (Before deposition both steps were smooth over large distances.) The kMC simulation performed with identical parameters as above and including the asymmetric jump probability assumption shows the same difference between the appearances of the two types of steps upon growth (fig. 12c) as the experiment. The reason for this can be recognized in fig. 12a. Decisive for the degree of roughness of the growing steps is the fate of the cross-marked atoms, which are onefold coordinated to the step atoms: while those near the B-step can reach the step by an "easy" jump via bridge sites, those near the A-step can reach it, without loosing in plane coordination, only via the unfavorable on-top sites. Accordingly, the latter atoms will stay for a longer time onefold coordinated to the step adatoms and thus will have a correspondingly higher probability to be stabilized in this position by an arriving atom. This is the beginning of the strongly branched growth observed at the A-steps.

In conclusion, the asymmetric jump probability of onefold coordinated atoms to twofold coordinated sites determines the specific features of the tem-

---

have the same structure as the close packed steps on Pt(111), lies between 270 K and 330 K [50,51].

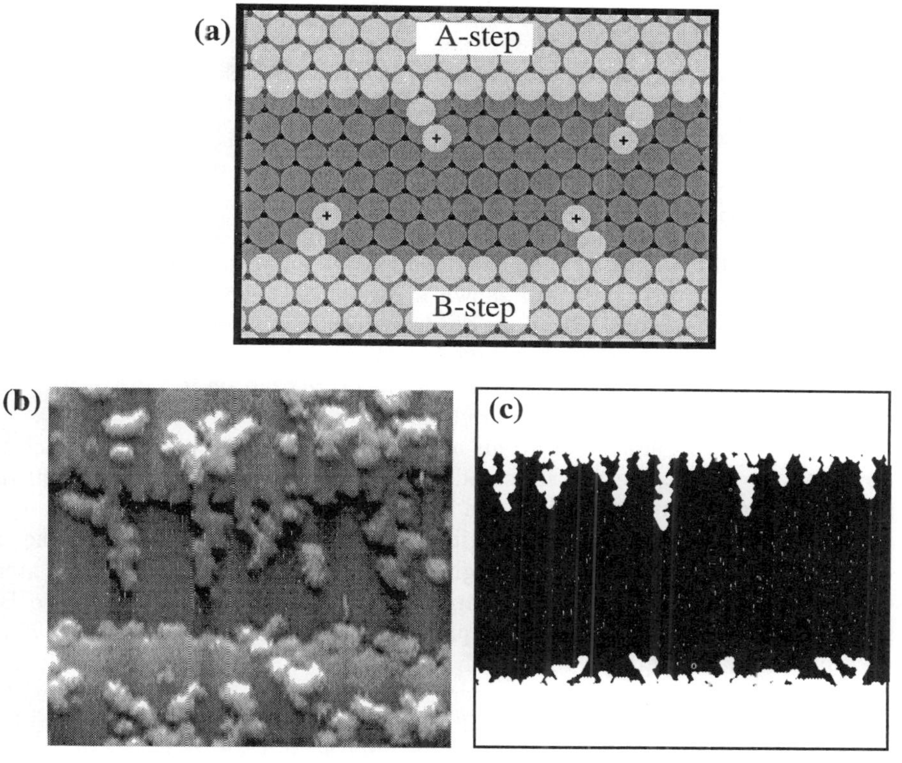

Figure 12: (a) Ball-model of the two types of close-packed steps on Pt(111). The nature of the path of the cross-marked onefold coordinated atoms towards higher coordinated sites at the adjacent step determines the shape of the growing step (see text). (b), (c) Growth shape of an A- and of a B-step after deposition at 180K with a rate of 6.7×10^{-4} ML/s: (b) experiment and (c) simulation.

perature dependence of the branch thickness and the overall shape of the dendritic Pt-islands grown on Pt(111) in the range 150 K-250 K, as well as the shape of the aggregation at the different types of close-packed step edges. This result has been obtained by a careful comparison between STM-data and kMC simulations. The uniqueness of this conclusion was possible because at these low temperatures only a limited number of processes are activated. The asymmetric jump probability is likely to induce similar effects on other surfaces with hexagonal symmetry. Indeed, the growth behavior of Ag deposited on Pt(111)- and Ag(111)-surfaces can be explained by a similar mechanism [52].

Island shapes obtained at higher temperatures, which will be discussed in the next paragraph, do not allow conclusions of the same degree of unambiguity, because of the much larger number of processes simultaneously active. The experimentally observed shapes can be simulated equally well by different sets of processes and parameters. In this case, the same degree of reliabilty of the conclusions asks for an increased number of data obtained under various experimental conditions.

3.2 The polygonal shapes

This paragraph focusses on the growth shapes obtained by MBE in the intermediate temperature range and shown in figs. 1b-d. This range extends for the Pt/Pt(111)-system, depending on deposition flux, between about 350 K and 650 K. At these temperatures more and more diffusion processes become active, and their individual as well as combined contribution to the growth process changes drastically. A large variety of island shapes is the consequence. In particular, due to diffusion of atoms along close-packed steps becoming active above about 260 K, the overall island shape becomes compact around 350 K. At somewhat higher temperature the island sides become more or less straight and the shape can be roughly described as polygonal. The sides are, like in the equilibrium case, monatomic high consisting mainly of close-packed step edges running along the $< 110 >$ directions. However, the ratio of A- to B-side length has, in contrast to the equilibrium case, not an almost fixed value determined by the ratio of the step free-energies, but may have any arbitrary value depending on deposition temperature and rate. Indeed, as illustrated in fig. 1 for $R \approx 1 \times 10^{-2}$ ML/s, at T = 455 K the shape is nearly a regular hexagon (side length ratio 1), while at T = 400 K and 640 K triangular, with A- and B-sides, respectively, with the corresponding side length ratio ∞ and 0. The low and high temperature triangular shapes will be denoted Δ_l and Δ_h, respectively.

As shown above, the origin of both the equilibrium (section 2) and the branched island shapes (paragraph 3.1) could be unequivocally traced back to the ratio of step edge free-energies and to a few kinetic processes, respectively. For the polygonal shapes observed in the intermediate temperature range, the situation is so far less satisfactory, in spite of many efforts and obvious successes. The reason, which has been already mentioned, is the vast number of diffusion processes, which individually as well as in combination contribute in varying extent to the overall process. Towards higher temperatures some processes become equilibrated so that besides kinetics (barriers) also thermodynamics (free-energies) come into play. In order to assess the individual contribution of the various processes, a faithful basis of activation energy and attempt frequency values - not yet available - is be necessary. So far only a few diffusion parameters have been determined experimentally for Pt/Pt(111) by FIM [48,50,51,53] and by STM [45,47], which in addition are relevant

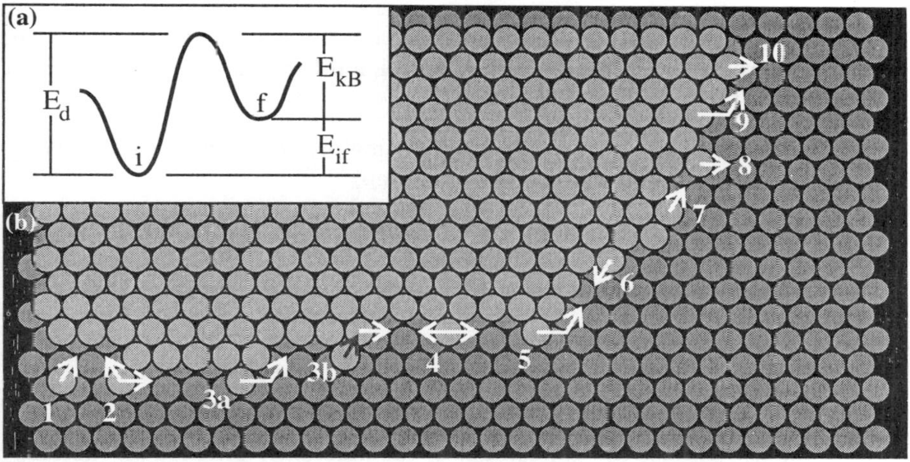

Figure 13: (a) The activation energy for an atomic jump process from state i to f, can be divided into the energy difference between the two states, E_{if}, and the additional kinetic barrier, E_{kb}. (b) Ball model of an fcc (111)-surface, with relevant in plane atomic processes at the island edges: (1) adatom attachment, (2) "umklapp"-processes of onefold coordinated atoms, (3a) corner-incorporation by jumping or (3b) by exchange, (4) step edge diffusion, (5) diffusion over corner, (6) step adatom creation by kink evaporation, (7) step dimer dissociation, (8) step adatom evaporation, (9) step adatom/step vacancy creation and (10) kink atom evaporation.

only at low temperature. The number of values obtained by ab initio calculations[29,30,48] is also modest and their coincidence with established experimental values rather limited. Activation energies for a large number of relevant diffusion processes have been calculated by the embedded atom method (EAM) and effective medium theory (EMT)[32,33,54,55,56,57,58,59,60,61,62,63], but in most cases they agree neither mutually nor with the experimental data. Accordingly, the presentation of the (probably) relevant concepts and atomic processes as well as the assessment of the so far published work on the polygonal Pt-island shapes on Pt(111) will be largely qualitative.

The approximate hierarchy of the onset of diffusion processes as it results from a nearest-neighbor-model, as well as from EAM- and EMT-calculations, and which corresponds to experimental data on fcc (111)-surfaces, may serve as a rough guidance for the following discussion. Let us consider the schematic potential along the diffusion path of an atom from an initial state i to a final one f plotted in fig. 13a. The following rules of thumb are useful:

1. If $E_{if} = 0$, the higher the coordination of the involved atoms, the larger E_{kb}. For instance, while the kinetic barrier for Pt-adatom diffusion on the Pt(111) terraces is 0.26 eV[47] it lies for diffusion along the closed

packed steps between 0.6 and 0.85 eV [50,51].

2. E_{if} increases with increasing coordination difference between the i and f state [59,63].

3. The values of E_{kb} and of E_{if} for the same process may depend substantially on the step structure (A- or B-step) [55,56]

The activation energies of the processes, shown on the ball model in fig. 13b, increase roughly with their indexing number.[f]

Let us analyse in the following in more detail two questions related to the island shapes in fig. 1:

1. The onset of which atomic mechanisms determines the appearance of the compact Δ_l shape instead of the dendritic one?

2. What determines the appearance of the differently oriented Δ_l- and Δ_h-shapes?

In this analysis the focus is on processes occuring at islands step edges. Most of the processes and objects involved are similar to those on surfaces, but have one dimension less. They are in general named also similarly: e.g. the hexagon side is a 1D-terrace, the analogon of the 2D-(111)-terraces; the 1D-nucleation on the hexagon side leads to an 1D-island, the analogon of 2D-nucleation on the 2D-terraces leading to 2D-islands; the corner-incorporation (3) defined in fig. 13, causing inter-row transport, is the analogon of the jumping down of adatoms over step edges, causing the inter-layer transport. The "growth" of step edges perpendicular to the step line is called step advancement.

The first question above can be decomposed in three partial ones: which are the atomic processes needed during deposition in order to 1a) maintain a hexagonal shape, 1b) transform an arbitrarily shaped (but compact) island into a hexagonal one and 1c) transform a hexagonally shaped island into a triangularly Δ_l-shaped one?[g]

In order to maintain a hexagonal island shape during deposition, besides the obvious processes (1) and (2), also the processes (4) and (3a or 3b) are necessary. The step edge diffusion (4) causes the lateral material transport parallel to a hexagon side, the corner-incorporation (3) the transport perpendicular to

[f] An exchange process, i. e. the correlated motion of several atoms [64] may have a substantially lower kinetic barrier than the corresponding jump process. Such a process should be taken into consideration especially, when an intermediate state with very low coordination is involved in the jump process, e. g. processes (3) and (5) in fig. 13b.

[g] This decomposition corresponds to the experimental observation (see e. g. fig 17 in the next paragraph) that during growth the islands undergo a transformation from small more or less hexagonal islands to triangular Δ-shaped islands.

it. At a given adatom attachment rate, a sufficiently large step edge diffusion hinders that more than one step edge island nucleates on a step edge island. However, even this leads, in principle, to a rough, 2D-step advancement (the analogon of 3D-growth). This can be hindered if corner-incorporation is also sufficiently large. Thus, if both (3) and (4) are active enough, only one step edge island is present on each hexagon side. It is only after this step edge island has grown laterally to become a full hexagon side, that a new step edge nucleus can be formed. This can be called "row-by-row" step advancement, the analogon of layer-by-layer growth. If the diffusion-over-corners (5) is effective enough, it can reduce, within this scenario, the number of step edge islands along the whole hexagon perimeter to a single one. While this is not a necessary condition for maintaining the hexagonal shape (question 1a), it appears that the effectiveness of (5) is needed to transform during growth an arbitrarly shaped, compact island in a hexagonal one (question 1b). Indeed, the diffusion-over-corners can hinder the appearance of notches at the island sides. Assuming that corner-incorporation (3) and step edge diffusion (4) are effective only on A-, but not (yet) on B-steps, the island shapes observed at 280K can be reproduced rather precisely by kMC simulations[65].

The answer to question 1c, concerning the transformation of a hexagonal into a Δ_l-shaped island, is approached by considering the atom fluxes to, from and along the island step edges of a regular hexagon. The first step of the answer is obtained, in analogy with the 3D-crystal case, simply by looking at fig. 14a. The development of a Δ_l-shaped island from a hexagonal one can be viewed as due to a larger advancement speed of the B-steps, v_B, compared to that of the A-steps, v_A. The second part of the answer to question 1c, why is $v_B > v_A$, is more involved. Assuming the incoming adatom fluxes from the (111)-terraces to be equal, $j_A = j_B$, this can be realized only by the effect of at least one of the diffusion currents j_{AB}, $j_{ev,A}$, and $j_{ag,A}$ defined in figs. 14b-d.

In the following, a few examples will illustrate how these currents can be effective. Note that their origin is both energetic and kinetic. These examples are neither exhaustive, nor fully independent. The prerequisite for the current j_{AB} to flow is the diffusion of atoms over corners (5). The current j_{AB} may result from the following mechanisms:

a.) Step-adatoms are stronger bound on B- than on A-steps. Accordingly, their dwell time will be larger on B-steps and, at equal incorporation probability, the incorporation rate at the kink sites of the B-steps will be also larger. Likewise, the probability to meet an incoming atom and to form a stable nucleus will be larger on the B-steps.

b.) Step nuclei (step adatom dimers) dissociate (7) easier on A- than on B-steps. Accordingly, step island growth is largely suppressed on A-steps. In view of the hierarchy of the processes (fig. 13b), kink evaporation is

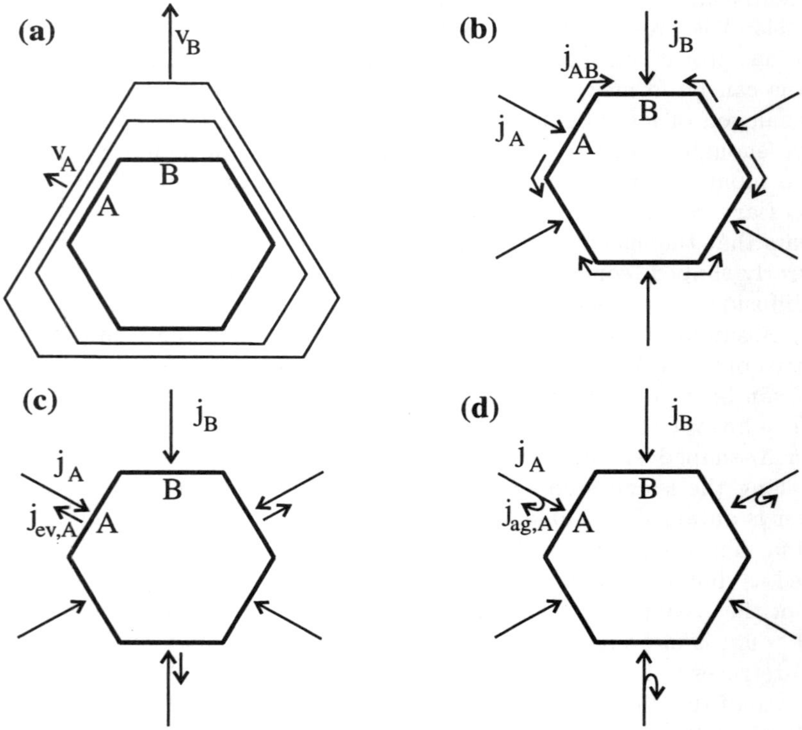

Figure 14: (a) The development of a Δ_l-shaped island from a hexagonal one is due to a larger advancement speed of the B-steps, compared to that of the A-steps: $v_B > v_A$. This can be realized, for identical adatom incoming fluxes from the (111)-terraces $j_A = j_B$ by (b) an edge diffusion current j_{AB} from A- to B-steps; (c) a current $j_{ev,A}$ corresponding to a preferred evaporation of adatoms from A-steps onto the (111)-terraces or (d) a lower effective attachement probability of the adatoms impinging on the A-steps due to an incorporation barrier in front of or at the A-step, corresponding to a reverse diffusion current $j_{ag,A}$.

also active and easier on A-steps and with it the thermal step adatom creation. This would further increase j_{AB}.

c.) The diffusion coefficient (4) along the B-steps is larger. Accordingly, if (6) and (7) are not yet active, a step adatom will find on a B-step more rapidly a kink position to be incorporated[32].

d.) Inversely, the diffusion coefficient (4) along the A-steps is larger and the kink creation probability (9) is so low that there are practically no thermally created kinks present. Accordingly, the dwell time on the B-steps and thus the probability to form a stable nucleus with an incoming atom is also larger[36].

Possible mechanisms leading to the currents $j_{eV,A}$ and $j_{ag,A}$ are:

e.) Step adatoms are stronger bound at B-steps. Accordingly, step adatom evaporation (8) from A-steps is stronger and thus the adatom incorporation rate in A-step islands is reduced.

f.) The attachment probability of adatoms impinging on A-steps might be reduced by a possible additional potential barrier located in front of the A-steps.

In spite of being incomplete (in particular, because combination effects have not been considered), the list shows how complex the action of the atomic processes in this temperature range is.

The analysis above, concerning the transformation during growth of a hexagonally shaped island into a Δ_l-shaped one, makes it easy to give an answer, at least a formally one, to question 2: what determines the appearance of Δ_l-shapes (A-step bounded, fig. 1b) and Δ_h-shapes (B-step bounded, fig. 1d) at low and high temperatures, respectively? Let us denote the above list, which leads to A-step bounded islands by the enhanced advancement of the B-steps, as list-A. By interchanging the A's and B's, one obtains the list-B, which trivially leads to B-step bounded islands, i.e., to the Δ_h-shape. The problem is now reduced to the choice of one (or several) mechanism from list-A, which is decisive at low temperatures, but insignificant at higher ones, and to the choice of one (or several) mechanism from list-B, which is decisive at higher temperature, but not yet activated at lower temperatures. Obviously, there will be a narrow temperature range in between, where the mechanisms chosen from list-A and -B have nearly equal contribution, leading to similar step advancement speeds ($v_A = v_B$) and thus to almost regular hexagonal shapes (fig. 1c).

The description of the mechanisms which lead to Δ_l-, hexagonally-, and Δ_h-shaped islands at the corresponding deposition temperatures is certainly somewhat formal. It reflects, however, the present understanding of these phenomena, in which a manifold of atomic processes play a role involving often only slightly differing energies. After all, certain mechanism combinations can be found to be more or less plausible by internal consistence considerations and by comparison with available experimental and theoretical results. This is illustrated by the following analysis of the papers related to the questions 1 and 2 published so far [32,33,36,66,55,44,67,68]:

I. Originally, the shape inversion (from Δ_l to Δ_h) seen in figs. 1b-d has been given a kinetic explanation, by assuming an inversion of the hierarchy of the step diffusion coefficients occuring around 450 K [36]: below 450 K the larger diffusion coefficient should be that along A-steps (mechanism d-A), while above 450 K that along B-steps (mechanism d-B). This assumption is plausible in view of some FIM-results [50,51], but later kMC simulations [32] suggest that with this assumption and in view of mechanism c, the orientation of the triangular islands should be exactly inverse to that found experimentally. Very recent ab initio kMC simulations [68], however, find that for the growth of Al on Al(111) the original explanation holds.

II. A combined kinetic-energetic explanation of the shape inversion was given in ref. 66. The Δ_l-shape would result from a higher diffusion coefficient along B-steps (mechanism c-A), while at 600 K an easier dissociation of step-dimers, trimers, etc., and of kink atoms at B-steps favors the formation of Δ_h-shapes (mechanism b-B). The step edge diffusion energies of 0.40 eV and 0.46 eV used in the simulations are certainly too low and the transition temperature is too high. Moreover, the instability of the simulated shapes in the transition region as well as the low number of kinks on the island edges seen in the simulations are clearly at variance with the experimental data.

III. In an analysis of the transition from the dendritic-fractal growth to the Δ_l-shape, the triangular shape is enforced by an ad-hoc hypothesis concerning the mobility of corner atoms [44]

IV. In a series of detailed simulations of Pt-island shapes on Pt(111) as a function of deposition temperature [32,55,33] a basically energetic explanation is given. The diffusion parameters used were derived from EMT calculations. The formation of the Δ_l-shape results from a stronger adatom binding at B-steps (mechanism e-A). At higher temperatures, both the step-adatom creation by kink evaporation and by step-dimer dissociation become active and are assumed to have a lower activation energy at the B-steps. This favors the formation of Δ_h-shapes (mechanism b-B). The result of the simulations shown in fig. 15 is the only one published so far which reproduces the STM-images in a realistic way in the whole temperature range. However, while the energy parameters chosen are consistent, the choice seems less plausible in view of fundamental considerations concerning the relation between step and kink energies. From the 2D-Wulff construction it follows that the kink creation energy is lower at the higher free energy close-packed step edge, i.e. at the A-step. In the approximation that the creation of one step adatom from a kink as well as the dissociation of a step dimer are equivalent with the creation a double kink, both processes have a consideravly lower energy difference E_{if} on A- than on B-steps, and a step adatom is considerably stronger bound at the A- than at the B-step [21,24]. There is also a minor, but clear discrepancy to the experimental data concerning the contribution of adatoms descending from the island tops to the island shape, which will be discussed in the next paragraph 3.3.

V. Another energetic explanation is based on a possible temperature dependence of the binding energy of adatoms at step edges [67]. The Δ_l-shape results again from a stronger binding of adatoms to B-steps. Due to entropy effects the binding to A-steps becomes stronger at higher temperatures leading to Δ_h-shapes. At least at low temperature the assumption is again at variance with the considerations based on the Wulff-construction.

VI. An, in principle, possible energetic explanation for the shape transition could be that the Δ_l-shape results from a lower activation energy for step dimer dissociation (as well as for kink evaporation) on A-steps (mechanism b-A), while due to an easier evaporation of step adatoms onto the (111)-terraces from B-steps (mechanism e-B), the formation of Δ_h-shapes is favored. These assumptions are seemingly not in contradiction to the experimental results known so far, and are compatible with the energetic considerations outlined above.

Due to the complexity of the atomic processes involved, precise and faithful answers to the questions presented here can hardly be given without further detailed experiments. FIM- and special STM-experiments[h] devised to uncover the behavior of individual step adatoms would be necessary. kMC-simulations making for instance predictions on the deposition rate dependence of the island shapes would be also very useful.

Two other ways should be also considered. One would be the analysis of homoepitaxy on Ir(111). The present qualitative and quantitative knowledge

[h] For instance, a continuous monitoring during the deposition of individual Pt-atoms on a prestructured Pt(111) surface could supply far-reaching information on the relevant atomic processes.

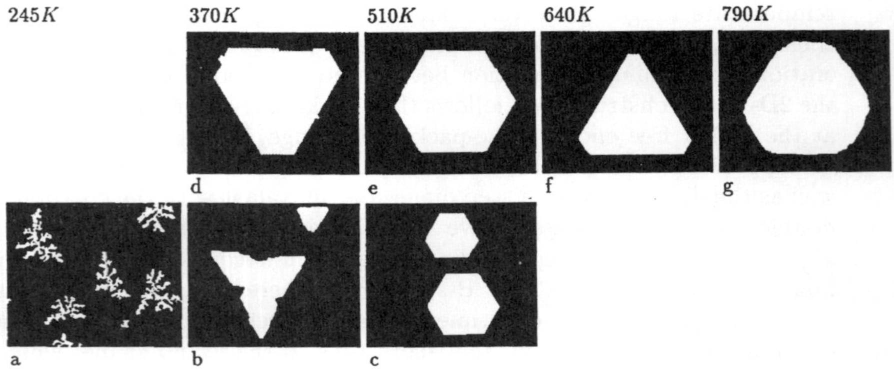

Figure 15: Simulated island shapes as a function of temperature after deposition of 0.016 - 0.1 ML Pt on Pt(111) (a-c) Random deposition on the (111)-terraces; (d-f) random deposition directly along the island edge. The surface orientaion is nearly the same as in fig. 1. We are indebted to the authors of ref. [33] for the permission to reproduce this figure.

on the atomic processes and diffusion parameters for this system from FIM-measurements is unique[49,69,70,71,72,73,74,75,76,77,78,79,80]. A comparison between kMC simulations based on this knowledge and STM-experiments could lead to substantial progress. The starting point for the other way is the large number of diffusion-pararameters and -energies, which have been calculated for the homoepitaxial Al(111)-system[31,68,81,82]. Ab initio calculations for this system are rather accurate. A comparison of kMC simulations based on these values with STM-experiments appears to be promising for a deeper understanding of complex growth phenomena on the ground of atomic processes.

3.3 The role of instabilities in determining the island shapes

In recent years a number of studies have focussed on the morphological instabilities which are due to the influence of the surrounding diffusion field during island or step edge growth[43,83,84]. These studies have revived the interest on this aspect and applied to surface steps the basic ideas of the Mullins-Sekerka work[85,86]. Indications for instabilities are obvious when looking at the island shapes obtained under various conditions. However, a quantitative interpretation of the observed shapes is not yet possible mainly because in these theories the anisotropies of the diffusional processes, which are essential in determining the island shapes, were not consistently considered. On the other hand, the diffusion field is, in principle, inherently built-in in kMC simulations, if the atoms are allowed to fall randomly in time and space onto a large enough surface (i.e., containing several islands). In particular at higher temperatures, however, due to computational time considerations this is not always possible: accordingly, the atoms are deposited directly at the island step edges, thus excluding the possible influence of the diffusion field. In view of this and of the fact that again at higher temperatures the comparison of experiment and kMC simulation does not always lead to a unique set of processes and parameters, it is worthwhile to try to explain experimentally observed growth shapes on the basis of instabilities. The fact that this is a continuum approach might contribute to a better understanding of the phenomena.

In this paragraph a number of examples of experimentally observed island shapes, in which morphological instabilities are particularly evident, will be presented. When explaining the shapes in the light of isotropic morphological stability theories the necessity to include anisotropic effects will become evident in most cases. (Conversely, the importance of the proper consideration of the diffusion field in kMC simulations will be also emphasized.) Of particular interest is the fact that specific island shapes resulting from anisotropic growth processes influence the actual diffusion field and thus may change the morphological stability criteria. For instance, the triangular shapes discussed in the preceding paragraph lead to instabilities at the triangle corners (apexes) under circumstances where the smooth (straight or circular) step stability conditions are fulfilled. This phenomenon will be called instability of type II in contrast to the instability I resulting from the usual morphology analysis of smooth steps.

To facilitate the discussion, we consider first the isotropic growth problem, more specifically the morphological stability of circular 2D-islands with isotropic specific step free energy γ in an adatom diffusion field and adatom incorporation at the ascending steps. The islands are assumed to grow due to the diffusion field of adatoms on the terrace, i.e., the contribution of the adatoms deposited on top of the islands is neglected. (This is certainly acceptable at low coverages in particular in the presence of a high Ehrlich-Schwoebel barrier.) Under these conditions and in the absence of stabilizing effects any accidental protuberance on the island contour, having otherwise a constant radius of curvature, leads to instability, because the protuberance grows more rapidly than the other regions of the island contour. The reason is that in front of the protuberance the adatom isoconcentration lines are bunched together, while being rarefied in front of the depressions adjacent to the protuberance. This leads to a focussing of the diffusion flux onto the protuberance (the so-called "point effect of diffusion") and hence - as the step advancement speed is proportional to the adatom concentration gradient normal to the step - to an enhancement of the protuberance growth speed [85,86,87]. As a consequence and as fluctuations always create accidental protuberances on growing islands, any island of any size should be unstable. The experimental fact that this is generally not the case and thus that stabilizing effects are also present, is demonstrated for instance by various polygonal island shapes observed in experiment.

The driving force of both stabilizing effects, considered to be active, can be traced back to the increase of the contour length and thus of the total step free energy due the growth of the protuberance. The system tries to relax to a state of lower free energy by a flow of atoms away from the protuberance. This may happen in two ways: by evaporation of atoms from the tip of the protuberance, i.e., by increasing the local equilibrium adatom concentration in front of the tip, and by step edge diffusion from the tip towards the adjacent depressions.

From an "order of magnitude" stability analysis, Pimpinelli et al. [43,84] have derived the following relation for the time evolution of the amplitude δ of an accidental protuberance of width (wavelength) λ:

$$\frac{\partial \delta}{\partial t} \approx \left(\frac{Fl^2}{r_0 \lambda} - K_1 \frac{D\gamma}{kT\lambda^3} - K_2 \frac{D_{st}\gamma}{kT\lambda^4}\right)\delta, \tag{4}$$

with r_0 and l the average island size and separation, respectively, F the deposition flux, D and D_{st} the adatom diffusion coefficient on terraces and along step edges, respectively, T the substrate temperature, k the Boltzmann constant, and K_1 and K_2 temperature dependent coefficients.[i] Obviously, when the first term outweighs the two stabilizing ones, the step becomes unstable. Eq. (4) leads to the following conclusions:

A. For $\gamma \to \infty$ both stabilizing effects vanish: the steps become unstable.

B. The wavelength of the protuberances being limited by the island size ($\lambda < r_0$) protuberances with increasing wave length become possible with increasing island size. Thus, as long as the islands are small, all protuberances are driven back. However, as soon as the islands become larger than a critical value r_c ($r_0 > r_c \approx \lambda_c$) the destabilizing term outweighs and the island growth becomes unstable.

C. The destabilizing effect of the "point effect of diffusion" increases with the deposition flux.[j] Thus with increasing flux the critical size r_c should decrease.

In the following we will discuss, based on experimental examples, the applicability and limitations of Eq. (4) and of the corresponding conclusions as well as, incidentally, of other approaches.

a) Pt-islands deposited on Pt(111) below 265K

[i] K_1 and K_2 contain the adatom and step-adatom concentrations respectively; their magnitudes increase with temperature.

[j] This is generally true, allthough l, r_0 and F are actually coupled. For instance, if the dimer is assumed to be the stable nucleus, the first term in Eq. (4) becomes $\frac{F^{2/3}D^{1/3}}{r_0\lambda}$.

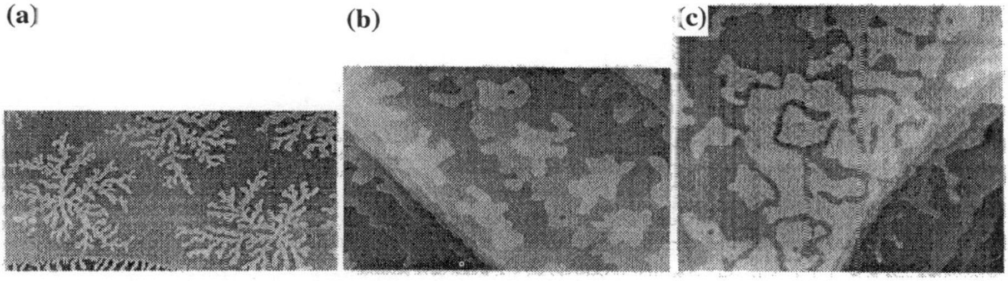

Figure 16: (a) Island shapes upon deposition of 0.3 ML of Au on Ru(0001) at room temperature (scale 10000 Å × 6500 Å); similarly prepared surface (except for $\Theta_{Au} = 0.37$ ML) but with additional annealing at (b) 650 K (scale: 8300 Å × 5400 Å) and c) 1100 K (scale: 7600 Å × 6600 Å) [34]. The permission to publish the images is gratefully acknowledged.

The fact that the sequence of STM-images in fig. 6, discussed in detail in paragraph 3.1, show branched islands, can be interpreted as a consequence of the instability predicted by Eq.(4). Indeed, for Pt/Pt(111) at these low temperatures both 2D-adatom evaporation and step edge diffusion are inactive (see section 2 and paragraph 3.1). Accordingly, the "point effect of diffusion" is outweighing for protuberances of wavelengths down to atomic dimensions. The growth is fully unstable. However, as Eq.(4) does not contain any asymmetric element, the predicted branched islands should be isotropic (i.e., fractal) at variance with the experiment and the corresponding discussion in paragraph 3.1.

b) Growth of fractal Au-islands on Ru(0001)

Fig. 16 shows that upon deposition of 0.3 ML Au on Ru(0001) at room temperature monatomic high, strongly ramified, isotropic (fractal) islands with typical branch widths of about 100 Å are formed [34]. The island contour has in all directions the same structure: the branch edges are rounded, exhibiting a waviness with radii of curvature of the same size as the branch width. Thus, step edge diffusion with a diffusion coefficient largely independent on orientation appears to be effective at room temperature. Island shape and step edge diffusion being isotropic Eq. (4) is applicable: as soon as during growth a critical radius r_0 is reached, the stabilizing effect of step edge diffusion is outweighed by the point effect of diffusion and the protuberances grow rampantly.

An inspection of figs. 16b and c points to another significant aspect: annealing of the Au-islands does not result in convex islands, but in meandering step edges. Annealing at 650 K (fig. 16b) leads, admittedly, to more compact islands with larger curvature radii, but these islands are by no means convex.

Although upon annealing at 1100 K (fig. 16c) the Au-adlayer is fully rearranged, it still exhibits meandering steps and many channel-like holes. The behavior is clearly different from that of the homoepitaxial systems investigated so far, which upon growth or annealing at high temperatures, exhibit always compact islands[35,88,89,90,91,92]. The minimization of the total step free energy being the driving force for the generation of compact islands, one may conclude that the specific step free energy γ for the Au/Ru(0001) system is small.[k] This makes the stabilizing terms in Eq. (4) ineffective even at room temperature. Accordingly, the fractal shapes observed (fig. 16a) for room temperature growth of Au on Ru(0001) are a consequence of the small γ, rather than of a limited step edge diffusion.

Unstable growth similar to that of Au on Ru(0001) has been reported for deposition of Ag on Au(111) at room temperature[95]. This is astonishing, because deposition of Au on Au(111) and of Ag on Ag(111) leads to compact islands at the same temperature and because Ag and Au have practically the same lattice constant. A relatively low step free-energy could be also in this case the cause for unstable growth.

One may speculate that in well wetting systems the stronger binding between deposit and substrate atoms compared to the mutual binding of deposit atoms might be the cause for the reduced step free energy. Indeed, the strong binding to the substrate might further reduce the mutual binding within the adsorbed layer and thus also the step free energy. The observed effects being strong, even approximate energy calculations may confirm or infirm this speculation.

c) Stability of polygonal growth shapes in the Pt/Pt(111) system

This and the next example demonstrate that anisotropy effects alone can largely determine the island stability. Figures 17a and b show various growth stages of islands obtained by deposition at 370 K and 640 K, respectively. Small islands are compact, either irregular or hexagonal (stage 1). The next stage 2 is represented by triangular islands, Δ_l for low and Δ_h for high temperature deposition, discussed in detail in paragraph 3.2.

The three corners of the Δ-shape are obviously protuberances of small wavelength. The essential aspect is, however, that the formation of the Δ-shape with its sharp corners is not a consequence of an accidental fluctuation or instability, but of the anisotropic growth kinetics. Once the triangle corners are

[k] A possible influence of the lattice mismatch on the shape of the structures resulting in *hetero*epitaxial systems cannot be excluded a priori. Indeed, the resulting strain energy can be relaxed by breaking of large islands into small ones[93,94]. However, the results in fig. 16 show that the structures do not become smaller by annealing, giving no evidence for such relaxation effects.

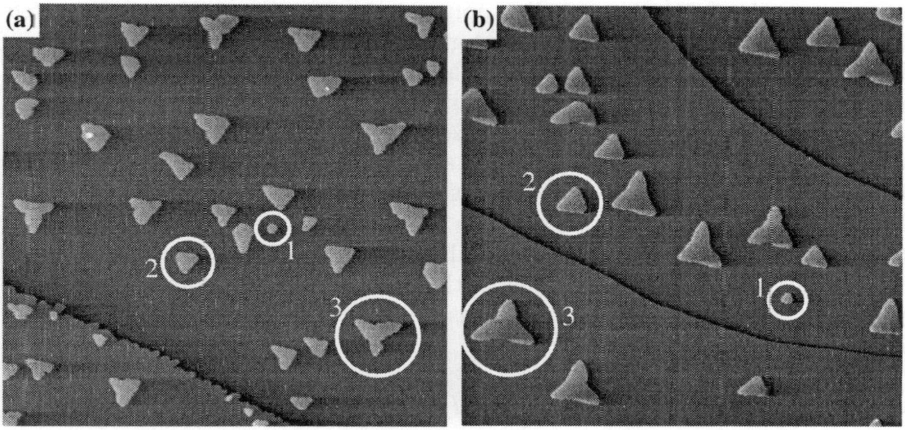

Figure 17: Growth stages of islands imaged upon deposition of Pt on Pt(111) at (a) 370 K and (b) 640 K. The small islands (stage 1) are compact, irregular or hexagonal; the intermediate ones (stage 2) have a Δ-shape and the larger ones (stage 3) are dendritic-skeletal. (a) flux: 3.5×10^{-3} ML/s, $\Theta = 0.1$ ML, scale: 3300 Å × 3300 Å, (b) flux: 1.4×10^{-2} ML/s, $\Theta = 0.2$ ML, scale: 8500 Å × 8500 Å.

developed, the point effect of diffusion leads to an increased material transport to these corners. As long as the side length L of the triangular shape does not exceed a critical length L_c, the enhanced material, supplied to the corners by the diffusion field, can be carried away by step edge diffusion: the Δ-shape stays stable (stage 2) as evidenced by the smooth edges. For $L > L_c$ this is no more the case. The material supplied to the corners, which is no more carried away, favors the growth at the corners, which - due to the point effect of diffusion - becomes self accelerated. This leads to tip formation and eventually to the dendritic-skeletal shape of stage 3. Note that this effect was present in the kMC simulations of Pt-island growth on Pt(111) [33] when the atoms were deposited randomly on the terrace at 370 K (fig. 15b), but absent upon random deposition directly along the island edge at the same temperature (fig. 15d). Because in the latter case the diffusion field has been artificially excluded, the simulations obviously confirm the determining role of the point effect of diffusion for the rampant growth of the triangle corners. The diffusion field may lead even at high temperatures to a clearly dendritic-skeletal island shape (see e.g. the islands in stage 3 obtained by deposition at 640 K - fig. 17b) and therefore cannot be neglected.

The instability evidenced by the shapes in fig. 17 is similar to the "skeletal coarsening" of polygonal growth shapes in 3 dimensions discussed in the literature since more than one century [87,96]. This instability originates in the incessant and enhanced material supply at given points, which are generated by specific anisotropies of the crystal growth process. This incessant and enhanced material supply leads to a substantially sharper stability condition than implied by Eq. (4), where the instability is due to the accidental appearance of a small protuberance on an otherwise smooth step. The difference between these two instabilities (we call the instability described by Eq. (4) instability I, the other one instability II), which is known in the 3D-case[96,97] appears to be important also in 2D-systems. The Pt/Pt(111)-system seems so far to be the only one in which the two instabilities can be observed clearly and well separated. The instability I leads to the dendritic-fractal shape at low temperatures, the instability II to the onset of dendritic-skeletal growth once a Δ-shape is formed.

d) The flux dependence of instability II

The island shapes shown in figs. 18a and b have been obtained under identical deposition conditions except for the deposition flux. The 100 times higher flux in fig. 18b leads to irregular hexagonal islands (stage 1 type, fig. 17) with a slight tendency to form Δ_l-shapes. In contrast, the low flux (fig. 18a) generates dendritic-skeletal islands (stage 3 type) even at comparable sizes. This astonishing observation is in apparent contradiction to Eq. (4), which predicts a decrease of the critical radius r_c with increasing deposition flux. The contradiction is only apparent because Eq. (4) describes instability I (of straight steps), while fig. 18a shows the effect of instability II.

What is the cause for the inverse flux dependency in the onset of instability II? As discussed above, instability II sets in because at given projected regions of the island contour (in the present case the corners of the Δ-shape), there is an incessant and enhanced material supply which, due to too long triangle sides $L > L_c$, cannot be transported fast enough away. The essential point is that such a contour - here the Δ-shape - is created first, due to anisotropic kinetics. At low deposition fluxes the anisotropy of the kinetics can fully develop: each atom arriving at the contour has enough time to find the most favorable site by means of the allowed kinetic processes and the net particle transport from A- to B-steps is fully active. Δ_l-shaped islands with sharp triangle tips develop at an early stage and the dendritic-skeletal growth sets in during further deposition. This growth is also favored by the larger inter-island distance (another consequence of the low deposition flux), i.e., by the larger diffusion field surrounding each island (see below under f)). In contrast, at high deposition flux the anisotropy of the kinetics has no chance to really develop. The net material flux from the A- to B-step is severely suppressed. For instance, the high impingement rate may lead even on steps with short adatom lingering time (here the A-steps) to nuclei formation and thus to adatom incorporation.

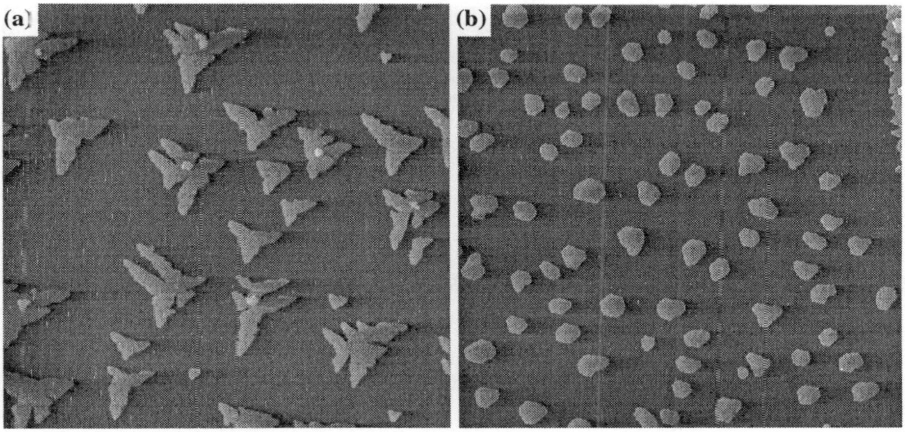

Figure 18: Deposition flux dependence of the island shape. Images taken after deposition of 0.1 ML Pt on Pt(111) at 445 K with very different fluxes: (a) 2.1×10^{-4} ML/s and (b) 2.1×10^{-2} ML/s but otherwise identical conditions (scales for (a) and (b): 2200 Å × 2200 Å).

As a consequence the A-steps will also advance substantially and the island contour will consist of both A- and B-steps, having thus a hexagonal to circular shape with blunt corners only (fig. 18b). Accordingly, there is no significant enhancement of adatom supply at any special section of the contour and thus no reason for the onset of instability II. This example shows how important the difference between the two instabilities is, when trying to understand at least qualitatively the growth behavior of islands.

The analysis above suggests a speculative conclusion concerning the impact of instability II as a function of the symmetry of the growth situation: the lower the surface symmetry, the sharper the island corners, and hence, the stronger the impact of instability II. Indeed, for the twofold symmetry situation, extreme 1D-island shapes are reported [98,99,100], while for fourfold symmetries no dendritic skeletal shapes have yet been observed [38,92,101].

e) The effect of adsorbed carbon

Previous experiments on Pt(111) have shown that in the presence of C-contamination the equilibrium shape of Pt-islands is no longer hexagonal with threefold symmetry, but practically circular [16]. This demonstrates that carbon adsorbs preferentially at steps and - of particular significance here - that the specific step free energy with adsorbed carbon is isotropic. The consequences for the onset of the instabilities are evidenced in fig. 19. The three images have

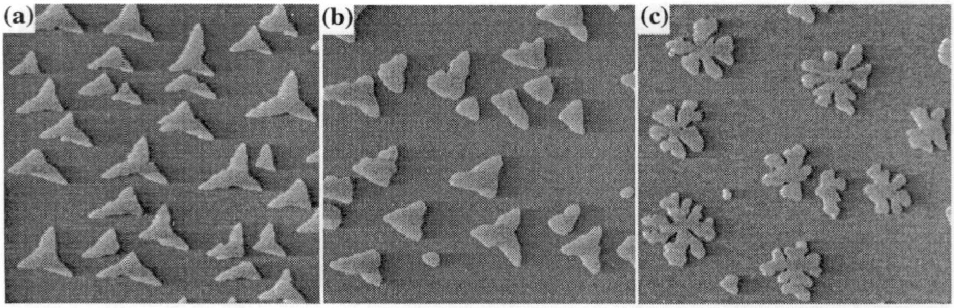

Figure 19: Influence of adsorbed carbon on the island shape. Images taken upon 0.2 ML Pt deposition on (a) a clean, (b) a slightly, and (c) a heavily ($\approx 0.03\ ML$) C-contaminated Pt(111) surface under otherwise identical conditions (except for a slight difference in temperature: (a) 600K, (b) and (c) 610K). Scale: 4400 Å × 4400 Å.

been taken after Pt deposition on increasingly C-contaminated Pt(111) surfaces, under otherwise identical conditions. Fig. 19a shows dendritic-skeletal islands based upon the Δ_h-shape, fig. 19c isotropic fractal ones. (A careful look at fig. 19c shows that "tip splitting" had taken place, which is a clear indication for the absence of anisotropy [102].) The observed shapes are obvious consequences of the effect of the instabilities II and I, respectively. Because carbon adsorption removes the anisotropy, instability II disappears. At the same time carbon adsorption seems to lower the step free-energy and therefore creates the conditions for the onset of instability I (Eq.(4)).

f) The influence of the diffusion field on island growth

Eq. (4) has been derived under the assumption that except for the influence of the island itself, the island is growing in a homogeneous diffusion field, i.e. that at a sufficiently large distance the adatom concentration in all direction is the same. This approximation is fairly well fulfilled in early stages of the monolayer growth. However, as soon as the distances between edges of neighboring islands become of the order of the island dimensions themselves, new effects appear, showing that the approximation no longer holds. The islands start to mutually "communicate" by their overlapping diffusion fields. Fig. 20, which shows a typical dendritic-skeletal island landscape upon deposition of 0.2 ML of Pt on Pt(111) at 640 K, illustrates such effects:

A. The islands indexed by A are located in areas relatively free of other islands. Of particular importance is the fact that there are no closely located islands or step edges in the directions of the triangle corners as seen from the island center. The island tips are surrounded by large diffusion fields and can develop undisturbed.

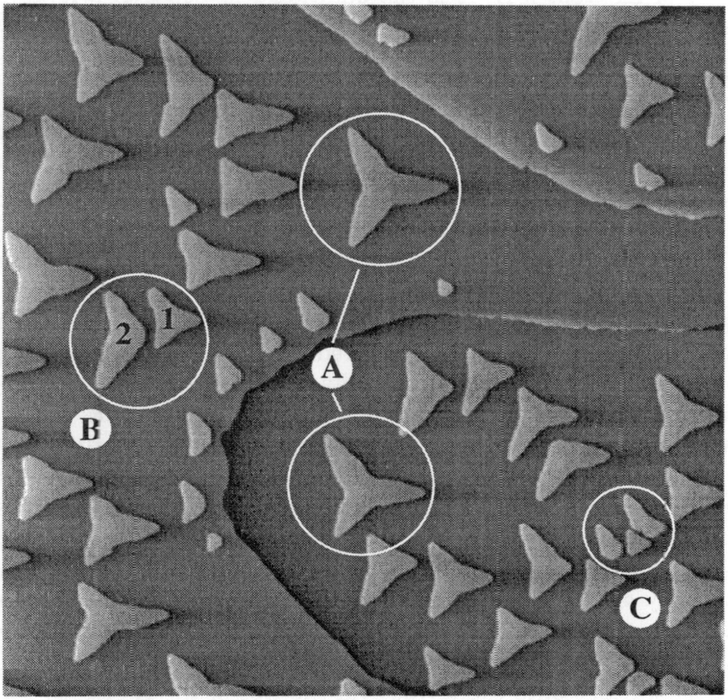

Figure 20: Influence of the diffusion field on the island growth. Image taken after deposition of 0.2 ML Pt on Pt(111) at 640 K with a flux of 7×10^{-3} ML/s. The shapes of the islands in the areas indexed by A, B, and C are discussed in the text.

B. In contrast, the situation of the islands indexed by B is typical for mutually influenced islands via their diffusion field. The tip of island 2 pointing to island 1 is severely shadowed by the vicinity of island 1 and cannot grow as rapidly as the other two tips of island 2 (the latter clearly exhibits the features of instability II) due to the substantially reduced adatom supply. Thus, the simple existence of a Δ-shape is not sufficient for the onset of instability II. It is the additional presence of a large diffusion field which warrants the enhanced adatom supply.

C. The islands indexed by C being closer, the mutual influence is even stronger. They evidence a surprising aspect: the islands have a propensitiy not to coalesce. As soon as their edges become closer, the islands seem to be mutually repelling. The narrower the space between islands, the lesser the number of atoms deposited in this space. The growth of an island towards a neighboring one slows down compared to the growth in open directions, where islands are further away. (The same effect is observed for triangular islands grown at lower temperatures.) The appearance and persistence of channels between islands show that the contribution of the atoms deposited onto the island top and jumping onto the terrace, is not particularly important at these coverages. As shown in the preceding paragraph, the kMC simulations performed for the growth of Pt-islands on Pt(111) mentioned previously[32,55,33] reproduce the observed phenomena remarkably well. However, there is a small but significant difference between the features in fig. 20 discussed here and the results shown in fig. 3 of ref. 54 and fig. 6 of ref. 33. Indeed, in the simulations island tips shadowed by a closely located island are still very sharp. These sharp tips are even protruding into the edges of neighboring islands. The contradiction to the experimental observation may originate in an overestimation of the number of atoms jumping from the island tops onto the terrace via the tips of the Δ-shaped islands.

Acknowledgments

We acknowledge the essential contributions in experiment and analysis to the presented work by Michael Hohage, Michael Bott, Stefanie Esch and Markus Morgenstern. Discussions with Zhenyu Zhang (section 3.1) and Dietrich Wolf (section 3.3) were useful as was the help of Christian Tölkes with TEX-processing.

References

1. Ichiro Sunagawa, *Morphology of Crystals* (Terra Scientific Publishing, Tokyo 1987) page V.

2. J. W. Gibbs *On the equilibrium of heterogeneous Substances, Collected works* (Longmans and Green, New York 1928) **1**.
3. P. Curie, Bull. Soc. min. de France **8**, 145 (1885).
4. excellent introductions to the subject are: C. Herring, in *Structure and Properties of Solid Surfaces* ed. R. Gomer and C. S. Smith (University of Chicago Press, Chicago 1953) p. 5 and M. Wortis, *Chemistry and Physics of Solid Surfaces, Vol. 7* ed. R. Vanselow and R. Howe (Springer, Berlin 1988) p. 367.
5. J. C. Heyrauld and J. J. Métois, J. Crys. Growth **50**, 571 (1980).
6. J. C. Heyrauld and J. J. Métois, Surf. Sci. **128**, 334 (1983).
7. J. C. Heyrauld and J. J. Métois, J. Crys. Growth **84**, 503 (1987).
8. J. C. Heyrauld, J. J. Métois, J. M. Bermond, J. Crys. Growth **98**, 355 (1989).
9. D. J. Eaglesham, A. E. White, L. C. Feldman, N. Moriya, D. C. Jacobsen, Phys. Rev. Lett. **70**, 1643 (1993).
10. A. Pavlovska, D. Dobrev, E. Bauer, Surf. Sci. **314**, 341 (1994).
11. A. Pavlovska, D. Dobrev, E. Bauer, Surf. Sci. **326**, 101 (1995).
12. H. P. Bonzel and S. Surnev in *Dynamics of Crystal Surfaces and Interfaces* ed. P. M. Duxbury and T. J. Pence (Plenum Press, New York 1997) p. 41.
13. Y. W. Mo, B. S. Swartzentruber, R. Kariotis, M. B. Webb, M. G. Lagally, Phys. Rev. Lett. **63**, 2393 (1989).
14. W. Święch and E. Bauer, Surf. Sci. **255**, 219 (1991).
15. N. C. Bartelt, R. M. Tromp, E. D. Williams, Phys. Rev. Lett. **73**, 1656 (1994).
16. Th. Michely, *Rastertunnelmikroskopuntersuchung der Morphologie von Pt(111) nach Ionenbeschuß, Jül-Bericht 2569* (Forschungszentrum Jülich, Jülich 1991).
17. Th. Michely and G. Comsa, Surf. Sci. **256**, 217 (1991).
18. Th. Michely, T. A. Land, U. Littmark, G. Comsa, Surf. Sci. **272**, 204 (1992).
19. G. Ehrlich, F. G. Hudda, J. Chem. Phys. **44**, 1039 (1966).
20. R. L. Schwoebel, E. J. Shipsey, J. Appl. Phys. **37**, 3682 (1966).
21. Th. Michely, Habilitationschrift *Atomare Prozesse bei der Pt-Abscheidung auf Pt(111)* (Bonn 1996).
22. G. Wulff, Zeitschr. f. Kristallog. **34**, 449 (1901).
23. W. Burton, N. Carbrera, F. C. Frank, Trans. Roy. Soc. **A 243**, 299 (1951).
24. M. Rost, Th. Michely, G. Comsa, Phys. Rev. **57**, in print.
25. The approximate identity of the step free energies of A- and B-steps on Ag(111) may be inferred from the published data on island shapes: J. A. Meyer and R. J. Behm, Phys. Rev. Lett. **73**, 364 (1994); G. Rosenfeld,

N. N. Lipkin, W. Wulfhekel, J. Kliewer, K. Morgenstern, B. Poelsema, G. Comsa, Applied Physics **A 61**, 455 (1995); K. Morgenstern, *Jül-Bericht 3278* (Forschungszentrum Jülich, Jülich 1996).
26. S. Surnev, P. Coenen, B. Voigtländer, H. P. Bonzel, P. Wynblatt, Phys. Rev. B, submitted.
27. R. C. Nelson, T. L. Einstein, S. V. Khare, P.J. Rous, Surf. Sci. **295**, 462 (1993).
28. S. V. Khare, T. L. Einstein, Surf. Sci. Lett. **314**, 857 (1994).
29. P. J. Feibelman, Phys. Rev. **B 52**, 16845 (1995).
30. G. Boisvert, L. J. Lewis, M. Scheffler, Phys. Rev. B, submitted.
31. R. Stumpf and M. Scheffler, Phys. Rev. Lett. **72**, 254 (1994).
32. J. Jacobsen, K. W. Jacobsen, J. K. Nørskov, Surf. Sci. **359**, 37 (1996).
33. J. Jacobsen, K. W. Jacobsen, J. K. Nørskov, Scanning Microscopy **11**, in print.
34. R. Q. Hwang, J. Schröder, C. Günther, R. J. Behm, Phys. Rev. Lett. **67**, 3279 (1991).
35. M. Bott, Th. Michely, G. Comsa, Surf. Sci. **272**, 161 (1992).
36. Th. Michely, M. Hohage, M. Bott, G. Comsa, Phys. Rev. Lett. **70**, 3943 (1993).
37. H. Brune, C. Romainczyk, H. Röder, K. Kern, Nature **369**, 469 (1994).
38. H. Brune, H. Röder, C. Romainczyk, C. Boragno, K. Kern, Appl. Phys. **A 60**, 167 (1995).
39. H. Röder, K. Bromann, H. Brune, K. Kern, Phys. Rev. Lett. **74**, 3217 (1995).
40. T. A. Witten, L. M. Sander, Phys. Rev. Lett. **47**, 1400 (1981).
41. P. Meakin, Phys. Rev. **A 27**, 1495 (1983).
42. Z. Zhang, X. Chen, M. G. Lagally, Phys. Rev. Lett. **73**, 1829 (1994).
43. A. Pimpinelli, J. Villain, D. E. Wolf, J. Phys. 1 (Paris) **3**, 447 (1993).
44. M. C. Bartelt, J. W. Evans, Surf. Sci. Lett. **314**, 829 (1994).
45. M. Hohage, M. Bott, M. Morgenstern, Th. Michely, G. Comsa, Phys. Rev. Lett. **76**, 2366 (1996).
46. J. Wintterlin, J. Wiechers, H. Brune, T. Grietsch, H. Höfer, R. J. Behm, Phys. Rev. Lett. **68**, 59 (1989).
47. M. Bott, M. Hohage, M. Morgenstern, Th. Michely, G. Comsa, Phys. Rev. Lett. **76**, 1304 (1996).
48. P. J. Feibelman, J. S. Nelson, G. L. Kellogg, Phys. Rev. **B 49**, 10548 (1994).
49. S. C. Wang, G. Ehrlich, Surf. Sci. **239**, 301 (1990).
50. D. W. Bassett, P. R. Webber, Surf. Sci. **70**, 520 (1978).
51. G. L. Kellogg, J. Phys. (Paris) **47**, C2-331 (1986).
52. H. Brune, H. Röder, K. Bromann, K. Kern, J. Jacobsen, P. Stoltze, K. Jacobsen, J. K. Nørskov, Surf. Sci. Lett. **349**, 115 (1996).

53. A. Gölzhäuser, G. Ehrlich, Phys. Rev. Lett. **77**, 1334 (1996).
54. J. Jacobsen, K. W. Jacobsen, P. Stoltze, J. K. Nørskov, Phys. Rev. Lett. **74**, 2295 (1995).
55. J. Jacobsen, Ph. D. thesis (University Lyngby, Lyngby 1996).
56. M. Villarba, H. Jónsson, Surf. Sci. **317**, 15 (1994).
57. M. Villarba und H. Jónsson, Phys. Rev. **B 49**, 2208 (1994).
58. S. L. Liu, Z. Zhang, J. K. Nørskov, H. Metiu, Surf. Sci. **321**, 161 (1994).
59. P. Stoltze, J. Phys.: Condens. Matter **6**, 9495 (1994).
60. C. L. Liu, J. M. Cohen, J. B. Adams, A. F. Voter, Surf. Sci. **253**, 334 (1991).
61. R. Wang, K. A. Fichthorn, Surf. Sci. **301**, 253 (1994.
62. Y. Li, A. E. DePristo, Surf. Sci. **319**, 141 (1994).
63. M. C. Fallis, M. S. Daw, C. Y. Fong, Phys. Rev. **B 51**, 7817 (1995).
64. G. Kellogg, P. J. Feibelman, Phys. Rev. Lett. **64**, 3143 (1990).
65. M. Hohage, *Jül-Bericht 3255* (Forschungszentrum Jülich, Jülich 1996).
66. S. Liu, Z. Zhang, G. Comsa, H. Metiu, Phys. Rev. Lett. **71**, 2967 (1993).
67. M. Villarba, Ph. D. thesis (University of Washington Seattle, Seattle 1995).
68. P. Ruggereone, C. Ratsch, M. Scheffler in *Growth and Properties of Ultrathin Epitaxial Layers, The Chemical Physics of Solid Surfaces, Vol. 8* ed. D. A. King and D. P. Woodruff (Elsevier, Amsterdam 1997).
69. S. C. Wang, G. Ehrlich, Phys. Rev. Lett. **62**, 2297 (1989).
70. S. C. Wang und G. Ehrlich, J. Chem. Phys. **94**, 4071 (1991).
71. S. C. Wang, G. Ehrlich, Phys. Rev. Lett. **67**, 2509 (1991).
72. S. C. Wang, G. Ehrlich, Phys, Rev. Lett. **68**, 1160 (1992).
73. S. C. Wang, G. Ehrlich, Phys. Rev. Lett. **70**, 41 (1993).
74. S. C. Wang, G. Ehrlich, Phys. Rev. Lett. **71**, 4174 (1993).
75. C. Chen , T. T. Tsong, Phys. Rev. **B 41**, 12403 (1990).
76. C. Chen, T. T. Tsong, Appl. Phys. **A 51**, 405 (1990).
77. T. Y. Fu, Y. R. Tzeng, T. T. Tsong, Phys. Rev. Lett. 76, 2539 (1996).
78. T. Y. Fu, Y. R. Tzeng, T. T. Tsong, Phys. Rev. **B 54**, 5932 (1996).
79. T. Y. Fu, Y. R. Tzeng, T. T. Tsong, Surf. Sci. Lett. **366** L691 (1996).
80. G. Ehrlich, Surf. Sci. **331**, 865 (1995).
81. R. Stumpf, Ph. D. thesis (University Berlin, Berlin 1994).
82. R. Stumpf, M. Scheffler, Phys. Rev. **B 53**, 4958 (1996).
83. G. S. Bales, A. Zangwill, Phys. Rev. **B 41**, 5500 (1990).
84. A. Pimpinelli, J. Villain, D. E. Wolf, J. J. Métois, J. C. Heyraud, I. Elkinani, G. Uimin, Surf. Sci. **295**, 143 (1993).
85. W. W. Mullins, R. F. Sekerka, J. Appl. Phys. **34**, 323 (1963).
86. W. W. Mullins, R. F. Sekerka, J. Appl. Phys. **35**, 444 (1964).
87. O. Lehmann, *Molekularphysik*, (Leipzig 1888/1889).

88. J. A. Stroscio, D. T. Pierce, R. A. Dragoset, Phys. Rev. Lett. **70**, 3615 (1993).
89. K. Meinel, M. Klaua, H. Bethge, J. Crys. Growth **89**, 447 (1988).
90. C. A. Lang, M. M. Dovek, J. Nogami, C. F. Quate, Surf. Sci. **224**, L947 (1989).
91. J. Vrijmoeth, H. A. van der Vegt, J. A. Meyer, E. Vlieg, R. J. Behm, Phys. Rev. Lett. **72**, 3843 (1994).
92. E. Kopatzki, S. Günther, W. Nichtl-Pecher, R. J. Behm, Surf. Sci. **284**, 154 (1993).
93. J. Tersoff, R. M. Tromp, Phys. Rev. Lett. **70**, 2782 (1993).
94. A. F. Becker, G. Rosenfeld, B. Poelsema, G. Comsa, Phys. Rev. Lett. **70**, 477 (1993).
95. D. Chambliss, R. J. Wilson, J. Vac. Sci. Technol. **B 9**, 928 (1991).
96. K. Spangenberg, in *Handwörterbuch der Naturwissenschaften, Bd. 10*, (Gustav Fischer Verlag, Jena 1934) 2. edition, p. 362.
97. L. M. Sanders in *Solids Far From Equilibrium*, ed.C. Godrèche (Cambridge University Press, Cambridge 1992), p. 433.
98. H. Röder, E. Hahn, H. Brune, J. P. Bucher, K. Kern, Nature **366**, 141 (1993).
99. Y. W. Mo, J. Kleiner, M. B. Webb, M. G. Lagally, Phys. Rev. Lett. **66**, 998 (1991).
100. Y. W. Mo, M. G. Lagally, Surf. Sci. **248**, 313 (1991).
101. J. M. Wen, J. W. Evans, M. C. Bartelt, J. W. Burnett, P. A. Thiel, Phys. Rev. Lett. **76**, 652 (1996).
102. D. Wolf, private communication.

Ramified Growth in Metal on Metal Epitaxy

Robert Q. Hwang
Sandia National Laboratories, Livermore, CA 94550

Abstract

Ramified and fractal growth patterns have been shown to exist in many metal on metal film systems. In this paper, the observation, verification and modeling of such behavior is reviewed and the nature of the diffusion processes that give rise to this phenomenon will be discussed. These diffusion processes govern growth in epitaxial systems and the ramified nature of islands points to specific relationships between them.

The study of metal film growth has a history spanning several decades. However, it has only been recently that a detailed understanding of the atomistic processes that govern growth has been addressed. With recent advances in both experimental and theoretical techniques, many traditional concepts have been validated, and one has witnessed the discovery of unexpected phenomena related to the dynamics of film growth. In particular, scanning tunneling microscopy (STM) has permitted the detailed structure and morphology of thin films to be examined on the atomic length scale. In this chapter, the phenomenon of ramified island growth and its evolution from atomistic processes will be discussed. This is especially important for islands in the submonolayer regime act as a "template" for further film growth. In particular, island shapes can have a great influence on the eventual morphology of thicker films.

As atoms are deposited onto a surface and form islands, information on atomic kinetics can be obtained by examining features of the nucleated islands and their dependence on deposition conditions. A simple way to appreciate this is to consider a highly idealized model of epitaxy in which atoms are deposited onto a smooth surface where they are free to diffuse randomly. Islands are formed probabilistically by adatoms colliding. Information related to the rate at which isolated adatoms diffuse, and the stability of the clusters they form can be obtained by studying the size distribution of nucleated islands (see chapter by Evans and Bartelt in this volume). More information can be obtained by examining not only the size but also the shape of the clusters. In particular, the mobility of atoms along the perimeter of islands can greatly affect their shape. For many systems of metal growth on fcc(100) metal surfaces, submonolayer islands were found to be compact, exhibiting the square substrate symmetry[1-3]. On close-packed (111) surfaces however, irregular ramified islands have been observed as a result of room temperature deposition. This was first observed by Chambliss and Wilson[4] studying the growth of Ag on Au(111). Broad finger-like islands were found in the first monolayer. Crystal growth models incorporating this effect have been extensively studied within diffusion limited aggregation (DLA) and related models[5-9]. In their original DLA model, Witten and Sander[5,6] simulated two-dimensional (2-D) growth by injecting atoms (or particles) onto a lattice very far from a seed island at its center. The atoms then performed random walks until they either walked off the lattice or encountered the island. If they did reach the island, the atoms became trapped so that they could neither leave the island nor diffuse along the island edge. Such islands formed highly dendritic shapes. They were in fact shown to be self-similar with a fractal dimension of 1.7. This result proved to be quite robust and independent of the details of the model. That is, relaxing the assumptions of the model in several ways, such as varying the symmetry of the lattice or allowing for non-unity sticking of atoms onto the island, had no effect on the fractal dimension. Therefore, this dimension appeared to be universal.

It is easy to appreciate the intuitive appeal of this simple model to metal epitaxy, since in the majority of cases, films are grown from atomic beams. Quantitative application of this model was first made in the study of Au on Ru(0001).[10] At room temperature, submonolayer Au islands form highly ramified structures as shown in the series of STM images in Fig. 1(a-d). In this series of experiments, the films were grown with a consistent deposition rate of 1.8 monolayers per minute (ML/min). For very low coverages of 0.03 ML (Fig. 1a), two types of islands were found: small compact islands about 100Å and larger islands made up of a core with arms extending outward. These arms also exhibited a width of about 100Å. At 0.15 ML (Fig. 1b), ramified monolayer islands continued with the same arm width of about 100Å. At the higher coverage of 0.37 ML (Fig. 1c), the ramified shapes are even more pronounced. The arms are now thicker. At 0.69 ML, the thickening of the arms was very apparent (Fig. 1d). This can be explained by the fact that at higher coverages, a larger fraction of the deposited Au occurred either between the arms or directly on top of existing islands. In the first case, these atoms can easily diffuse to the sides of the arms thus widening them. Atoms landing on top of islands can also easily find the edge of arms where they can diffuse off the islands and condense on the arm edge, again leading to the thickening of the structure. Note that even at the high coverage of 0.69 ML, there are still deep "fjords" in the islands which can act as traps for atoms diffusing on top of the islands. In this way, the ramified structure of the film acts to enhance layer-by-layer growth by reducing the probability that second layer atoms collide to form islands before they encounter a first layer step edge[11].

These ramified structures are not energetically favorable because of their high edge to area ratio. They are produced by kinetic limitations. A check of this was

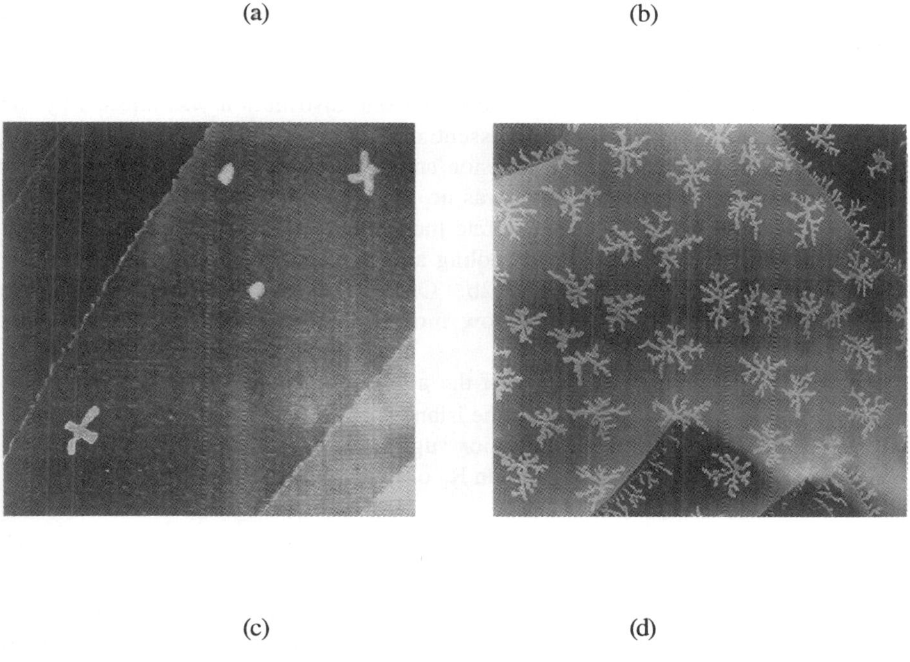

Fig. 1: STM images of various coverages of Au islands deposited on Ru(0001) at room temperature with a flux rate of 1.8 ML/min. (a) 0.03 ML (0.60 μm X 0.60 μm), (b) 0.15 ML (1.20 μm X 1.20 μm), (c) 0.37 ML (0.96 μm X 0.96 μm), (d) 0.69 ML (1.13 μm X 1.13 μm) [10].

performed by a simple annealing experiment. Fig. 2a shows the film of Fig. 1c after a momentary heating to 650K. The islands restructured into much more compact and energetically favorable configurations. However, it should be noted that neither the number density nor the island size distribution was affected by the annealing, indicating that each island essentially collapsed onto itself. The heating process only enhanced edge atom diffusion and did not result in substantial transfer of atoms between islands, i.e. there was no measurable island coarsening. Flash heating to 1100K however, did evaporate the Au back into a 2-D gas from which islands are able to re-condense upon cooling at a higher temperature. The resulting island morphology is shown in Fig. 2b. One can conclude that the ramified Au islands are indeed stabilized by the low mobility of edge atoms as in the DLA model.

A more quantitative evaluation of the applicability of DLA to this growth is to perform a dimensional analysis on the island shapes[12] and check its compatibility with DLA predictions. Fractal behavior suggests a power-law dependence of the island size N with the radius of gyration R_g of the form,

$$N \sim R_g^D, \qquad (1)$$

were D is the fractal dimension (D=2 for compact islands). In applying this analysis, larger islands are desirable to access a larger range of R_g. To achieve this, Au islands were grown at a deposition rate of 0.2 ML/min, resulting in islands similar to that shown in Fig. 3. Power dependence of the size of the island can be seen by plotting $\log_{10}N$ versus $\log_{10}R_g$ as shown in Fig. 4. The linearity of the plot implies that there is in fact scaling of the form described by Eqn. (1) with a fractal dimension of $D=1.72\pm0.07$. This value is in excellent agreement with that predicted from DLA, and served to identify the importance of edge diffusion to island morphology.

Because edge diffusion is thermally activated, the development of non-equilibrium island shape instabilities should be generally important in epitaxy at low temperatures. In fact, dendritic island growth has been found using low temperature STM for Pt on Pt(111) [13], Ag on Pt(111) [14] and Ag on Ag(111).[15] In general, ramified growth is expected when the mobility of edge adatoms is much less than that of isolated adatoms on the surface, so that the arrival rate of adatoms at the edge of an island is faster than the relaxation rate of edge atoms around the island perimeter (i.e. the rate at which they find kink sites). However, for Au/Ru(0001) and Ag/Au(111), the arm widths are far from the single atomic widths of a pure DLA process. Therefore, relaxations of the model are needed to account for the finite thicknesses found in the experiments. Generalizations of the basic DLA mechanism have been considered by several groups, including models where the island edge sticking probability less than one[12, 16, 17] and models with non-zero edge

(a)

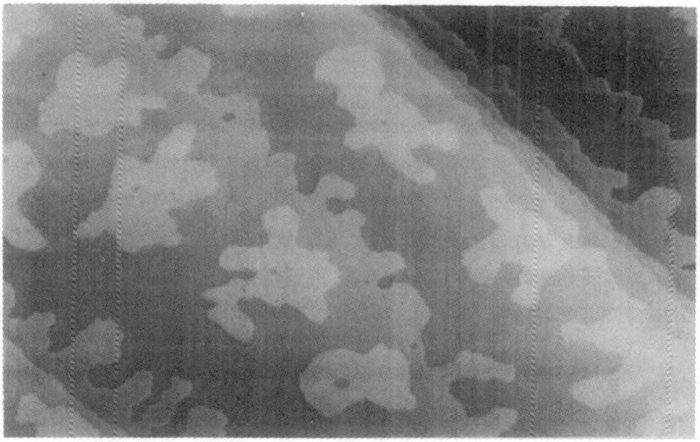

(b)

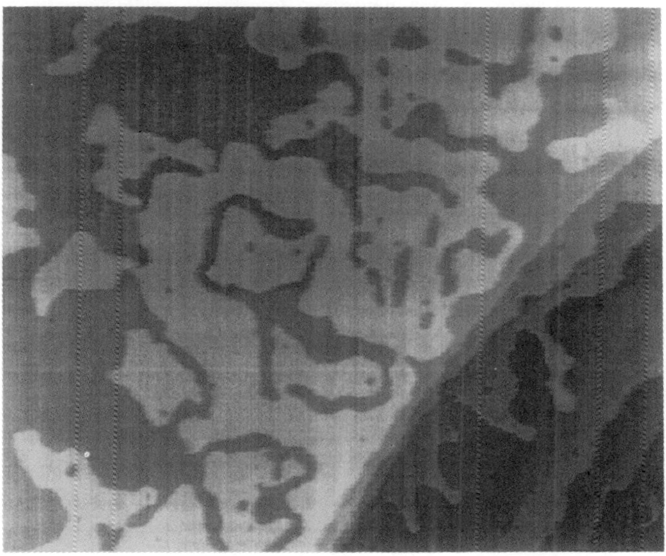

Fig. 2: STM images of the identical film shown in Fig. 1c after annealing to (a) 650K (0.83μm X 0.54 μm) and (b) 1100K (0.76μm X 0.66μm) [10].

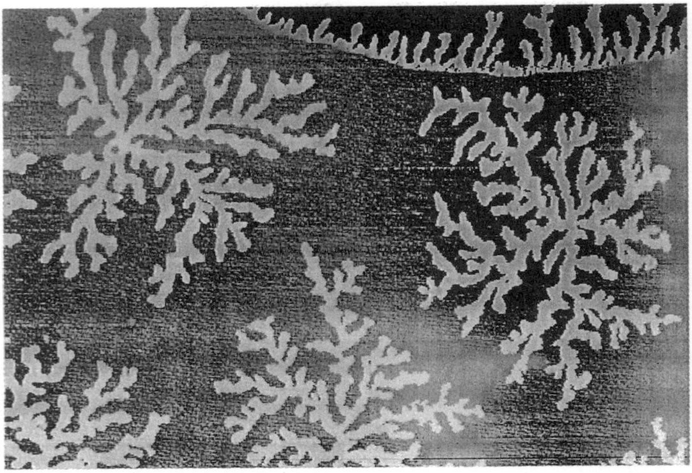

Fig. 3: STM image of ramified islands of Au on Ru(0001) (0.3 ML) formed at room temperature with a flux rate of 0.2 ML/min (1.00μm X 0.65μm) [10].

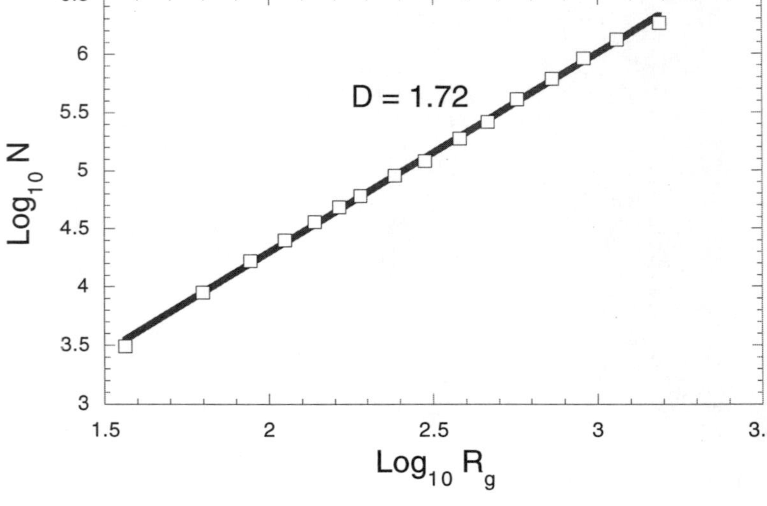

Fig. 4: Scaling nature of island size, N, versus radius of gyration, R_g for islands of the type shown in Fig. 3. Slope of log vs. log plot indicates fractal dimension D=1.72±0.07 [10].

diffusion[18-22]. In particular, Bartelt and Evans[19] have simulated the effects of a non-zero edge hopping rate, h_e, on the thickness of the island arms. The important parameter is the ratio of h_e to the impingement rate of adatoms, I. For sufficiently large h_e/I, compact islands initially grow. As the islands become larger, the mobility of the edge atoms is not large enough for them to "sample" the entire perimeter of the island. Thus an instability occurs at some island size L_c, leading to dendritic growth[18]. From their simulations, they found that the average arm width of dendritic islands, L_w, and L_c depended on this ratio as,

$$L_w \sim L_c \sim 1.7(h_e/I)^{1/2}. \tag{2}$$

Diffusion processes are generally assumed to be thermally activated, so that the edge hop rate should have a temperature dependence of the form

$$h_e = \upsilon \exp[-E_e/(k_B T)], \tag{3}$$

where E_e is the energy barrier for adatom edge diffusion, υ is the effective attempt frequency, k_B is Boltzmann's constant and T is the temperature. Furthermore, the impingement rate, I, can be obtained from the island density, N_{island}, and the known deposition flux, F. In the steady-state regime of growth in which no further islands are nucleated,

$$N_{island} \approx F/I. \tag{4}$$

Therefore by measuring the temperature dependence of L_w for controlled fluxes, an effective energy barrier for edge diffusion can be extracted from Arrhenius plots of $L_w\sqrt{I}$. This was performed by Röder et al, for the dendritic growth of Ag on both Pt(111) and Ag(111) from which they extracted the values of 125±10 meV and 65±10 meV, respectively[15].

Other forms of ramified growth have also been found which can provide even more detailed information regarding island edge diffusion barriers. For instance on Pt(111), Pt and Ag ramified islands grow in highly anisotropic shapes. This issue has been investigated in detail by Hohage et al.[23] and Brune et al.[24], using low temperature STM. Shapes of Ag and Pt islands on Pt(111) reflect the three-fold symmetry of the substrate, with branches growing preferentially in the $[\bar{1}\bar{1}2]$ directions. To understand how the anisotropy develops, consider Fig. 5 which shows a schematic of a hexagonal island on an fcc (111) surface. Two types of step edges are present and labeled A and B. They differ in that they form small facets of different orientation and atomic structure, (111) and (100) for A and B respectively. This structural difference leads to differing edge diffusion barriers along the two step types. The observed branching in the $[\bar{1}\,\bar{1}2]$ directions implies that atoms are preferentially trapped at A steps. Using effective medium theory (EMT), Brune et al.[24] calculate E_A=187meV and E_B=389meV for the edge diffusion barrier along the A and B steps respectively, for Ag/Pt(111). In addition, they find different diffusion barriers from the corner site to either the A or B step as illustrated in Fig. 5. Note that the atom in position C can diffuse to site A through the intermediate high coordination hcp site. No such site exists between C and B, so the barrier for this latter process is expected to be larger. Indeed, EMT calculations find that $E_{C\text{-}A}$=80meV while $E_{C\text{-}B}$=248meV.[24] The value of $E_{C\text{-}A}$ is in fact very close to the EMT value for the energy barrier of a freely diffusing Ag atom on Pt(111). This lower barrier preferentially allows more corner atoms to diffuse to A steps, thus promoting branching in the $[\bar{1}\,\bar{1}2]$ directions.

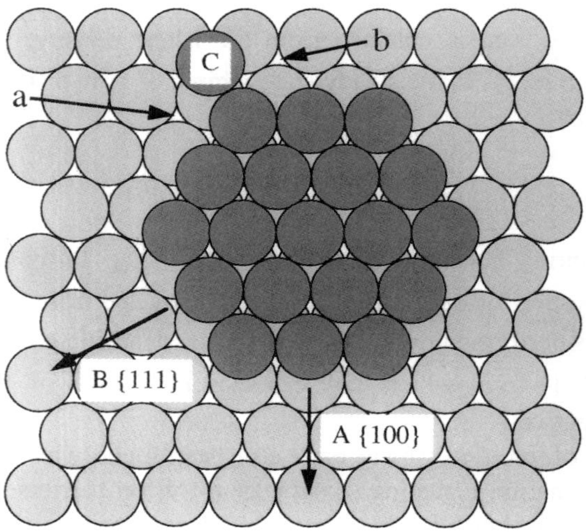

Fig. 5: Geometric differences encountered by an Ag adatom diffusing around an Ag cluster on Pt(111). A- and B- directions are indicated in the ball model. The diffusion path with the lowest barrier is the one from the corner (C) and the hcp site close to it, at an A step [24].

Fig. 6: (a) 7500 X 7500 Å image of a Co-Ag alloy grown at 70°C as described in the text. (b) 500 X 500 Å image of a type A dendritic island[25].

For Pt and Ag on Pt(111) described above, the edge diffusion barriers leading to ramified growth was due to structural differences along island edges. Similar effects can have chemical origins as well. This is particularly important in the growth of alloy films as shown in the case of bi-metallic films of Ag-Co on Ru(0001).[25] At room temperature, dendritic islands form on the surface (Fig. 6a) despite the fact that neither metal grown individually on Ru(0001) produces such shapes. These islands are in fact made up of a mixture of both metals with a specific structure. Alternating veins of Ag and Co run along the primary directions of the dendritic growth (Fig. 6b). The link between this internal structure and the island shapes can be seen in the series of STM images of Fig. 7. Deposition at elevated temperatures continues to produce dendritic islands with the internal vein structure up to about 170°C (Fig. 7a and Fig. 7b). Above this temperature, the dendritic shape and the internal structure simultaneously disappear (Fig. 7c and Fig. 7d). The connection between the vein structure and the edge diffusion barriers around an alloy island can be understood schematically by considering a small Co-Ag (Co-gray and Ag-white) cluster diagrammed in Fig. 8. Also plotted is the energy diagram for a Co atom at various positions along the edge of the cluster. For diffusion along each metal individually, a small barrier is encountered due to the close-packed structure of the island edge. However, for a Co atom to cross from a Co region to a Ag region along an edge, an additional energy barrier, δ, must be crossed, which depends on the energy difference between a Co-Co bond and a Co-Ag bond. Thus it is more difficult for atoms to diffuse across these interfaces, leading to a reduced edge mobility. (A similar energy diagram exists for Ag.) In this way, the interface energetics lead to diffusion mechanisms that favor the continued growth of like-metal species and encourages dendritic island shapes.

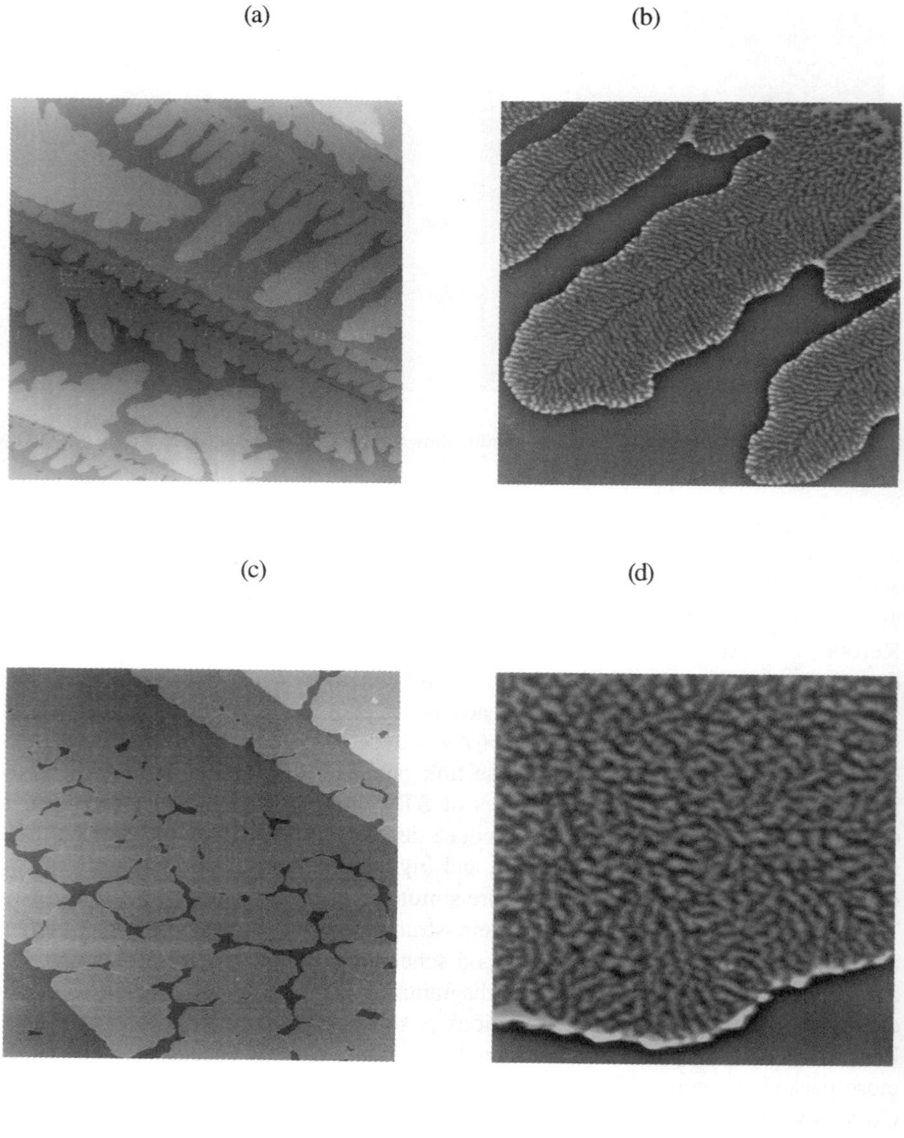

Fig. 7: (a) 6000 X 6000 Å image of a Co-Ag alloy grown at 138°C as described in text. (b) 1750 X 1750 Å image of the vein structure with the finger islands of fig. 2a. The image is contrast enhanced, so that the intensity difference between the Co and Ag represents a corrugation height of about 0.2Å. Note the integrity of the veins. (c) 10,000 X 10,000 Å image of a Co-Ag alloy grown at 170°C. Islands now exhibit compact shapes as described in text. (d) 600 X 600 Å image of the internal structure with the islands of fig. 2c. Continuous vein structure no longer exists at this growth temperature[25].

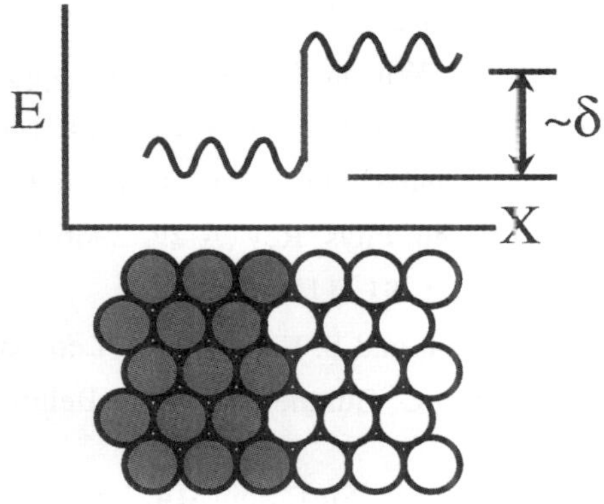

Fig 8: Schematic of the energy diagram experienced by a Co at the edge of a vein structure. Ag and Co atoms are represented as white and black balls respectively [25].

In conclusion, the appearance of ramified growth in metal epitaxy is a generic phenomenon related to the difference in atomistic diffusion processes along an island edge as compared to free adatom mobility. Since the first observation and identification of DLA-like growth, careful measurements coupled with detailed modeling has resulted in the ability to measure these mobilities and extract energy barriers. One question has resulted from the observation that on fcc(100) surfaces, ramified growth is uncommon and in fact has only recently been observed[26,27]. This points to the very different diffusion mechanisms that occur on these surfaces as compared to those on the close-packed (111) surfaces[28-30]. FIM measurements have indicated that the primary difference leading to different growth structure on the two types of surfaces is primarily due to the large difference in isolated adatom diffusion and not to differences in edge mobilities[31].

ACKNOWLEDGEMENTS: The author is grateful to Dr. Maria C. Bartelt for her critical reading of this manuscript. The preparation of the review was supported by the Office of Basic Energy Science, Division of Materials Science, of the U.S. Dept. of Energy under Contract No. DE-AC04-94AL85000.

References

1 E. Kopatzki, S. Günther, W. Nichtl-Pecher, and R. J. Behm, Surf. Sci. **284**, 154 (1993).

2 J. A. Stroscio, D. T. Pierce, and R. A. Dragoset, Phys. Rev. Lett. **70**, 3615 (1993).

3 J. A. Stroscio and D. T. Pierce, Phys. Rev. B **49**, 8522 (1994).

4 D. D. Chambliss and R. J. Wilson, J. Vac. Sci. Technol. B **9**, 928 (1991).

5 T. A. Witten and L. M. Sander, Phys. Rev. Lett. **47**, 1400 (1981).

6 T. A. Witten and L. M. Sander, Phys. Rev. B **27**, 5686 (1983).

7 R. C. Ball and T. A. Witten, Phys. Rev. A **29**, 2966 (1984).

8 P. Meakin, Phys. Rev. Lett. **51**, 1119 (1983).

9 D. Bensimon, B. Shraiman, and S. Liang, Phys. Lett. **102A**, 238 (1984).

10 R. Q. Hwang, J. Schröder, C. Günther, and R. J. Behm, Phys. Rev. Lett. **67**, 3279 (1991).

11 M. C. Bartelt and J. W. Evans, Surf. Sci. **314**, L835 (1994).

12 P. Meakin, Phys. Rev. A **27**, 1495 (1983).

13 T. Michely, M. Hohage, M. Bott, and G. Comsa, Phys. Rev. Lett. **70**, 3943 (1993).

14 H. Brune, C. Romainszyk, H. Röder, and K. Kern, Nature (London) **369**, 469 (1994).

15 H. Röder, K. Bromann, H. Brune, and K. Kern, Phys. Rev. Lett. **74**, 3217 (1995).

16 P. Wynblatt, J. J. Métois, and J. C. Heyraud, J. Cryst. Growth **102**, 618 (1990).

17 J. G. Amar and F. Family, Phys. Rev. Lett. , 2066 (1995).

18 A. Pimpinelli, J. Villain, and D. E. Wolf, J. Phys. (Paris) **13**, 447 (1993).

19 M. C. Bartelt and J. W. Evans, Surf. Sci. **314**, L829 (1994).

20 J. Villain, A. Pimpinelli, and D. E. Wolf, Comments Condensed Mater. Phys. **16**, 1 (1992).

21 J. M. García-Ruiz and F. Otálora, Physica A **178**, 415 (1991).

22 T. Viscek, Phys. Rev. Lett. **53**, 2281 (1984).

23 M. Hohage, M. Bott, M. Morgentstern, Z. Y. Zhang, T. Michely, and G. Comsa, Phys. Rev. Lett. **76**, 2366 (1996).

24 H. Brune, H. Röder, K. Bromann, K. Kern, J. Jacobsen, P. Stoltze, K. Jacobsen, and J. Norskov, Surf. Sci. **349**, L115 (1996).

25 R. Q. Hwang, Phys. Rev. Lett. **76**, 4757 (1996).

26 B. Müller, L. Nedelmann, B. Fischer, H. Brune, and K. Kern, Phys. Rev. B **54**, 17858 (1996).

27 Z. Y. Zhang, X. Chen, and M. G. Lagally, Phys. Rev. Lett. **73**, 1829 (1994).

28 G. L. Kellogg, Surf. Sci. Rep. **21**, 1 (1994).

29 J.-M. Wen, J. W. Evans, M. C. Bartelt, J. W. Burnett, and P. A. Thiel, Phys. Rev. Lett. **76**, 652 (1996).

30 J.-M. Wen, S.-L. Chang, J. W. Burnett, J. W. Evans, and P. A. Thiel, Phys. Rev. Lett. **73**, 2591 (1994).

31 G. L. Kellogg, Prog. Surf. Sci. **53**, 217 (1996).

MORPHOLOGY AND ENERGY BARRIERS IN HOMOEPITAXIAL GROWTH AND COARSENING: A CASE STUDY FOR Cu(100)

J. F. Wendelken[1], Anna K. Swan[1], Woei-Wu Pai[1] and J.-K. Zuo[2]
[1]*Solid State Division, Oak Ridge National Laboratory, Oak Ridge, Tennessee 37831-6030, USA*

[2]*Department of Physics and Astronomy, Southwest Missouri State University, Springfield, Missouri 65804, USA*

Abstract

Kinetic limitations during homoepitaxial growth result in a well defined, inherently nonequilibrium, morphology. The observed submonolayer morphology of Cu/Cu(100) with $T = 200-300$ K consists of nearly square islands whose distribution is a function of temperature and flux. Scaling relationships indicate critical cluster sizes of 1, 3, and 8 atoms for initiation of island growth. After deposition, island coarsening is observed to result from island diffusion and coalescence. The island diffusion results from periphery diffusion of atoms mediated by the kink density.

1. Introduction:

With the confluence of technological need, new experimental tools, and theoretical advances, epitaxial growth of thin films has been the subject of an ever increasing number of studies in recent years at both a fundamental and applied level. The field of epitaxial growth divides into subfields depending on film thickness, and the combination of film and substrate material properties (metallic, semiconducting, magnetic, etc.). Essential elements in growth processes are the extrinsic controllable factors of deposition rate and temperature and the intrinsic system properties of diffusion barriers, binding energies, lattice mismatch stress, stoichiometry, and electronic structure. In all cases, a basic knowledge of the initial growth processes at the submonolayer level is essential to the understanding and manipulation of subsequent growth. This understanding can yield improvements and innovations in technologically important thin films while also providing a test of basic physical understanding.

It is easiest to begin a study of epitaxial growth with a homoepitaxial system because this involves the fewest parameters. This chapter focuses on the case of homoepitaxy employing Cu(100) as a model system with discussion limited to compact, 2D growth. Both the mechanisms responsible for the morphology during growth and for the evolution of that morphology after growth are examined.

2. Theoretical Background: Submonolayer Growth

The basic physical processes in submonolayer homoepitaxy are nucleation, island growth, and island coalescence, as well as the onset of second-layer nucleation. These processes determine the ultimate morphology of both thin and thick films. Each process is governed by a balance between the deposition rate and the surface diffusion, association, and dissociation of adatoms and small clusters [1,2]. When adatoms are deposited onto a clean surface, they diffuse in a random walk and, upon supersaturation, nucleate to form small clusters. This diffusion process may involve an exchange mechanism of the adatom with the substrate, as first predicted by Feibelman [3] for Al(100) and observed by Kellogg *et al.* for Pt(100) [4] and also by Chen and Tsong for Ir(100) [5]; but in most cases including Cu(100), embedded atom method, EAM, calculations [6] predict that the diffusion should involve discrete adatom hopping in directions determined by the crystal lattice. Recent *ab initio* calculations [7,8] confirm this for Cu(100). When the clusters, or nuclei, reach a *critical size i*, the addition of one more atom results in a stable island, at least on a time scale proportional to the time interval between the arrival of adatoms near the island. This may be represented as $1/F$, where F is the deposition flux in units of monolayers per second (ML/s). As deposition proceeds, the stable islands act as sinks for deposited atoms, thus limiting the density of adatoms which have a diffusion length determined by their diffusion coefficient D, F, and the existing densities of islands and critical nuclei [9]. Eventually, a balance between arrival of adatoms and losses to islands results in a steady state adatom density, N_1, marking the beginning of the *steady-state growth regime*. At almost the same time, the island density, N, which is determined by the competition between D and F, may become nearly stable with increasing coverage as most adatoms diffuse to existing islands rather than nucleate new ones. Because N is more easily observed than N_1, observation of a stable N vs. coverage is often used as a sufficient indication of steady-state growth, although it is not a necessary condition. Often, the less restrictive term *aggregation regime* is used to describe this period of island growth [2,10]. With further deposition, the *coalescence regime* begins as islands grow larger and begin to touch. This causes a reduction of N until a monolayer is complete in the rare case of perfect layer-by-layer growth. Normally, second-layer growth begins in the coalescence regime, enhanced by a step-edge barrier [11,12] which resists the step-down diffusion of deposited atoms, leading to a pyramid-like multilayer morphology in most systems [13-19]. As is shown below, the steady state regime is most amenable to the study of basic physical parameters since it typically occurs in a broad coverage range from < 0.1 ML to nearly 0.5 ML where diffraction signals are strong and the islands are strictly single layer.

Island shapes during growth are a function of the relative energy barriers for diffusion along different step-edge directions and around island corners. These barriers depend on the crystal face and temperature, and examination of the equilibrium island shapes provides qualitative information about the relative magnitudes of these energy barriers [20]. In the case of metal (100) surfaces, the islands are typically found to be compact and square as shown by studies of Cu, Ni, Fe, and Ag [21-24]. Even vacancy islands and the mounds that occur during multilayer growth [13] show the same square tendency. Exceptions to this behavior can occur when the (100) substrate displays an anisotropic reconstruction as in the case of the Au(100) and Pt(100)-hex reconstructed surfaces on which strongly rectangular growth patterns are observed [25,26], the interpretation of which is not simple [27]. A (110) surface generally displays growth which is compact with elongated rectangular islands while islands on the (111) surface may be compact, fractal, or dendritic with threefold symmetry [28–33]. The compact shapes are generally associated with higher growth temperatures or low-energy barriers for diffusion along close-packed step edges with the result that square islands on (100) surfaces are aligned with sides along the close-packed [110] directions [1,6,7,34]. When the barrier for atom diffusion around a corner is high compared with kT, as in the case of (111) surfaces at low temperatures, noncompact growth forms are observed [28]. Only the case of compact growth on unreconstructed metal(100) surfaces will be discussed below.

Although a variety of diffusion pathways and energy barriers have been isolated and studied [2,6,9,10,35–43], the fundamental growth behavior, when adatom diffusion is isotropic and two-dimensional, is typically described by a simple scaling relationship involving N, F, θ, $D \sim \exp(-E_D/kT)$, and the growth temperature T:

$$N \propto \theta^z (1/F)^{-p} \exp[p(E_D + E_i/i)/kT], \tag{1a}$$

where $\theta = Ft$ is the coverage, E_i is the binding energy of critical nuclei of size i, E_D is the adatom diffusion barrier, and z and p are scaling exponents which contain information about the number of atoms in a critical nucleus, i, but which also depend on the detailed physical growth processes. Manipulation of the growth parameters T and F may be used to control growth morphology [44,45], while measurement of the exponents may be used to test the limits of existing models. Thus far, the coverage dependent factor θ^z is much less studied, and in general it is believed that the factor should be replaced with a less restrictive term $f(\theta)$.

Because the island separation L is proportional to $N^{-1/2}$, Eq. (1a) can be recast in the form

$$L \propto \theta^{-z/2}(1/F)^{p/2} \exp[-p(E_D + E_i / i) / 2kT]. \tag{1b}$$

Both experiment [1,46,47] and Monte Carlo simulations [2,48,49] have confirmed the flux-dependent scaling, while the coverage-dependent scaling is more problematic. Eq. (1) and the exponents z and p have been derived from model-dependent rate equations [9,35–37,50–53], which in general can be written as

$$\frac{dN_1}{dt} = F(1-\theta) - 2K_1N_1^2 - N_1\sum_{s\geq 2} K_sN_s + \sum_{s\leq i}\gamma_sN_s \tag{2a}$$

and $$\frac{dN}{dt} = K_iN_1N_i - 4\pi R\frac{dR}{dt}N^2. \tag{2b}$$

In Eq. (2a), $F(1-\theta)$ is the flux received by the substrate surface (for low θ, $F(1-\theta) \sim F$), K_s is the capture rate of adatoms by islands of size s ($K_1 \sim D$), and γ_s is the detachment rate of atoms from islands of size $s \leq i$, since larger clusters are assumed to be stable. It is also assumed here that there is no barrier for adatom attachment to island edges. In Eq. (2b), the first term in the right side represents the creation rate of stable islands, and the second term represents the loss rate of stable islands by coalescence [35]. In both the steady-state growth and coalescence regimes, the second and fourth terms in Eq. (2a) can be neglected. In the steady-state growth regime, the second term in Eq. (2b) can also be dropped. As long as K_s is given, the rate equations can be solved either analytically or numerically.

In classic nucleation theory, the capture rate K_s is assumed to be independent of island size (i.e., $K_s \sim D$). This is the so-called point-island model. For this simple model, the exponents z and p in the steady-state growth regime can be solved from Eq. (2) to be

$$z = 1/(i+2) \qquad \text{and} \qquad p = i/(i+2). \tag{3}$$

This simple scaling model, often without consideration of any scaling with coverage, has long been used to discuss nucleation and growth. By considering the finite-size effect of islands and the distribution of islands, Ratsch *et al.* [2] suggested a capture rate dependence, $K_s \sim Ds^xN^y$. The physical meaning of this model is that the capture rate K_s is proportional to the island perimeter sites when $x = 1/2$ and is inversely proportional to the island separation when $y = 1/2$. With this model, if $x \leq 1/(i+1)$, the exponents z and p in the steady-state growth regime can be solved from Eq. (2) to be

$$z = \frac{1 - x(i+1)}{i+2+(y-x)(i+1)} \qquad \text{and} \qquad p = \frac{i}{i+2+(y-x)(i+1)} \,. \tag{4}$$

As can be seen, when $x = y = 0$, Eq. (4) reduces to Eq. (3) (i.e., the point-island model). For the realistic case with $x = y = 1/2$, the exponents become

$$z = (1-i)/2(i+2) \qquad \text{and} \qquad p = i/(i+2) \,. \tag{5}$$

Only the exponent z is different from the point-island model. In particular, z is always positive for the point-island model and zero or negative in this finite size model. A recent rate equation and Monte Carlo analysis [54] show a gradual reduction in N vs. θ for Cu/Cu(100) at 350 K and for coverages between 0.1 and 0.4 ML; however, this result includes the effect of small island mobility which is not a part of the above model. The expression for z must be regarded with caution, and a very recent study indicates that it may not be possible to compute an exponent z because a simple power law dependence on coverage probably does not exist [55]. The data discussed below are only consistent with this expression for z when $i = 1$. It should be noted that Eq. (1) will hold in the anisotropic diffusion case, but that the exponent p may be expected to change [38,42,43]. Likewise, in the case where adatom attachment to island edges is hindered by an energy barrier, as when surfactants are present, the value of p has been shown to change [55,56].

These simple relationships of z and p to i in the steady-state regime facilitate experimental tests of the above models because it is generally not feasible to observe directly the initial nucleation events. According to Eq. (1), a coverage-, flux- or temperature-dependent measurement of N or L allows determination of the exponents z and p or the energies E_D and E_i, respectively. The coalescence regime, while interesting in itself, is not included in Eq. (1) and is not as useful for determination of these fundamental parameters. This regime is also complicated by a continuously changing length scale and the initiation of second-layer growth in many systems. Equation (1) is derived for two-dimensional growth under the assumptions of complete condensation (no desorption from the surface) and immobile islands. While the first assumption will normally hold at typical growth temperatures, the latter assumption often will not.

While, in principle, critical sizes of nuclei of any magnitude could be encountered, the detailed energy barriers associated with a particular surface lattice dictate a limited set of critical sizes. This can easily be understood by simple bond counting arguments, where the only cluster sizes to become critical are those whose next size up are particularly stable. While this is often the case it does not have to be [57]. The simplest case for a surface of any symmetry is simply $i = 1$, which occurs at temperatures low enough that the thermal energy is insufficient to break the single bond of a dimer on a relevant time scale, but high enough so that single adatoms can still diffuse. At a sufficiently high temperature, this single bond will break. On a (100) surface at the higher temperature, if three atoms form a cluster long enough for a fourth atom to arrive, a stable island could be formed which would require breaking two nearest neighbor (nn) bonds to dissociate the island giving a second critical size of $i = 3$. This configuration would be stable up to some higher temperature at which double bonds would break and no islands would be stable. Recently, it was shown that a third critical nucleus size of $i = 8$ is also possible [58]. This results from island breakup not only from sequential atom motion, but also including the possibility of two atoms moving together in a dimer shearing motion so that compact islands of size < 9 can dissociate via single-bond scission rather than by breaking 2 bonds. Thus, minimum stable island sizes of 2, 4, and 9 atoms could be encountered on (100) surfaces. The occurrence of any of these possibilities under given conditions depends on finding the lowest free energy of the system that is kinetically accessible. In particular, a low initial energy barrier toward dissociation is necessary but not sufficient to determine the stability of clusters. Bartelt *et al.* defined an effective dissociation barrier to consist of the diffusion barrier plus the cohesive energy difference between the initial and final state [41]. Thus, a given system could display any combination of these three minimum stable island sizes as a function of changing conditions. It must be remembered that the term "stable" is used with a relative time scale and that no island is totally stable in the long term. The scaling relationships associated with the idea of critical sizes also suffer limitations which are beginning to be addressed [55,56,59]. Nevertheless, these concepts serve well to describe the behavior observed in many experiments and simulations.

In most discussions of island nucleation and critical nucleus sizes, it is assumed not only that islands grow irreversibly after a cluster of size $i+1$ occurs, but also that they are immobile. In fact, the motion of small clusters has been observed on Rh(100) [60] and the dimer shearing mechanism reduces the diffusion barrier for islands of fewer than 9 atoms [58]. There have been several theoretical efforts to predict the effect of island motion on the growth process [37,54,61–64]. In an early study, Villain [37] assumed that all islands are mobile up to a size i^* with the island diffusion coefficient decreasing monotonically with island size. The resulting scaling exponent in the steady-state regime was then found to be

$$p = i^*/(2i^*+1)\,. \tag{6}$$

This result is exact if all mobile islands have the same diffusion coefficient [62]. While this condition is not valid, Eq. (6) was found to be a good approximation if the diffusion coefficients are the same order of magnitude and their ratios are independent of temperature when islands of 3 atoms or less are allowed to move [62]. If the islands are less compact (i.e. more fractal in nature), the value of p is expected to increase [61]. To date, there has been no experimental confirmation of these results; however, simulations in which dimers [62,63] and trimers [62] were allowed to move are generally consistent with Eq. (6), though clearly lower values of p are reported in one case [64]. The effect of island mobility must be viewed in terms of the relevant time scales. If the cluster grows before it can collide with another cluster because of the adatom arrival rate, then its diffusion rate does not affect the growth scaling. As is discussed later in this chapter, there is an expected discontinuity in island motion at the cluster size of 9 atoms in addition to an oscillatory behavior in the diffusion rates of small islands which could impact this problem.

Following growth, the resulting array of islands constitutes a nonequilibrium morphology that is observed to coarsen in time so that if one waited a very long time, the final morphology would consist of one island if the substrate were perfectly flat. It had been believed that the observed coarsening behavior was the result of Ostwald ripening in which large islands grow at the expense of smaller ones as a function of time through an evaporation-condensation process. However, with the advent of STM studies of epitaxial growth, it has become clear that mass transport also occurs by the movement of whole islands [24,65–68] even when the sizes become very large. Recent STM studies of Ag/Ag(100) and Cu/Cu(100) [24,65] showed that while Ostwald ripening may occur when islands are very far apart, the dominant mechanism for coarsening, at least at room temperature, is diffusion of whole islands with random collisions and coalescence.

3. Experimental Background:

Most experimental studies of submonolayer growth on metal surfaces in recent years have utilized the real space imaging technique of scanning tunneling microscopy (STM) [21–26,28–33,69] or high-resolution diffraction techniques such as helium atom scattering (HAS) [46,70,71] and spot profile analysis low energy electron diffraction (SPA-LEED, also commonly referred to as HRLEED for high resolution LEED) [1,72–76]. Important contributions using the field ion microscopy (FIM) technique have also been made with regard to detailed motion of individual adatoms on terraces and along step edges [4,77]. While the STM technique directly reveals the general form of the island growth with regard to island shapes and compactness, extraction of scaling properties from STM images can be tedious. Also, studies, which require variable temperatures, particularly if the temperature must be varied quickly, have been difficult in the past with STM. Using diffraction techniques, it is easy to determine average island separations, and temperature dependent studies present no difficulty. In the case of compact growth, the island shapes are also easily determined. Diffraction is also inherently faster for these studies because of the vastly greater sampling area of the diffraction beam vs. the typical STM scan area; however, as shown below, STM is essential both as a check on the diffraction and to elucidate mechanistic details which are not apparent in diffraction data.

The system of Cu/Cu(100) has been a popular system for the study of metal (100) homoepitaxy. In general, observations made on Cu will be the same as those made for other metal (100) systems except for differences in temperature- dependent energy barriers. Growth of Cu/Cu(100) may be considered as imperfect layer-by-layer growth between 200 and 305 K, as determined by the well-known diffraction intensity oscillations observed by SPA-LEED [74], RHEED [78], and HAS [70] which gradually decay with increasing coverage. This decay indicates the initial stages of rough 3D growth. These oscillations are gone at a temperature of 400 K where step-flow growth occurs. Surprisingly, as the temperature is decreased below 200 K, the oscillations decrease at first, but then are found to be stronger again near 100 K indicating the return of layer-by-layer growth [70]. Diffraction rings following a deposition of ~9 ML of Cu [72] suggest some type of adatom mobility at these very low temperatures. Explanations ranging from "transient mobility" [72] to "downward funneling" [79] have been proposed to explain this phenomenon. These low-temperature phenomena will not be dealt with further here.

In this chapter, both high-resolution diffraction and real space STM measurements of island morphologies and energy barriers as a function of F and T in the steady state growth regime for Cu/Cu(100) are reviewed. From these, both the scaling behavior, Eq. (1), and energy barriers which determine the growth character are examined. The diffraction technique of SPA-LEED, with a transfer width >1000 Å for the data discussed here, was used for the determination of average island separations for temperatures ranging from 213 to 305 K. STM measurements at the two extremes of this temperature range confirm the morphological information contained in the diffraction data and also show a high degree of island mobility at room temperature. The data indicate that the smallest stable island at low temperatures is the dimer ($i = 1$); while at an elevated temperature, the data are indicative of a tetramer ($i = 3$) as the smallest stable island. This observation is further supported by dissociation energy calculations for single-atom removal using the embedded-atom method (EAM) [1], but questions have been raised in the literature [39–41]. At extremely low fluxes, observation of an apparent large increase in the smallest stable island size to something near 9 atoms, a nonomer ($i = 8$) has been reported only for Cu(100) to date [80]. From the temperature dependence of the island separation in the $i = 1$ regime, the adatom diffusion barrier on Cu(001) was determined to be 0.36 ± 0.03 eV. In addition, STM observations of the post-deposition coarsening process which illustrate the importance of island diffusion to the coarsening process and elucidate the mechanism that enables and controls island diffusion at room temperature are discussed.

4. Experimental Details: Cu/Cu(100)

The measurements discussed below were all obtained in ultrahigh vacuum chambers (base pressures $< 1 \times 10^{-10}$ Torr) equipped with commercial SPA-LEED or STM systems, Auger electron spectroscopy (AES), and Cu evaporators. The Cu(100) surface was cleaned by Ne^+ sputtering and subsequently annealed at ~750 K for a few minutes. Deposition of a few ML of Cu and an additional anneal improve the surface quality further, as indicated by diffraction measurements. After preparation, no impurities were detected with AES or STM. The average terrace width on the Cu(100) surface was estimated to be ~850 Å with STM which revealed many terraces as large as 4000 Å wide. Copper was evaporated from a thermocouple monitored high-purity (6N) Cu disk heated by electron bombardment from the backside of the disk or from a flux-monitored e-beam evaporator. The samples could be heated by radiation or electron bombardment from the backside and cooled with liquid nitrogen to 110 K. Coverage, θ, was determined in the diffraction measurements using the intensity oscillations of the specular beam with the first minimum corresponding to $\theta = 0.5$ ML [81], while for the STM measurements, θ was determined by direct measurement of the island areas.

One-dimensional angular profiles centered on the (00) beam were obtained with the SPA-LEED system within ~2 min after each deposition during which time no change in profile shape was detected. Two-dimensional angular profiles require much longer, so the sample temperature was cooled below 150 K, where adatom thermal mobility is negligible [46,72,82], in order to avoid coarsening effects during data acquisition. Above a coverage of $\theta \sim 0.2$ ML, a first-order diffraction ring appears around the (00) beam in an out-of-phase diffraction condition at 141.6 eV. Figure 1(a) shows a three-dimensional view of the (00) beam, and Fig. 1(b) shows one-dimensional angular profiles scanned along the [110] and [100] directions. The surface was covered with 0.42 ML of Cu at a sample temperature of 263 K and a flux of 3.21×10^{-4} ML/s. A symmetric array of four lobes with intensity centered ~ 0.1 Å^{-1} from the central (00) beam and lying in the [110] and [110] directions [73,74] is also evident. The combination of circular and fourfold symmetries at different k values reflects two separate contributions to the diffraction pattern, where k is the scattering vector parallel to the surface. At the smaller k value, the circular ring comes from a narrow, nearly isotropic, distribution of island separations peaked at a distance L. At larger k values, the four-fold pattern comes from a narrow island

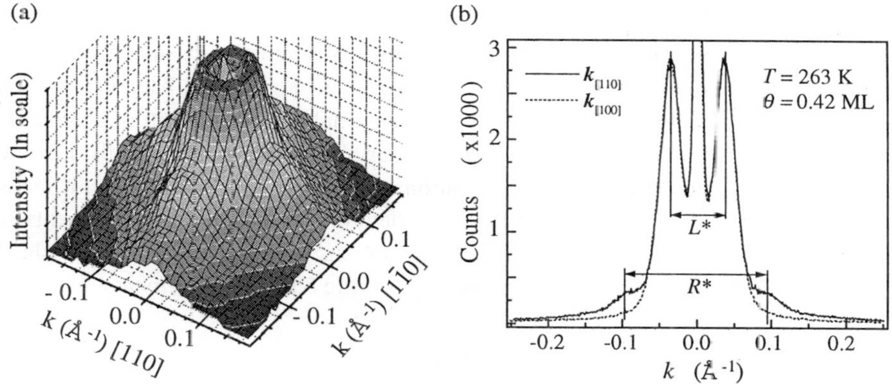

Fig. 1. (a) A two-dimensional angular profile of the (00) beam obtained by SPA-LEED following the deposition of 0.42 ML of Cu on Cu(100) at $T = 263$ K with a flux of 3.21×10^{-4} ML/s. The intensity is plotted with a logarithmic scale. (b) One-dimensional angular profiles from the same data plotted on a linear scale which were scanned in the (110) and (100) directions. L^* and R^* are the k-space separations which were determined by fitting the profiles with power Lorentzians for each peak.

size distribution with a peak size of R for the side length [73]. The diameter L^* of the circular ring is related to the mean island separation L via $L \sim 4\pi/L^*$, which is consistent with optical multiple slit diffraction [83,84]. With the necessarily larger length scale of the island separation, it is natural that the associated diffraction effect occurs at smaller k. The fourfold pattern at large k indicates square islands that have edges aligned along the close-packed [110] and [$1\bar{1}0$] directions. The optical diffraction analogy can be continued in order to relate the separation R^* of the lobes in Fig. 1(b) to R. If the islands were truly square, $R \sim 5.72\pi / R^*$ would be obtained. On the other hand, round islands would result in $R \sim 6.54\pi / R^*$. Because the island shapes are between these two extremes, only an approximate value of R can be obtained through this approach. Nevertheless, the validity of the assignment of L and R was shown by the relationship [73]

$$\theta \propto (R^2 / L^2)\,, \tag{7}$$

where R grew with increasing coverage proportional to $\theta^{1/2}$ up to a coverage of 0.5 ML, consistent with the experimentally determined region for steady-state growth [85] as discussed below. It should be noted that under some conditions, L^* and R^* may be so close together that the diffraction pattern has the appearance of a four-lobed Henzler ring [85,86].

5. Submonolayer Growth Results: Cu/Cu(100)

Diffraction measurements on Cu(100) always indicate the existence of square islands for temperatures above 180 K. Below 150 K, no ordered island formation was apparent, indicating significantly reduced thermal adatom motion in agreement with the results of others [46,72,82]. The existence of these square islands down to such low temperatures is a further indication of the very low barrier for diffusion along [110] step edges. This value has been determined by STM measurements to be 0.1 eV [87] in agreement with EAM calculations [88]. Examination of island growth below 150 K will require variable temperature STM measurements.

5.1. Coverage-Dependent Measurements

Figure 2 shows results of a coverage-dependent SPA-LEED measurement of L for various values of F at $T = 223$ K. No data appear below ~0.2 ML because of insufficient intensity in the diffraction ring. For a given F, L appears approximately constant up to a certain coverage θ_c, after which it starts to increase. The coverage invariant L below θ_c, indicates steady-state growth and a value of $z = 0$ in Eq.(1). This cannot occur in the point island model, but is consistent with the realistic model if a critical nucleus size $i = 1$ is assumed in the z expression in Eq. (5). As is discussed below, the same conclusion for $i = 1$ is also obtained from a flux-

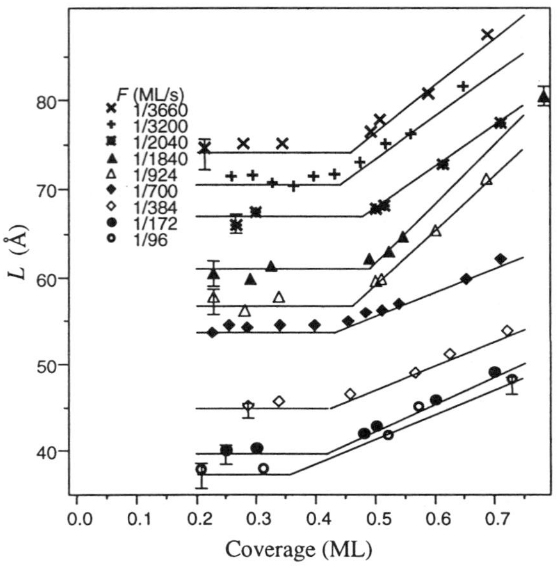

Fig. 2. Island separation vs. coverage for a series of fluxes at T = 223 K showing the aggregation (steady state) and coalescence regimes. Solid lines are fits of two connected line segments to the data.

dependent measurement (i.e., the smallest stable island is the dimer at $T = 223$ K). This result supports the realistic model of $K_s \sim D s^{1/2} N^{1/2}$ from which Eqs. (5) are derived and, not surprisingly, disagrees with the point-island model. However, the limitations of this model become apparent when the expression for z in Eq.(5) is compared with our results when $i > 1$. When $i = 3$ (see next section), $z \approx 0.16$ and 0.19 at temperatures of 253 and 267 K, respectively, are found in the coverage range of 0.17–0.4 ML. When $i = 8$, the coverage-dependent data [69] suggest $z \approx 0.26$. Thus the observed behavior, a slow increase in the island density, is just opposite to the prediction of this model. Above θ_c, the increase in L reflects the occurrence of coalescence among islands. Within experimental accuracy, θ_c ranges between 0.45 and 0.50 ML except for an apparent shift to lower coverages at high flux.

The STM is not handicapped at low coverages by low intensity as in the diffraction case. Figure 3 shows STM images obtained at room temperature for a series of coverages also deposited at room temperature, and it is observed that the island density at 0.08 ML has only slightly decreased at 0.45 ML where the beginning of coalescence can be observed as well as the first sign of second- layer growth. At 0.7 ML, the coalescence regime is well developed. Both diffraction and STM measurements indicate that a coverage of 0.3 ML is safely within the constant island density regime which is taken as an indication of steady- state growth, so all diffraction measurements discussed below were made at that coverage.

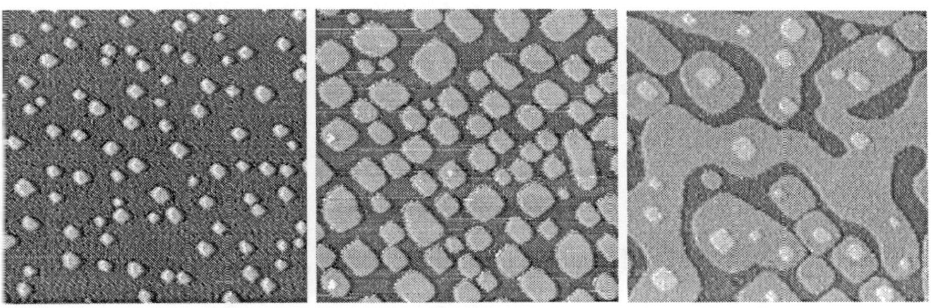

Fig. 3. STM images following the deposition of 0.08, 0.45, and 0.75 ML of Cu on Cu(100) at room temperature. Each image is 100×100 nm^2.

5.2. *Flux-Dependent Measurements*

By measuring the flux dependence of L^* with a constant $\theta = 0.3$ ML and using Eq. (1b), the scaling exponent p may be extracted without concern for the complexities which appear in the θ scaling. The results for three growth temperatures are shown in Fig. 4, where L is plotted vs. $1/F$ on a log-log scale and the slope gives $p/2$. Figure 5 shows the same type of data at a temperature of 213 K over a much larger flux range. The data in Fig. 4 at 223 K, which divides into three regimes, are discussed first. For the lower flux range of $1/F > 900$ s/ML, the exponent obtained from the slope is $p \sim 1/3$, giving $i = 1$ according to Eq. (5), which is consistent with the coverage-dependent measurement above. For higher fluxes the situation becomes complex and can only be roughly understood in terms of changing time scales where the scaling changes or breaks down. First, for $1/F <$ 900 s/ML, the exponent crosses over to $p \sim 1/2$. One explanation for this crossover of p is as follows: In the realistic model of $K_s \sim D s^{1/2} N^{1/2}$, the factor $N^{1/2}$ implies that the capture rate K_s is inversely proportional to the island separation L $(= N^{-1/2})$ since a deposited atom must travel a net distance $(L-R)/2$ by diffusion to reach an existing island, the travel time is proportional to L^2. At low deposition fluxes, islands are well separated; thus, travel time is significant and the inclusion of the factor $N^{1/2}$ in K_s gives $p = 1/3$ with $i = 1$. However, as L decreases with increasing flux, the time for an adatom to reach an island may become so insignificant that the factor $N^{1/2}$ in K_s may become negligible. This would correspond to a model first proposed by Blackman and Wilding [89] in which the capture rate K_s depends only on the perimeter sites ($\propto s^{1/2}$) of islands. In this model, $x = 1/2$ and $y = 0$ in Eq. (4), leading to $p = 1/2$ and $z = 0$ when $i = 1$. Ernst *et al.* [46] studied this same flux range

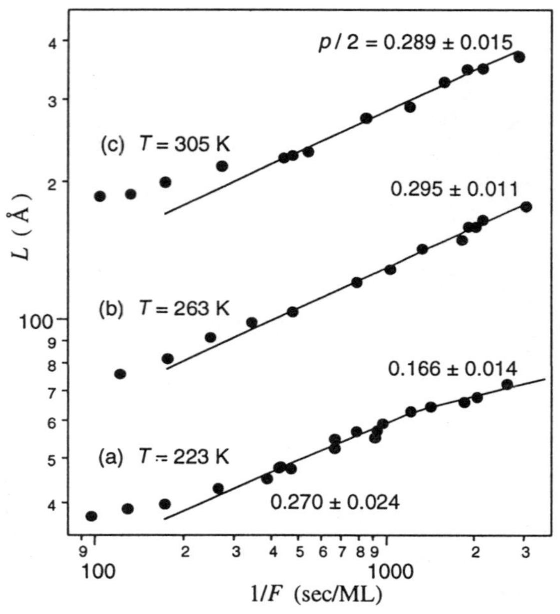

Fig. 4. Island separations vs. inverse flux in log-log format for deposition of Cu on Cu(100) at (a) 223 K, (b) 263 K, and (c) 305 K. The line slopes give p/2 with the value marked in each temperature regime

at temperatures of 220 and 230 K and also reported a value of $p \sim 1/2$, but the coverage in their case was $\theta \sim 0.5$ ML. As shown below, the data obtained at 223 K were very close to the point at which a change in i is observed; hence an alternative explanation could involve some instability at 223 K. As the flux is further increased so that $1/F < 350$ s/ML, the slope of the data for 223 K in Fig. 4(a) is reduced. The cause of this effect is most likely again a question of time scales. With extremely high deposition rates, the island density in the early nucleation stage may become so high and the separations so small that the newly created small and mobile islands can coalesce via collisions on the time scale of the deposition process so that the apparent nucleation density dependence on flux is weakened. In addition, early coalescence may also occur because of the small spacing as suggested by Fig. 2 where the coalescence regime begins earlier as the flux increases. Interestingly, a deviation from the scaling behavior similar to that seen here has also been found in Monte Carlo simulations in the $i = 1$ regime with island motion not considered. From a detailed consideration of length and time scales, Tang [53] obtains a logarithmic correction to the scaling. In a simulation which compares the island density as a function of the adatom hopping rate for Si on Si(100), Mo [47] finds that when the time between arriving adatoms is short compared with the time required to diffuse to an existing island (high-flux regime), most new islands are formed after this deposition time interval, and hence, N becomes independent of F.

Figure 4(b) shows the data obtained at a temperature of 263 K and Fig. 4(c) shows a very similar data set taken at 305 K. For both cases in the lower flux regime, $p = 3/5$ within experimental uncertainty, which suggest $i = 3$ according to Eq. (5). This implies that the smallest stable island at the higher temperatures is the tetramer instead of the dimer and that dimers and trimers are unstable and dissociate at elevated temperatures while the tetramer remains stable at least to 305 K. For the high-flux regime of $1/F < 470$ s/ML, a deviation from $p \sim 3/5$ which can be rationalized with the same argument of experimental time scale as discussed for the low-temperature case occurs. As a test of how reasonable the proposed transition from $i = 1$ to 3 is, the dissociation barrier for removal of one atom from various size clusters on Cu(100) were calculated [88] using the EAM at 0 K, which gives an upper limit estimate for the true dissociation barriers [90]. In this calculation, the dissociation barrier was found to be similar for the dimer (0.88 eV), trimer (0.96 eV), and pentamer (0.93 eV). However, a large jump for the tetramer (1.42 eV) was found, which can be understood through bond counting arguments; atom removal requires the breaking of two nn bonds in the square array. In this approach, which looks at removal of 1 atom at a time, clusters of 4, 6, 8, and all higher sizes would have a similar dissociation barrier [91]. As was mentioned earlier, a low initial dissociation barrier is important for dissociation to occur, but the difference in cohesive energy [41] and the entropy determines the stability of the different cluster sizes.

The measurements shown in Fig. 5 present a new challenge to our understanding. Here, flux measurements are made at 213 K which is below the transition temperature to $i=3$ (see Fig. 7). A striking increase in the slope of the L vs. $1/F$ data is observed for fluxes well below those studied for the data of Fig. 4. In the higher flux range, the slope is again consistent with $i = 1$ but with no apparent crossover to $p = 1/2$ with increasing flux. This may result from having fewer data points in this region in this data set or may be an indication of an instability in the other data as suggested above. The large increase in slope at extremely low fluxes indicates a new and unexpected growth behavior. If the usual relationship of Eq. (5) is applied to this slope, a critical nucleus size of approximately 7 to 12 is indicated. This is too large to be explained with the conventional bond counting picture of critical island sizes, but as discussed above, the possibility of a critical nucleus size of 8 has been demonstrated theoretically [58]. However, increases in the critical nucleus size are usually not expected without an increase in temperature to promote

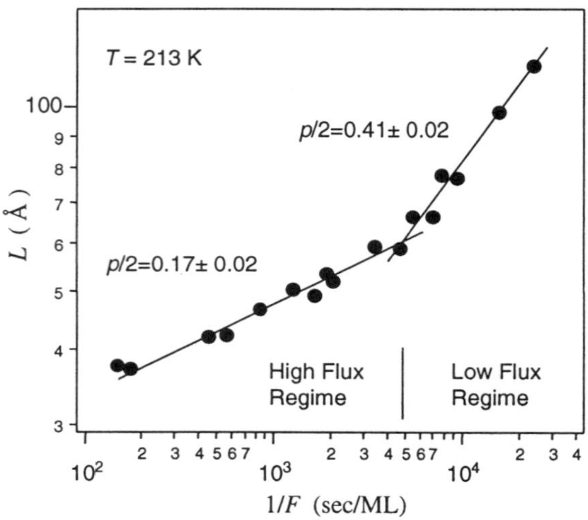

Fig. 5. Island separation vs. inverse flux at a temperature of 213 K for a much wider flux range than the data of Fig. 4.

thermally activated bond breaking. Several possible experimental artifacts were considered to explain this data. First, the effect of steps is discussed. Step-up edges acts as sinks for the adatoms leading to a lower adatom density below a step. This could conceivably increase the over-all island separation on the terrace. STM micrographs show the depletion zone is approximately equal to the island separation. Considering the distribution of island locations relative to terrace widths, the average site is a 1850-Å-wide terrace. Since the maximum depletion zone in this experiment is ~120 Å, the effect of steps should be minimal. Contamination due to adsorbates during the lengthy exposures (up to 2 h) to the Cu flux was ruled out because contaminants will generally have the opposite effect on the data because of their tendency to nucleate additional sites. Only atomic hydrogen, which is not present without special efforts, can promote the mobility of islands with subsequent coarsening [92]. Coarsening of the island distribution during the lengthy deposition process is a final possibility. However, monitoring of the diffraction profiles for extended periods after deposition showed that the coarsening rate following growth was negligible at 213 K.

Two possibilities exist to explain this behavior, both of which depend on the dimer shear mechanism. First, it has been noted that one of the key effects of the dimer shear process is to reduce the diffusion barriers for islands of 4 to 8 atoms [58]. For sufficiently low flux, motion and coalescence of small islands could take place before capturing arriving adatoms and thereby changing the exponent p according to Eq. (6). However, the large value of the measured slope is inconsistent with the existing theory [37] that predicts a maximum slope of $p = 0.5$ as the mobile cluster size increases to infinity. An estimate of the time required for the small islands to move a necessary distance also suggests that this process would be too slow for the time scale determined by the incident flux.

The second possible process considers clusters made unstable by an increasing time scale, thus allowing onset of dimer shearing and single-bond scission. The notion of critical island size is a thermodynamic concept, and a change of i is typically associated with an increase of T, causing thermally activated bond breaking. Whether a cluster is more likely to grow than fall apart is determined by the number of possible configurations and the cohesive energy (i.e., the free energy) of all the kinetically accessible states [93]. Lowering the flux changes the time scale so that more states can become kinetically accessible. It has been shown that the dimer shearing process lowers the maximum dissociation barrier of tetramers as well as other islands up to 8 atoms in size relative to the dissociation barrier for islands of size 9 atoms and above [58]. The stability of a sheared island is reduced, since one or more atoms is only attached to the cluster by a single nn bond (Fig. 6). Thus a change in the time scale resulting from a change in the flux could lead to a change in the value of i. A crude estimate of the time scale involved for activation of the dimer shear compared with the time scale imposed by the flux [80] suggests that this scenario is, in fact, reasonable. This estimate was made by first comparing the measured activation barrier for single adatom diffusion of 0.36 eV to the value of 0.50 eV obtained from EAM calculations. The EAM calculation overestimate of

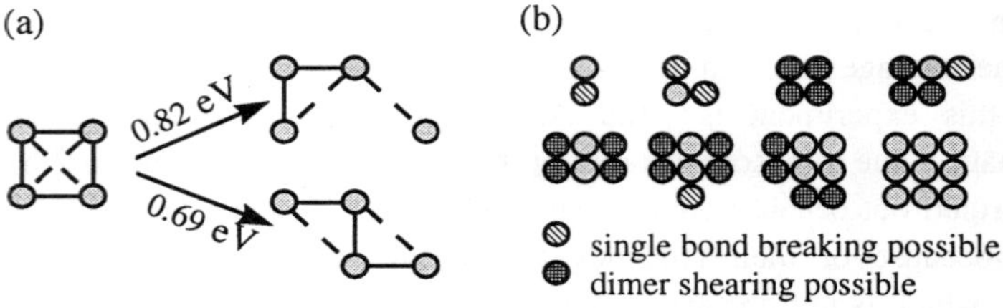

Fig. 6. The dimer shear mechanism allows shearing of a tetramer or any island with two atoms on a side to shear in a single step with a lower energy barrier than that required for a two-step sequential motion of single atoms.

0.14 eV in the case of the diffusion barrier was assumed to be the same for all calculated energy barriers, and in particular, for the dimer shearing barrier. With an assumed attempt frequency of 10^{-12} s^{-1}, the time required for the dimer shear was estimated at 0.2 min. This compares favorably with the estimated 1.5 min between arrival of adatoms in the area occupied by one island at the beginning of the low-flux regime. This illustrates that onset of dimer shearing is likely to take place at lower flux rates. But in order for dimer shearing to impact the stability of islands, a sizable fraction of the clusters (of size < 9) has to be in a sheared, or noncompact, configuration [Fig. 6(a)]. The sheared configuration provides the starting point for dissociation, and the effective dissociation energy of the sheared configuration is comparable to that of dimers and trimers [41]. However, a detailed model calculation is still needed to sort out the importance of the dimer shearing process relative to the observed phenomenon.

5.3. *Temperature-Dependent Measurements*

The temperature dependence of the island separation for fixed flux conditions is a function of the energy barriers according to Eq. (1). Once p or i are known, these energy barriers can be obtained [46]. Figure 7 displays an Arrhenius plot of such data, covering a temperature range between 180 and 265 K at a fixed coverage of 0.3 ML and a deposition flux of $F = 3.21 \times 10^{-4}$ ML/s. The data in Fig. 7 cannot be described by a single straight line but are best fit by two line segments with a fit intersection at 223 K. For temperatures below 223 K, where $p = 1/3$, $i = 1$, and $E_{i=1} = 0$, a diffusion barrier is obtained on terraces of $E_D = 0.36 \pm 0.03$ eV. This value is generally consistent with calculated values ranging from 0.42 to 0.69 eV [6–8,88,90,94] which assume a hopping mechanism rather than the exchange mechanism where recent calculations [6–8] indicate a much higher value. Other experimental groups have also measured E_D and obtained values 0.39 eV using low-energy ion scattering to observe adatoms created by ion bombardment [82], 0.4 eV using HAS to study multilayer growth [95], and 0.28 eV using HAS to study submonolayer epitaxial growth [46].

For temperatures above 223 K, the change in slope indicates the effect of additional energetic terms associated with the change from $i = 1$ to $i = 3$, but the exact nature of the energetic contributions is not yet clear. Adatom detachment from islands becomes more probable with increasing temperature. By assuming the change in slope to be purely due to the transition of $i = 1$ to $i = 3$ in this temperature range, an effective binding energy per atom, $E_i/i = 0.04$ eV, was determined [73] with $E_D = 0.36 \pm 0.03$ eV as above. If the trimer is in a linear form with only two nn bonds, then the average binding energy per bond is (3/2)(0.04) = 0.06 eV. In any case, this value is small as compared with the values of 0.18 and 0.26 eV calculated by EAM for dimers and trimers [88]. Attributing the sudden change in slope in Fig. 7 to the change between $i = 1$ and $i = 3$ has been questioned by Evans *et al.* [39–41]. In their model, the change in slope above 223 K is the result of a gradual transition from $i = 1$ to $i = 3$, thus giving a smaller slope and an erroneous estimate of the binding energy. An experimental indication of such a gradual transition has been

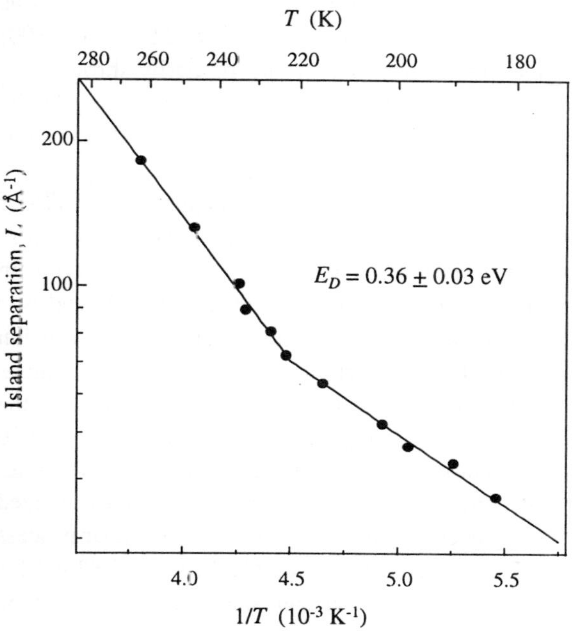

Fig. 7. An Arrhenius plot of the island separation vs. inverse temperature $1/T$.

reported [69] recently by the same group for Ag/Ag(100). However, a first-principles calculation by Feibelman [96] suggests that such a small apparent binding energy is plausible due to compensation between the cost of breaking an adatom-island bond and the gain by the simultaneous strengthening of adatom-surface bonds. A calculation for Al adatom pairs on Al(100) gives a small separation energy of only 0.07 eV. The same principle may be expected to reduce the apparent binding energy of small adatom pairs on Cu(100) below the normally calculated values.

5.4. Post-deposition Equilibration and Island Mobility

When considering the growth of thin films, it is necessary to consider the stability of the thin-film morphology after deposition. Deposition and growth processes produce an inherently nonequilibrium morphology, a condition which is self-rectifying given sufficient time. In the case of submonolayer depositions, what is often observed is a coarsening of the island distribution in which small islands tend to disappear over time and the large islands get larger. Based on the assumption that clusters of many atoms do not move, the coarsening effect was generally ascribed to the process of Ostwald ripening in the past. Several authors have shown for a fact via STM that both vacancy [66,67] and adatom [65,68] islands do move, and it was shown conclusively by Wen *et al.* [24] for Ag(100) and Pai *et al.* [65] for Cu(100) that adatom island coarsening at room temperature occurs primarily through the diffusion of islands with islands as large as 700 atoms observed to move. At the other end of the spectrum, the motion of small islands with fewer than 20 atoms has also been observed using FIM [60,97,98].

While the observation of island motion is now easy with an appropriate STM or FIM system, understanding the mechanism by which they move is far more complex. In the case of very small islands, a clearly oscillatory behavior in diffusion barrier with island size has been predicted for Cu/Cu(100) [54] and was observed for Rh on Rh(100) [60]. In the case of Rh on Rh(100), the diffusion activation energy for islands with sizes 4, 6 and 8 atoms was observed to be similar and to be higher than that for islands sizes 3, 5, and 7 atoms. Oddly though, the activation energy for an island with 9 atoms was higher than for islands with 8 or 10 atoms. This was explained very recently with the prediction [58] that the dimer shear motion discussed above will reduce the diffusion barrier for any even sized island with 2 atoms on a side. While the energy required for the dimer shear process in a tetramer is calculated via the EAM to be 0.69 eV on Cu(100) vs. 0.82 eV for double-bond breaking, the energy required for a trimer shear process would be greater than 0.82 eV. The effect on mobility of small islands is striking. A tetramer could move one lattice space with two steps having a maximum barrier of 0.69 eV vs. the four steps required for sequential single atom motion with a barrier of 0.82 eV. For larger islands, the dimer slide mechanism may still play a role in their motion, but its effect is reduced because there is always a core of atoms that will require double-bond breaking to induce motion. Thus the motion of islands may be expected to divide into two size regimes, with the small islands having less than 9 atoms diffusing significantly faster than the larger islands. The fast diffusion of the small islands provides the basis for one explanation of the observed apparent large critical island size associated with very slow deposition rates on Cu(100) discussed above. The large island, slow diffusing regime, is discussed in detail below.

For the larger islands, whose motion is easily observed via STM, most studies have attempted to associate the diffusion mechanism with a scaling behavior for the diffusion coefficient as a function of island size [24,67,99–102]. This approach associates the exponent α in the expression,

$$D \sim R^{-\alpha}, \tag{8}$$

with several different microscopic mass transport mechanisms, each of which gives a unique, integer value of α. For the evaporation-condensation (EC) mechanism, wherein the islands exchange adatoms with a thermodynamically balanced sea of adatoms, $\alpha = 1$. In the terrace diffusion (TD) limited process, in which atoms detaching from an island have a high probability of reattaching to the same island, $\alpha = 2$. Finally, the process of periphery diffusion (PD), which has been modeled as atoms diffusing around the structureless edge of essentially round islands, gives a value of $\alpha = 3$. These integer values for α are called into question by several Monte Carlo simulations [103–105] and more detailed theoretical treatments [99–101] which suggest that a wide range of non-integer values are possible.

The first study of the diffusion of large islands came from Wen et. al. [24] who found little variation in the diffusion rates of large islands on Ag(100) in the size range from 100 to 700 atoms. They concluded that the diffusion was most likely associated with the EC mechanism. Recently, the same group has made a more detailed study of coarsening kinetics and finds that [106] the diffusion coefficient is significantly larger for smaller islands (~150 atoms) than for larger islands (350 atoms), in agreement with our own results which are discussed below [65]. Morgenstern *et. al.* [67] looked at the diffusion of vacancy islands (100 – 5000 atoms) on the Ag(111) surface and concluded that $\alpha \approx 2$, suggesting the TD case. However, the same group has recently pointed out that the evaporation rate from the island edges is too small to account for the observed island diffusion coefficient [107] and a recent molecular dynamics simulation [108] finds PD to be the dominant mechanism in this case. Diffusion behavior suggesting $\alpha = 2$ has also been observed for vacancy islands on the partially Co-covered Cu(111) surface [109]. However, in a study with implications for the question of island diffusion, observations of step fluctuations on a vicinal Cu(100) surface in the temperature range 290–600 K [110] showed that these fluctuations could be associated almost entirely with mass transport along the step edge. This result suggest that islands on the Cu(100) surface might be expected to diffuse via the PD mechanism.

Following the background of studies above, comprehensive measurements of the diffusion and coarsening behavior of islands on Cu(100) were performed and then the same issue on Ag(100) was revisited. This study provided the first clear evidence for two cases in which the PD mechanism is the dominant factor in the island diffusion process. In this study, an STM was used first to make movies of the island motion at room temperature with between 800 and 2000 frames in each sequence covering a period of time up to 28 hours [65]. Figure 8 shows STM images of 0.08 ML of Cu on Cu(100) immediately following deposition and also 3 and 6 h later for the same location. As previously observed for island coarsening on Ag(100) [68], the diffusion, collision, and subsequent coalescence of the islands on

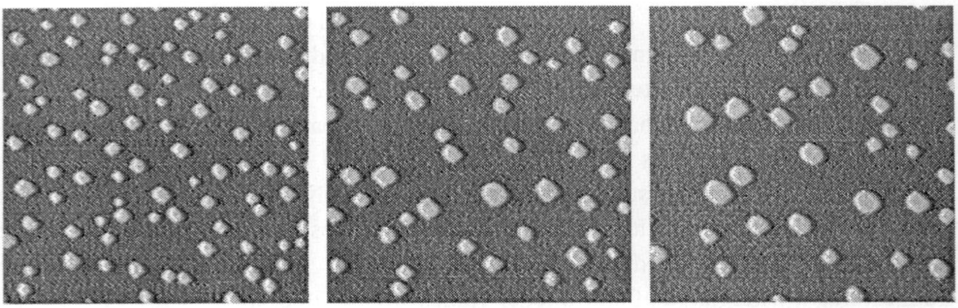

Fig. 8. STM images of square islands on Cu(100) in the same area just after deposition of 0.08 ML, 3 h after deposition, and 6 h after deposition going from left to right. Each image is 100 × 100 nm^2.

Cu(100) as shown in Figure 9 is also the principal coarsening mechanism. The movies show clearly that there is a strong size dependence in the island diffusion for both Cu(100) and Ag(100) with smaller islands moving faster. To determine the diffusion coefficient as a function of island size, the island motion was tracked by a computerized procedure which first determines the size and center of mass position of every island in every frame. Islands were then tracked through the sequence of frames and the procedure of Morgenstern [67] was used to eliminate drift effects by measuring the relative displacement, $\delta r = r_i - r_j$, between island pairs of nearly equal sizes. The relative diffusion which is thus measured is the sum of the individual diffusivities, $D_{i,j} = D_i + D_j = \langle [\delta r(t) - \delta r(0)]^2 \rangle / 4t$. Islands may disappear through coalescence or by drifting out of the field of view. In this way, new sequences begin and end throughout the set of frames. Data from an island pair

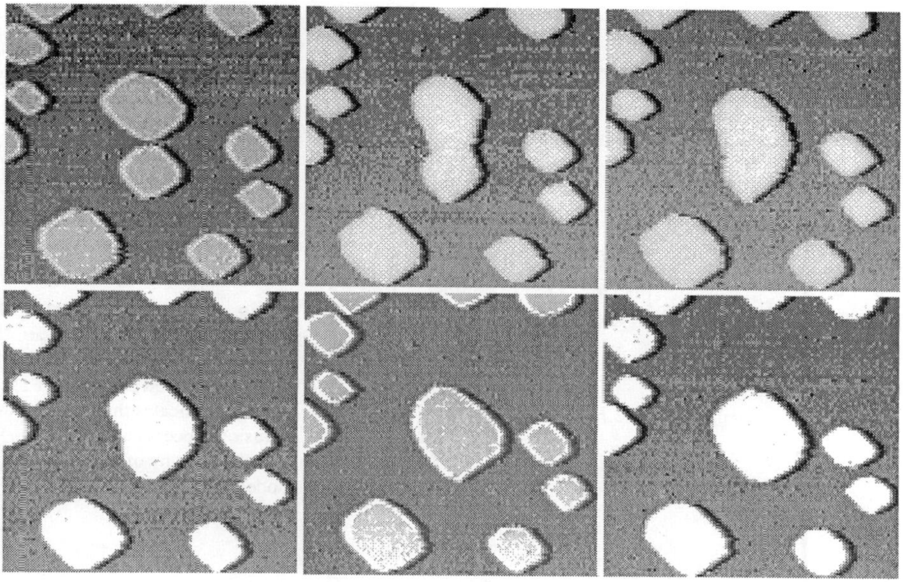

Fig. 9. Time sequence (upper left to lower right) of large island (~400 atoms) collision on Cu(100) with subsequent coalescence and reshaping. Each image is 50×50 nm^2 with a time interval of 3.5 min.

are only accepted when the pair can be tracked for a minimum of 50 frames. Figure 10 shows the culmination of this analysis where log D is plotted vs. log L and L^2 is the number of atoms in the cluster. A least-squares fit to the data yields clearly non-integer values of $\alpha = 2.49 \pm 0.09$ for Cu(100) and 2.28 ± 0.10 for Ag(100). These exponents can be shown to be associated with the PD mechanism as discussed below.

Further support for these non-integer values of α comes from the coarsening which follows a power law, $R \sim t^{\beta}$. Using Monte Carlo simulations and scale invariance, Sholl [103] has shown that when coarsening is due to island coalescence, the coarsening exponent β may be related to α by

$$\beta = 1/(\alpha + 2). \tag{9}$$

Figure 11 shows coarsening data obtained for islands on the Cu(100) surface in the form of log $\langle L \rangle$ vs. log t. The slope of the data following an initial period gives a value of $\beta = 0.227$, which is consistent with $\alpha = 2.49$ in Eq. (9). For Ag(100), a value of $\beta = 0.233$ which is likewise consistent [65] was found. An explanation for the initially slow coarsening behavior observed with both systems is as follows: After deposition, all islands are separated by an average length scale determined by

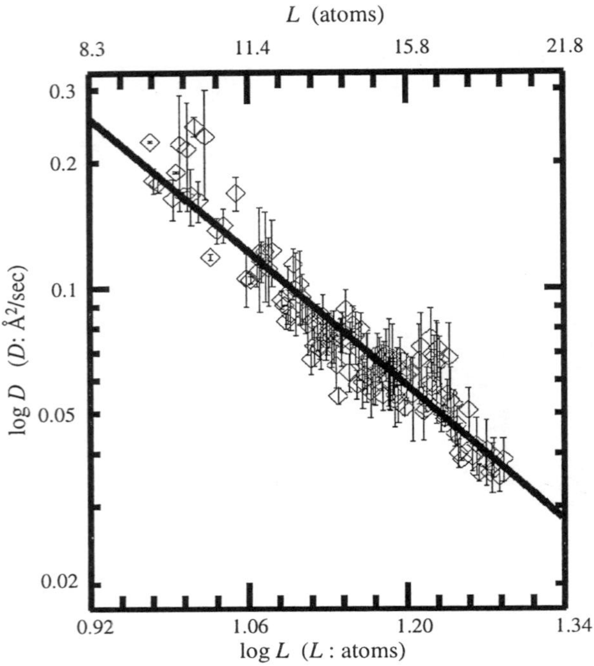

Fig. 10. Size dependence of island diffusion rates plotted as log D vs. log L. From the fitted slope, the derived exponent of $\alpha = -2.5$ indicates that the diffusion process cannot be described by any of the simple models which yield integer exponents.

Eq. (1). The theoretical treatment assumes an initially random distribution of island positions, and it takes a significant time after deposition before the islands become "randomized" by diffusion. With the non-integer values of α, the choice of diffusion mechanisms is not obvious. However, both the EC and TD mechanisms will result in some degree of island size decay (i.e., at least a limited degree of Ostwald ripening should occur). In the case of Cu(100), no detectable decay of the island sizes was observed during the lifetimes of the islands. For Ag(100), a maximum decay rate of ~ 0.125 atom/min was observed for an individual island. Assuming the EC and TD mechanisms to be active and forbidding the PD mechanism, the island diffusion coefficients were estimated to be two to five orders of magnitude smaller than the experimental value of $D \approx (3-6) \times 10^{-2}$ Å^{2}s^{-1} determined for Ag(100) [65]. Thus the ED and TD mechanisms cannot have a significant impact on the island diffusion process.

The observed non-integer exponents can be understood with the PD mechanism by considering the detailed structure of island edges and may not be unexpected. The Langevin analysis developed by Khare *et al.* [99] shows that even when PD is the only allowed mechanism, the "effective" exponent $\alpha(R)$ changes smoothly from 1 to 3 as a dimensionless quantity R/R_{st} changes from ~ 0.1 to ~ 10. It is likely that

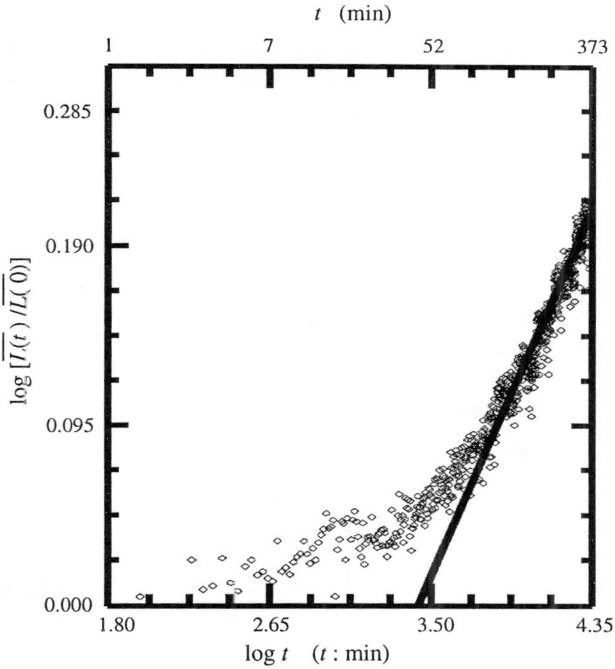

Fig. 11. Observed coarsening for 0.08 ML Cu/Cu(100) plotted in a log-log plot of island separation vs. time. The observed slope in the later stage is consistent with the noninteger value of α if the islands diffuse via the PD mechanism.

our systems lie in such a "crossover" regime, giving $\alpha \sim 2.3$ or 2.5. It is proposed that the detailed structure of the island edges must affect the scaling behavior. The observed rounding of corners, which is more pronounced on Cu(100) than on Ag(100), and the transient indentations which appear on the island edges away from the corners make it clear that there is always a certain density of kinks which is material dependent and, no doubt, temperature dependent. There are many possible atomic processes along such kinked island edges. The "rate-limiting" process controlling the island diffusion is more likely atom detachment from kink sites when compared with the very rapid atom diffusion along straight island edges [105,111]. The effect of kinks can be modeled through a scaling analysis based on uncorrelated random atom transport between kinks as follows:

$$D = \frac{\left\langle \delta^{2}_{c.m.} \right\rangle}{4t} \propto (2\pi R)\frac{\left\langle n_k f_k l_k^2 \right\rangle}{N^2} \propto \frac{\left\langle n_k f_k l_k^2 \right\rangle}{R^3} \tag{10}$$

where $N \sim R^2$, n_k is the line concentration of kink sites, l_k is the jump displacement between adjacent kinks, and f_k is the frequency at which atoms detach from a kink site and attach to an adjacent kink site through a random walk. This analysis was performed assuming a constant kink density, a random distribution of kinks, and a f_k with weak dependence on l_k because of the presence of an island corner barrier. Two interesting aspects can be clearly identified. First, $\alpha(R)$ changes continuously from 1 to 3 as R increases and closely matches the calculated form in Ref. [99] if their parameter R_{st} is interpreted as the average separation $\bar{l}$ between adjacent kinks. This result is shown in Fig. 12. It is also noted that in contrast to the comments made in Ref. [99], it should be quite plausible to observe the crossover regime experimentally with a reasonable value of the angular kink separation $\approx \bar{l}/R$. Second, $\alpha(R)$ varies inversely with $\bar{l}/R$. This is consistent with the observation that the corners of Cu islands are more rounded than those of Ag islands (i.e., smaller $\bar{l}$ for the same R), leading to $\alpha(\text{Cu}) > \alpha(\text{Ag})$.

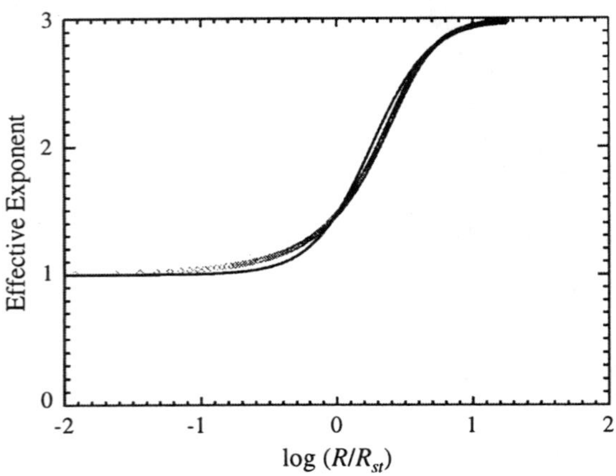

Fig. 12. Plot of the effective exponent α as a function of the dimensionless ratio R/R_{st}. The solid line is the prediction of Khare [99], and the diamonds represent the result of a scaling analysis where R_{st} is taken to be the average kink separation, $\bar{l}$.

A fundamental question that is not answered by the simple model above concerns the process referred to as core breakup. Every island may be described as having a rectangular core with some number of extra atoms on the edges which give the island its particular shape and which may diffuse easily around the periphery. However, for an island to move, its core must move. This requires removal of atoms from the core (i.e., core breakup) [112]. Corner atoms, having only two nn bonds should be easier to remove than edge atoms with three nn. If the core breakup is dependent on detachment of the corner atoms of the central core, then there should be little size dependence for the island diffusion. The strong size dependence shown in Fig. 10 indicates that the diffusion is not limited by removal of the four corner atoms of the central core; hence, kinks must be generated on the straight sides of the core between the four corners to provide more sources for mobile atoms. Figure 13 illustrates one possible mechanism that is presented without proof as a suggestion

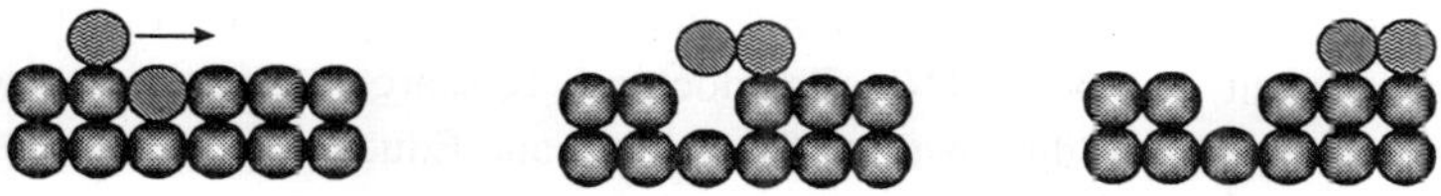

Fig. 13. A suggested mechanism for core breakup. A single atom diffusing along the edge may assist with the removal of an atom by replacing one of the three bonds holding the edge atom.

for future theoretical and experimental work. An atom anywhere on the edge having 3 nn might be induced to break out of the core by the passage of an adatom diffusing along the island edge. In this process, which may be referred to as "adatom-assisted core breakup," an atom with 3 nn would transfer a bond to a diffusing atom on the edge, effectively breaking only two nn bonds with an energy requirement similar to that for corner atom detachment. The resulting edge dimer could break apart or diffuse as a dimer.

6. Summary

In this chapter one of the simplest cases of epitaxial nucleation, growth, and coarsening (i.e., submonolayer homoepitaxy on Cu(100)) has been dealt with. From both electron diffraction and STM, critical island sizes and energy barriers have been predicted. After deposition, progression of the non-equilibrium island structure toward equilibrium through diffusion and coalescence of large islands has been observed and it has been shown that this diffusion process consists almost exclusively of atoms diffusing around the edges of the islands. Not surprisingly, the scaling data are more consistent with a model that includes finite island sizes rather than the point island model. However, the reader should not feel that everything is understood for such systems. While in most cases, the growth behavior can be represented by the continuum analysis presented in equations 1–6, the detailed atomistic behavior is only beginning to be understood. Questions such as the following are yet to be answered: What role does the diffusion of very small islands play in the growth process and when might that become important? Can the critical island sizes of 2, 4 and 9 atoms discussed for Cu(100) be confirmed by direct observation? Can the detailed dependence of the island diffusion on the structure of island edges be determined? How does the island core breakup process occur? In each case the challenge will be to match the physical time scales to experimental time scales. The proliferation of variable-temperature STM systems holds the promise that at least some of these questions can be answered in the near future.

Acknowledgments

This work was supported by the Oak Ridge National Laboratory, which is managed by the Lockheed Martin Energy Research Corp. for the U.S. Department of Energy under Contract No. DE-AC05-96OR22464, and was also supported in part (WWP) by an appointment to the ORNL Postdoctoral Research Associates Program administered by the Oak Ridge Institute for Science and Education.

---

1. J.-K. Zuo, J. F. Wendelken, H. Dürr, and C.-L. Liu, *Phys. Rev. Lett.* **72** (1994) 3064.
2. C. Ratsch, A. Zangwill, P. Smilauer, and D. D. Vvedensky, *Phys. Rev. Lett.* **72** (1994) 3194.
3. P. J. Feibelman, *Phys. Rev. Lett.* **65** (1990) 729.
4. G. L. Kellogg and Peter J. Feibelman, *Phys. Rev. Lett.* **64** (1990) 3143; G. L. Kellogg, A. F. Wright, and M. S. Daw, *J. Vac. Sci. Technol. A* **9** (1990) 1757.
5. Chonglin Chen and Tien T. Tsong, *Phys. Rev. Lett.* **64** (1990) 3147.
6. Chun-Li Liu, *Int. J. Mod. Phys.* **9** (1995) 1.
7. Ghyslain Boisvert and Laurent J. Lewis, *Phys. Rev. B* **56** (1997) 7643.
8. Changyol Lee, G. T. Barkema, M. Breeman, Alfredo Pasquarello, and Roberto Car, *Surf. Sci.* **306** (1994) L575.
9. A. Zangwill, in *Evolution of Surface and Thin film Microstructure*, ed. by H. A. Atwater, E. Chason, M. Grabow, and M. Lagally (Materials Research Society, Pittsburgh, 1993).
10. G.S. Bales and D.C. Chrzan, *Phys. Rev. B* **50** (1994) 6057.
11. R. L. Schwöbel and E. J. Shipsey. *J. Appl. Phys.* **37** (1966) 3682; R. L. Schwöbel, *J. Appl. Phys.* **40** (1969) 614.
12. G. Ehrlich and F. Hudda, *J. Chem Phys.* **44** (1966) 1039; S. C. Wang and G. Ehrlich, *Phys. Rev. Lett.* **67** (1991) 2509.
13. J.-K. Zuo and J. F. Wendelken, *Phys. Rev. Lett.* **78** (1997) 2791.
14. M. D. Johnson, C. Orme, A. W. Hunt, C. Graff, J. Sudijono, L. M. Sander, and B. G. Orr, *Phys. Rev. Lett.* **72** (1994) 116.
15. H.-J. Ernst. F. Fabre, R. Folkerts, and J. Lapujoulade, *Phys. Rev. Lett.* **72** (1994) 112.
16. J. E. Van Nostrand, S. Jay Chey, M.-A. Hasan, David G. Cahill, and J. E. Greene, *Phys. Rev. Lett.* **74**, (1995) 1127.
17. J. A. Stroscio, D. T. Pierce, M. D. Stiles, A. Zangwill, and L. M. Sander, *Phys. Rev. Lett.* **75** (1995) 4246.
18. K. Thürmer, R. Koch, M Weber, and K. H. Rieder, *Phys. Rev. Lett.* **75** (1995) 1767.
19. F. Tsui, J. Wellman, C. Uher, and R. Clarke, *Phys. Rev. Lett.* **76** (1996) 3164.
20. M. Bott, Th. Michely, and G. Comsa, *Surf. Sci.* **272** (1992) 161.
21. J. C. Girard, Y. Samson, S. Gauthier, S. Rousset, and J. Klein, *Surf. Sci.* **302** (1993) 73.
22. E. Kopatzki, S. Günther, W. Nichtl-Pecher, and R. J. Behm, *Surf. Sci.* **284**(1993) 154.
23. J. A. Stroscio, D. T. Pierce, and R. A. Dragoset, *Phys. Rev. Lett.* **70**, 3615 (1993); J. A. Stroscio and D. T. Pierce, *Phys. Rev. B* **49** (1994) 8522.
24. J.-M. Wen, S.-L. Chang, J. W. Burnett, J. W. Evans, and P. A. Thiel, *Phys. Rev. Lett.* **73** (1994) 2591.
25. S. Günter, E. Kopatzki, M.C. Bartelt, J.W. Evans, and R.J. Behm, *Phys. Rev. Lett.* **73** (1994) 553.
26. T. R. Linderoth, J. J. Mortensen, K. W. Jacobsen, E. Laegsgaard, I. Stensgaard,and F. Besenbacher, *Phys. Rev. Lett.* **77** (1996) 87.
27. Lutz Bönig, Shudun Liu and Horia Metiu, *Surf. Sci.* **365** (1996) 87.
28. Zhenyu Zhang and Max G. Lagally, *Science* **276** (1997) 377.
29. Th. Michely, M. Hohage, M. Bott, and G. Comsa, *Phys. Rev. Lett.* **70** (1993) 3943.
30. R. Q. Hwang, J. Schröder, C. Günther, and R. J. Behm, *Phys. Rev. Lett.* **67** (1991) 3279.
31. Z. Y. Zhang, X. Chen, and M.G. Lagally, *Phys. Rev. Lett.* **73** (1994) 1829.
32. M. Hohage, M. Bott, M. Morgenstern. Z. Y. Zhang, Th. Michely, and G. Comsa, *Phys. Rev. Lett.* **76** (1996) 2366.
33. H. Brune, H. Röder, K. Bromann, K. Kern, J. Jacobsen. P. Stoltze, K. Jacobsen, J. Nørskov,

Surf. Sci. **349** (1995) L115.
34. Chun-Li Liu and James B. Adams, *Surf. Sci.* **294** (1993) 211.
35. S. Stoyanov and D. Kashchie, in *Current Topic in Materials Science*, Vol. **7**, edited by E. Kaldis (North-Holland, Amsterdam, 1981), pp. 69–141.
36. J. A. Venables, G. D. T. Spiller and M. Hanbucken, *Rep. Prog. Phys.* **47** (1984) 399.
37. J. Villain, A. Pimpinelli and D. E. Wolf, *Comments Cond. Mat. Phys.* **16** (1992) 1.
38. J. W. Evans and M. C. Bartelt, *J. Vac. Sci. Technol. A* **12** (1994) 1800.
39. J. W. Evans and M. C. Bartelt, *Langmuir* **12** (1996) 217.
40. M. C. Bartelt, S. Günther, E. Kopatzki, R. J. Behm, and J. W. Evans *Phys. Rev. B* **53** (1996) 4099.
41. M. C. Bartelt, L. S. Perkins, and J. W. Evans, *Surf. Sci.* **344** (1995) L1193.
42. M. C. Bartelt and J. W. Evans, Phys. Rev. *B* **46** (1992) 12675.
43. M. C. Bartelt, M. C. Tringides, and J. W. Evans, *Phys. Rev. B* **47** (1993) 13891.
44. G. Rosenfeld, R. Servaty, C. Teichert, B. Poelsema, and G. Comsa, *Phys. Rev. Lett.* **71** (1993) 895.
45. Wulf Wulfhekel, Ingo Beckmann, Nuphar N. Lipkin, Georg Rosenfeld, Bene Poelsema, and George Comsa, *Appl. Phys. Lett.* **69** (1996) 3492.
46. H.-J. Ernst, F. Fabre and J. Lapujoulade, *Phys. Rev. B* **46** (1992) 1929.
47. Y. W. Mo, J. Kleiner, M. B. Webb, and M. Lagally, *Phys. Rev. Lett.* **66** (1991) 1998; Y. W. Mo, J. Kleiner, M. B. Webb, and M. Lagally, *Surf. Sci.* **268** (1992) 275.
48. G. T. Barkema, Ofer Biham, M. Breeman, D. O. Boerma, and G. Vidali, *Surf. Sci.* **306** (1994) L569.
49. Jacques G. Amar and Fereydoon Family, *Phys. Rev. Lett.* **74** (1995) 2066.
50. M. J. Stowell, Phil. Mag. **21** (1970) 125.
51. B. Lewis and D. S. Campbell, *J. Vac. Sci. Technol.* **4** (1967) 209.
52 Irisawa, Y. Arima, and T. Kuroda, *J. Cryst. Growth* **99** (1990) 491.
53 L. H. Tang, *J. Phys. I* (France) **3** (1993) 935.
54 Ofer Biham, G. T. Barkema, and M. Breeman, *Surf. Sci.* **324**, (1995) 47.
55. G. S. Bales and A. Zangwill, *Phys. Rev. B* **55** (1997) R1973; A. Zangwill, private communication.
56. D. Kandel, *Phys. Rev. Lett.* **78** (1997) 499.
57. J.A. Venables. *Phys. Rev B* **36** (1987) 4153.
58. Zhu-Pei Shi, Zhenyu Zhang , Anna K. Swan, and John F. Wendelken, *Phys. Rev. Lett.* **76** (1996) 4927.
59. C. Ratsch, P. Smilauer, A. Zangwill, and D. D. Vvedensky, *Surf. Sci.* **329** (1995) L599.
60. G. L. Kellogg, *Phys. Rev. Lett.* **73** (1994) 1833.
61. M. Schroeder and D. E. Wolf, *Phys. Rev. Lett.* **74**, 2062 (1995); D. E. Wolf in *Scale Invariance, Interfaces, and Non-Equilibrium Dynamics*, Vol. **344** of Nato Advanced Study Institute, Series B: Physics, ed. by M. Droz, A. J. McKane, J. Vannimenus and D. E. Wolf, (Plenum, New York, 1994).
62. Itay Furman and Ofer Biham, *Phys. Rev. B* **55** (1997) 7917.
63. Shudun Liu, Lutz Bönig, and Horia Metiu, *Phys. Rev.* **B 52** (1995) 2907.
64. Ofer Biham, Itay Furman, Majid Karima, Gianfranco Vidali, Rosemary Kennett and Hong Zeng, *Surf. Sci.* **400** (1998) 29.
65. Woei Wu Pai, A. K. Swan, Zhenyu Zhang, and J. F. Wendelken, *Phys. Rev. Lett.* **79** (1997) 3210.
66. J. de la Figuera, J. E. Prieto, C. Ocal and R. Miranda, *Solid State Commun.* **89** (1994) 815.
67. Karina Morgenstern, Georg Rosenfeld, Bene Poelsema, and George Comsa, *Phys. Rev. Lett.* **74** (1995) 2058.
68. J.-M. Wen, J. W. Evans, M. C. Bartelt, J. W. Burnett, and P. A. Thiel, *Phys. Rev. Lett.* **76** (1996) 652.
69. C.-M. Zhang, M. C. Bartelt, J.-M. Wen, C. J. Jenks, J. W. Evans, and P. A. Thiel, *J. Crystal*

Growth **174** (1997) 851.
70. H. -J. Ernst, F. Fabre and J. Lapujoulade, *Surf. Sci.* **275** (1992) L682.
71. H.-J. Ernst, F. Fabre, and J. Lapujoulade, *Phys. Rev. Lett.* **69** (1992) 458.
72. G. L. Nyberg, M. T. Kief, and W. F. Egelhoff, Jr., *Phys. Rev. B* **48** (1993) 14509.
73. H. Durr, J. F. Wendelken and J.-K. Zuo, *Surf. Sci.* **328** (1995) L527.
74. J.-K. Zuo and J. F. Wendelken, *Phys. Rev. Lett.* **70** (1993) 1662.
75. G.-C. Wang and T.-M. Lu, *Phys. Rev. Lett.* **50** (1983) 2014.
76. J. W. Evans, D. K. Flynn-Sanders, and P. A. Thiel, *Surf. Sci.* **298** (1993) 378.
77. J. D. Wrigley and G. Ehrlich, *Phys. Rev. Lett.* **44** (1980) 661.
78. W. F. Egelhoff, Jr. and I. Jacob, *Phys. Rev. Lett.* **62** (1989) 921.
79. J. W. Evans, D.E. Sanders, P. A. Thiel, and A. E. DePristo, *Phys. Rev. B* **41** (1990) 5410; J. W. Evans, *Phys. Rev. B* **43** (1991) 3897.
80. Anna K. Swan, Zhu-Pei Shi, John F. Wendelken, and Zhenyu Zhang, *Surf. Sci.* **391** (1997) L1205.
81. J. M. Pimbley and T.-M. Lu, *J. Appl. Phys.* **58** (1985) 2184.
82. M. Breeman and D. O. Boerma, *Surf. Sci.* **269/270** (1992) 224.
83. P. Hahn, J. Clabes, and M. Henzler, *J. Appl. Phys.* **51** (1980) 2079.
84. see for example: E. Hecht, Optics, (Addison-Wesley, 1987).
85. J.-K. Zuo and J. F. Wendelken, *Phys. Rev. Lett.* **66** (1991) 7777.
86. M. C. Bartelt and J. W. Evans, *Surf. Sci.* **298** (1993) 421; *MRS Proceedings* No. **312** (Materials Research Society, Pittsburgh, 1993) 255.
87. M. Poensgen, J. F. Wolf, J. Frohn, M. Giesen, and H. Ibach, *Surf. Sci.* **274** (1992) 430.
88. C.-L. Liu, *Surf. Sci.* **316** (1994) 294..
89. J. A. Blackman and A. Wilding, *Europhys. Lett.* **16** (1991) 115.
90. L. B. Hansen, P. Stolze, K. W. Jacobsen, and J. K. Nørskov, *Surf. Sci.* **289** (1993) 68.
91. M. Breeman, G. T. Barkema, and D. O. Boerma, *Surf. Sci.* **323** (1995) 71.
92. C. Walters, *thesis*, University of Pennsylvania (1997).
93. C. Kittel and H. Kroemer, *Thermal Physics*, (Freeman and Company, 1980), p.29.
94. L. S. Perkins and A. E. DePristo, *Surf. Sci.* **294** (1993) 67.
95. J. J. DeMiguel, A. Sánchez, A. Cebollada, J. M. Gallego, J. Ferrón, and S. Ferrer, *Surf. Sci.* **189/190** (1987) 1062.
96. Peter J. Feibelman, *Phys. Rev. Lett.* **58** (1987) 2766.
97. S. C. Wang and G. Ehrlich, *Surf. Sci.* **239** (1990) 310.
99. S. V. Khare, N. C. Bartelt, and T. L. Einstein, *Phys. Rev. Lett.* **75** (1995) 2148.
98. Tien T. Tsong and Chong-lin Chen, *Surf. Sci.* **246** (1991) 13.
100. S. V. Khare and T. L. Einstein, *Phys. Rev. B* **54** (1996) 11752.
101. J. M. Soler, *Phys Rev. B* **53** (1996) 10540.
102. C. D. Van Siclen, *Phys. Rev. Lett.* **75** (1995) 1574.
103. D. Sholl and R. T. Skodje, *Phys. Rev. Lett.* **75** (1995) 3158; *Physica A* **231** (1996) 631.
104. A. F. Voter, *Phys. Rev. B* **34** (1986) 6819.
105. A. Bogicevic, S. D. Liu, J. Jacobsen, B. Lundqvist and H. Metiu, *Phys. Rev. B* **57** (1998) R9439.
106. L. Bardottik, M. C. Bartelt, C. J. Jenks, C. Stoldt, J.-M. Wen, C.-M. Zhang, P. A. Thiel and J. W. Evans, *Phys. Rev. B* **57** (1998) 12544.
107. G. Rosenfeld, K. Morgenstern, and G. Comsa, ed.., M. C. Tringides and M. Sheffler, NATO-ASI (Plenum Press, 1997), p. 361.
108. Ulrike Kürpick, Peter Kürpick and Talat S. Rahman, *Surf. Sci.* **383** (1997) L713.
109. J. de la Figuera, J. E. Prieto, C. Ocal, and R. Miranda, *Solid Stat. Commun.* **89** (1994) 815.
110. M. Giesen-Siebert, R. Jentjens, M. Poensgen, and H. Ibach, *Phys. Rev. Lett.* **71** (1993) 3521.
111. S. V. Khare, private communication.
112. J. W. Evans, P. A. Thiel, and R. Wang, preprint (1997).

THE CONCEPT OF TWO MOBILITIES FOR GROWTH MANIPULATION

GEORG ROSENFELD, GEORGE COMSA
Institut für Physikalische und Theoretische Chemie der Universität Bonn, Wegelerstraße 12, D-53115 Bonn, Germany

BENE POELSEMA
Faculty of Applied Physics and Centre of Materials Research (CMO), University of Twente, P.O. Box 217, NL-7500 AE Enschede, The Netherlands

In this chapter recent work on manipulation of growth modes in molecular beam epitaxy of metals is reviewed. The basic idea (the "concept of two mobilities") is described, a mathematical formulation of this concept is sketched, and examples for its experimental realization are given.

1 The basic idea

Controlled growth in molecular beam epitaxy (MBE) requires a detailed understanding of growth kinetics because in many cases the desired film morphology is metastable. The concept of two mobilities is an example of how the identification and experimental control of the important atomic processes during far from equilibrium growth can be turned into recipes for growth manipulation. The concept was introduced for the simple case of homoepitaxy, [1] i.e., substrate and film consisting of the same kind. We will stick to this simple case in the following description of the basic idea behind the concept. A generalization to heteroepitaxy is straightforward but does not lead to further enlightening at this stage. We will, however, present heteroepitaxial examples in the experimental section.

The concept of two mobilities is based on the simple idea that the interlayer mass transport necessary for layer-by-layer growth can be enhanced if the mobility of atoms on top of a growing layer is increased. For *a given island size* and a given probability $s = exp(-E_S/kT)$ of surmounting a step edge in a single jump (where E_S is the excess barrier required to cross the step edge, [2,3] cf. Fig.1), the interlayer mass transport will be the more effective the higher the diffusivity of atoms *on top* of monolayer islands is. However, the island size is in general not a free parameter: it is determined by the diffusion length of atoms on the *lower layer* during nucleation. The higher the atom mobility during nucleation, the lower the island density and the larger the islands at a given coverage. A change in growth parameters that increases the atom

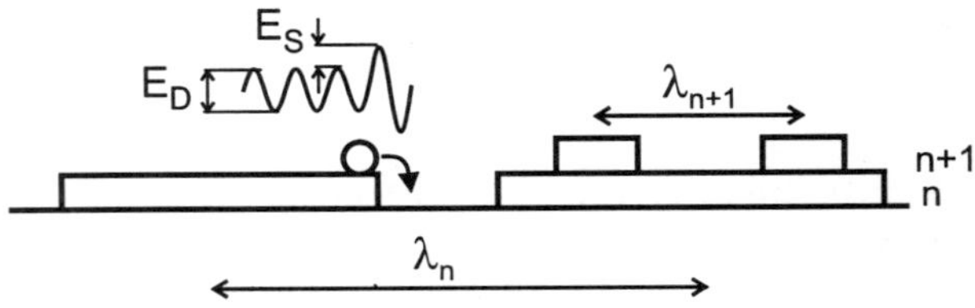

Figure 1: Schematic representation of a multi-level surface with potential energy diagram near a step edge. The step-edge barrier E_S (Ehrlich-Schwoebel barrier [2,3]) is defined as the barrier additional the terrace diffusion barrier E_D. Note, that the island separation on level $n+1$, λ_{n+1} is drawn *smaller* than that on level n. For growth improvement by the concept of two mobilities, λ_{n+1} must be chosen *larger* than λ_n.

mobility on top of islands, e.g., an increase in substrate temperature, increases also the atom mobility on the lower level leading to larger island sizes.

A good way to effectively increase the interlayer mass transport, and hence, to suppress second-layer nucleation before coalescence of islands[a] is therefore to *decouple* the atom mobility in the different levels of the growing surface so that the atom mobility on top of islands can be increased without increasing the island size at the same time. This is the concept of two mobilities: it requires two different sets of parameters controlling the diffusivity of atoms during layer growth, one in the level of the growing layer and one on top it (cf. Fig. 1). This *spatial* decoupling can also be a *temporal* decoupling. The island density (and hence the island size) at a given coverage is fixed during the first stage while second-layer nucleation is a problem only during a later stage of monolayer growth. Therefore, the concept of two mobilities can be realized by choosing a low mobility during the nucleation of a layer and a high mobility later during the rest of monolayer growth. In this way the mobility is the same at any moment on both levels of the growing film but it changes from low to high while the layer grows.

The main characteristic of the concept of two mobilities is that it takes the step-edge barrier as a given quantity and tries to improve the growth by changing the effect of this barrier but not the barrier itself nor the probability of surmounting this barrier in a single jump. In this sense, the two-mobilities concept is an indirect method. As the step-edge barrier is the key quantity in kinetically controlled epitaxy, the direct way for growth improvement would be to reduce this barrier (e.g., by suitable additives, or changes in the structure of island edges) or the probability s defined above (e.g., by increasing the

[a]Second-layer nucleation *before* or *after* coalescence is the physically meaningful criterion to distinguish between 3D- and 2D-growth, respectively.

substrate temperature). However, the direct way is not always feasible, e.g., if the substrate temperatures needs to be low in order to suppress interdiffusion. In such cases the concept of two mobilities provides a powerful means of growth manipulation as we will see in this chapter.

2 Mathematical formulation

The concept of two mobilities can be described mathematically by using standard nucleation theory [4] and its application to nucleation on top of growing islands.[5,6,7,8] We define a nucleation length (or mean free path for nucleation), λ, as the average distance an adatom travels before it forms a stable nucleus. In a description using average quantities, λ is the mean separation of islands on a flat terrace. This quantity increases with increasing temperature and decreasing deposition rate :[4]

$$\lambda = \lambda_0 \cdot \left(\frac{\nu_0}{R}\right)^{\frac{i}{2(i+2)}} \exp\left(-\frac{E_i + iE_D}{2(i+2)kT}\right), \tag{1}$$

where E_D is the adatom diffusion barrier, E_i the binding energy and i the size of the critical cluster, R the deposition rate (in ML/s), and ν_0 the attempt frequency for adatom diffusion. The factor λ_0 depends on the critical cluster size and contains capture numbers.[4] In the following, λ is regarded as dimensionless quantity (i.e., expressed in units of the lattice constant). We distinguish between this nucleation length on a level n of the growing surface, $\lambda_n = \lambda(T_n, R_n, E_n, i_n)$, and its value on the level $n+1$, i.e, on top of 2D islands growing on level n (see Fig. 1), $\lambda_{n+1} = \lambda(T_{n+1}, R_{n+1}, E_{n+1}, i_{n+1})$.

To predict growth modes one can calculate a critical coverage for nucleation on top of level $n-1$, Θ_C, and compare this value to the coverage Θ_{coal}, at which island coalescence typically sets in. If $\Theta_C < \Theta_{coal}$, one has 3D-growth while for 2D-growth, $\Theta_C > \Theta_{coal}$ must be fulfilled. The critical coverage for nucleation can be determined by assuming circular islands of radius r and computing the steady-state radial adatom density profile $n_1(r')$ on top of islands subject to the boundary condition given by the transmission factor s at the island edge. The nucleation rate $\omega(r') \propto n_1(r')^{i+1}$,[9] is integrated across the island top:

$$\Omega = \int_0^r 2\pi\omega r' dr', \tag{2}$$

and the fraction of islands on top of which second-layer islands have nucleated, $f(\Theta)$, is calculated:

$$f(\Theta) = 1 - \exp\left(-\int_0^\Theta \Omega \, d\Theta\right). \tag{3}$$

A critical coverage for second-layer nucleation, Θ_C, is defined by stating that at this coverage, f has a certain value f_C (e.g., $f_C = 0.1$), i.e.:

$$f(\Theta_C) = 1 - \exp\Big(-\int_0^{\Theta_C} \Omega \, d\Theta\Big) \equiv f_C. \tag{4}$$

In the end, this leads to an equation for the critical coverage which, expressed in terms of the nucleation lengths λ_n and λ_{n+1} as defined above, reads: [10]

$$\sum_{k=0}^{i+1} \beta_k(i)\, (s \cdot \lambda_{n+1})^{-k} \Big(\frac{\lambda_{n+1}}{\lambda_n}\Big)^{2(i+2)-k} \Theta_C^{i+3-k/2} = -\ln(1 - f_C), \tag{5}$$

where s is the transmission factor defined above, and the $\beta_k(i)$ are coefficients depending on capture numbers and on the critical island size $i = i_{n+1}$ on level $n+1$ (here and in the following, we drop the index n+1 for the critical island size). [10] Equation (5) takes a much simpler form for two limiting cases. If $s \cdot \lambda_{n+1} \gg 1$, e.g., in case of a vanishing step-edge barrier, only the term with $k = 0$ needs to be considered in the sum in Eq. (5), and one obtains:

$$\Theta_C = \Theta_{C,0} \Big(\frac{\lambda_{n+1}}{\lambda_n}\Big)^{\frac{2(i+2)}{i+3}} = \Theta_{C,0}\, \xi^{\frac{i+2}{i+3}}, \tag{6}$$

where $\Theta_{C,0}$ denotes the critical coverage during conventional (i.e., not manipulated) homoepitaxy with $\lambda = \lambda_{n+1}$ $(= \lambda_n)$ and $\xi = (\lambda_{n+1}/\lambda_n)^2$ is the enhancement factor of the island density. In this case of a vanishing step-edge barrier, $\Theta_{C,0}$ does not depend explicitely on substrate temperature and deposition rate: [10]

$$\Theta_{C,0} = A(i), \tag{7}$$

where the constant $A(i)$ depends on capture numbers and the critical island size and is of order of magnitude one but greater than unity. [11] Hence, one has $\Theta_C > \Theta_{coal}$, i.e., the condition for 2D-growth is fulfilled.

Similarly, for a high step-edge barrier, $s \cdot \lambda_{n+1} \ll 1$, only the term with $k = i + 1$ needs to be considered in the sum of Eq. (5), and one obtains:

$$\Theta_C = \Theta_{C,0} \Big(\frac{\lambda_{n+1}}{\lambda_n}\Big)^{\frac{2(i+3)}{i+5}} = \Theta_{C,0}\, \xi^{\frac{i+3}{i+5}}. \tag{8}$$

Here, the expression for $\Theta_{C,0}$, i.e., conventional homoepitaxy, reads : [10]

$$\Theta_{C,0} = B(i)\, (s \cdot \lambda_{n+1})^{\frac{2(i+1)}{i+5}}. \tag{9}$$

Again, the $B(i)$ are of order of magnitude one,[12] and as $s \cdot \lambda_{n+1} \ll 1$, one has $\Theta_C < \Theta_{coal}$, the condition for 3D-growth.

From these equations, the effect of two mobilities, or equivalently, an artificial enhancement of the island density, is obvious. For instance, for an enhancement factor of $\xi = 10$ and $i = 1$, Eq. (8) leads to an increase of the critical coverage by a factor of 4.6. For the case of a vanishing step-edge barrier, the critical coverage is greater than the coverage for coalescence already in conventional homoepitaxy. An artificial enhancement of the island density has thus only a quantitative effect, improving further the 2D-growth.[13] In contrast, in the other limit of a high step-edge barrier there can be a qualitative effect: depending on the value of ξ, the critical coverage may shift from $\Theta_C < \Theta_{coal}$ to $\Theta_C > \Theta_{coal}$ so that a transition from 3D- to 2D-growth can be induced.

In principle, a change in any kinetic parameter during monolayer growth that leads to $\lambda_{n+1} > \lambda_n$, and hence, to $\xi > 1$, can be used to decrease the roughness of a growing film. This inequality can be realized by reducing the temperature during the nucleation period on a layer n ($T_n < T_{n+1}$), or by increasing the deposition rate during nucleation ($R_n > R_{n+1}$). However, also changes in the critical cluster size, i, may have a similar effect, or for a given critical cluster size, a difference in characteristic energies (e.g., $E_{D,n} > E_{D,n+1}$) enhances the interlayer mass transport. While temperature and deposition rate modulation are the direct way to realize the concept of two mobilities, the experimental tuning of energies require more sophisticated methods, e.g., the use of surfactants with different concentrations in the different layers of the growing film. Another means to influence the nucleation without temperature or deposition rate variation is ion bombardment during the early stages of monolayer growth. The effect of the ions is to produce so-called target adatoms on the surface which nucleate close to the ion impact position and form small clusters.[14] These clusters act as nucleation centres for the deposited atoms, and as their density can be made higher than the island density during conventional growth at the same temperature, an enhanced island density is produced.[15] In the next sections we will present examples of metal homo- and heteroepitaxy where temperature modulation, deposition rate modulation, and pulsed ion beam assisted deposition (PIBAD) were successfully applied to improve the quality of thin layers.

3 Experimental realization

3.1 Temperature variation

In our first experiments we chose the substrate temperature as the parameter to be changed during monolayer growth. As a test system we first chose Ag(111) which has a substantial step-edge barrier E_S as it grows three-dimensionally rough at all temperatures below the transition to step flow.[19,20,21,22,23] Pars pro toto, we concentrated on the growth of one monolayer and, according to our general concept, prepared an enhanced island density by depositing 0.05 monolayers (ML) of Ag at a reduced temperature, T_1. Then, we increased the temperature to the growth temperature, T_2, and continued the deposition.[24,25,26]

To monitor the growth mode during these manipulation experiments we used the scattering of thermal energy helium atoms (TEAS).[16,17] Like in diffraction techniques using electrons or x-rays, the intensity of the specularly reflected He beam shows oscillations for layer-by-layer growth. The oscillation period corresponds to the deposition of one monolayer. For multilayer growth, the intensity decreases monotonically.

A series of such experiments is shown in Fig. 2 for a growth temperature $T_2 = 260$ K and T_1 as indicated. For conventional growth at 260 K (Fig. 2 a) a monotonically decaying intensity is observed indicating three-dimensional growth. For manipulated growth (Figs. 2 b-d) a maximum in intensity is observed at a time roughly corresponding to the deposition of one monolayer. The height of the maximum increases with increasing island density, i.e., decreasing substrate temperature during nucleation, T_1, indicating improved quality of the grown monolayer. The decreasing intensity in the minimum is a direct signature of the increasing island density. To get a more quantitative idea of the remaining surface roughness after deposition of one monolayer, we calculated the height distribution of the deposited layer from the height of the intensity maximum. Assuming an ideal instrument and a three-level system (substrate, first and second layer level) we obtained, e.g., for the curve 2 d that at a coverage of one monolayer the first layer is filled to 94 percent and only 6 percent of the deposited material is in the second layer.[27]

The set of experiments shown in Fig. 2 is only a selection of the data available. We performed similar experiments at a variety of different temperature combinations with T_2 between 125 K and 260 K. In all cases we found improved growth compared to conventional growth at constant temperature, and the quality of the grown layer improved with increasing difference between T_1 and T_2. We have also tried the same recipe for a different system, namely Cu(111), which similarly to Ag(111) grows three- dimensionally rough in a broad temperature range in conventional homoepitaxy.[28,29,30,31] As exemplified by the experiments shown in Fig. 3, 2D growth can be induced in the same manner.

The highest growth temperatures T_2 used in these experiments (260 K for Ag and 300 K for Cu) were chosen subject to the condition that the high

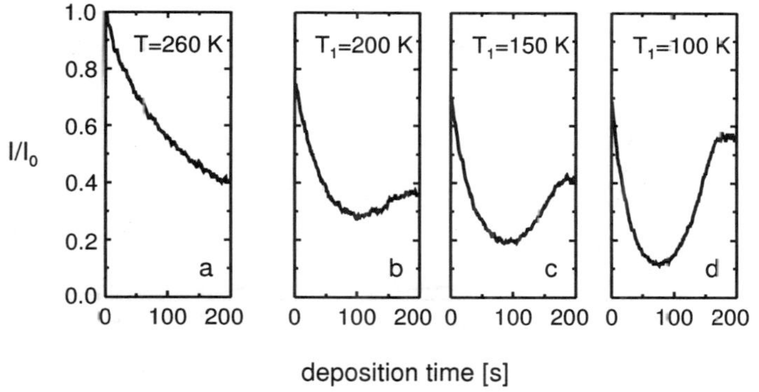

Figure 2: Normalized anti-phase He intensity during deposition of Ag onto Ag(111) at a substrate temperature of 260 K and a deposition rate of $4.2 \cdot 10^{-3}$ ML/s. a) Conventional growth. b) - d) Substrate prepared with high island density by pre-depositing 0.05 ML of Ag at the indicated preparation temperature T_1.

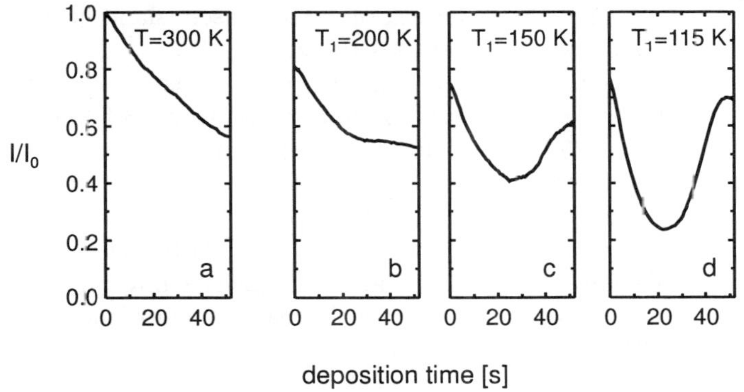

Figure 3: Normalized anti-phase He intensity during deposition of Cu onto Cu(111) at a substrate temperature of 300 K and a deposition rate of $1.8 \cdot 10^{-2}$ ML/s. a) Conventional growth. b) - d) Substrate prepared with high island density by pre-depositing 0.05 ML of Cu at the indicated preparation temperature T_1.

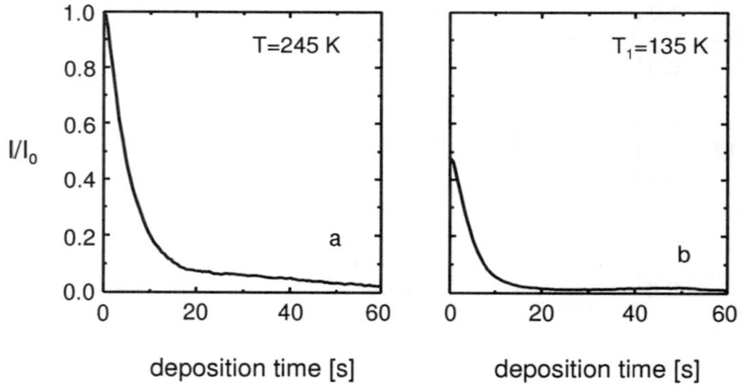

Figure 4: Normalized anti-phase He intensity during deposition of Ni onto Cu(111) at a substrate temperature of 245 K and a deposition rate of $2 \cdot 10^{-3}$ ML/s. a) Conventional growth. b) Substrate prepared with high island density by pre-depositing 0.1 ML of Ni at 135 K.

island density prepared at T_1 is not destroyed by Ostwald ripening during the time scale of the experiments when the substrate temperature is raised from its low value during nucleation to the growth temperature. From annealing experiments we have concluded that Ostwald ripening on the time scale of the experiment sets in near room temperature for Ag/Ag(111) and around 350 K for Cu/Cu(111).[25,31,32] Therefore, an increased island density is indeed maintained for the growth temperatures used.

In order to test whether the concept of two mobilities could also be applied to heteroepitaxial systems, we tried to improve the growth of Ni on Cu(111) at temperatures below 375 K, i.e., below the temperature at which intermixing at the interface is supposed to set in.[33] Under these conditions, Ni grows pseudomorphically on Cu(111) up to coverages of seven monolayers,[37,38,39] but not smoothly: during conventional growth of Ni/Cu(111) between 300 K and 375 K only one oscillation is observed in the He intensity curves; for temperatures below room temperature, the intensity is decreasing monotonically,[34,35] in line with previous RHEED-studies.[36] Obviously, the growth is rough in this temperature regime. Flat films can be grown at elevated temperatures (around 470 K)[37] but at these temperatures interdiffusion is active.

Unlike the homoepitaxial case, attempts to improve the growth via temperature variation did not lead to convincing results. Fig. 4 shows a typical result of such an experiment where 0.1 ML of Ni were deposited onto Cu(111) at 135 K and subsequent deposition was carried out at 245 K.[34] For comparison, also a deposition curve at constant temperature of 250 K is presented.

The manipulated growth curve shows a shallow maximum at about 50 s, but the low intensity indicates that the film after deposition of 1 ML is not smooth.

The origin of the failure of these experiments could be related to an unusual annealing behaviour of the 0.1 ML Ni-film during heating of the sample from T_1 to T_2. Instead of becoming smoother during this annealing step, the surface appears to roughen as concluded from a decreasing He reflectivity. [34] This anomalous behaviour presumably destroys the positive effect of an enhanced density of 2D islands. We could, however, successfully apply the concept of two mobilities for this system by using deposition rate variation as discussed in the next section.

The experiments discussed in this section showed that temperature variation may be successful in improving the layer growth unless additional processes counteract the positive effect of island density enhancement. The general disadvantage of this technique, however, is that in order to grow many layers in a reasonable amount of time, a fast temperature modulation is required which is impractical if large temperature differences are needed (i.e., if the step-edge barrier is high). The techniques discussed in the next two sections overcome this problem as they work at constant substrate temperature.

3.2 Deposition rate variation

In principle, instead of a variation of the substrate temperature a modulation of the deposition rate can also be used to improve the layer quality. In this case the depositon rate has to be high during nucleation and low during the rest of monolayer growth. The advantage of this technique is the better control and the straightforward realization (e.g., by using two evaporators operating at different fluxes) without the problems associated with fast changes of the substrate temperature. Compared to methods using energetic particles (as pulsed ion beam assisted deposition which will be discussed in the next section) it has also the advantage that radiation damage of the film is excluded. The disadvantage is that it is much less efficient. For instance, an increase of the deposition rate by a factor of 10 during the first five percent of monolayer growth of Ag/Ag(111) at a temperature of 260 K has hardly any effect on the growth behaviour (see Fig. 5). [25] The reason for this inefficiency is that the deposition rate has a weaker influence on the island density than the substrate temperature (cf. Eq. (1)): assuming the critical cluster size to be $i = 2$, the enhancement in island density is only $\xi = 3.2$, and the critical coverage increases according to Eq. (8) by a factor of 2.3. Hence, for the case of high step-edge barriers, a change in deposition rate by many orders of magnitudes is necessary in order to shift the growth mode from 3D to 2D.

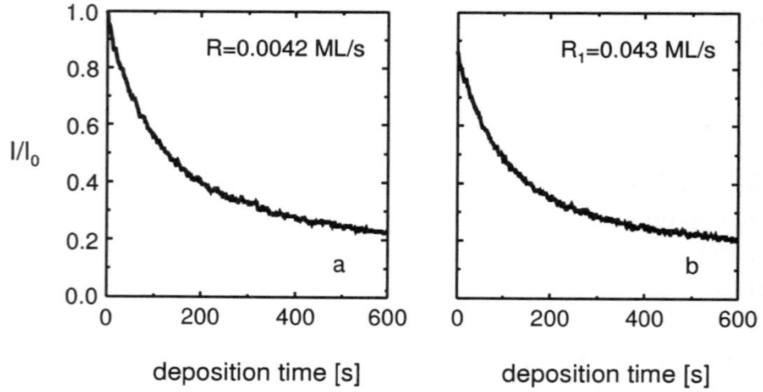

Figure 5: Normalized anti-phase He intensity during deposition of Ag onto Ag(111) at a substrate temperature of 260 K and a rate of $4.2 \cdot 10^{-3}$ ML/s. a) Conventional growth. b) After pre-depositing 0.05 ML of Ag at the same temperature and a tenfold higher rate ($4.3 \cdot 10^{-2}$ ML/s).

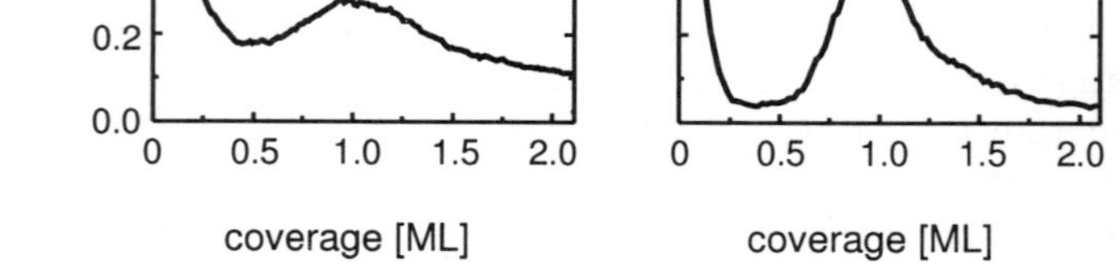

Figure 6: Normalized anti-phase He intensity during deposition of Ni onto Cu(111) at a substrate temperature of 300 K and a deposition rate of $2 \cdot 10^{-2}$ ML/s. a) Conventional growth. b) After pre-depositing less than 0.05 ML of Ni at 300 K using a "shot-evaporator" (see text).

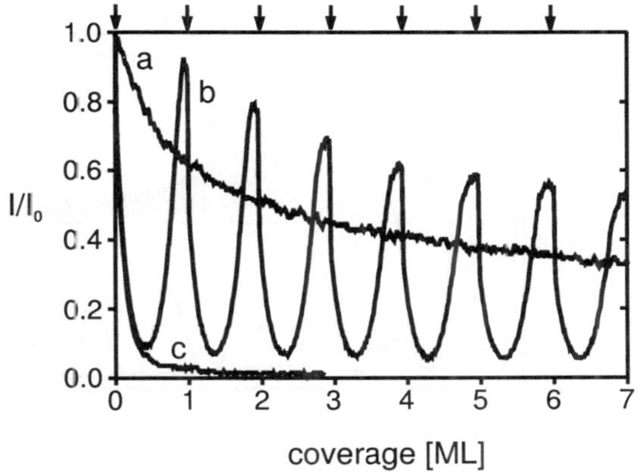

Figure 7: Normalized anti-phase He intensity during deposition of Ag onto Ag(111) at a substrate temperature of 300 K. a) Continuous deposition at a rate of $4.7 \cdot 10^{-3}$ ML/s (conventional growth). b) Same as for a), but in addition at the times marked by the arrows a short ion pulse was given (600 eV, $6 \cdot 10^{16} ions/m^2$). c) Continuous deposition at a rate of $1.3 \cdot 10^{-2}$ ML/s combined with continuous ion bombardment (600 eV) correponding to a removal rate of $7.7 \cdot 10^{-4}$ ML/s.

However, using conventional evaporators with mechanical shutters, the variation in deposition rates is limited by the control of the shutter for the fast deposition rate, and for the slow one by the constraint that the time to deposit the rest of the monolayer should not exceed reasonable experiment times. In any case, variations of more than three orders of magnitude are difficult to achieve. (E.g., three orders of magnitude difference correspond roughly to a situation where ten percent of a monolayer are deposited from a fast evaporator (deposition rate = 1 ML/s) opened for 0.1 seconds, while the slow evaporator deposits at a rate of 0.001 ML/s.) For test purposes, we therefore built a simple "shot-evaporation" source which consists of very thin wires of the evaporant fixed between the pins of an electrical feedthrough. By discharging a capacitor across a single wire, the evaporant explodes and an amount corresponding to a small fraction of a monolayer (< 5%) is instantaneously deposited onto the surface. Using this shot-evaporator we could significantly improve the growth of Ni/Cu(111) at room temperature as shown in Fig. 6.[35]

3.3 Pulsed Ion Beam Assisted Deposition

The method described in this section has the advantage that both substrate temperature and deposition rate remain constant during growth. In order to

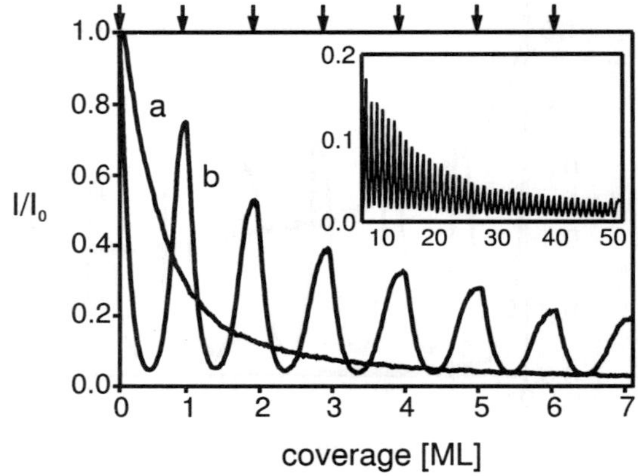

Figure 8: Normalized anti-phase He intensity during deposition of Cu onto Cu(111) at a substrate temperature of 300 K. a) Continuous deposition at a rate of $1.7 \cdot 10^{-2}$ ML/s (conventional growth). b) Same as for a), but in addition at the times marked by the arrows a short ion pulse was given (1.2 keV, $1 \cdot 10^{17} ions/m^2$).

enhance the island density, a pulsed ion beam is used to expose the surface to ions at the beginning of the nucleation period of each layer. The effect of these ions is to produce target adatoms on the surface which nucleate close to the ion impact position. The cluster produced in this way act as nucleation centres for the deposited atoms and the number of these clusters can exceed the number of islands produced during conventional epitaxy at the same temperature by a large factor. For the metals considered here this factor is typically between 10 and 100 for ion energies of several 100 eV and ion fluences of the order of 10^{16} to 10^{17} $ions/m^2$.[15,27,34,40,41] Figures 7 to 9 show examples where we used this method to improve the growth of Ag/Ag(111), Cu/Cu(111), and Ni/Cu(111). The method was also recently applied to improve the growth of Pt/Pt(111) and the result was monitored by STM,[40] see Fig. 11.

Although straightforward to use, this method bears the danger of damage produced by the energetic particles, or gas atoms implanted near the growing surface. Moreover, for heteroepitaxial systems, this method may produce slightly rough interfaces. The formation of clusters of target atoms and the creation of holes in the substrate layer may lead to a scenario as sketched in Fig. 10: the deposit fills the holes created by the ion beam and hence, an unsharp interface is the price to be paid for the smoother film surface. Note, however, that Fig. 10 exaggerates the situation. During typical conditions only a few percent (or less) of the surface layer atoms are removed. Moreover,

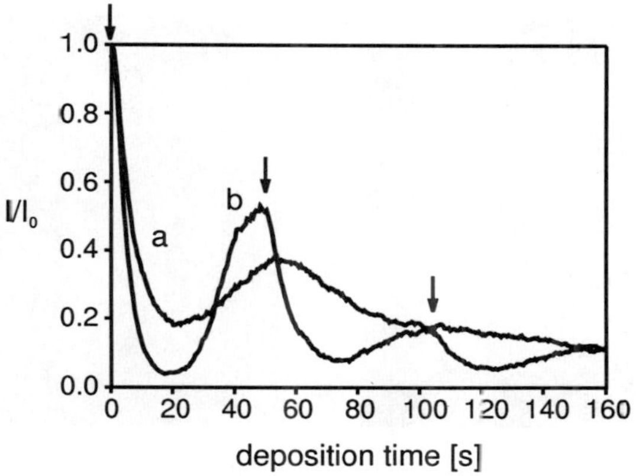

Figure 9: Normalized anti-phase He intensity during deposition of Ni onto Cu(111) at 300 K. a) Continuous deposition. b) Sputter pulses (1.2 keV Ne^+, $4 \cdot 10^{17} ions/m^2$) were given at times indicated by arrows.

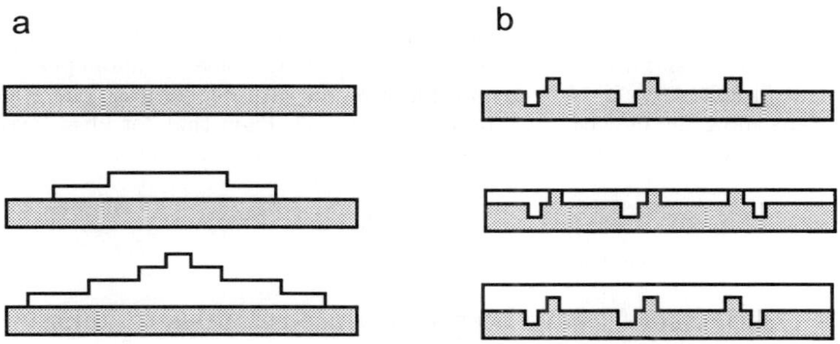

Figure 10: Schematic sketch of a growing heteroepitaxial film. a) Conventional growth. b) Film grown with pulsed ion beam assisted deposition.

in the case of Ni/Cu(111) the interface roughness produced in this way is likely to be lower than that present if the film were grown at an elevated temperature where the film grows smoother but at the same time, massive interdiffusion takes place.

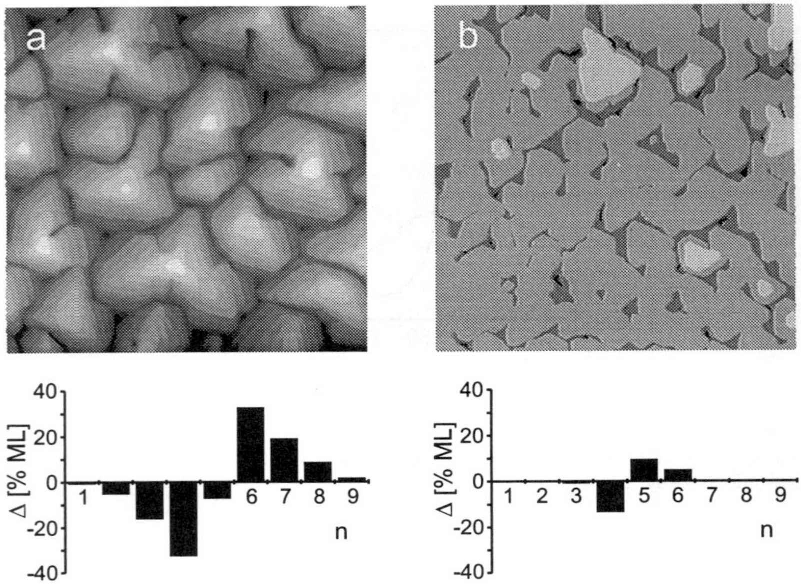

Figure 11: STM-topographs (2200Å×2200Å) of $\simeq$ 4.5 ML Pt grown on Pt(111) at a substrate temperature of 400 K and a rate of $\simeq 5 \cdot 10^{-3}$ ML/s).[40] a) Conventional deposition. b) Ion bombardment (4 keV Ar^+, $2 \cdot 10^{16} ions/m^2$) alternated with deposition. The graphs below the STM-scans show the deviation of the layer filling, Δ, from that for ideal layer-by-layer growth versus the layer number n. With kind permission by T. Michely.

3.4 Surfactant-mediated growth

Additives ("surfactants") have been used since centuries to improve growth. (Regardless of its original meaning the term surfactant is used for any adsorbate that leads to improved growth.) In quasi-equilibrium growth the improvement is due to the reduction of the surface and interface free energy. In the case of kinetically controlled epitaxy discussed here, it is due to different mechanisms. (Note, however, that in both cases in order to be really useful, the surfactant atoms have to leave the growing film by floating on its surface). The most direct way in which surfactants may improve growth is by reducing the height of the step-edge barrier. This has been demonstrated for the action of oxygen in the homoepitaxial growth of Pt(111).[42]

However, surfactants could also work in a different way. Many surfactants are found (both experimentally and theoretically) to limit the lateral mobility of adatoms of the growing film.[43,44] If the transfer to the upper layer ("the floating") is delayed so that a concentration gradient develops with a relatively high concentration of surfactant atoms in the growing layer but a low or even vanishing concentration on top of the growing islands, the concept of two mobilities is realized. The surfactants induce an enhanced island density in the growing layer but they hardly effect the mobility on top of growing islands. Especially, this mechanism might be effective if a small amount of surfactant (corresponding to submonolayer coverages) is pre-dosed onto the substrate before deposition rather than being supplied continuously during growth. In this case the surfactant atoms might stay in the growing layer until they are pushed up during coalescence of islands in a late stage of monolayer growth.[24] Recent experiments and theoretical studies on the surfactant action of Sb during growth of Ag/Ag(111) show that at least the 2D growth of the first monolayer can be explained by this mechanism.[45,46]

However, the identification of suitable surfactants is still a matter of trial and error, it is hard to tell in advance whether a certain candidate for a surfactant acts in the desired way. In this context, the concept of two mobilities serves more as a possible explanation for an observed effect than as a manipulation recipe. Besides surfactants, there are other examples where the concept of two mobilities may be realized during epitaxial growth without the active intervention of the experimentalist. A particularly nice example is highlighted in the next section.

3.5 Two mobilities realized by nature

Nature appears to have already used the concept of two mobilities. Indeed, the experimental observation of an exceptionally smooth growth of Pt/Pt(111) at higher temperatures may be directly related to the concept of two mobilities. It was found by He-scattering that growth at temperatures above 600 K is remarkably smooth: oscillations in the specular He intensity were observed to persist at coverages exceeding 150 monolayers.[47,48] If the growth temperature is lowered, a gradual transition to an almost perfect 3D-growth regime is observed (below $T = 450$ K), demonstrating the existence of a substantial step-edge barrier E_S. This is followed by a regime of re-entrant layer-by-layer growth (below $T = 300$ K).[16,47,49]

As revealed by STM-studies on this system, the high-temperature 2D-growth mode is caused by an unexpected reconstruction of the Pt(111) surface which typically forms upon deposition of Pt at temperatures above 450 K and consists of fcc and hcp domains separated by a network of dislocation lines.[50,51] In the absence of this reconstruction, 3D-growth is observed.[51] The combination of two properties of this reconstruction directly leads to improved growth according to the concept of two mobilities: a) adatom diffusion on a reconstructed layer is reduced leading to an enhanced island density, and b) on top of the islands the adatom mobility is not affected since the island tops do not reconstruct until a large terrace is formed by coalescence. Reconstruction phenomena which occur on many other surfaces, in particular on semiconductor surfaces, are believed to influence the growth in a similar way.[51]

Finally, we note that different mobilities on different levels of a growing film occur naturally during heteroepitaxy. As a consequence, 2D-growth may be favoured *kinetically* (in contrast to the classical energetic arguments),[52] if the island density on the foreign substrate is much higher than the island density observed if the substrate were of the same kind than the film.

Acknowledgements

The contributions of Ingo Beckmann, Louis C. Jorritsma, Jörg Kliewer, Nuphar N. Lipkin, Karina Morgenstern, Christian Teichert, and Wulf Wulfhekel to the work presented in this chapter are gratefully acknowledged. We are also indebted to Thomas Michely and his co-workers for stimulating discussions and for the permission to present their data in Figure 11.

References

1. G. Rosenfeld, B. Poelsema, and G. Comsa, J. Cryst. Growth. **151**, 230 1995.
2. G. Ehrlich and F.G. Hudda, J. Chem. Phys. **44**, 1939 (1966).
3. R.L. Schwoebel and E.J. Shipsey, J. Appl. Phys. **37**, 3682 (1966).
4. J. A. Venables, G. D. T. Spiller, and M. Hanbücken, Rept. Prog. Phys. **47**, 399 (1984).
5. S. Stoyanov and I. Markov, Surf. Sci. **116**, 313 (1982).
6. J. Tersoff, A. W. Denier van der Gon, and R. M. Tromp, Phys. Rev. Lett. **72**, 266 (1994).
7. A. A. Chernov, J. Cryst. Growth **42**, 55 (1977).
8. I. Markov, Phys. Rev. B **54**, 17930 (1996).
9. D. Walton, J. Chem. Phys. **37**, 2182 (1962).
10. G. Rosenfeld, B. Poelsema, and G. Comsa, in *Growth and Properties of Ultrathin Epitaxial Layers*, The Chemical Physics of Solid Surfaces, Vol. 8, eds. D.A. King and D.P. Woodruff (Elsevier, Amsterdam 1997), p. 66
11. $A(i) = (-\ln(1-f_C)/\beta_0(i))^{2/(i+5)}$. For $f_C = 0.1$, typical values are: $A(1) = 1.2$ and $A(2) = 1.6$.[10]
12. $B(i) = (-\ln(1-f_C)/\beta_{i+1}(i))^{1/(i+3)}$. For $f_C = 0.1$, typical values are: $B(1) = 0.84$ and $B(2) = 1.5$.[10]

13. Nice examples for growth improvement of layer-by-layer systems by the concept of two mobilities (although the original interpretation of the experiments is different) are the growth experiments by Markov et al., who modulated the substrate temperature and deposition rate during MBE of Si/Si(111) and Ge/Ge(111) (V.A. Markov, O.P. Pchelyakov, L.V. Sokolov, S.I. Stenin, and S. Stoyanov, Surf. Sci. **250**, 229 (1991)).
14. T. Michely and G. Comsa, Phys. Rev. B **44**, 8411 (1991), T. Michely and C. Teichert, Phys. Rev. B **54**, 11156 (1994).
15. S. Esch, M. Bott, T. Michely, and G. Comsa, Appl. Phys. Lett. **67**, 3209 (1995).
16. B. Poelsema, R. Kunkel, N. Nagel, A.F. Becker, G. Rosenfeld, L.K. Verheij, G. Comsa, Appl. Phys. A **53**, 369 (1991).
17. B. Poelsema and G. Comsa, *Scattering of Thermal Energy Atoms*, Springer Tracts in Modern Physics **115**, (Springer, Berlin, Heidelberg 1989).
18. K.H. Besocke, Surf. Sci. **181**, 145 (1987).
19. Y. Suzuki, H. Kikuchi, N. Koshizuka, Jpn. J. Appl. Phys. **27**, L1175 (1988).
20. K. Meinel, M. Klaua, H. Bethge, J. Cryst. Growth **89**, 477 (1988), and phys. stat. sol. **110**, 189 (1988).
21. H.A. van der Vegt, H.M. van Pinxteren, M. Lohmeier, E. Vlieg, J.M.C. Thornton, Phys. Rev. Lett. **68**, 3335 (1992).
22. W.C. Elliott, P.F. Miceli, T. Tse, and P.W. Stephens, Phys. Rev. B **54**, 17938 (1996).
23. M. Stanley, C. Papageorgopoulos, K.R. Roos, M.C. Tringides, Surf. Sci. **355**, L264 (1996).
24. G. Rosenfeld, R. Servaty, C. Teichert, B. Poelsema, G. Comsa, Phys. Rev. Lett. **71**, 895 (1993).
25. G. Rosenfeld, PhD-Thesis, Jül-Report No. 2914, (Jülich 1994).
26. To further enhance the island density, the deposition rate was chosen one order of magnitude higher during deposition at T_1. However, it turned out that the latter had only a minor influence on the results.
27. G. Rosenfeld, N. N. Lipkin, W. Wulfhekel, J. Kliewer, K. Morgenstern, B. Poelsema, G. Comsa, Appl. Phys. A **61**, 455 (1995).
28. M. Henzler, Prog. Surf. Sci. **42**, 297 (1993).
29. P. C. Dastoor, J. Ellis, A. Reichmuth, H. Bullman, B. Holst, and W. Allison, Surf. Sci. Lett. **1**, 509 (1994).
30. H. A. van der Vegt, J. Alvarez, X. Torrelles, S. Ferrer, and E. Vlieg, Phys. Rev. B **52**, 17443 (1995).
31. W. Wulfhekel, N. N. Lipkin, J. Kliewer, G. Rosenfeld, L. C. Jorritsma,

B. Poelsema, and G. Comsa, Surf. Sci. **348**, 227 (1996).
32. K. Morgenstern, G.Rosenfeld, G. Comsa, Phys. Rev. Lett. **76**, 2113 (1996).
33. S. Yang, M. Lu, G. Meigs, X.H. Feng, and E. Garfunkel, Surf. Sci. **205**, L777 (1988).
34. W. Wulfhekel, I. Beckmann, N.N. Lipkin, G. Rosenfeld, B. Poelsema, and G. Comsa, Appl. Phys. Lett. **69**, 3492 (1996).
35. W. Wulfhekel, I. Beckmann, G. Rosenfeld, B. Poelsema, and G. Comsa, Surf. Sci., in print
36. F. Huang, M.T. Kief, G.J. Mankey, and R.F. Willis, Phys. Rev. B **49**, 3962 (1994).
37. U. Gradmann, Ann. Phys. (Leipzig) **13**, 213 (1964).
38. D.W. Gidley, Phys. Rev. Lett. **62**, 811 (1989).
39. J. Zhang, Z.L. Han, and S. Varma, Surf. Sci. **298**, 351 (1993).
40. S. Esch, M. Breeman, M. Morgenstern, T. Michely, and G. Comsa, Surf. Sci. **365**, 187 (1996).
41. K. Morgenstern, PhD-Thesis, Jül-Report No. 3278, (Jülich 1996).
42. S. Esch, M. Hohage, T. Michely, and G. Comsa, Phys. Rev. Lett. **72**, 518 (1994).
43. B. Voigtländer and A. Zinner, Surf. Sci. **292**, L775 (1993).
44. S. Oppo, V. Fiorentini, and M. Scheffler, Phys. Rev. Lett. **71**, 2437 (1993).
45. J. Vrijmoeth, H.A. van der Vegt, J.A. Meyer, E. Vlieg, and R.J. Behm, Phys. Rev. Lett. **72**, 3842 (1994).
46. V. Fiorentini, S. Oppo, M. Scheffler, Appl. Phys. A **60**, 399 (1995).
47. R. Kunkel, B. Poelsema, L.K. Verheij, G. Comsa, Phys. Rev. Lett. **65**, 733 (1990).
48. J.E. Parmeter, R. Kunkel, B. Poelsema, L.K. Verheij, G. Comsa, Vacuum **41**, 467 (1990).
49. B. Poelsema, A.F. Becker, R. Kunkel, G. Rosenfeld, L.K. Verheij, G. Comsa, in *Surface Science: Principles and Applications*, Springer Proc. in Physics Vol. 115, eds. R.F. Howe, R.N. Lamb, K. Wandelt, (Berlin 1993), p. 95.
50. M. Bott, M. Hohage, T. Michely, G. Comsa, Phys. Rev. Lett. **70**, 1489 (1993).
51. T. Michely, M. Hohage, S. Esch, G. Comsa, Surf. Sci. Lett. **349**, L89, (1996).
52. E. Bauer, Z. Kristallographie **110**, 372 (1958).

Tailoring epitaxial growth of low-dimensional magnetic heterostructures by means of surfactants

J.J. de Miguel*, J. Camarero, J. de la Figuera, J.E. Prieto, and R. Miranda*
Dpto. de Física de la Materia Condensada, C-III, Univ. Autónoma Madrid, Spain
* also: Instituto de Ciencia de Materiales *"Nicolás Cabrera"*

Abstract

The preparation of low-dimensional, epitaxial, magnetic heterostructures requires the use of sophisticated growth techniques to allow the experimentalist an always increasing degree of control over the atomic processes that finally determine the sample quality and properties. In this article, we review the application of surfactant agents to custom-modify the growth of magnetic ultrathin films, multilayers and quantum wires.

The steady increase in the sensitivity of surface probes, and specially the appearance of Scanning Tunneling Microscopy (STM) in recent years have revolutionized our views of epitaxial growth at the atomic scale. Far from the naïve pictures based on simple solid-on-solid models, we now know that good-quality, layer-by-layer growth is not the rule but the exception, even in those cases fulfilling the basic thermodynamic requirements [1]. The outcome of the growth process results from a complicated interplay of multiple atomic processes, such as inter- and intra-layer diffusion, segregation, interdiffusion, etc. All these processes are governed by different activation energies and proceed at different rates. The challenge we face is to find ways to control them efficiently.

Resulting from this improvement in our understanding of the mechanism of epitaxial growth, the physics of thin magnetic films and superlattices has been explored with renewed interest in the past few years. As a result, magnetic materials are becoming increasingly controllable at the nanometer scale and new properties such as surface or interface-induced perpendicular anisotropy [2], oscillatory magnetic coupling [3], and giant magnetoresistance (GMR) [4] have been discovered. Applications of these phenomena to perpendicular recording media [5], heads for digital compact cassette players [6] or GMR reading heads for computer hard disks [7] are already available.

In spite of it, there is still ample room for improvement in our capability to control the microstructure, and thus, the properties of epitaxial magnetic films. Most of the artificial materials that we are interested in, such as heterostructures (e.g., *crystalline superlattices*) or systems with reduced dimensionality (*quantum wires* or *dots*) have structures that do not appear naturally: they are therefore metastable. They can only exist thanks to kinetic constraints that prevent them from reaching their equilibrium configurations. For this reason, experimentalists usually try to eliminate unwanted effects by manipulating the growth kinetics. While this procedure might be effective in some cases, such as preventing 3D islanding of materials with a tendency to grow following the Volmer-Weber or Stranski-Krastanov mechanisms, it is not sufficient to solve the wide variety of problems that appear when attempting to grow exotic magnetic materials. For example, spin valve multilayers showing the largest (80%) GMR values at room temperature (RT) [8], require high magnetic fields (~1 T) to display this effect. A better control of the sample microstructure should reduce the contributions of magnetostatic coupling, coercivity and anisotropy to the switching field, opening the way to fabricate spin valves with large values of GMR at low magnetic fields. Clearly, a flexible method of altering the growth process is needed.

The use of suitable *surfactant* agents holds promise to become such a method. A surfactant could be defined as a catalyst for growth: it is a substance that, incorporated to the growth process, alters its characteristics and the resulting film structure without being incorporated into it. A necessary condition for a substance to be a surfactant is thus that it efficiently segregates to the surface of the growing film, so that it maintains its activity without introducing any impurities. Surfactants have long been known in the field of crystal growth [9], but they have only recently become a subject of common interest for the growth of thin films and artificial structures, mostly of semiconductors [10]. In the field of metal growth there have been a few examples on the use of surfactants to assist homoepitaxial growth (Sb/Ag(111) [11], O/Pt(111) [12], In/Cu(111) [13]). We will concentrate in what follows in describing how one can use Pb as a surfactant to improve the growth of low-dimensional magnetic heterostructures. The system that we investigate is Co on Cu(111). There exists a great number of studies on the growth and magnetic properties of Co on Cu(100), due to the nearly perfect epitaxial growth of these materials for that particular orientation [14]. As a consequence of the structural simplicity, there is general agreement on the magnetic properties of thin Co films on Cu(100) [15] and (100)-oriented {Co/Cu} superlattices. Oscillatory magnetic coupling across Cu spacers of increasing thickness, for instance, has been repeteadly observed by different groups [3,16] and explained by theory [17]. On the other hand, similar studies on the {Co/Cu}(111) system have produced disappointing and often contradictory results: while {Co/Cu} multilayers grown by sputtering display oscillatory coupling and GMR [18], neither of these properties could be observed in samples grown by Molecular Beam Epitaxy (MBE) [19].

The solution to this puzzling situation came from a detailed STM study performed by our group characterizing Co layers grown on Cu(111) and show-

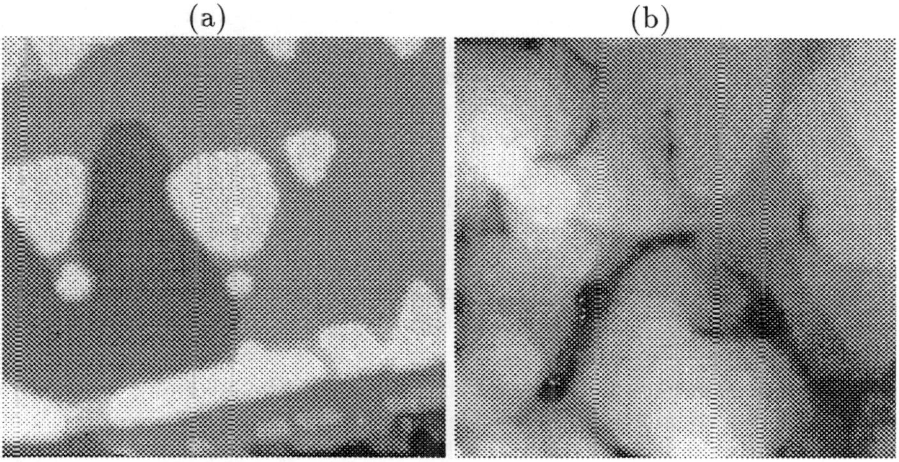

Figure 1: STM images of Co films grown on clean Cu(111). (a) 0.3 ML; (b) 5 ML. The size of the images is 500×500 Å^2.

ing that the films prepared by MBE contain a very large amount of defects [20]. A detailed structural characterization of Co films grown on Cu(111) without the aid of a surfactant lies beyond the scope of this paper, and is done elsewhere [21]. Here we will briefly summarize the main results to emphasize the modifications brought about by the use of Pb. During the early stages of Co deposition at RT, a considerable amount of interdiffusion takes place; the irregularly shaped clusters that decorate the natural steps of the substrate (see Fig. 1a) and the islands on the first layer (all of them having double atomic height) contain a mixture of Co and Cu. The interdiffusion of Co into the substrate is accompanied by etching of Cu, resulting in the appearance of monolayer-deep vacancy islands and an increase of the total step density [22]. The islands have a typical size of ~150 Å and a density of $\sim 3 \times 10^{11}$ cm^{-2}. At some point during the completion of the first bilayer, interdiffusion stops and the growth of pure Co goes on. Fig. 1b shows the typical aspect of a Co film grown on clean Cu(111) at RT, as observed by STM. Although the equivalent coverage is only 5 ML, one can observe as many as 12 atomic levels simultaneously exposed. The film morphology corresponds to a multilayer growth mode with a Poisson distribution of layer occupations. The corresponding LEED pattern shows the sixfold symmetry characteristic of the hcp crystal structure.

The huge roughness that can be seen in the image is caused by two characteristics of the (111) orientation: the existence of energetic barriers at the atomic steps, hindering interlayer diffusion and resulting in 3D growth (*"Schwoebel barrier"*), and the symmetry of the fcc-(111) substrate, offering two different, but very similar, three-fold adsorption sites for the incoming Co atoms. Since the equilibrium crystal structure of Co at RT is hcp, it has a strong tendency to abandon the fcc stacking sequence dictated by the substrate and adopt its own. This structural transformation occurs gradually during the growth of the first 5-6 ML, until the whole film has become hcp[23]. Due to their different stacking, fcc and hcp fractions do not match laterally; therefore, their coexistence during the first stages of growth forces the Co film to break up into regions separated by empty ditches that are very hard to fill in. Summarizing, a single Co film is *not* continuous and shows a 3D pyramidal morphology with mixed fcc-hcp stacking.

In a situation like this, reducing the mobility of the incoming Co adatoms can only partially alleviate the problem. When the islands formed are small and irregularly shaped, the effective Schwoebel barrier is reduced and interlayer diffusion is enhanced[24]. However, it is impossible to select the adsorption site of the deposited Co atoms simply by applying kinetic constraints: in the limit of no diffusivity, the adatoms will just stick to the point where they first hit the surface. Fortunately, using a single layer of Pb as surfactant seems to be a much more efficient method of controlling the growth and achieving the desired results.

We chose Pb because its low surface energy and large atomic volume makes it a good candidate for a surfactant: the main requirement that such an agent must fulfill is to segregate efficiently to the surface of the growing film; in this way, no impurities are left behind during the growth and the surfactant maintains its activity indefinitely. Pb grows on clean Cu(111) in the Stranski-Krastanov mode, forming first a flat monolayer with a $p(4 \times 4)$ superstructure followed by the growth of 3D clusters. Below 1 ML coverage, the distance between Pb atoms forming the superstructure (3.51Å) is close to that of bulk fcc-Pb (3.50Å). This distance is incommensurate with the Cu substrate, so one cannot speak properly of a $p(4 \times 4)$ superstructure. When the surface is completely covered, additional Pb atoms are inserted into the $p(4 \times 4)$ layer, which becomes commensurate with the substrate and laterally compressed ($\sim$ 3%) with respect to the bulk Pb-Pb distance. This compression is accommodated by vertical displacements of the Pb atoms within the superstructure unit cell (buckling). We have been able to assess, using several different experimental techniques, that the Pb-$p(4 \times 4)$ layer does indeed segregate efficiently to the surface upon either Co or Cu deposition. First of all, we do not detect any significant loss of Pb with Auger Spectroscopy (AES), after having grown several alternated layers of Co and Cu adding up to a total of 20 ML. But still more

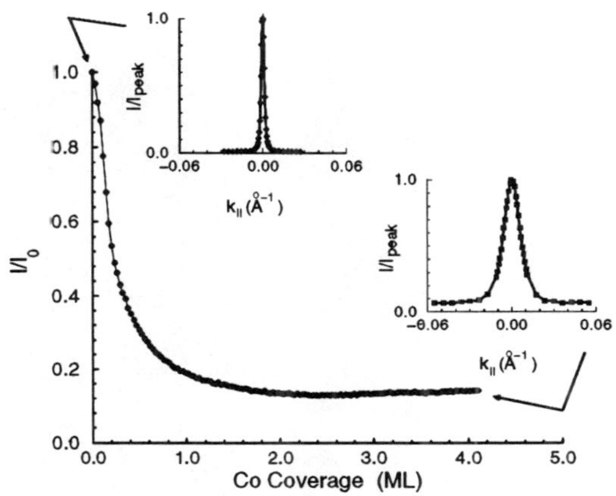

Figure 2: Segregation of Pb to the surface of a growing film: The curve shows the peak intensity of the S-XRD (3/4,0) diffracted beam, characteristic of the Pb-$p(4 \times 4)$ superstructure, during Co growth. Also shown are the peak profiles before and after deposition.

remarkable is the fact that the superstructure is maintained during and after the segregation process. This is demonstrated by the Surface X-Ray Diffraction (S-XRD) data presented in Fig. 2. Here we show in the insets the angular profiles of one of the fractional order diffracted beams, characteristic of the $p(4 \times 4)$ superstructure, before (upper left) and after (right) growing a 4 ML-thick Co film on top of the Cu substrate covered by Pb. The curve shows the evolution of the peak intensity of this diffracted beam measured in real time during Co deposition. After the growth of Co, the FWHM of the diffracted beam has increased, indicating that the average size of the ordered domains is reduced; however, integrating the peak profile to obtain the total diffracted intensity we find that after segregation we still have up to 84% of the initial value. This means that at least 84% of the Pb layer initially prepared not only has segregated to the surface, but has also kept its superstructure. The time evolution during Co deposition is also important: the fall of the diffracted peak intensity (due to the broadening) is completed after the first ML of Co is evaporated, and from this point on, no further changes are observed; the diffracted peak remains visible all through the growth experiment. Our interpretation of these results is as follows: rather than Pb segregating to the surface, it is the individual Co (or Cu) atoms deposited on top that exchange position *individually* with the Pb and so get buried beneath the Pb layer. The broadening of the diffracted peak signals a reduction of the average domain size of the

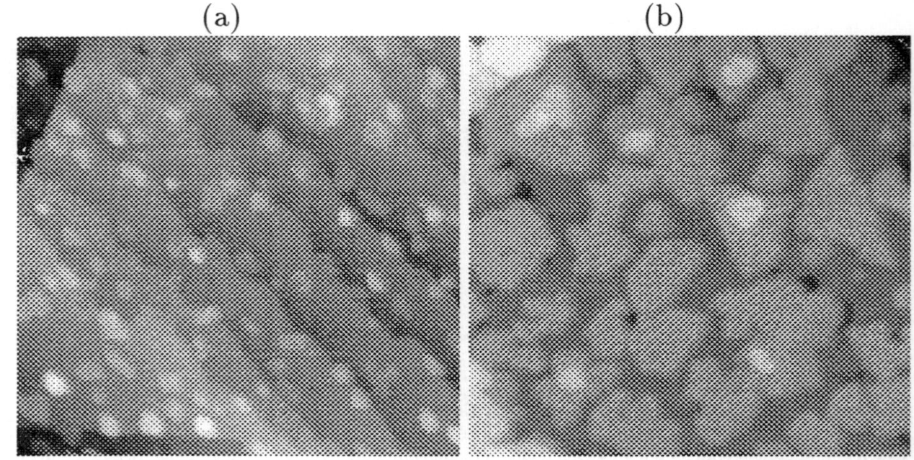

Figure 3: STM images of Co films grown on Pb-$p(4 \times 4)$/Cu(111). (a) 0.3 ML; (b) 4 ML. No etching of the surface is observed in this case, and the thicker film shows only three levels exposed. The size of the images is 500×500 Å^2.

reconstruction caused by the appearance of additional steps during the layer-by-layer growth of the incoming material. The build-up of new steps takes place during the first stages of evaporation and then the film keeps growing maintaining a stable level of roughness. The amount of intensity "lost" from the fractional-order diffracted peak must therefore reflect the fraction of surface where the superstructure is destroyed by the presence of those new steps. We believe that a real-loss of Pb can be ruled out by the fact that no relaxation of the compressed, commensurate superstructure to its relaxed, incommensurate counterpart is observed, neither during nor after growth. As a consequence of this segregation process, the growth of Co is essentially modified. Fig. 3a reproduces a STM image corresponding to 0.3 ML of Co deposited at RT on a Pb-$p(4 \times 4)$-precovered Cu(111) substrate. The Co islands that have grown underneath the Pb overlayer do not decorate the steps of the substrate, are smaller (average size ~ 25 Å) than those grown without Pb, and have an order of magnitude larger density ($n \simeq 3.6 \times 10^{12}$ cm^{-2}). No holes on the surface nor etching of the steps can be observed. Furthermore, most (at least 2/3) of the islands now have single atomic height. Thus, the amount of interdiffusion seems to be drastically reduced by the Pb overlayer. Another obvious effect of the surfactant is to enhance the nucleation density. At larger Co thicknesses, the surfactant produces a smoothening of the growth front. This is proven by

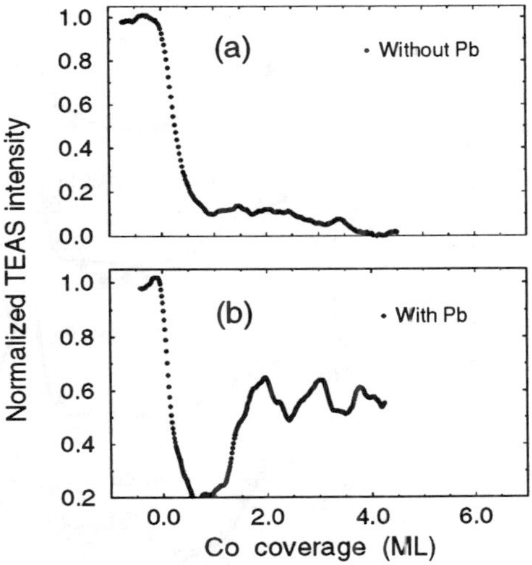

Figure 4: Growth of Co on Cu(111) observed by TEAS: (a) Without surfactant, a rough, multilayer film is formed, containing a large amount of defects; (b) With Pb, the appearance of diffracted intensity oscillations indicates layer-by-layer growth.

Fig. 3b, where the STM image of a 4 ML-thick film of Co grown with Pb shows that only three atomic levels are now exposed. The islands (with a density of 1.4×10^{12} cm^{-2} and an average size of ~75 Å) are now interconnected and the film grows flat on the substrate. A second, important modification of the growth process is thus the appearance of layer-by-layer growth. The direct experimental proof of this mode of growth is the existence of periodic oscillations in the diffracted intensity measured during the deposition. We have observed these intensity oscillations with a variety of diffraction techniques: MEED [27], S-XRD [25] and Thermal Energy Atom Scattering (TEAS). These latter data are shown in Fig. 4, and they are specially relevant because of the high sensitivity of He scattering for surface defects. Starting from a level of diffracted intensity determined by the reflectivity of the soft and corrugated Pb layer, the fact that so little of it is lost during Co deposition indicates that a very small amount of new defects is introduced by its growth. It deserves being mentioned that the first intensity oscillation is absent, indicating that the growth of the first monolayer is still somewhat different from the rest (in agreement with the S-XRD results shown in Fig. 2), but after that, the growth proceeds in a layer-by-layer fashion with good structural quality. This overall picture

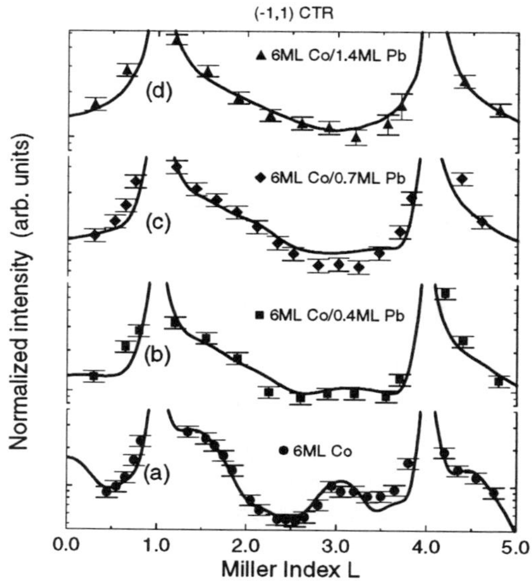

Figure 5: S-XRD characterization of the growth of Co on Cu(111). On a clean Cu(111) substrate, the Co film rapidly transforms to hcp, as revealed by the features developing at L=1.5 and L=3 (curve *a*); these features are reduced when submonolayer coverages of Pb are used (curves *b* and *c*) and absent when the whole substrate surface is covered by Pb (curve *d*).

also agrees with the STM data presented in Fig. 3. On the contrary, when no surfactant is used, the specularly reflected intensity rapidly falls down to zero indicating a steady accumulation of defects and the growth of a 3D, rough film. Besides the above mentioned effects, the most important consequence of using the surfactant is the *delay of the structural transformation* leading the Co to adopt the hcp structure. Fig. 5 shows S-XRD scans along the a Crystal Truncation Rod (CTR), the $(\bar{1},1)$ in surface notation, after growing 6 ML of Co on Cu(111) substrates prepared with different Pb coverages. These scans depict the dependence of the integrated diffracted intensity with perpendicular momentum transfer, and thus carry information on the stacking structure of the films studied. Together with the very intense peaks located at values 1 and 4 of the Miller index L (corresponding to the bulk fcc structure), the film grown without using Pb (curve labeled *a*) shows additional structures appearing at L=1.5 and L=3, corresponding to a developing hcp fraction. These structures are absent in the film grown on a substrate fully covered by Pb (curve *d*),

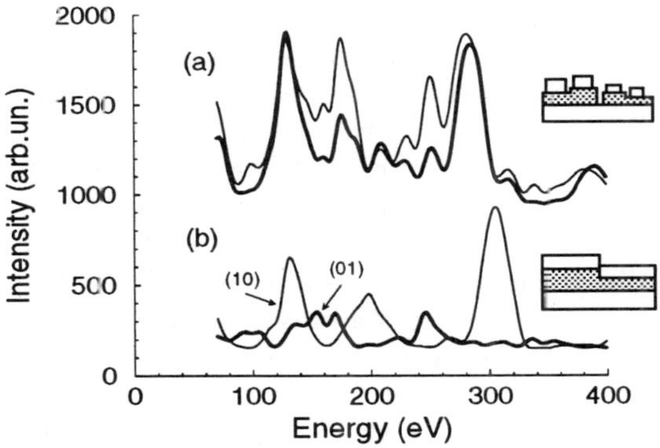

Figure 6: LEED I/V curves on {Co/Cu} bilayers: (a) Without Pb, the two sets of diffracted beams behave identically, indicating that the upper Cu layer is composed of fcc-twins. (b) Using Pb, the fcc crystal structure is maintained, and the two sets of diffracted beams, {10} and {01}, remain non-equivalent.

while they are still visible in those cases in which the Cu surface was partially covered with Pb. A thorough structural characterization of the Co films grown with the aid of Pb is presented elsewhere[25]; here we can summarize our results by saying that, up to the highest coverage that we have studied (about 6 ML, where the transition to hcp was almost completed when no Pb was used), the presence of Pb seems to preserve the fcc crystal structure of the Co film. The data shown in Fig. 5 further indicate that this is a local effect: in the patches not originally covered by Pb, the growth of Co proceeds following the normal pattern (curves *b* and *c*). This effect is of utmost importance, because islands of fcc Co can now match laterally with their neighbors and form continuous layers. This, together with the promotion of layer-by-layer growth mentioned above, is the origin of the higher structural quality of the films grown with the aid of the surfactant. Forcing Co layers to remain fcc is important not only for the improvement of its own structural quality, but also because they act as the substrate for the growth of Cu during the preparation of bilayers and superlattices. When Co abandons the three-fold fcc stacking sequence and adopts the two-fold hcp one, the transmission of structural information from the substrate to the surface of the growing film is interrupted. If a capping Cu layer is then grown on top of the Co, there will be more than a single choice of fcc stacking for the Cu. Specifically, terraces of hcp Co separated by a monoatomic step present specularly symmetric terminations (like, f.i., -AB and -BA). Cu islands nucleated on these two terraces will most likely choose C adsorption sites, thus acquiring immediately fcc stacking. In this way, these Cu islands will be fcc twins of each other, and again they will not be able to coalesce at higher coverages. The LEED intensity vs. energy (I/V) data displayed in Fig. 6a confirm this hypothesis: the two families of diffracted beams, labeled {1,0} and {0,1} should be inequivalent due to the three-fold symmetry of the fcc structure, but those measured on a {Co/Cu} bilayer grown without Pb behave identically. These curves result from the incoherent addition of intensities originating from twin crystallites of average size smaller than the instrument transfer width (typically, $\sim$ 100 Å). This problem adds to the already mentioned sources of roughness and imperfections in the Co films, and constitutes the likely explanation of the repeated experimental failures to observe the expected magnetic properties. In particular it seems plausible in the light of our results that the lack of antiferromagnetic (AFM) coupling between Co layers separated by thin Cu spacers is due to direct contacts through pinholes and structural imperfections, forcing ferromagnetic (FM) alignment of the magnetic layers through the whole sample. At larger Cu thicknesses the absence of AFM coupling is probably due to Néel's "orange peel" effect [28], originated by conformal roughness. On the other hand, samples grown with the aid of Pb are free of these types of defects: as we have seen, they consist of flat, continuous layers of fcc Co and Cu separated by sharp interfaces with a minimum amount of roughness. The LEED I/V curves of one such sample (a bilayer made of 3ML Cu and 2ML Co, grown with Pb) are presented in Fig. 6b: in this case, the two sets of diffracted beams are clearly different, as expected from crystalline layers of fcc structure [29]. We consider therefore that samples grown following this procedure constitute excellent candidates to show the desired magnetic behavior.

To ascertain this assumption, we have determined *in-situ* the magnetic character of our samples by means of Magneto-Optic Kerr Effect (MOKE) measurements. Our experimental setup allows us to record hysteresis curves for both longitudinal and polar Kerr effect, so we can probe the existence of in-plane and out-of-plane magnetization, respectively. We have found that Co films grown with the aid of Pb and sandwiched between 2 Cu layers exhibit an easy axis of magnetization perpendicular to the surface for thicknesses up to $\sim$ 4 ML; beyond this point, shape anisotropy dominates over interfacial anisotropy and the sample magnetization vector lies in the surface plane [27]. This is important because it allows us to freely choose the direction of the easy axis of magnetization simply by selecting the Co layer thickness. In practice, this is only possible thanks to the layer-by-layer mode of growth induced by the surfactant. We also observe out-of-plane magnetization in samples grown

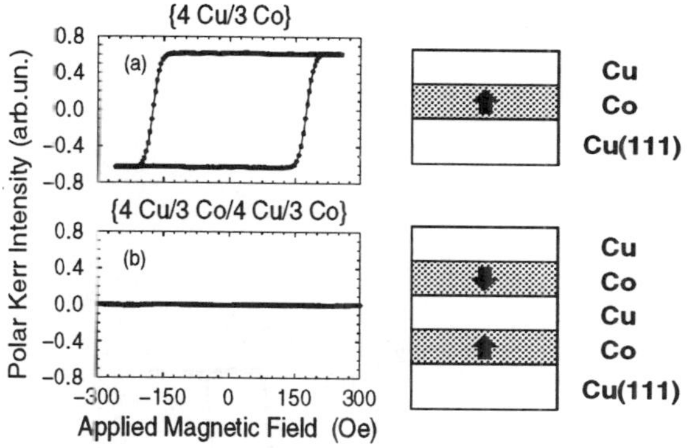

Figure 7: Polar MOKE measurements for one (a) and two (b) {Co/Cu} bilayers grown with Pb; in the second case, the hysteresis curve of the lower Co layer is exactly compensated by the opposite contribution of the upper one, resulting in a cancelation of the total magnetization.

without Pb, but only for a reduced range of Co coverages because of the large thickness inhomogeneities existing in those rough samples. Also, growing two bilayers of {Co/Cu} we can observe the effect of exchange coupling between the magnetic layers transmitted across the non-magnetic spacer. For the first time in {Co/Cu}(111) samples grown by MBE, we have been able to observe *complete* AFM coupling between consecutive magnetic layers. Fig. 7 shows an example of our results, presenting data obtained on a perpendicularly magnetized sample. The penetration depth of the laser light used in these measurements can be estimated to be several hundreds of Å's; thus, the MOKE intensity results from the addition of all atomic magnetic moments existing within this length. Panel (a) shows the polar Kerr effect hysteresis curve obtained after growing a single bilayer, composed of 3 ML Co and 4 ML Cu. When a second bilayer, equal to the first, is deposited on top of that, the total magnetization vanishes as a result of the combination of the two identical contributions with opposite orientation. The existence of AFM coupling can be observed in more detail with in-plane magnetization. The magnetic properties of the two Co layers can only be compared if they are exactly equivalent. In the previous example, this means that the upper one must also be covered with Cu, as the lower one is, so that both of them feel the same interfacial anisotropy and display perpendicular magnetization. For this reason, it is impossible to observe the evolution of the total sample magnetization for varying thickness of the upper Co layer in a single experiment. On the other hand, increasing

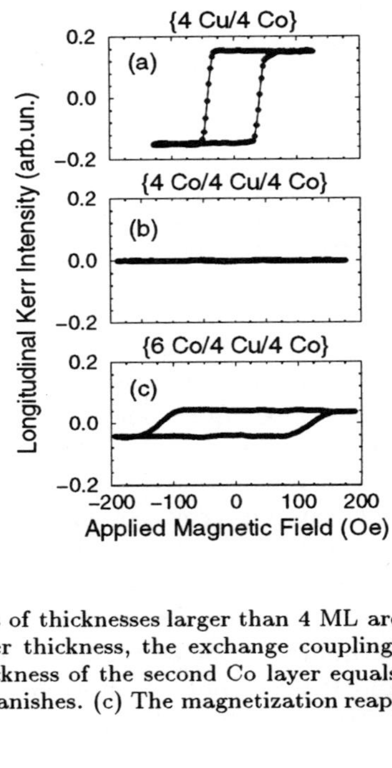

Figure 8: (a) Individual Co films of thicknesses larger than 4 ML are magnetized in-plane. (b) With an adequate Cu spacer thickness, the exchange coupling between Co layers is antiferromagnetic; when the thickness of the second Co layer equals that of the first one, the total sample magnetization vanishes. (c) The magnetization reappears when the second layer thickness exceeds the first.

the Co thickness beyond 4 ML forces the magnetization to be in the surface plane. In this case, the contribution of the uppermost capping Cu layer to the interfacial anisotropy is not relevant for the magnetic behavior of the sample, and after growing a first {Co/Cu} bilayer one can monitor the change of the total magnetization during the deposition of the second Co layer. The results of this experiment are summarized in Fig. 8. Starting with a single magnetic film (*a*) covered by a Cu layer of adequate thickness, we grow a second Co film and observe how the total magnetization gradually decreases and finally vanishes completely (as in the previous case) when both magnetic layers contain the same amount of material (*b*). If the deposition of Co is continued, the balance is broken and a net, non-zero magnetization reappears (*c*). This result demonstrates unambiguously that the thickness of the Cu spacer is homogeneous at least over a length scale of the size of the laser beam used (about 1 mm^2) and that the disappearance of the net magnetization is due to the exact cancelation of two opposed contributions.

In summary, the fine degree of control that we can attain by means of the surfactant allows us to choose independently all the magnetic properties of these heterostructures: the individual Co layer thickness determines the direction of the easy axis of magnetization while the Cu layer thickness controls the sign of the exchange coupling. This is therefore a clear example of how the surfactant method can be used to solve a specific growth problem and to prepare samples with custom-chosen properties. In fact, it has been recently shown that the properties of {Co/Cu} spin valves can be improved substantially by using Pb as surfactant [30], with a reported reduction by a factor of ten in the switching field (~10 Oe) with respect to spin valves grown without the surfactant, and a GMR at RT of 10%. Obviously, if the surfactant allows us to grow crystalline films layer-by-layer, with low roughness and well-defined interfaces, it is possible to think of other artificial structures with potentially interesting properties. The prototype of 1D systems is the growth of magnetic wires along the steps of a vicinal surface. The concept has been demonstrated by growing Co wires on step bunches of Cu(111) with a proper adjustment of evaporation rate and substrate temperature [31]. Fig. 9 shows an STM image of the wires produced in this way. There are however several problematic aspects of the "spontaneous" step decoration that we would like to control and modify: first of all, the irregular shape of the islands indicates that they are composed of some mixture of Co and Cu, as a result of rather strong interdiffusion even at RT. Intermixing seems to occur preferentially at the steps, and it also controls cluster size: when they become saturated with Co, their growth stops and islands start to appear across the terraces. Evidently, it would be desirable to be able to grow Co stripes parallel to the steps, with no width

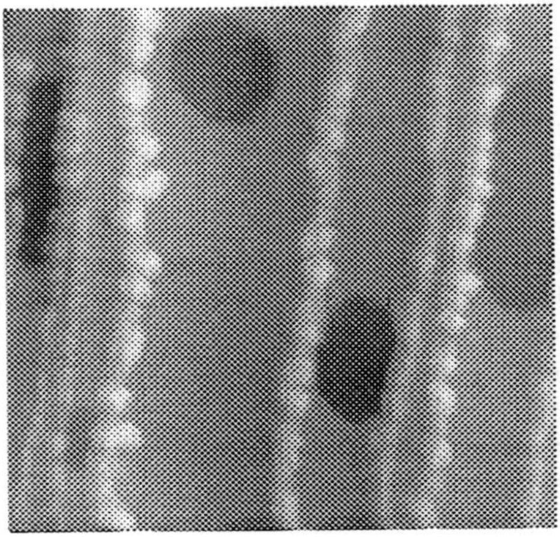

Figure 9: STM image showing arrays of clusters decorating atomic steps after Co deposition on clean Cu(111).

restrictions and without interdiffusion: the ideal step-flow growth mode. Also in this case Pb can help us obtain the desired results. As we have seen above, the presence of the surfactant prevents interdiffusion during the growth of 2D layers on nominally flat surfaces at RT. It should be possible therefore to use higher substrate temperatures during growth on a Pb-covered vicinal surface: the idea is to promote diffusion of the deposited Co adatoms on top of the Pb layer, so that they can reach the substrate steps and exchange position there instead of nucleating islands across the terrace. We expect this exchange to be favored at the substrate steps, because Pb atoms are less coordinated there and it should be easier for Co to displace them. On the other hand, the Pb layer is quite corrugated, so diffusion of Co on top of it must be very slow. This means that one should use slow deposition rates and substrate temperatures as high as possible. For this experiment we have used a 2.5°-miscut Cu(111) crystal. The best way to quantify the growth mode is to use a surface-sensitive diffraction technique such as TEAS. The evolution of the specularly diffracted intensity in real time during deposition can be directly translated to concentration of defects (e.g., island edges) on the surface. Thus, starting with a well-ordered layer, a decrease of intensity indicates that the formation of islands is increasing the surface roughness, while a constant level of diffracted intensity reveals that growth is proceeding by the step-flow mechanism. Fig. 10 summarizes our experimental results. In this series of Co depositions on the vicinal Cu surface covered with Pb, the level of diffracted intensity increases with increasing sub-

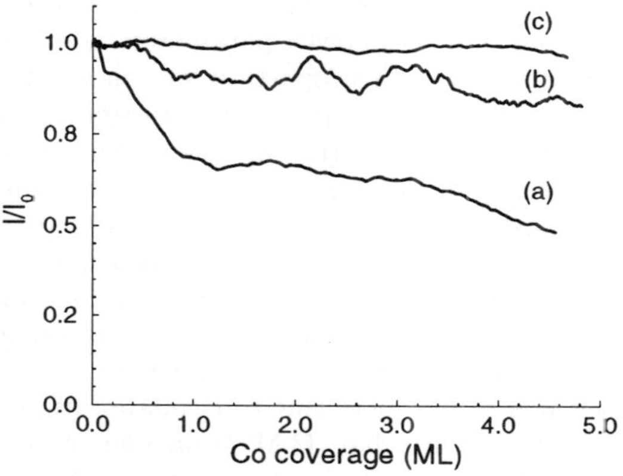

Figure 10: Deposition of Co on a Pb-covered, 2.5°-miscut Cu(111) surface at different temperatures. (a) 25°C; (b) 40°C; (c) 60°C. The presence of Pb suppresses interdiffusion of Co into the substrate; thus, it is possible to raise the deposition temperature to attain the step-flow growth mode.

strate temperature, as expected for a diffusion-driven growth process. In all cases the deposition rate was 200 sec/ML, and for a substrate temperature of 60°C the intensity remains constant indicating that no additional defects are appearing on the surface: the deposited material must therefore be sticking to the existing steps in such a way that the growing surface replicates itself. At the end of deposition the total Co coverage was cross-checked by AES, with no indications of interdiffusion or Cu segregation, and the Pb-$p(4 \times 4)$ superstructure was still visible. Furthermore, LEED I/V measurements (not shown) on these Co films several ML's thick indicate that they are purely fcc. This demonstrates that *Pb can effectively be used to induce step-flow growth*; with an adequate control of the amount of deposited material, it should be possible to produce highly-ordered arrays of quasi-1D magnetic wires. As a final conclusion, it seems clear that using surfactant agents to assist growth is a very promising method opening a rich variety of opportunities for the preparation of artificial materials with exotic magnetic properties.

In summary, we have demonstrated the application of surfactants to controlling epitaxial growth at the atomic scale. In the system {Co/Cu}, covering the Cu(111) substrate with a full ML of Pb prevents the formation of stacking-faults and fcc-twins, and delays the transition fcc-hcp of Co. It also induces good-quality layer-by-layer growth and reduces interdiffusion. Multilayers grown by this method have very little roughness and sharp interfaces. Our improved control of all the growth parameters allows us to prepare samples with predetermined magnetic properties: we can independently choose the direction of easy magnetization (in-plane or out-of-plane) and the sign of the exchange coupling (ferro- or antiferromagnetic). The surfactant can also be used to grow magnetic quantum wires by promoting step-flow.

Acknowledgments: We are pleased to thank all our collaborators during the realizations of these experiments. The S-XRD measurements were done at the ESRF with the help of J. Alvarez, X. Torrellas and S. Ferrer. The LEED I/V data were obtained and analyzed with the help of G. Kostka, S. Müller C. Rath and Prof. K. Heinz, from the Univ. of Erlangen-Nürnberg. The MOKE experiments were performed in collaboration with W. Kuch, M. Jarnikov, C. Teichert, A. Dittschar, C.M. Schneider, and Prof. J. Kirschner, from the MPI-Halle. The stepped Cu crystal was kindly provided be Dr. J.E. Ortega. This work has been supported in part by the CICyT through Grants PB93-0271 and PB94-1527. Financial support for scientific exchange from "Acciones Integradas" HA-95-38 and HA-95-39 is also acknowledged.

References

1. E. Bauer and J.H. van der Merwe, Phys. Rev. B, **33**, 3657 (1986).
2. U. Gradmann and J. Müller, Phys. Stat. Sol. **27**, 316 (1968).
3. J.J. de Miguel, A. Cebollada, J.M. Gallego, R. Miranda, C.M. Schneider, P. Schuster, and J. Kirschner, J. Magn. Magn. Mater., **91**, 1 (1991).
4. M.N. Baibich, J.M. Broto, A. Fert, F. Nguyen Van Dau, F. Petroff, P. Etienne, G. Creuzet, A. Friederich, and J. Chazelas, Phys. Rev. Lett. **61**, 2472 (1988).
5. S. Iwasaki and K. Takemura, IEEE Trans. Magn. **11**, 1173 (1975).
6. A. Barthelemy, A. Fert, R. Morel, and L. Steren, Physics World, Nov., 34 (1994).
7. D.E. Heim, IEEE Trans. Magn. **30**, 1453 (1994).
8. H. Kano, K. Kagawa, A. Suzuki, A. Okabe, K. Hayashi, and K. Aso, Appl. Phys. Lett. **63**, 2839 (1993).
9. N. Cabrera and D.A. Vermilyea in *"The Growth of Crystals from Solution"* Ed. D. Turnball (Wiley, New York, 1958) p. 393.
10. M. Copel, M.C. Reuter, E. Kaxiras, and R.M. Tromp, Phys. Rev. Lett. **63**, 632 (1989).
11. H.A. van der Vegt, H.M. van Pinxteren, M. Lohmeier, E. Vlieg, and J.M.C. Thornton, Phys. Rev. Lett. **68**, 3335 (1992).
12. S. Esch, M. Hohage, T. Michely, and G. Comsa, Phys. Rev. Lett. **72**, 518 (1994).

13. H.A. van der Vegt, J. Alvarez, X. Torrelles, S. Ferrer, and E. Vlieg, Phys. Rev. B **52**, 17443 (1995).
14. J.J. de Miguel, A. Cebollada, J.M. Gallego, S. Ferrer, R. Miranda, C.M. Schneider, P. Bressler, J. Garbe, K. Bethke, and J. Kirschner, Surf. Sci. **211/212**, 732 (1989).
15. C.M. Schneider, P. Bressler, P. Schuster, J. Kirschner, J.J. de Miguel, and R. Miranda, Phys. Rev. Lett. **64**, 1059 (1990).
16. Z.Q. Qiu, J. Pearson, and S.D. Bader, Phys. Rev. B **46**, 8659 (1993).
17. P. Bruno and C. Chappert, Phys. Rev. Lett. **66**, 2152 (1991).
18. S.S.P. Parkin, R. Bahdra, and K.P. Roche, Phys. Rev. Lett. **66**, 2152 (1991).
19. M.T. Johnson, R. Coehoorn, J.J. de Vries, N.W.E. McGee, J. aan de Stegge, and P.J.H. Bloemen, Phys. Rev. Lett. **69**, 969 (1992).
20. J. de la Figuera, J.E. Prieto, C. Ocal, and R. Miranda, Phys. Rev. B **47**, 13043 (1993).
21. J. Camarero, J. de la Figuera, J. Alvarez, L. Spendeler, X. Torrellas, S. Ferrer, J.J. de Miguel, and R. Miranda, in preparation.
22. J. de la Figuera, J.E. Prieto, C. Ocal, and R. Miranda, Surf. Sci., **307-309**, 538 (1994).
23. S. Müller, G. Kostka, T. Schäfer, J. de la Figuera, J.E. Prieto, C. Ocal, R. Miranda, K. Heinz, and K. Müller, Surf. Sci. **352-354**, 46 (1996).
24. M. Bott, T. Michely, and G. Comsa, Surf. Sci. **272**, 161 (1992).
25. L. Spendeler, J. Camarero, J. Alvarez, X. Torrellas, S. Ferrer, J.J. de Miguel, and R. Miranda, in preparation.
26. G. Meyer, M. Michailov, and M. Henzler, Surf. Sci. **202**, 125 (1988).
27. J. Camarero, T. Graf, J.J. de Miguel, R. Miranda, W. Kuch, M. Jarnikov, A. Dittschar, C.M. Schneider, and J. Kirschner, Phys. Rev. Lett. **76**, 4428 (1996).
28. L. Néel, Comp. Ren. Acad. Sci. (France) **255**, 1545 (1962); *ibid.* **255**, 1676 (1962).
29. J. Camarero, L. Spendeler, G. Schmidt, K. Heinz, J.J. de Miguel, and R. Miranda, Phys. Rev. Lett. **73**, 2448 (1994).
30. W.F. Egelhoff, Jr., P.J. Chen, C.J. Powell, M.D. Stiles, R.D. McMichael, C.-L. Lin, J.M. Sivertsen, J.H. Judy, K. Takano, and A.E. Berkowitz, J. Appl. Phys. **80**, 5183 (1996).
31. J. de la Figuera, M.A. Huerta-Garnica, J.E. Prieto, C. Ocal, and R. Miranda, Appl. Phys. Lett. **60**, 1006 (1995).

CLUSTER DIFFUSION, COALESCENCE, AND COARSENING IN METAL(100) HOMOEPITAXIAL SYSTEMS

P.A. Thiel and J.W. Evans
Departments of Chemistry and Mathematics, and Ames Laboratory
Iowa State University, Ames, Iowa 50011 USA

Abstract

We describe recent studies of the diffusion of large 2D metal homoepitaxial islands or clusters, of their coalescence or restructuring upon "collision", and of the role of these processes in the subsequent coarsening of adlayers formed by nucleation and growth of islands during deposition. These phenomena are illustrated with results from STM observations of the evolution of Ag/Ag(100) adlayers at room temperature.

1. Introduction

Only a few years ago, the first reports appeared concerning the surface diffusion of two-dimensional (2D), compact metal islands or clusters containing hundreds of atoms. Motion was observed on time scales of minutes to hours at room temperature.[1,2] These first studies were followed quickly by a report of the diffusion of metal vacancy clusters with analogous 2D (i.e., a single atom deep) compact structure, containing hundreds or thousands of vacancy sites.[3] At the time of these experimental observations, there had been some earlier theoretical work on the diffusion of such large clusters.[4-6] The theoretical work had received little attention because its. relevance to experimental phenomena had been unclear. However, the experimental results served to renew interest in this older work, and to foster the development of many new theoretical studies.[7-12] Interest also turned quickly to the role which cluster diffusion might have on the evolution of adlayer structure following deposition, both in theory and in experiment.[13,14] Indeed, all these observations of diffusion and coarsening were quite intriguing, because they suggested a degree of fluidity, with an appearance of large-scale concerted behavior, which had not been part of the usual mental image of metal surface dynamics. In this article, we survey the status of this rapidly-changing field, with emphasis on the results from our own laboratory, and, hence, on the homoepitaxial Ag(100) system.

The adlayers which serve as the starting points in these studies are formed by submonolayer nucleation and growth of islands during deposition, typically at around room temperature. Our focus is primarily on low coverage regime (below about 0.3ML), where the adlayer consists of a distribution of isolated, near-square 2D adatom islands or clusters. However, we shall also comment on the intermediate coverage regime characterized by a connected network of filled regions, as well as high coverage regime of isolated vacancy islands. In all cases, there is a finite characteristic length determined by the surface diffusivity and the details of the nucleation process, and which can be controlled by the deposition conditions (flux and temperature).[15,16] Since temperatures of interest are below critical value for 2D phase separation, the equilibrium state consists of a single large region of condensed island phase coexisting with a (very) dilute 2D gas phase. Thus the equilibration process will involve coarsening of the adlayer, which is associated with an increase in its characteristic length scale.[17]

The process of coarsening during phase separation has been subject of extensive experimental and theoretical studies for decades in a variety of two-phase systems including, e.g., binary alloys.[18] In general, partial concentrations play the role of coverage in surface systems. For the extremes of low and high partial concentration, the mechanism of coarsening is expected to be Ostwald Ripening which involves diffusive transfer of "material" from smaller to larger domains of the minority phase.[18] (The "material" is vacancies in surface systems at near-monolayer coverages.) This transfer is driven by the higher step edge free energy of the smaller domains. However, here we emphasize that traditional "generic" theoretical treatments[18] cannot predict the competition between different kinetic pathways, which in general determines the nature of coarsening of far-from-equilibrium states.[13] Indeed, experimental observations for Ag/Ag(100) adlayers,[13] and more recently for Cu/Cu(100) adlayers,[19] reveal that coarsening at low coverages is dominated by diffusion and coalescence of (large) islands or clusters, rather than by the anticipated Ostwald ripening. This is particularly surprising since the 2D clusters are strongly bound to the substrate. By way of contrast, coarsening dominated by cluster diffusion has been observed long ago in some heteroepitaxial systems with 3D clusters, where dislocations at the cluster-surface interface have been proposed to play an important role.[20]

Given the recent interest in the surface diffusion of large 2D clusters, in Sec.II, we first review relevant theoretical ideas and experimental observations for this phenomenon. Next, we consider the ramifications for coarsening. In Ag/Ag(100) adlayers at low coverages, we observe "fairly rapid" coalescence and restructuring of pairs of clusters following "collision" to form a single larger square cluster. Thus, in Sec.III, we briefly describe such coalescence phenomena. Finally, in Sec.IV, we present a "dynamic phase diagram" giving an overview of the coarsening process for Ag/Ag(100) adlayers. Some "tailored" studies of the coarsening kinetics at low coverages are also described.

2. Diffusion of Large 2D Clusters

ATOMIC MECHANISMS AND REGIMES OF CLUSTER DIFFUSION

The collective motion of large 2D clusters is achieved as a result of the hopping between adjacent sites of individual constituent atoms (or perhaps dimers, etc.). Here we describe various possible mechanisms and associated mass-transport regimes controlling cluster diffusion.[8-14] See Fig.1. Generally, all mechanisms will be active to some degree, and transitions could occur between regimes dominated by different mechanisms.

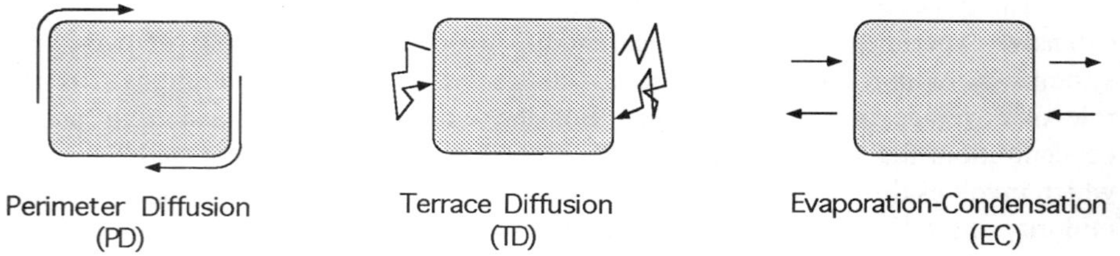

Figure 1: Schematic of various mass transport mechanisms leading to diffusion of large 2D clusters.

Perimeter Diffusion (PD)

PD involves hopping of perimeter atoms around the edge of the cluster, without detaching. There are a variety of atomically distinct PD processes: diffusion of single atoms along straight edges with no net change in bonding; escape from and attachment at kinks; etc..[1,5] A key distinguishing feature is that all PD processes conserve cluster size.

Terrace Diffusion (TD)

In the TD regime, cluster diffusion is limited by the terrace diffusion of single adatoms or vacancies which detach from and reattach at different points on the cluster perimeter. Cluster diffusion is not limited by the rates for detachment or reattachment. We distinguish between "interior" and "exterior" terrace diffusion. In the former case, advacancies (adatoms) diffuse across the interior of the adatom (vacancy) cluster.[3,9] In the latter case, adatoms (advacancies) diffuse exterior to the adatom (vacancy) cluster. For exterior diffusion, reattachment will likely occur to "close" to the detachment point if there is no extra barrier to reattachment (as is expected in metal homoepitaxy), at least for isolated clusters.[8] This is a consequence of the recurrence property of 2D random walks, as exemplified by Polya's theorem.[21]

This has been described as correlated evaporation-condensation.[8] For TD, one expects some degree of fluctuation in cluster size.

Evaporation-Condensation (EC)

In the EC regime, cluster diffusion is limited by "interface-transfer" of adatoms or vacancies detaching and reattaching at the cluster perimeter.[9] EC applies if there is a significant barrier to reattachment. EC via exterior terrace diffusion should also be more likely for a significant equilibrium density of the 2D dilute gas phase, or where adatom clusters are not completely isolated, enhancing flow of adatoms between other clusters and steps. For EC, one expects the most extreme fluctuations in cluster size to occur. One could also make the distinction between interior and exterior EC, as for TD.

Finally, for EC or TD, it is appropriate to identify the effective activation barrier for detachment. For a metal(100) system, one might imagine that this is the barrier to detach an isolated adatom from a straight (110) edge. However, one the population of such adatoms is low since they gain energy upon binding at kink sites. Thus, instead, one associates the effective barrier for EC with that for the detachment of doubly coordinated corner or kink atoms from the cluster.[1] This barrier equals the sum of the binding energy for such corner atoms and the terrace diffusion barrier.[1] This is consistent with the exact Ising model result for the density of the 2D dilute gas phase in equilibrium with the island phase.[22] For further discussion, see for example Ref.23.

THEORETICAL TREATMENTS OF CLUSTER DIFFUSION

Below, we let S denote number of atoms or vacancies in the cluster (i.e., its size), and R denote its linear dimension or "radius" (so $S \sim R^2$ for the compact clusters). We now briefly describe a number of treatments leading to predictions for the size-dependence of the cluster diffusion coefficient, $D \sim S^{-\alpha} \sim R^{-2\alpha}$, for large sizes. While exact analytic treatment is possible in principle, at least for PD, it is not practical for larger clusters.[24-26]

Classical Analyses

Following Binder *et al.*,[4] one adopts the heuristic formula, $D \sim H_{eff}(\delta d_{cm})^2$. Here $\delta d_{cm} = \delta d_a / S$ is the change in position of the center of mass (CM) of the entire cluster for each "atomic event", involving a displacement of an individual adatom or vacancy by δd_a. H_{eff} is the effective total rate of those events. For the different mechanisms above, one has

PD: $\delta d_a = 1$ and $H_{eff} \sim$ perimeter length $\sim \rho_e R$, so $D \sim R^{-3}$,
Interior TD: $\delta d_a = 1$ and $H_{eff} \sim$ area $\sim \rho_a S$, so $D \sim R^{-2}$,
Exterior TD: $\delta d_a \sim R^{1/2}$ and $H_{eff} \sim$ perimeter length $\sim R$, so $D \sim R^{-2}$,

EC: $\delta d_a \sim R$ and $H_{eff} \sim$ perimeter length $\sim R$, so $D \sim R^{-1}$.

Here ρ_e denotes the linear density of edge atoms active in PD, ρ_a denotes the areal density of diffusing adatoms or vacancies, and one has $\delta d_a \sim R^{1/2}$ (rather than $\delta d_a \sim R$) for exterior TD since the detaching adatom or vacancy tends to reattach a distance $O(R^{1/2})$ from the detachment point.[8] The limitations of these analyses are discussed below.

Continuum Langevin Analyses

This mesoscopic approach describes the cluster perimeter by a continuous curve, and views cluster diffusion as a result of fluctuations of this perimeter location.[12] The diffusion coefficient is determined by the decay rate of fundamental fluctuations, which in turn depends on the dominant pathway for mass transport (PD, TD, or EC). Perimeter evolution is described quantitatively by a Langevin equation, incorporating mesoscopic equilibrium parameters such as step stiffness, diffusivity, and step mobilities. Analysis reveals the same scaling of D with R as described above.[9] The precise regime of validity of this formalism is difficult to assess. Clearly R cannot be so small that lattice effects modify continuum predictions for evaporation rates, etc.. Also, the temperature cannot be so low that anisotropy in the mesoscopic step edge energy leads to faceted cluster shapes, or to specific atomic processes (e.g., at corners) controlling cluster diffusion.

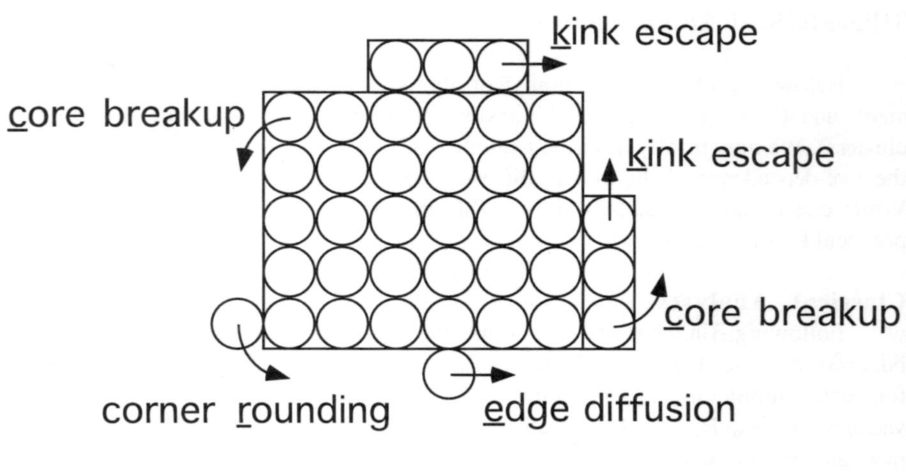

Figure 2: Schematic of various atomistic perimeter diffusion processes for metal(100) homoepitaxial clusters: edge diffusion at rate h_e; corner rounding at rate h_r; escape from kink sites at rate h_k; and "core breakup" at rate h_c, where $h_e >> h_r \approx h_k >> h_c$. The rectangular "core" of the cluster is also indicated.

Specialized Atomistic Models and Simulation Analyses

The classical analyses above are oversimplistic by comparison with atomic scale models of cluster diffusion. Pioneering studies by Voter[5] of PD-mediated cluster diffusion in metal(100) systems emphasized the variety of possible edge diffusion processes. These are illustrated in Fig.2, and include rapid diffusion of adatoms along straight edges, moderately fast kink escape and corner rounding, and slow "core breakup" (using the terminology of Ref.1). Note that the rapid and moderately fast periphery diffusion processes cannot alone lead to true diffusion, since the cluster is tethered by its rectangular core (see Fig.2). Thus core breakup is required for cluster diffusion, and is the rate determining step.[1,5] This was confirmed in simulations noting that the activation barrier for overall cluster diffusion corresponds to that for core breakup.[5] Our work emphasized an important consequence of these observations: while one might expect PD to be much more energetically favorable than TD or EC, in fact there may be little advantage since the effective barrier for evaporation (described above) is likely only slightly higher than the substantial barrier for core breakup.[1] See Ref. 5.

The scaling of D with S has often been observed to deviate from classical predictions. For PD-mediated cluster diffusion, Voter found $D \sim S^{-1.75 \pm 0.1}$, for $S \leq 100$, in a model for metal(100) homoepitaxy with roughly square clusters.[5] Large deviations found by Soler for EC and TD-mediated cluster diffusion on a triangular lattice were attributed to step edge energy effects.[7] Deviations were also found in simulations of EC- and PD-mediated Xe-cluster diffusion on Pt(111),[10] where it was emphasized that the effective number of active edge atoms scales differently than $S^{0.5}$. A detailed simulation study of size scaling in PD-mediated cluster diffusion on metal(111) surfaces found that the scaling exponent generally depends on cluster size and temperature.[12] Molecular dynamics simulations for metal(111) systems[27] reveal the possibility of dislocation mechanisms for smaller clusters. In no case is there a definitive understanding at present!

Next, we remark on an extreme situation of PD-mediated diffusion of near-square faceted clusters in metal(100) systems at low T,[11] noting that the barrier for core breakup should be much larger than for kink escape, etc.. Thus, after each core breakup event, atoms remaining on that initially complete edge of the cluster can "quickly" peel off via kink escape, and potentially be reincorporated on another side of the cluster to complete and "lock in" that edge. For such an event $\delta d_{cm}=1$. However, one must account for the possibility of reincorporation on the same edge, for which $\delta d_{cm}=0$. A simplified analysis[11] suggests that the probability, P, of successful completion of a new edge decreases like $P \sim R^{-1}$, so that $D \sim H_{eff} \sim P \sim R^{-1}$.

Finally, we note that in general the motion of the center of mass of a cluster does not constitute a pure random walk, but rather a walk with "backward correlation".[1,6] This is a consequence of the connectivity constraint on atoms in the cluster,[1,6] and can be demonstrated explicitly for small clusters.[26]

EXPERIMENTAL MEASUREMENT OF CLUSTER DIFFUSION

In our studies for the Ag/Ag(100) systems,[1] we examine only configurations with a low density of well separated islands to avoid any potential interference effects. The position of the cluster center of mass is measured relative to an impurity, or impurity-pinned step edge. This position, R_n, is monitored at times t_n, n=1,2,.., where $t_{n+1}-t_n \approx \tau$=10-15 minutes, thus mapping out the cluster's trajectory typically for several hours. To extract the maximum information from this data, we average values of $|R_{n+1}-R_n|^2/(t_{n+1}-t_n)$, $|R_{n+2}-R_n|^2/(t_{n+2}-t_n)$, etc.. We thus obtain $D(\delta t)=\langle (d_{cm})^2/(4\delta t)\rangle$ versus δt, where d_{cm} is the displacement of the cluster CM. This quantity tends to plateau for large δt, and one obtains $D=\lim_{\delta t\to\infty} D(\delta t)$. Next we comment on two practical issues. Firstly, we typically have 30-50 independent measurements for $\delta t=\tau$, effectively less for $\delta t=2\tau$, etc., and these values become strongly correlated for progressively larger δt. This produces a smooth variation of $D(\delta t)$ with δt, and makes estimates of uncertainties non-trivial. Secondly, there is a significant, but random uncertainty in the measurement of each R_n, of magitude E, say. Thus, one can show that the estimated (est) and true (true) values of $D(\delta t)$ are related by $D(\delta t)_{est} = E/(2\delta t) + D(\delta t)_{true}$. This decrease in $D(\delta t)_{est}$ due to E>0 gives the appearance of a back correlation, and we expect that it will dominate the decrease due to any true back correlation on the cluster's walk.[28] In other words, we attribute our previous reports of an apparent back correlation in cluster diffusion (on the time scale of tens of minutes) to experimental uncertainty in measuring the cluster's position.

Another approach suited to higher island densities is to monitor the relative separation of pairs of clusters. This has been utilized more often than the first.[3,14,19] Here one exploits the feature that for two clusters with diffusion coefficients D_1 and D_2, the relative separation will evolve according to diffusion with coefficient D_1+D_2. Using this method, one cannot directly extract D for a single size other than by choosing equally sized pairs. Interference effects are typically not apparent, and are assumed negligible.

Observations for Ag/Ag(100)

Fig.3 shows positions of a few islands after deposition, and about 6 hours later. Fig.4a shows a typical trajectory of the CM of a cluster over several hours. The corresponding $D(\delta t)$ versus δt is shown in Fig.4b. Similar studies of the motion of 12 clusters produces the behavior for D versus S from about 100 to 700 atoms, shown in Fig.4c. While the overall variation with S appears weak, "tailored" studies of adlayer coarsening described below,[28] as well as more comprehensive studies by Wendelken *et al.*[19], indicate a strong decrease of D with S for the regime of smaller sizes up to about 250 atoms. Perhaps there is a transition from PD, for small S, to EC-dominated diffusion, for large S, although we caution that the data for larger S is very limited. Although the clusters chosen for analysis in Fig.4c have nearly constant size, in general there are significant size fluctuations, so there clearly is some level of mass transport between the diffusing cluster and other clusters and steps. For large enough sizes, EC might dominate contributions from other mechanisms due to weaker decrease with S (but cf. Ref.9)

We performed analogous studies for the diffusion of vacancy clusters in Ag/Ag(100) adlayers with sizes ranging from about 500 to 4000 atoms.[29] Behavior observed for the trajectory of the center of mass, and for the variation of D(δt) with δt, was qualitatively similar to that for adatom clusters. Large uncertainties in data for D versus S make it difficult to assess a trend; however, for a given size, the value of D appears lower than that for adatom clusters by a factor of about two or three. Slow variations in cluster size were observed consistent with Ostwald ripening (see below).

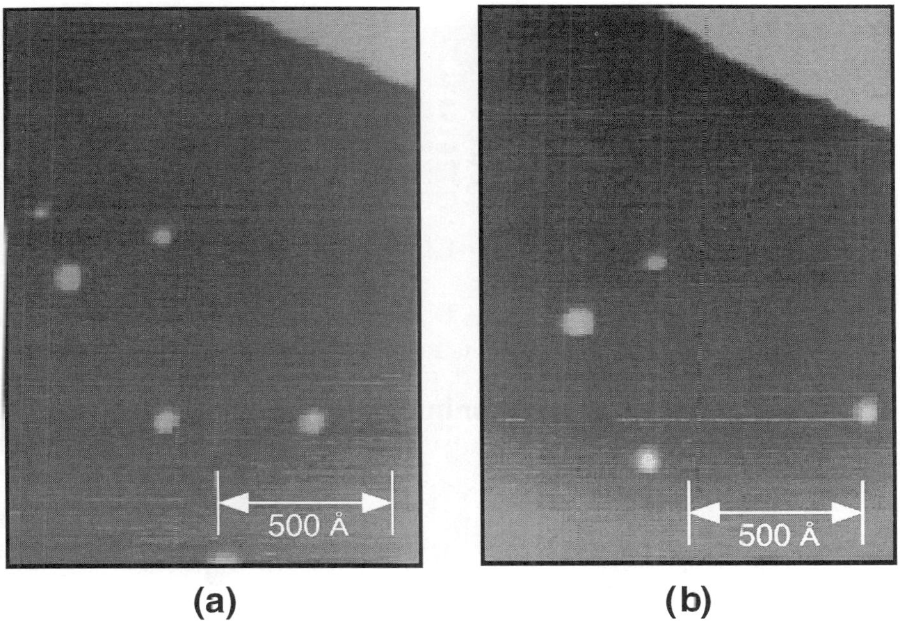

Figure 3: STM image of a few islands: (a) just after deposition; and (b) 5.7 hours later. Note the change in relative positions, and the disappearance of one island due to OR.

Other Systems

For contrast, we note that studies of the diffusion of Ag/Ag(111) vacancy clusters[3] find that $D \sim S^{-1.0 \pm 0.2}$, for 500<S<10000. This was first interpreted as TD-mediated diffusion associated with diffusion across the cluster interior of adatoms [which are much more mobile than vacancies in Ag/Ag(111)]. However, recent estimates of evaporation rates, supplemented by simulations with realistic parameters for this system, suggest cluster diffusion might actually be PD-mediated,[14] emphasizing the point that deriving mechanisms from simple scaling-law predictions can be dangerous. Ag/Ag(111) adatom clusters also diffuse, but they decay too quickly for reliable estimates of diffusion coefficients.[14] We also mention recent studies of Cu/Cu(100) adatom cluster diffusion which find $D \sim S^{-1.5}$, for $S<250$, thus suggesting PD-mediated diffusion.[19] Finally, we note an earlier study of vacancy cluster diffusion in the Cu/Cu(111) system, but here diffusion is induced only by adding submonolayer amounts of Co.[2] Rather than Co acting as an immobile pinning impurity, it tends to undergo place-exchange with the Cu substrate. This was proposed to create a strained 2D alloy, which has the effect of lowering certain activation barriers controlling diffusion.

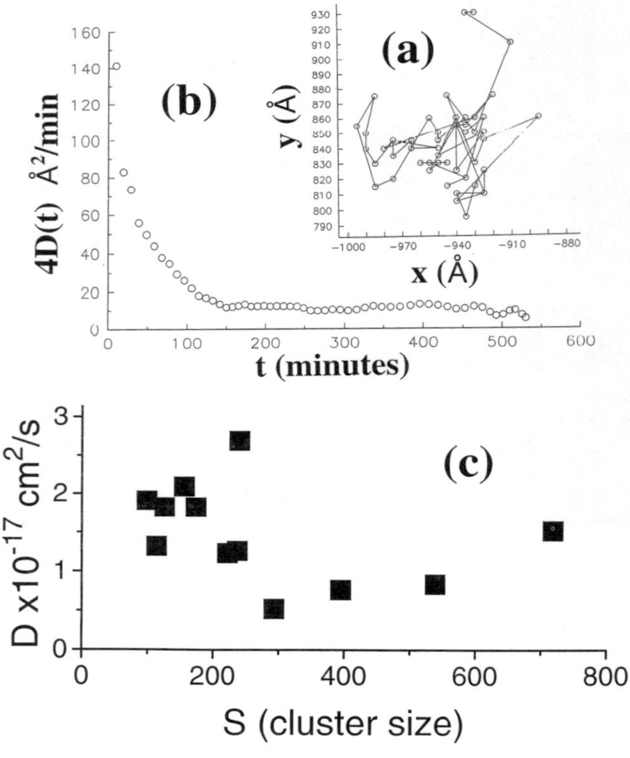

Figure 4: (a) Trajectory of a cluster of about 300 atoms followed for about 9 hours. (b) Behavior of D(t) versus t for this trajectory. (c) Estimated values of diffusion coefficients, D, for twelve clusters with sizes from 100 to 720 atoms.

3. Cluster-Cluster Coalescence

As mentioned in the Introduction, the overall coarsening process for Ag/Ag(100) adlayers at low coverages around 295K involves "fairly rapid" coalescence and restructuring of pairs of clusters after their diffusion-mediated collision. STM studies reveal the following specific events.[29] Near-square clusters can meet "side on" after which they "quickly" restructure to form a rectangular shape (Fig.5). Significant shape changes of this type occur on a time scale of 5-10 minutes, followed by substantially slower restructuring to achieve a near-square equilibrium shape. Alternatively, clusters can meet "corner-to-corner" creating an overlapping square configuration, which then restructures slowly on a time scale of several tens of minutes to form a convex shape (Fig.6), and ultimately a square shape. It has already been noted that such restructuring, between say 0.3ML and 0.5ML, can have a significant affect on spatial correlations, and thus on surface sensitive diffraction profiles, for the adlayer.[30] Similar observations were made for the restructuring of colliding vacancy clusters in Ag/Ag(100) adlayers,[29] as well as in Co-modified Cu/Cu(111) adlayers.[2]

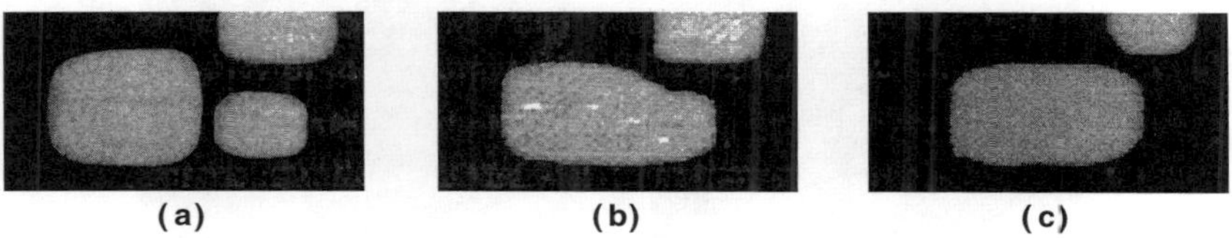

Figure 5: Side-on collision and subsequent coalescence of two unequal sized clusters. Configurations shown are at: (a) t=0 min. (before collision); (b) t=191 min. (<10 min. after collision); (c) 214 min. All images have a size of 25nm x 50nm.

Observation of step edge evolution naturally leads to consideration of the underlying mass transport mechanisms. Mass transport at step edges has been studied extensively through the analysis of the corresponding fluctuations in step edges about their equilibrium configuration. Langevin-type equations (cf. Sec.II) are utilized to determine, e.g., effective step mobilities, or mass transport parameters.[31] In contrast, we emphasize that the above coalescence processes create various "geometric" far-from-equilibrium step edge configurations. The relaxation of these can potentially provide much more information about, e.g., various specific edge diffusion processes, than the study of near-equilibrium fluctuations. By analogy, the analysis of far-from-equilibrium island nucleation processes provides more insight into terrace diffusion, bond breaking, etc., than can be obtained from a study of near-equilibrium coarsening.

In future work, we shall present a quantitative analysis of the relaxation of a variety of far-from-equilibrium step edge nanostructures, which are specially created to emphasize different step orientations, etc.. Observed evolution is compared against predictions of atomistic models with various activation barriers for core breakup, kink escape, etc. (see Sec.II), which also play a crucial role in PD-mediated cluster diffusion.

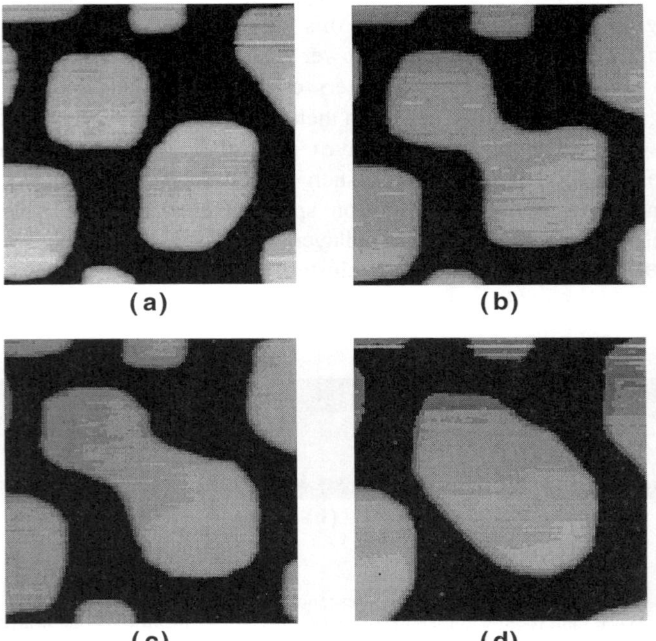

Figure 6: Corner-to-corner collision and subsequent coalescence of two clusters. Configurations shown are at: (a) t=0 min. (before collision); (b) t=46 min. (<10 min. after collison; (c) t=91 min.; (d) t=271 min.. All images have a size of 50nm x 50nm.

4. Coarsening of the Adlayer

Here, we consider the time-evolution following deposition of such basic quantities as the mean island density, N_{av}, and the mean separation between island centers, $L_{av}=N_{av}^{-1/2}$. These quantities are simply related to the mean linear cluster size, R_{av}, or the mean number of atoms per cluster, $S_{av}\propto R_{av}^{2}$. More complete information is provided by the densities, N_S, of islands of specific sizes, S, where $N_{av}=\Sigma_S N_S$. If we define $S_{av}=\Sigma_S SN_S/\Sigma_S N_S=\theta/N_{av}$, then one writes $N_S \sim \theta S_{av}^{-2} f(S/S_{av})$, for large S_{av}, where the shape 'f' of the normalized island size distribution satisfies the normalization conditions $\int_0^\infty dx\, f(x) = \int_0^\infty dx\, xf(x) = 1$. We note that the definition of S_{av} and R_{av} is not unique; also the size distribution is often presented as a function of "radius", R, making comparison non-trivial.[32] Finally, we note that coarsening leads to evolution of the shape 'f' towards some characteristic asymptotic form, dependent on the mechanism of coarsening.

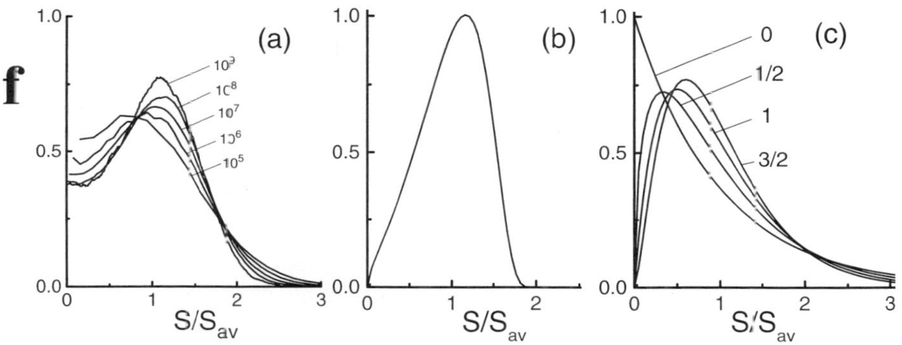

Figure 7: Shape of the normalized island size distribution, f(x), where $x=S/S_{av}$, at low coverages: (a) immediately following deposition for irreversible island formation, for various h/F indicated; (b) for long times after coarsening due to Ostwald ripening; (c) for long times after coarsening due to cluster diffusion and coalescence, where $D \sim S^{-\alpha}$ for various α indicated.

The initial distribution of islands is determined by the process of nucleation and growth during deposition.[15,16] A detailed characterization is available for the irreversible formation of square islands.[33] Here one has $N_{av} \sim (h/F)^{-1/3}$, for large h/F, where F is the deposition flux, h is the hop rate for adatoms diffusing across the terrace, with a weak θ-dependence up to the onset of island coalescence. The island size distribution for lower precoalescence coverages has a characteristic monomodal shape shown in Fig.7a. An analytic expression was recently obtained for the function f(x).[34] Here we just note that f(x) assumes a value of about 0.4, as $x \to 0$, rising to a maximum of about 0.8, around x=1. We caution that significant mobility of dimers and other small clusters on the time scale of aggregation modifies this form, e.g., lowering f(0).[35] See also below. Another key feature of the nucleation process is that the islands are not randomly distributed, there being a depletion in the population of nearby pairs of islands.[15,16] This feature delays the onset of island coalescence, and also the percolation threshold which occurs around

0.75-0.8ML, for large h/F (cf. around 0.7ML for random island distributions).[16,33] Irreversible island formation applies to Ag/Ag(100) homoepitaxy below 320K, where we find for room temperature deposition that $N_{av} \approx \kappa F^{0.31}$ (per site) with $\kappa=(2.0\pm0.2)\times10^{-3}$. See Ref.36.

THEORIES OF COARSENING

We consider two possible coarsening mechanisms, restricting discussion to 2D islands. In both cases, the average linear size of clusters is expected to scale like $R_{av}(t) \sim R_0(1+ t/\tau)^{\beta}$, so the average cluster size satisfies $S_{av}(t) \sim S_0(1+ t/\tau)^{2\beta}$, and the average cluster density satisfies $N_{av} \sim N_0(1+ t/\tau)^{-2\beta}$. Here τ represents a characteristic time. The two mechanisms predict different forms of β and τ, a point which will be emphasized below.

Ostwald Ripening

In Ostwald ripening, transfer of atoms from islands smaller than some "critical size", $R<R_c$, to larger islands, $R>R_c$, is driven by a gradient in adatom density. This is in turn due to higher chemical potential at step edges of smaller, more curved clusters. One has the classic result $\beta=1/3$, and explicit expressions for τ are available.[17] If one sets $\rho=R/R_c \propto x^{1/2}$ then the asymptotic form of island size distribution satisfies (cf. Ref.17)

$f(x)=0.2325\rho(1+ \rho/3)^{-7/3}(1- 2\rho/3)^{-11/3}\exp[-\rho/(3/2 -\rho)]$, for $\rho<3/2$, and $f(x)=0$, for $\rho>3/2$.

See Fig.7b. Normalization determines the numerical prefactor, as well as the relation $\rho=1.023(R/R_{av} = x^{1/2})$. Finally, we note that recent more sophisticated treatments of coarsening have accounted for subtle correlations which develop in the island distribution.[37]

Cluster Diffusion and Coalescence

Analysis of this coarsening mechanism can be based on the classic Smoluchowski theory of coagulation kinetics. A basic result from a mean-field rate equation analysis[38,39] is that $\beta=1/(2+2\alpha)$. The characteristic time τ scales like $(a^2N_0)^{-1-\alpha}(a^{-2}D_0)^{-1}$, for surface lattice constant 'a'.[28] The classic rate equations do not account for finite extent of islands, so some modification is required to obtain the correct θ-dependence of coarsening.[13,38] Simple theories ignoring finite island extent, but including a scaling of $D \sim S^{-\alpha}$, predict that the asymptotic form of island size distribution satisfies[40]

$f(x)=A(\alpha)x^{\alpha}\exp[-C(\alpha)x]$, where $C(\alpha)=\alpha+1$, and $A(\alpha)=(\alpha+1)^{\alpha+1}/\Gamma(\alpha+1)$.

See Fig.7c. The values of $A(\alpha)$ and $C(\alpha)$ follow from normalization. Some modification to this form of f(x) is apparent from simulations for realistic islands at finite coverage.[38]

From the above, it is clear that both coarsening kinetics, and evolution of the shape of the island size distribution, provide information on the S-dependence of D. We noted above that, e.g., dimer mobility on a short time scale can modify the initial form of f(x). For cluster diffusion on a slower time-scale, even a decrease of D with S only for smaller clusters (so effectively $\alpha>0$) will have the effect of "quickly" lowering f(0) towards zero, perhaps before the first STM image can be taken. Finally, we note a complication due to the initial correlations and, specifically, the "separation" of island pairs. This feature is not included in the above mean-field rate equation theory, so one might expect an "additional" transient initial period of little change in N_{av}.[19,41]

DYNAMIC PHASE DIAGRAM FOR Ag/Ag(100)

Next we provide an overview of coarsening in the submonolayer Ag/Ag(100) system at room temperature,[1] which is summarized by the STM images shown in Fig.8. For both high and low θ, we exploit the STM technique to monitor the actual events leading to a decrease in the number of clusters. For the regime of isolated adatom clusters at "low" θ below about 0.6ML, one finds that coarsening is typically dominated by cluster diffusion and coalescence. In contrast, for the regime of isolated vacancy clusters at high θ above about 0.85ML, coarsening occurs almost exclusively due to Ostwald ripening. There is an "intermediate" θ-regime between about 0.65 and 0.8ML characterized by a ramified, interconnected network of filled regions. Here an appropriate measure of the characteristic length might be the mean chord length or terrace width (in some specified direction), which clearly increases with time, while the overall pattern remains self-similar. In this regime, one might anticipate that perimeter diffusion or "edge running" along the extensive network of step edges could dominate the coarsening process (although we do not have any explicit evidence that this is the dominant pathway).

It is instructive to compare these observations with a previous He-beam scattering study of coarsening in Cu/Cu(100) adlayers at 0.5ML where edge running was inferred to be the dominant mechanism.[42] In this study, low temperature deposition was used to create a quasi-random initial adlayer. At 0.5ML, this initial quasi-random state is much closer to the "random percolation" threshold of about 0.6ML, compared to an adlayer with large islands created by deposition of 0.5ML at, say, room temperature. We also note that recent STM studies of coarsening in Cu/Cu(100) at lower coverages demonstrate that coarsening is dominated by diffusion and coalescence of large clusters,[19] just as for Ag/Ag(100). Finally, we note that coarsening behavior for Ag/Ag(111) seems "opposite" to Ag/Ag(100), in that Ostwald ripening of adatom islands presumably dominates at low θ, and diffusion and coalescence of vacancy islands dominates at high θ.[3,14]

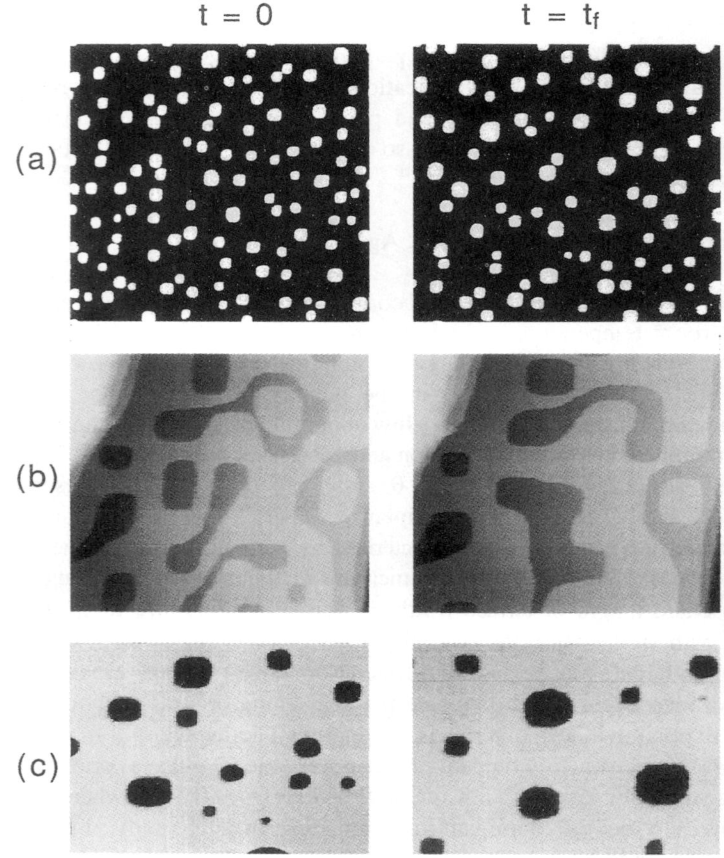

Figure 8: STM images following the coarsening of Ag/Ag(100) adlayers at room temperature. The left frames show the starting point, and the right frames show the adlayer several hours later. The full horizontal image width is 150 nm. Bright areas are the deposited film, one layer deep, and the dark areas show the substrate. Conditions are: (a) 0.11 ML, $N_{av}(t{=}0){=}4.9\times10^{-3}$ nm^{-2}, t_f =520min. (b) 0.69 ML, t_f =400 min.; and (c) 0.87 ML, vacancy cluster $N_{av}(t{=}0){=}1.5\times10^{-3}$ nm^{-2}, t_f =390 min..

TAILORED STUDIES OF COARSENING FOR Ag/Ag(100)

We now describe preliminary data from two comparative studies of coarsening "tailored" to elucidate specific aspects of the kinetics, or of the underlying cluster diffusion dynamics.[28] Since a linear increase of $1/N_{av}$ with time is predicted for size-independent cluster diffusivity (α=0 and β=1), we plot the data shown below in this form.

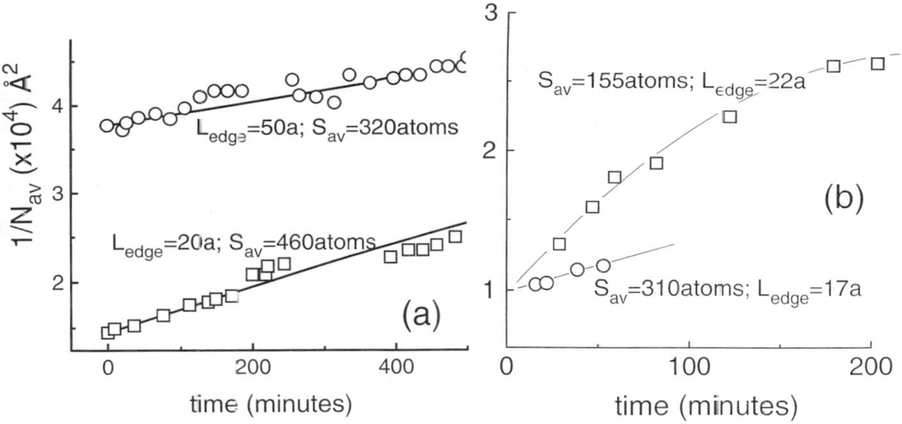

Figure 9: Tailored studies of the coarsening kinetics following $1/N_{av}$ versus t: (a) varying the mean edge-to-edge island separation, at roughly fixed mean island size; (b) varying the mean island size, at roughly fixed mean edge-to-edge separation.

For a fixed initial mean cluster size, S_{av}, coarsening should be inhibited with an increase in the mean separation between clusters, L_{av}. Such an increase in L_{av} produces an increase in the mean separation between cluster edges, $L_{edge} \approx (1-\theta^{1/2})L_{av}$, which controls the rate at which clusters coalesce. By simultaneously decreasing the deposition flux, F, and decreasing the deposited coverage, θ, one can create island distributions maintaining a constant $S_{av}=\theta/N_{av} \sim \theta F^{-1/3}$, while at the same time increasing $L_{edge} \sim (1-\theta^{1/2})F^{-1/6}$. In this way, one can test these ideas experimentally. The results of such a comparison, shown in Fig.9a, reveal the anticipated decrease in the rate of coarsening (shown by the smaller slope of the upper curve) with a significant increase in L_{edge}. This behavior is reproduced by rate equation modeling if one first modifies the standard Smoluchowski equations to account for the finite extent of the islands.[13] Specifically, the rate of coalescence is increased since the cluster center of mass needs only to travel a distance $\sim L_{edge}$, rather than $\sim L_{av}$, before coalescence. The solid curves in Fig.8a are the results of such rate equation analyses[13] using the average value of $D=2x10^{-17}cm^2/s$ for the cluster diffusion coefficient. (Note that t=0 in these plots is about _ hour after deposition.)

For fixed initial mean edge-to-edge separation of clusters, L_{edge}, the rate of coarsening should be controlled by the cluster diffusion coefficient, $D(S \approx S_{av})$, for cluster sizes $S \approx S_{av}$. Thus one can investigate the size dependence of the cluster diffusion coefficient by creating island distributions with fixed L_{edge} and increasing S_{av}. This is done by slightly decreasing F (and thus slightly increasing L_{av}), while significantly increasing the deposited coverage, θ. Results of such a comparison are shown in Fig.9b for initial mean size S_{av} equal to 155 and 310 atoms. Coarsening occurs much more quickly for the smaller mean size, even though the initial L_{edge} is slightly larger. This can only be due to a larger cluster diffusion coefficient, by a factor of roughly three, for "small" sizes around 150 atoms compared to larger sizes around 300 atoms. This is consistent with the recent observations by Wendelken *et al.*[19] of a strongly size-dependent D in the range of cluster sizes below about 300 atoms. Notice also that the only case where there are noticeable deviations from a linear variation of $1/N_{av}$ with time (consistent with $\beta<1$ and $\alpha>0$) is for the case of the smallest initial $S_{av}=155$ atoms.

5. Summary

In this article, we have discussed recent experimental observations and theoretical investigations of the diffusion of large 2D metal homoepitaxial clusters, a topic in which understanding is still limited, and subject to rapid change. We have also described the dominance of diffusion and coalescence of large 2D adatom clusters in the coarsening of Ag/Ag(100) adlayers. This dominance was not anticipated since, although coarsening during phase separation has been studied for decades in various two-phase systems, such generic studies cannot predict the competition between pathways to coarsening in a specific system or class of systems. We have noted, however, that the more "traditional" distinct mechanism of Ostwald ripening can dominate coarsening under different conditions, e.g., in Ag/Ag(100) adlayers at high θ, or in Ag/Ag(111) adlayers at low θ.

ACKNOWLEDGEMENTS: This work was supported by NSF Grant CHE-9700592. It was performed at Ames Laboratory, which is operated for the US Department of Energy by Iowa State University under Contract No. W-7405-Eng-82.

References

1. J.-M. Wen, S.-L. Chang, J.W. Burnett, J.W. Evans, and P.A. Thiel, Phys. Rev. Lett. **73**, 2591 (1994); J.-M. Wen, J.W. Evans, S.-L. Chang, J.W. Burnett, and P.A. Thiel, MRS Proc. **355**, 15 (1995).
2. J. de la Figuera, J.E. Pristo, O. Ocal, and R. Miranda, Solid State Commun. **89**, 815 (1994).
3. K. Morgenstern, G. Rosenfeld, B. Poelsma, and G. Comsa, Phys. Rev. Lett. **74**, 2058 (1995).
4. K.Binder and M.H. Kalos, J. Stat. Phys. **22**, 363 (1980); M. Rao, M.H. Kalos, J.L. Lebowitz, and J. Marro, Phys. Rev. B **13**, 4328 (1976).
5. A.F. Voter, SPIE **821**, 214 (1986); Phys. Rev. B **34**, 6819 (1986).
6. H.C. Kang, P.A. Thiel, and J.W. Evans, J. Chem. Phys. **93**, 9018 (1990); in "Structure of Surfaces III", edited by S.Y. Tong et al. (Springer, Berlin, 1991).
7. J.M. Soler, Phys. Rev. B **50**, 5578 (1994); **53**, R10540 (1996).
8. C. DeW. van Siclen, Phys. Rev. Lett. **75**, 1574 (1995).
9. S.V. Khare, N.C. Bartelt, and T.L. Einstein, Phys. Rev. Lett. **75**, 2148 (1995); S.V. Khare and T.L. Einstein, Phys. Rev. B **54**, 11752 (1996). For significant edge hopping, a transition is predicted from EC- to PD- to TD-dominated diffusion, for increasing size, S. However, lattice effects could modify small-S behavior, and the presence of other clusters and steps could modify large-S behavior.
10. D.S. Sholl and R.T. Skodje, Phys. Rev. Lett. **75**, 3158 (1995).
11. J.W. Evans, P.A. Thiel, and R. Wang, Phys. Rev. B (1997).
12. A. Bogicevic, S. Liu, J. Jacobsen, B. Lundqvist, and H. Metiu, , Phys. Rev. B **57**, R9459 (1998).
13. J-M. Wen, J.W. Evans, M.C. Bartelt, J.W. Burnett, and P.A. Thiel, Phys. Rev. Lett. **76**, 652 (1996).
14. K. Morgenstern, G. Rosenfeld, and G. Comsa, Phys. Rev. Lett. **76**, 2113 (1996); G. Rosenfeld, K. Morgenstern, and G. Comsa, in "Surface Diffusion: Atomistic and Collective Processes", M.C. Tringides, M. Scheffler, ed.s, NATO ASI Series (Plenum, New York, 1997).
15. S. Stoyanov and K. Kashchiev, Curr. Topics Mater. Sci. **7**, 69 (1981); J.A. Venables, Philos. Mag. **27**, 697 (1973).
16. J.W. Evans and M.C. Bartelt, in this Volume.
17. M. Zinke-Allmang, L.C. Feldman, and M.H. Grabow, Surf. Sci. Rep. **16**, 377 (1992).
18. J.D. Gunton and M. Droz, "Introduction to the Theory of Metastable and Unstable States", Springer Lecture Notes in Physics, Vol. 183 (Springer, Berlin, 1983)
19. Woei Wu Pai, A. K. Swan, Zhenyu Zhang, and J. F. Wendelken, Phys. Rev. Lett. **79**, 3210 (1997).

20. J.J. Metois, J.C. Zanghi, R. Erre, and R. Kern, Thin Solid Films **22**, 331 (1974).
21. B.D. Hughes, "Random Walks and Random Environments" Vol. 1 (Oxford University Press, Oxford, 1995).
22. R.J. Baxter, "Exactly Solved Models in Statistical Mechanics" (Academic Press, London, 1982).
23. H. Shao, P.C. Weakliem, and H. Metiu, Phys. Rev. B **53**, 16041 (1996).
24. U.M. Titulaer and J.M. Deutch, J. Chem. Phys. **77**, 472 (1982).
25. J.R. Sanchez and J.W. Evans, to be published.
26. For PD, diffusion can be described exactly by master equations with finite dimension equal to the number of cluster configurations. Transition rates between configurations satisfy detailed balance, producing conventional equilibrium cluster shapes. For EC, D cannot be defined rigorously since a cluster embedded in a dilute 2D gas is metastable.
27. Hamilton, M.S. Daw, and S.M Foiles, Phys. Rev. Lett. **74**, 2760 (1995).
28. L. Bardotti, M.C. Bartelt, C.J. Jenks, C. Stoldt, J.-M. Wen, C.-M. Zhang, P.A. Thiel, and J.W. Evans, Langmuir, to appear (Proc. 1997 ACS Symp. in memory of Brian Bent).
29. J.-M. Wen, Ph.D. Thesis (unpublished, Iowa State University, 1995).
30. J.W. Evans and M.C. Bartelt, in "Surface Diffusion". See Ref.14.
31. E.D. Williams, Surf.Sci. **299/300**, 502 (1994).
32. The island size distribution as a function of R can be written as $N_R \sim \theta S_{av}^{-1} R_{av}^{-1} F(R/R_{av})$, where $\int_0^\infty dy\, F(y)=1$. Since $N_R dR = N_S dS$, one has $F(y)=2yf(y^2)$ and $\int_0^\infty dy\, y^2F(y)=1$ for compact islands, if one defines $R_{av}^2=\Sigma_R R^2 N_R/\Sigma_R N_R$ so $S/S_{av}=(R/R_{av})^2$. Note that modified definitions $S_{av}'=\Sigma_S S^2 N_S/\Sigma_S S N_S$ ($>S_{av}$) and $R_{av}'=\Sigma_R R N_R/\Sigma_R N_R$ ($<R_{av}$) are sometimes used.
33. M.C. Bartelt and J.W. Evans, Surf. Sci. **298** 421 (1993); MRS Proc. **312**, 255 (1993).
34. M.C. Bartelt and J.W. Evans, Phys. Rev. B **54**, R17359 (1996); MRS Proc. **440** (1997).
35. M.C. Bartelt, S. Gunther, E. Kopatzki, R.J. Behm, and J.W. Evans, Phys. Rev. B **53**, 4099 (1996).
36. C.-M. Zhang, M.C. Bartelt, J.-M. Wen, C.J. Jenks, J.W. Evans, and P.A. Thiel, J. Crystal Growth **173**, in press (1997); Surf. Sci. (1997).
37. W. Theis, N.C. Bartelt, and R.M. Tromp, Phys. Rev. Lett. **75**, 3328 (1995).
38. P. Meakin, Physica A **165**, 1 (1990).
39. D.S. Sholl and R.T. Skodje, Physica A **231**, 631 (1996).
40. M. Villarica, M.J. Casey, J. Goodisman, and J. Chaiken, J. Chem. Phys. **98**, 4610 (1993).
41. L. Bardotti, C. Stoldt, C. Song, C.J. Jenks, J.W. Evans, and P.A. Thiel, to be published.
42. H.J. Ernst, F. Fabre, and J. Lapujoulade, Phys. Rev. Lett. **69**, 458 (1992).

METAL-ON-SEMICONDUCTOR GROWTH

Ag on Si-Surfaces: From Insulator to Metal

Michael Horn-von Hoegen and Martin Henzler
Institut für Festkörperphysik, Universität Hannover, Appelstrasse 2, 30167 Hannover, Germany

Gerhard Meyer
Institut für Experimentalphysik, Freie Universität Berlin, Arminallee 14, 14195 Berlin, Germany

1. Introduction

Epitaxial growth of nonreacting metals on semiconductors is of strong interest due to its potential for a variety of applications in semiconductor technology and in basic research. In contrast to metals forming silicides or those doping the semiconductor both the bulk properties of the semiconductor and of the metal overlayer are unaffected. This allows to study solely the influence of the surface and/or interface on the electronic transport or the lattice accommodation between metal and semiconductor.

In this paper we present three different examples from the wide field of metal deposition on Si surfaces, covering so different aspects as electronic transport in Ag monolayers at low temperatures and high magnetic fields, film morphology and lattice accommodation of thicker Ag films and the reversible selforganised nanostructuring of the Si substrate by metal induced facetting.

2. Experimental

Experimental results on surface and film morphology were obtained by *in situ* examination *during deposition* with Spot Profile Analysing-LEED (SPA-LEED) [1,2] and room temperature STM, and with low temperature STM [3]. Magnetotransport has been studied in a unique apparatus combining UHV-sample preparation and film deposition with *in situ* Hall measurements in Van der Pauw geometry at temperatures as low as 4 K and magnetic fields up to 4 Tesla [4].

Clean Si samples were prepared under ultrahigh vacuum conditions by degassing followed by a short flash to 1250°C to remove the native oxide. Ag was deposited from an e-beam heated tantalum cruicible, All evaporators are mounted in a water cooled Cu shroud equipped with a shutter and quartz microbalance. Deposition took place at temperatures ranging from 800°C down to 15 K, depending on the kind of experiment.

3. Lattice Accommodation of Ag(111) on Si(100)

Heteroepitaxial growth of metals on semiconductors is governed by a variety of problems which arise from the difference in lattice constant, surface and interface free energies and the kinetics of growth. Due to the high mobility of metal atoms (even at room temperature) they usually form islands in the shape of 3D-clusters (Stranski-Krastanov mode) [5,6]. In general the systems are also lattice mismatched which again increases the tendency towards islanding in order to relieve lattice strain either by elastic deformation or by generation of dislocations. As an additional complication metal surfaces usually tend to roughen during growth due to a barrier for inter-layer diffusion (Ehrlich-Schwoebel barrier) at the step edges [7,8,9].

The difference in the type of lattice (diamond versus fcc) and in the lattice constant is also the reason for the occurence of metal 3D-clusters with a different crystal orientation than the substrate, as observed for room temperature formation of Ag(111) 3D-clusters on Si(001) [10,11]. Obviously the free energy of a system with a Ag(111)/Si(001) interface is lower than those with a Ag(001)/Si(001) interface. The question arises: How fit these two completely different lattices together?

3.1. Kinetic Pathway

Here we demonstrate that use of the kinetic pathway (growth at low temperatures) overcomes most of these problems. Growth of Ag at 130 K on Si(100) substrates proceeds in a layer growth and finally results in continuous Ag(111) films. Although the Ag/Si system is highly mismatched (3:4), the two different interface lattices with (100)- and (111)-orientation could be adjusted with very small remaining mismatch of 2.2%. This mismatch is partially compensated by a linear array of interfacial dislocations which are observed by Low Energy Electron Diffraction (LEED) and Scanning Tunneling Microscopy (STM) in a weak height undulation of the surface.

Instead of formation of 3D-clusters (as observed for growth at room temperature), deposition at 130 K results in an epitaxial layer growth mode. This is proved in fig. 1 with the intensity oscillations of the (00)-spot as function of coverage in units of Ag(111) monolayers (1 ML= $13.9 \times 10^{14}\mathrm{cm}^{-2}$). The integer order- and (2×1)-spots from the Si(100)-substrate vanish at a Ag coverage of 4 ML, which reflects the growth of a continuous metal film which completely covers the Si substrate. The decrease and damping of the oscillations (note the logarithmic scale) reflects the kinetic roughening of the growth front with coverage. In contrast to this behaviour oscillations do not occur for homoepitaxy of Ag on Ag(111) [8,12], as shown in section 3.4. *Surface Roughness and Electronic Transport* of this contribution.

This random growth mode could be changed by introducing Sb as a surfactant, which increases the density of islands and causes a layer-by-layer growth [12,13]. Here we expect a similiar reason: Nucleation and growth on the Si(100) surface may result in an increased island density compared with growth on Ag(111).

Annealing of this 20 ML Ag film to 300 K results in the peculiar LEED pattern (electron energy 109 eV) shown in Fig. 2. The (00)-spot at the origin of the figure

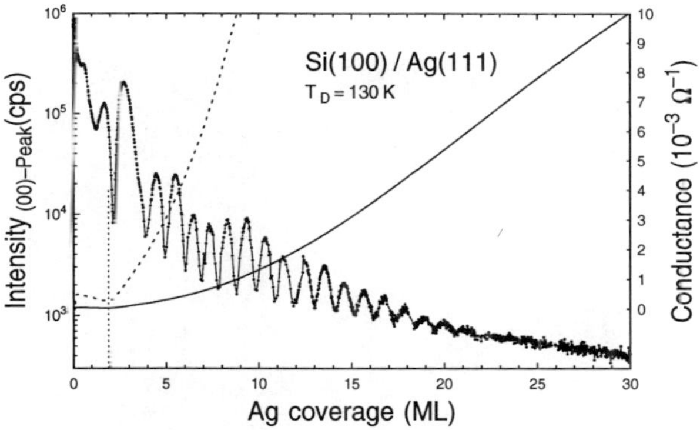

Figure 1. Intensity oscillations during growth of Ag on Si(100) at 130 K reflect epitaxial growth of a continuous film. The decrease and damping is caused by a kinetic roughening of the Ag film due to the barrier for interlayer diffusion.

is surrounded by twelf first order spots and many higher integer order spots. This pattern is easily explained as superposition of two LEED patterns of Ag(111)-surfaces which are rotated by 90° with respect to each other. For all electron energies all the first order spots have identical intensity reflecting the equal occurrence of twinned domains in the (111)-Ag film [14].

A fine structure in the LEED spots reduces the symmetry of the LEED-pattern: The (00)-spot has the shape of a plus-sign and the integer order spots are elongated in either the [011]- or [01$\bar{1}$]-directions. At higher magnification, the nature of the elongated spots becomes apparent: Each of the integer order spots is split into short chains of satellite spots with a separation of 1.4% of the Brillouin zone (100% BZ is the distance between the integer order spots), i.e. a periodicity length $L \simeq 180$ Å. This is shown for the (00)-spot and one (10)-spot in the insets of Fig. 2

3.2. Interfacial Dislocation Network

These chains of satellite spots are caused by a weak and periodic height undulation of the surface, which has the form of plane waves for each of the two (111)-crystal orientations. The undulation originates from a periodic arrangement of lattice matching dislocations at the interface. The associated strain fields (surrounding each dislocation) elastically distort the lattice up to the surface of the Ag film [15]. During diffraction the phase of the electrons is modulated due to the surface height undulation. The result is the same as expected for a phase grid for electrons [16,17,18]. As usual in optics, this leads to a splitting of all spots. The amplitude of the height undulation could be estimated from this dependence of the satellite intensities on the scattering condition [19,20].

With STM the size, orientation and morphology of the Ag(111)-crystallites of the

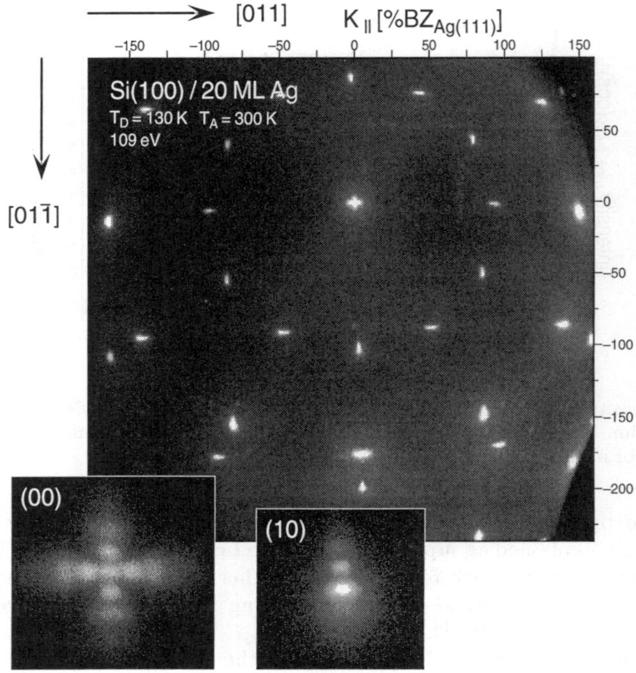

Figure 2. LEED pattern of 20 ML Ag on Si(001) after annealing to 300 K. The (00)-spot at the origin is surrounded by twelf first order spots and many higher integer order spots. The spots are elongated in either the [011]- or [01$\bar{1}$]-directions. The pattern is a superposition of two hexagonal patterns rotated by 90°. The insets show the splitting of the (00)-spot and one first order spot into satellites.

film has been detemined. Figure 3 shows a 40 ML thick Ag film on a Si(001) substrate which has been deposited at 90 K and annealed at 300 K. The steps of the underlying Si(001) substrate are still visible at the Ag surface: The azimuthal orientation of the Ag(111) film rotates from Si-step to Si-step by 90°. This is observed in the alternating orientation of the edges (indicated by arrows) of the monoatomic high Ag islands on larger terraces. This behaviour reflects the strict determination of the azimuthal orientation of the Ag(111) film by the direction of the Si-bonds on the Si(001) surface. The Ag grows only in one orientation on each of the alternating Si steps. However, twinning with Ag(111)-crystallites rotated by 180° is equally observed because there is no preference for the orientation perpendicular to the dimer rows.

The weak surface height undulation is observable also with STM for this 40 ML thick Ag film. The plane wave like height undulation could be clearly seen in Fig. 4 where the image contrast has simply been enhanced (without filter) that much that

Figure 3. Low temperature STM image (5000×5000 Å^2 wide, measured at 90 K) of 40 ML of Ag deposited at 90 K and annealed at 300 K. The alternating orientation (arrows) of the monoatomic high islands reflects the azimuthal rotation by 90° of the Ag(111)-crystallites for neighboured steps of the underlying Si-substrate.

the height undulation becomes clearly visible. A line-cut profile through this terrace yields an amplitude of 0.18 Å and a periodicity length of 185 Å which agrees perfectly with the LEED results. The amplitude of the undulation is a bit smaller, because the Ag film has the double thickness than in the LEED experiment.

The reason for this nearly perfect growth (only few threading dislocations have been observed with STM) is the almost exact 4:3 ratio of the extremely mismatched lattice constants of Ag (4.09 Å) and Si (5.43 Å). In Fig. 5 the surface lattices for the Si(001)-substrate and the Ag(111)-face have been drawn. In the [$\bar{1}$10] direction the distance between three Si atoms (11.52 Å) matches nearly exactly the distance between four Ag atoms (11.56 Å). The remaining compressive strain is only 0.36%. The mismatch in the [110]-direction is larger: the separation of the dimer rows of 7.68 Å has to fit with the distance of three dense packed rows of 7.51 Å. The interface, however, has a periodicity of two Si dimer row distances or six dense packed Ag rows. In this direction a tensile strain of 2.2% would remain. This anisotropic misfit is relieved by arrays of parallel dislocations in the interface plane, giving rise to the spot splitting in LEED.

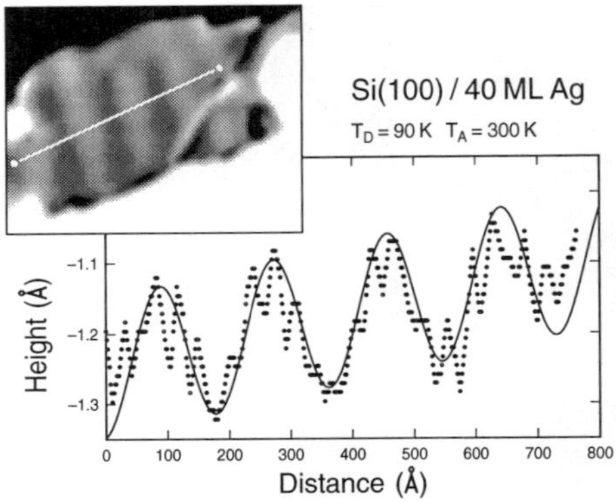

Figure 4. Increasing the image contrast, the plane wave height undulation due to the interfacial dislocation array becomes visible. Image is $1000 \times 800\,\text{Å}^2$ wide. From the linear cut an amplitude of 0.18 Å and a periodicity length of 185 Å is estimated.

3.3. Manipulation of Ag(111) in-plane Orientation

The strict relationship between the orientation of the surface Si bonds and the azimuthal orientation of the Ag(111) crystallites was proved by Ag deposition on single domain (2×1) structures on 4° vicinal Si(100). The single domain structure allows relief of the strong dimer induced anisotropic surface stress by the formation of double steps and only one type of (2×1) domain.

The LEED pattern of a 40 ML thick Ag film grown on this substrate and annealed at 300 K is shown in Fig. 6(a). The most prominent feature is the sixfold symmetry of the Ag film. Here the close-packed rows of Ag atoms run parallel to the dimer rows of

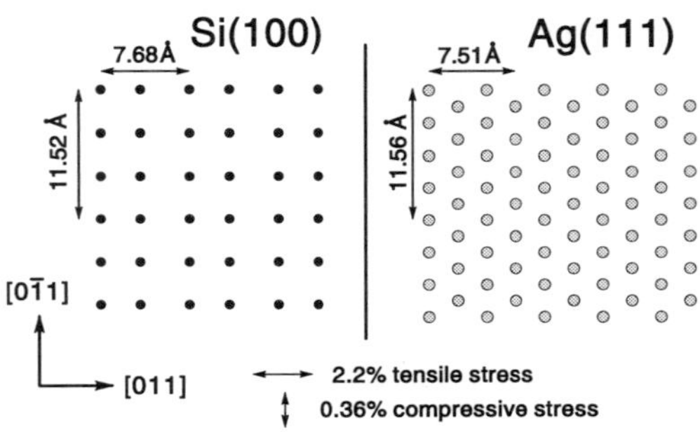

Figure 5. The lattice parameters of the Si(100)-(2×1) surface and the Ag(111) match nearly exactly in [0$\bar{1}$1]-direction. In the [011]-direction the Ag film has to be extended by 2.2% in order to match the Si dimer distance.

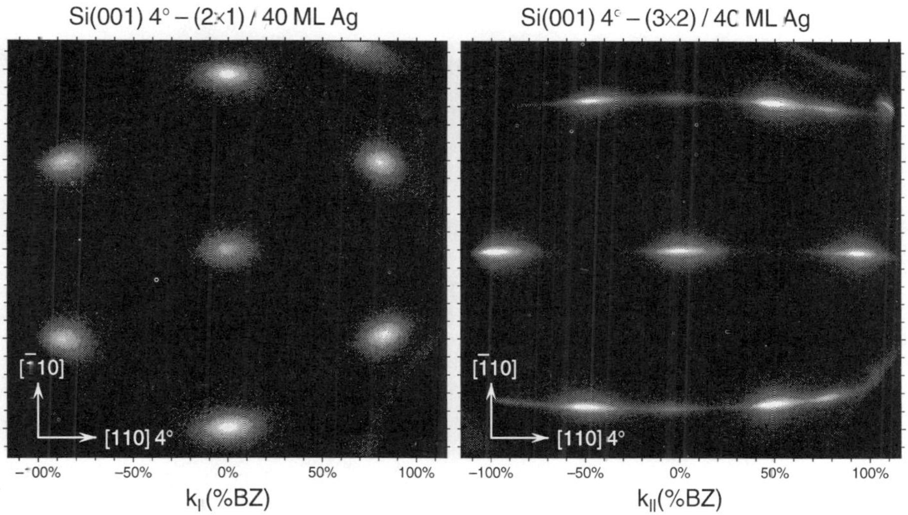

Figure 6. The single domain reconstructions of 4° vicinal Si(100) surfaces determine the in-plane orientation of the Ag film. (a) LEED pattern of a 40 ML thick Ag film deposited on single domain Si(100)-(2×1) substrate at 130 K and annealed at 300 K. The close-packed rows of Ag atoms run perpendicular to the step edge direction. (b) LEED pattern of a 40 ML thick Ag film deposited on single domain Si(100)-(3×2)/Ag substrate at 130 K and annealed at 300 K. The in-plane orientation of the Ag film is rotated by 90° and the close-packed rows of Ag atoms run parallel to the step edge direction

the Si-substrate. This orientation is easily understood on the basis of row-matching arguments, since every third close-packed Ag row fits with the twofold periodicity of the dimer rows with a mismatch of only 2%. The close packed row of the Ag atoms run *perpendicular* to the step edges. Because the Si substrate has a double step any 40 Å also the Ag film exhibits a high density of steps. Due to surface roughness the spots are broadened. Therefore it is impossible to determine, if a lattice matching dislocation network is present at the interface.

This in-plane orientation of the Ag(111) film could be drastically manipulated by altering the dimer row direction via a change of the reconstruction of the substrate. As observed for the flat Si(100) surface, a (3×2) reconstruction is induced by monolayer adsorption of Ag at elevated temperatures. Deposition of 40 ML Ag leads to the LEED pattern shown in Fig. 6 (b). The striking feature is the rotation of the in-plane orientation by 90°. As a result, the close packed Ag rows run *parallel* to the step edges of the Si substrate. The inclination of the macroscopic film orientation due to the 4° miscut of the substrate is accommodated by the formation of [335]-facets. This is apparent from facet spots which run along horizontal lines between the integral-order spots in Fig. 6 (b).

3.4. Surface Roughness and Electronic Transport

The LEED pattern of a 60 ML thick Ag film annealed at 300 K exhibit only very sharp spots without any detectable mosaic spread. Due to the larger thickness the intensity of the satellite spots has almost disappeared, because the height undulation smoothes out with increasing film thickness. The distance of dislocations in the Ag film is much larger than the mean free path of electrons. Therefore the thin Ag(111) films on flat Si(100) produced by the recipe described above are the ideal template to study further deposition of metal and the influence of surface and interface roughness on the electronic transport.

Deposition of Ag on this film at 130 K results in the formation of a very rough film. The Ehrlich-Schwoebel step edge barrier hinders step down diffusion of mobile Ag atoms resulting in a random growth mode [21]. The evolution of the surface roughness as function of Ag coverage has been studied *in-situ during the deposition* with SPA-LEED.

Fig. 7(a) shows the strong decrease of the 00-spot central spike intensity as expected for a random growth mode [22]. The lack of pronounced intensity oscillations is a direct hint for the formation of an increasingly rough surface morphology. This is also confirmed by the increasing half width of the spot profile as plotted in Fig. 7(b) as function of $\sqrt{\theta}$ for different scattering conditions. The linear increase with $\sqrt{\theta}$ is expected for a random growth behaviour.

During deposition at 130 K also the resistance of the Ag-film has been measured as shown in Fig. 7(a). A significant increase is observed, which is attributed to an increased diffuse scattering of the conduction electrons at the rough surface. The specularity parameter describing the scattering process decreases from 1 to zero, i.e. the surface is no longer a perfect reflecting mirror for electrons.

From the width of the spot at different scattering conditions and the rms-value of the roughness the parameters describing the surface roughness could be determined. With this knowledge the influence of surface roughness on the resistance on a thin metal film could be estimated. Following the description by Kaser and Gerlach [23] the contribution of conduction electron scattering at the rough surface is given by

$$\rho_{\mathrm{sr}} = \frac{9\pi\hbar\Delta^2 d^2}{2e^2 D} F(k_f\xi) \quad ,$$

with Δ the rms value of surface roughness, D the thickness of the Ag film, d the atomic step height, and ξ the correlation length describing the surface roughness, which could be expressed by [24]:

$$\xi = 2\Delta^2 \langle L \rangle$$

with $\langle L \rangle$ as average terrace length, determined from the spot profile at the out of phase condition. For the relevant range $k_f\xi \gg 5$ the function $F(k_f\xi)$ could be approximated by

$$F(k_f\xi) = \frac{2}{3}\frac{1}{k_f\xi} \quad .$$

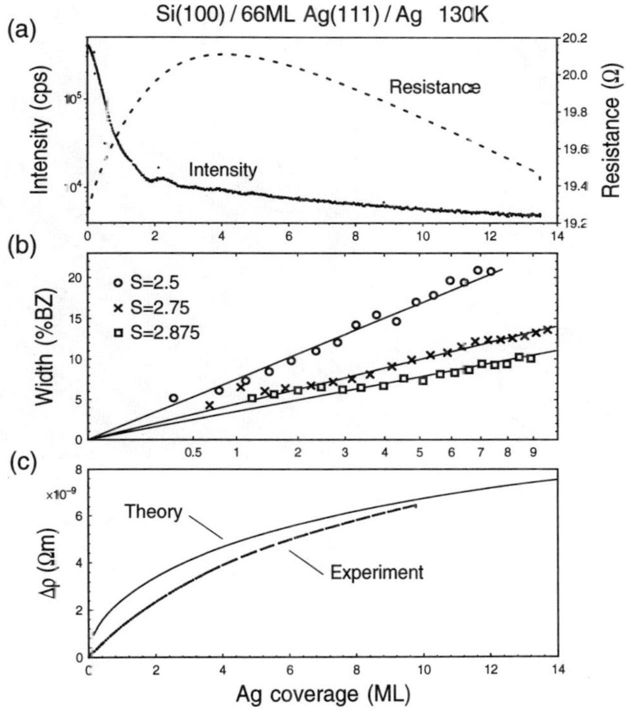

Figure 7. A 60 ML thick Ag(111) film on Si(100) is used as template for Ag growth at 130 K. (a) The exponential decrease of 00-spot intensity reflecting random growth mode due to a Ehrlich-Schwoebel step edge barrier. The resistance of the thin Ag film increases due to the additional scattering at the rough growth front. (b) The increase of the full width at half maxima of the broadened part of the spot profile indicates a built up of roughness during growth. Measured for different scattering conditions. (c) Comparison between measured change of conductance and theoretical expected behaviour due to increased contribution of surface scattering.

The experimental and theoretical behaviour is compared in Fig. 7(c). It has to be pointed out, that the theoretical curve is derived without any fitting parameter, using only the experimentally determined surface roughness as input parameter.

4. Ag induced Facetting of vicinal Si(100)

Controlled manipulation of surface morphology is one of the big goals for future applications of surface science as for example the self organised formation of quantum dots or wires without lithographic processes or heteroepitaxial growth and lattice accommodation for lattice mismatched systems. One way to change the surface mor-

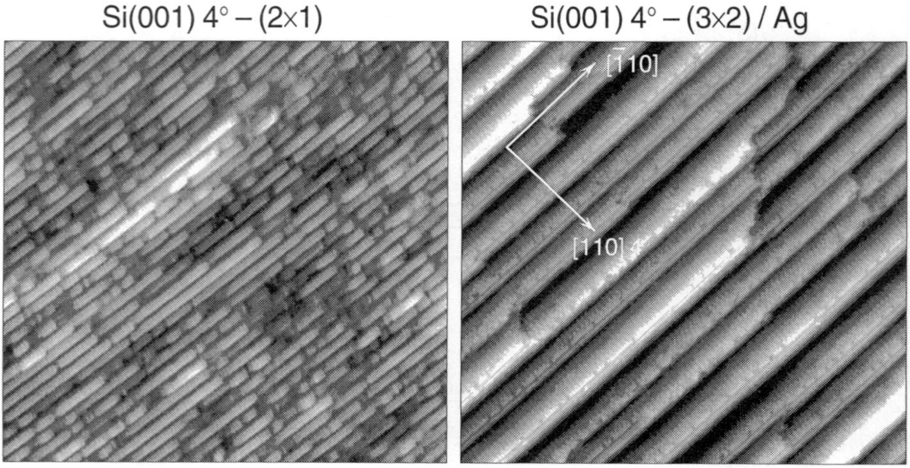

Figure 8. (a) STM image (140×140 nm scan) of the bare Si(100)-(2×1) surface measured at 90 K. Terraces with 4 nm width from are well ordered and separated by doublesteps. (b) STM image (140×140 nm scan) measured at 90 K after Ag monolayer adsorption at elevated temperature. Larger terraces with an Ag induced (3×2) reconstruction have formed which are separated by fourfold, sixfold and eightfold steps.

phology is by introduction of adsorbates or surfactants, i.e. manipulating the surface free energy. This will be demonstrated here with the change of step morphology of vicinal Si surfaces upon adsorption of Ag.

Adsorption of monolayer coverages of Ag at elevated temperatures changes the

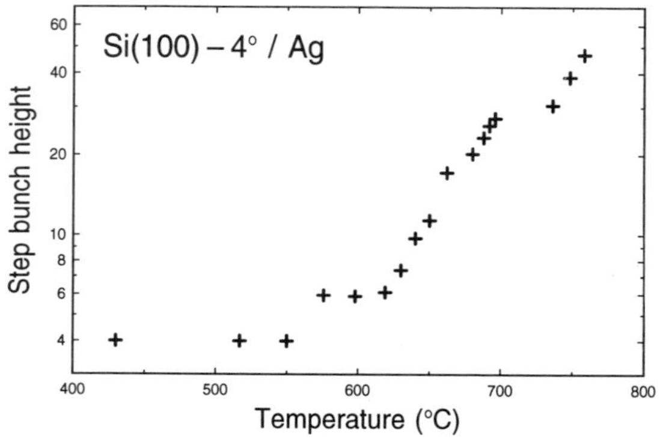

Figure 9. Multi step height as function of adsorption temperature. Equilibrium has not been reached due to kinetic limitations.

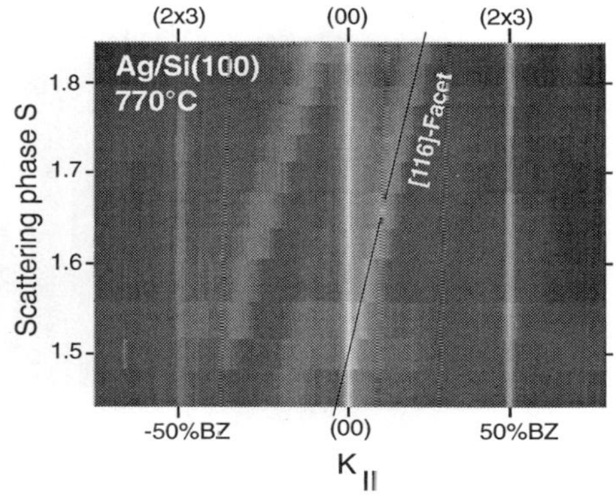

Figure 10. (a) Room temperature STM image of a 10 fold step. From the slope a [117]-facet orientation is deduced. (b) LEED spot profiles are plotted horizontal line by line as function of k_z as a cut through reciprocal space. From the slope of the moving facet spot an average facet orientation of [116] could be derived, indicating the existence of both [117]- and [115]-facets.

(2×1) dimer reconstruction of a Si(100) surface towards a (3×2) structure. On 4° vicinal Si(100) substrates this process is accompanied with a dramatic change of the step morphology: Formation of step bunches ranging from fourfold to 16fold step heights occurs. Figure 8 shows the change from the perfectly stepped single domain (2×1) Si substrate which exhibits a double step structure towards the formation of a (3×2) surface with multilayer terraces separated by fourfold, sixfold and eightfold steps. This process of multistep formation is completely reversible with the desorption of the Ag layer [25]. From the time constant of the facet decay the activation energy for Ag desorption is determined to 2.76 eV.

With increasing temperature the average step height increases from fourfold steps at 450°C via six-fold steps (as shown in Fig. 8) towards 50-fold steps at 750°C as revealed by spot splitting by means of SPA-LEED [26].

The height of the multi step bunches is plotted in Fig. 9 as function of adsorption temperature. With increasing temperature kinetic limitation for the Si atom mass transport becomes less inportant and results in larger, well ordered superterraces and very high step bunches. At higher adsorption temperature the step bunches form distinct facets with an average [116] orientation, as revealed by SPA-LEED and STM as shown in Fig. 10.

In Fig. 11 the uniform step morphology of ten-fold steps on a larger area of 1000 by 750 nm is shown. During the facetting process the macroscopic slope of the vicinal Si surface is also locally conserved: a shorter terrace is separated by smaller step bunches from the neighbouring terraces. This conservation of long range order is derived from the special shape of the diffraction profile [27].

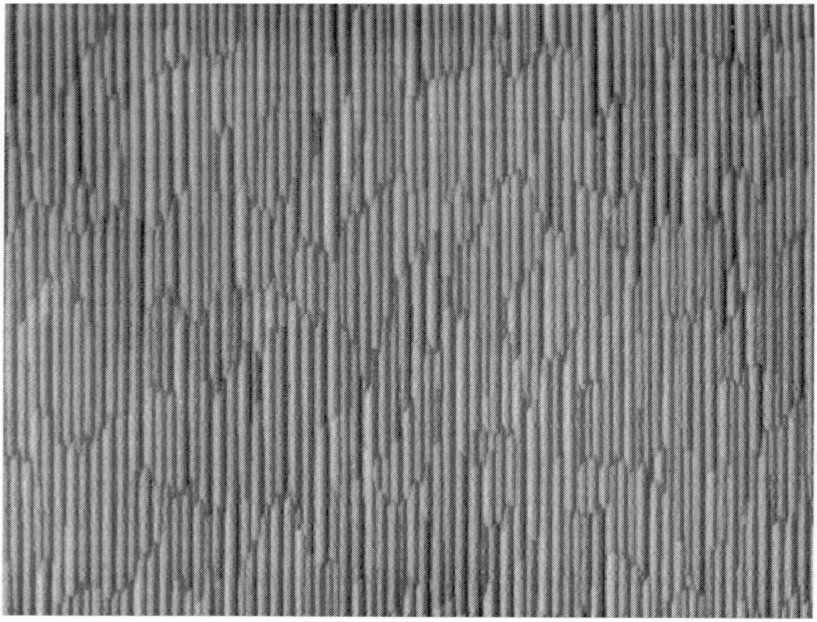

Figure 11. Large room temperature STM image (1000×750 nm scan) after Ag adsorption at 640°C resulting in a mean of 10 fold step heights.

5. Conductance of Ag Monolayers: a two-dimensional percolation problem

Single crystal Si surfaces are the ideal substrate to study the electronic transport quasi 2 dimensional, ultra thin Ag films. At low temperatures Si is (semi)insulating because no free carriers are left in the bulk, which allows to solely measure the conductivity of Ag films even for a coverage below one monolayer.

The Si(111) surface, however, provides an ideal substrate with very well defined crystaline order over the whole sample. Deposition at temperatures below 130 K hinders Ag cluster formation and allows 2 dimensional layer growth [28,29,30]. The layer growth mode is attributed to the decreased mobility and an increased island density, enabling a higher interlayer exchange rate. Therefore percolation of the Ag-film, i.e. the first current path through the incomplete film, is already observed for coverages around and below one monolayer [31,32,33]. For lower temperatures (deposition at 15 K) the percolation threshold θ_c increases to 1.4 ML (Fig. 12).

This percolation thresholds between 0.75 and 1.4 ML are much higher than the value 0.5 ML expected for random percolation on a triangular lattice. This increased value is attributed to the special binding sites of the Ag atoms on the (7×7) reconstruction of the Si(111) surface. STM investigations (Fig. 13) reveal preferential

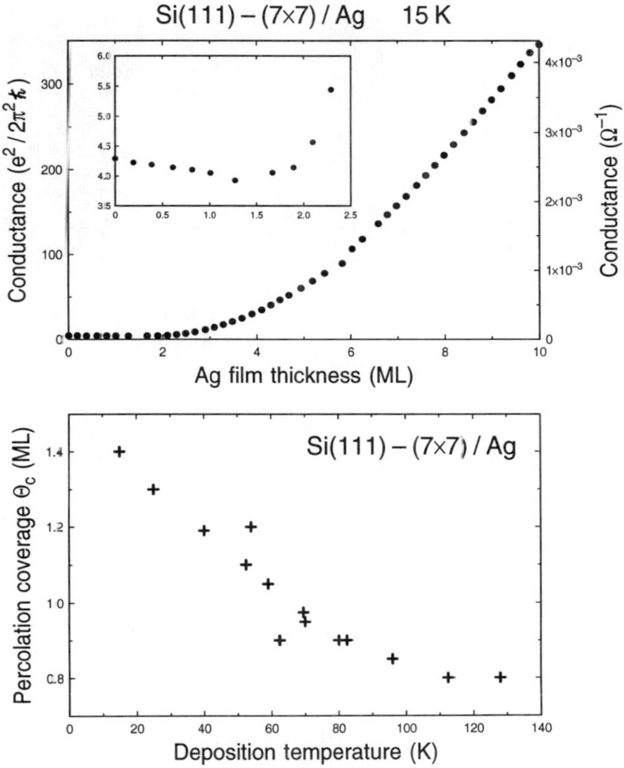

Figure 12. Upper plot: Conductance during Ag deposition at 15 K. The critical coverage for percolation is $\theta_c = 1.4$ ML as shown in the inset. Lower plot: Temperature dependence of the percolation threshold θ_c i.e. the coverage at which conduction through the Ag film starts.

nucleation and formation of very small Ag clusters inside *both* halves of the (7×7) reconstruction without occupation of the corner holes.

At low temperatures both halves, faulted and unfaulted, are equally occupied, as can be seen from Fig. 13(a). The diffusion of Ag atoms seems to be limited to within half of the unit cell, since always only one cluster is observed in each unit cell and the number of empty halves at low coverages can be simply explained with statistical arguments. This behaviour is attributed to a diffusion barrier at the edges of the halves of the unit cell, i.e. the dimers, so that the Ag atoms are captured in each separate half.

After deposition of 0.3 ML percolation is not observed and the islands have merely increased in size as seen in Fig. 13(b). Only a few islands are touching and the corner

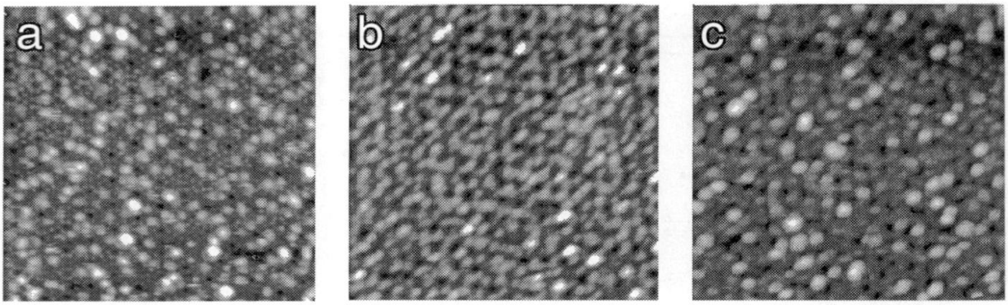

Figure 13. Nucleation and percolation behaviour of Ag on Si(111)-(7×7) at 90 K. (a) 0.08 ML Ag, 300 Å× 300 Å, (b) 0.3 ML Ag, 510 Å× 510 Å, (c) 0.85 ML Ag, 510 Å× 510 Å. The nucleation occurs in each half of the (7×7) with the corner holes remaining visible to the highest coverage.

holes are still not occupied. For a 0.85 ML thick film [Fig. 13(c)] the first layer is almost complete with a small amount of Ag already nucleated in the second layer. The periodic pattern of the corner holes can still be recognized: This determines the symmetry of the percolation problem. Percolation occurs through contact between the nucleated 2D Ag islands. This results in a honeycomb random site percolation problem with an critical coverage for percolation $\theta_c = 2/3$ of a monolayer [30]. Therefore the percolation threshold $\theta_c = 0.8$ ML deduced from the conductivity measurements as shown in Fig. 12 [31,32,33] for higher temperatures can be explained with this model.

With decreasing deposition temperature the percolation threshold increases almost by a factor of two at 15 K as plotted in Fig. 12. This behaviour is yet not understood and may be explained by cluster formation at very low temperatures due to non-thermal diffusion of the deposited Ag atoms [34]: It has to be assumed, that the arriving atom with its kinetic energy of several eV (condensation energy to the substrate) has to exite substrate phonons in order to equilibrate to the substrate temperature, which becomes more unlikely with decreasing temperature. With increasing equilibration time the atom should perform several or many diffusion hops until its kinetic energy is consumed. The attachment to an already resting Ag atom is energetically prefered, because energy transfer to an atom of the same mass should be favorable. Such a mechanism favors island growth especially at low temperatures.

The specific conductivity σ measured at 4 K for Ag films is shown in Fig. 14 in a log-log plot as function of film thickness d as deposited (circles) and annealed to maximum σ (dots). The extreme change of conductivity with thickness is explained by using different transport and scattering mechanism.

Applying the Drude formula $\sigma = ne^2\tau/m$ (a carrier density of one elctron per atom, free electron mass m, and a scattering time $\tau = l/v_{Fermi}$) an elastic mean free path l could be derived as plotted on the right axis. The value of l should always be larger than the interatomic distances. This condition is usually expressed by the Ioffe-Regel criterion $k_F l > 1$ as a metal-insulator transition [35] and is indicated in Fig. 14 by the horizontal dashed line.

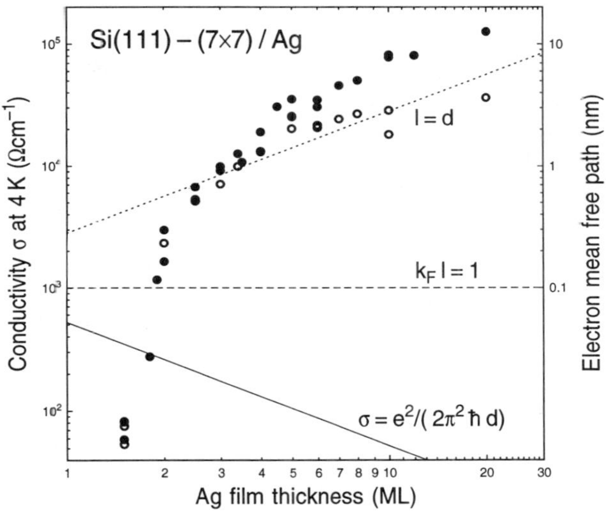

Figure 14. Conductivity and elastic mean free path l vs film thickness d measured at 4 K. Shown are two sets of data: ∘ after deposition at 15 K; • same film after annealing to optimum conductivity. The horizontal dashed line indicates $k_F l = 1$, a lower limit for reasonable l related to the Ioffe-Regel criterion. The dotted line $l = d$ indicates a mean free path equal the film thickness, with diffuse interface scattering as dominant mechanism.

The electronic transport for films thicker than 2 ML can be understood in terms of a metallic film with reduced elastic mean free path l of the electrons, which is limited by bulk defects and the finite thickness d of the Ag film. It is assumed, that the corrugation of the (7×7)-structure at the interface may be the reason for diffuse scattering of the electrons [32]. This limit due to finite thickness is expressed by the dotted line in Fig. 14 for the criterion $l = d$. Hall effect measurements of films thicker than 2 ML shows the Ag bulk charge density. The magnetoresistance and temperature dependence of resistance have classical behaviour for films with thickness above 50 ML. The magnetoresistance and resistance are modified for films between 2.5-16 ML due to weak localization of the electrons [34].

Films thinner than 2.5 ML show an elastic mean free path which is smaller than the atomic distances and below the value of the Ioffe-Regel criterion. The transport in this regime is no longer of diffusive nature, rather it is via localized states, which is consistent with the conductance lower than the universal quantum conductance $L_{00} = e^2/2\pi^2\hbar$ as the lower limit of extended electron states: Anderson localization is present.

Financial support by the Deutsche Forschungsgemeinschaft is gratefully acknowledged.

References

1. U. Scheithauer, G. Meyer, and M. Henzler, Surf. Sci. **178**, 441 (1986)
2. M. Horn, J. Falta, and M. Henzler, Thin Solid Films **183**, 213 (1989)
3. G. Meyer, Rev. Sci. Instrum. **67**, 2960 (1996)
4. F. Jentsch, H. Froitzheim, and R. Theile, J. Appl. Phys. **66**, 5901 (1989)
5. R.M. Tromp, A.W. Denier van der Gon, F.K. LeGoues, and M.C. Reuter, Phys. Rev. Lett. **71**, 3299 (1993)
6. J. Tersoff, and R. Tromp, Phys. Rev. Lett. **70**, 2782 (1993)
7. R. Kunkel, B. Poelsema, L.K. Verheij, and G. Comsa, Phys. Rev. Lett. **65**, 733 (1990)
8. G. Rosenfeld, R. Servaty, C. Teichert, B. Poelsema, and G. Comsa, Phys. Rev. Lett. **71**, 895 (1993)
9. C. Ammer, T. Schaefer, C. Teichert, K. Meinel, and M. Klaua, Surf. Sci. **307-309**, 570 (1994)
10. M. Hanbücken, and H. Neddermeyer, Surf. Sci. **114**, 563 (1982)
11. D. Winau, H. Itoh, A.K. Schid, and T. Ichinokawa, J. Vac. Sci. Technol B **12**, 2082 (1994)
12. H.A. van der Vegt, J.M.C. Thornton, H.M. van Pinxteren, M. Lohmeier, and E. Vlieg, Phys. Rev. Lett. **71**, 895 (1993)
13. J. Vrijmoeth, H.A. van der Vegt, J.A. Meyer, E. Vlieg, and R.J. Behm, Phys. Rev. Lett. **72**, 3843 (1994)
14. A Ag(111)-surface has a threefold symmetry and therefore in general also threefold LEED-patterns are expected, i.e. the first integer order spots usually show different intensity. If they have for all electron energies the same intensity (sixfold symmetry) stacking faults are equally present
15. F.R.N. Nabarro, Theory of Crystal Dislocations (Dover, New York, 1987)
16. U. Gradmann, and G. Waller, Surf. Sci. **116**, 539 (1982)
17. M. Horn-von Hoegen, M. Pook, A. Al-Falou, B.H. Müller, and M. Henzler, Surf. Sci. **284**, 53 (1993)
18. M. Horn-von Hoegen, A. Al-Falou, H. Pietsch, B.H. Müller and M. Henzler, Surf. Sci. **298**, 29 (1993)
19. M. Horn-von Hoegen, T. Schmidt, M. Henzler, G. Meyer, D. Winau and K.H. Rieder , Surf. Sci. **331-333**, 575 (1995)
20. M. Horn-von Hoegen, T. Schmidt,G. Meyer, D. Winau and K.H. Rieder, Phys. Rev. B **52**, 10764 (1995)
21. G. Meyer, J. Wollschläger and M. Henzler, Surf. Sci. **231**, 64 (1990)
22. E.Z. Luo, J. Wollschläger, F. Wegner, M. Henzler, Appl. Phys. A **60**, 19 (1995)
23. A. Kaser, E. Gerlach, Z. Phys. B **97**, 139 (1995)
24. E.Z. Luo, S. Heun, M. Kennedy, J. Wollschläger, M. Henzler, Phys. Rev. B **49**, 4858 (1994)
25. A. Meier, R. Vockenroth and M. Horn-von Hoegen, Surf. Sci. (submitted)
26. A. Meier, P. Zahl, R. Vockenroth and M. Horn-von Hoegen, Appl. Surf. Sci. (submitted)
27. S. Fölsch, G. Meyer, K.H. Rieder, M. Horn-von Hoegen, T. Schmidt and M. Henzler, Surf. Sci. (submitted)
28. M. Jalochowski and E. Bauer, Phys. Rev. B **37**, 8622 (1988)
29. K.R. Roos and M.C. Tringides, Europhys. Lett. **23**, 257 (1993)
30. G. Meyer and K.H. Rieder, Appl. Phys. Lett. **64**, 3561 (1994)
31. R. Schad, S. Heun, T. Heidenblut, and M. Henzler, Appl. Phys. A **55**, 231 (1992)
32. R. Schad, S. Heun, T. Heidenblut, and M. Henzler, Phys. Rev. B **45**, 11430 (1992)
33. S. Heun, J. Bange, R. Schad, and M. Henzler, J. Phys.: Condens. Matter **5**, 2913 (1993)
34. M. Henzler, T. Lüer, A. Burdach, Phys. Rev. B submitted
35. A.F. Ioffe and A.R. Regel, Prog. Semicond. **4**, 237 (1960)

METAL ON SEMICONDUCTOR GROWTH AT LOW TEMPERATURES

M.C. Tringides
Department of Physics, Iowa State University, Ames, IA 50011

Abstract

The search of the optimal growth conditions in epitaxy requires exploring the full space of growth parameters (i.e. temperature, flux rate, annealing cycles, etc.) since epitaxy is a non-equilibrium phenomenon. For the growth of metal on a semiconductor for several systems (Ag/Si(111), Pb/Si(111), etc.), it is found that high quality films can be grown at low temperatures (T=18-300 K) and high flux rates ($F > \frac{1}{20} ML/sec$). Evidence of the improved growth is based on the presence of diffraction intensity oscillations, sharp diffraction profiles, low roughness extracted from STM images and higher conductivity measured with in situ 4-probe techniques. Since most likely thermal diffusion does not operate at these low temperature / high flux rates possible mechanisms (interlayer diffusion enhancement due to higher island density, inefficiency of the energy transfer from the incoming atom to the substrate, etc.) will be discussed.

1 Introduction

The goal of ultrathin film growth is to attain the ability to fabricate defect free structures with atomic scale control. The quality of the grown films can be characterized by the degree of vertical and lateral perfection as the film thickness increases. In turn the structural perfection of the film (i.e. sharpness of the interface, low step density, absence of dislocations, etc.) critically determines its electronic properties. It is of great practical and fundamental interest to determine the range of growth parameters (i.e. temperature, flux rate, annealing cycles, etc.) that can produce films with ideal morphology. Commonly such film growth is carried out with MBE techniques under highly non-equilibrium conditions (since low flux rates and temperatures are usually used, etc.) so the final product depends on the interplay between thermodynamic and kinetic factors. The optimal growth conditions cannot be easily deduced from information describing equilibrium properties but it can be determined with systematic search over the complete space of growth parameters.

In the present chapter we would like to review recent work on ultrathin film growth of metal on semiconductor substrates at low temperatures [1-3] that surprisingly (and still for not fully understood reasons) has shown that films of high structural quality can be grown at low temperatures (or high flux rates). At a first glance this contradicts the intuitive expectation that thermal diffusion is necessary for atoms to move to the correct places and eliminate defects and imperfections.

Two main probes have been used to characterize the quality of the film by measuring complementary properties: (i). RHEED diffraction measurements that provide information about the film structure (ii). In situ resistivity measurements that provide information about electronic transport. Most of the discussion will be on the Ag/Si(111) system but other examples in the literature of similar effects observed at low temperatures will be briefly reviewed. Although no clear understanding of all the low temperature growth phenomena is currently available, possible explanations and operating mechanisms will be presented at the end.

2 Methods

Diffraction methods are very convenient in describing the quality of the grown film. They can be easily installed in an MBE chamber to allow measurements at the growth temperature and in real time. Several models have been developed to easily translate diffraction measurements into real space information related to the order in the film. Routinely the time dependence of the peak intensity vs. time can be used to identify the growth mode i.e. whether it is 2-dimensional (indicated by the presence of diffraction intensity oscillations), 3-dimensional (indicated by the monotonic decrease of the diffracted intensity) or step flow (when the diffracted intensity is practically unchanged with time). Although in practice several complications can be present (i.e. what scattering mechanism operates, whether dynamic effects are important, what is the role of step density, etc.) the presence of the oscillations is a necessary condition to confirm the growth of a smooth film. Information about the interface width (i.e. number of exposed layers) is obtained from the decay of the envelope of the oscillations, while the lateral order in the film (average terrace size, step density etc.) is obtained from spot profile analysis (intensity vs. $S_{\|}$ at fixed $S_{\perp} = (odd)\pi$, where $S_{\|}$, $S_{\perp}$ are the parallel and vertical components of the momentum transfer; for kinematic scattering $S_{\perp}$ is chosen at the out of phase condition to have maximum sensitivity to imperfections) [4]. It is common to use the full-width-at half maximum (FWHM) as a measure of the average island size. More detailed analysis within specific models can reveal additional information about the terrace and step size distribution. For a 2-level system [5] in the kinematic approximation (i.e. only the top atoms contribute to the scattering) the shape of the profile can be decomposed in two contributions, a narrow and a diffuse one.

$$I(S_{\parallel}) = \{\theta(1-\theta)(1-\cos(S_{\perp}d))L(S_{\parallel})\} + \{1-2\theta(1-\theta)(1-\cos(S_{\perp}d))\delta(S_{\parallel})\}$$

where θ is the coverage of the top layer, d the step height, $L(S_{\parallel})$ the Lorentzian diffuse component and $\delta(S_{\parallel})$ the narrow instrumentally limited component. The FWHM of the diffuse component measures the harmonic mean of the island size and the island separation. In our setup RHEED patterns are formed on a phosphor screen and the images are collected with a high gain video camera at acquisition speeds higher than 10 frames/sec for digitization and further analysis.

In situ resistivity measurements have been carried out systematically only recently. Reasons of the reluctance to perform such experiments is the difficulty to make well defined contacts to the film and at the same time to reproducibly prepare a clean surface after heating to higher temperatures (at least 500 C for the desorption of Ag from Si(111)). Depending on the quantity of interest, whether only relative changes or the absolute value of the resistivity is measured, different contact methods of the electrode probes have been used. Simple mechanical contact by the probes on the film can be used to measure the voltage drop as the resistivity decreases, [6-8]while pre-deposited Ni-pads [9, 10] which provide low and constant contact resistance to the film are more appropriate for absolute resistivity measurements when the equivalent electric circuit is analyzed. In our measurements mechanical contact by the probes and a constant current I=10 μA was used with voltage drops in the meV range with Si(111) substrates of initial resistance ranging from 1-3 k .

3 Results

The growth of Ag/Si(111) has been extensively studied at temperatures higher than room temperature especially in connection to the formation of the Si(111)-Ag($\sqrt{3}\times\sqrt{3}$) superstructure at T>480 K. For the range between room temperature and T=480 K the film grows in the Stranski-Krastanov growth mode; after the first layer is completed which consists of small Ag(111) crystallites (of average size 30 A) and different orientations (with respect to the Si substrate) large 3-dimensional clusters are formed.

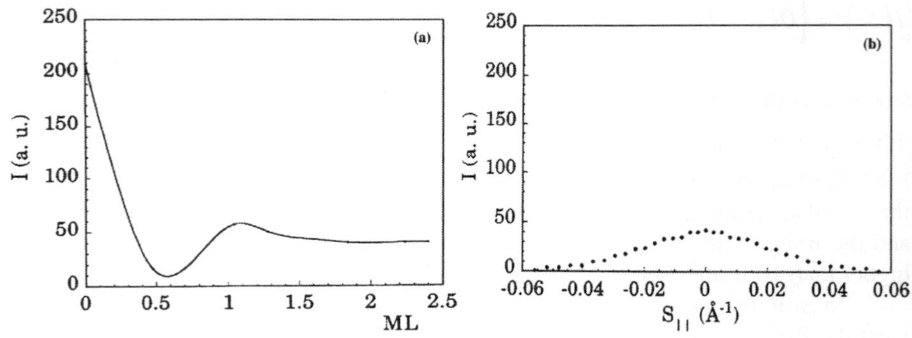

Fig 1: (a) Specular RHEED intensity vs. deposited amount for growth of Ag/Si(111) at room temperature at an out-of-phase condition. (b) Spot profile of the specular spot after 4 ML of Ag have been deposited. The low intensity indicates 3-dimensional growth.

Fig 1(a) shows the specular beam intensity vs. time during Ag/Si(111) growth at room temperature and at an out of phase condition. After one weak oscillation which corresponds to the formation of the first layer the intensity saturates to a value 20% its initial value which indicates the formation of a rough film. Fig 1(b) shows a corresponding specular beam profile taken at the same diffraction condition and after 4 ML of Ag have been deposited. The profile has a FWHM=0.05 A^{-1} and low peak intensity which indicates that the film has considerable lateral and vertical disorder (i.e. 3-dimensional cluster formation).

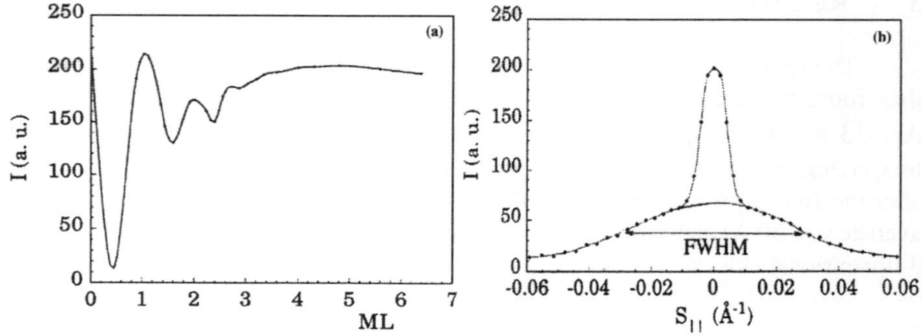

Fig 2: (a) Specular RHEED intensity vs. deposited amount for growth of Ag/Si(111) at T=150 K at an out-of-phase condition, showing oscillations and high saturation intensity. (b) Corresponding profiles after 5 ML showing two distinct

components. Both results are consistent with quasi layer-by-layer growth at low temperatures.

Fig 2(a) and Fig 2(b) show the dependence of the peak intensity during growth and the corresponding specular spot profile at a low temperature T=150 K at the same out of phase condition. The results suggest that the growing film is smoother and grows in a quasi-layer-by layer growth mode. This is indicated by the weak short-lived intensity oscillations that saturate at a much higher value of the initial intensity 80% and the 2-component shape of the spot profile. Both observations suggest that the film has a sharp interface (i.e. only 2-levels are exposed). In addition the diffuse component of the profile is narrow (FWHM=0.06 A^{-1} of the Brillouin zone) which implies that the film also is laterally ordered with average terrace size 20 A.

The comparison of the diffraction results at the two temperatures show clearly that better Ag films can be grown on Si(111) at lower than room temperature. Despite the numerous studies of Ag on Si(111) no direct measurement of the diffusion coefficient has been carried out and no information is available about the diffusion barrier. It is known from STM measurements at room temperature [11] that islands of Ag are formed at submonolayer coverage which are confined to the faulted part of the 7×7 unit cell. This suggests finite but limited mobility with possible different barriers for inter-unit and intra-unit (7×7) cell diffusion. Information about the origin of the mobility necessary for the quasi layer-by-layer growth discussed before can be obtained if the growth is studied as a function of the flux rate to distinguish thermal (i.e. the Ag atom hopping over the diffusion barrier because of its vibrational energy) from non-thermal (i.e. transient mobility [12], local heating effects related to the equilibrium energy [13], etc.). If the diffusion barrier is low enough so the atom motion is thermal then one expects strong flux dependence (with better films growing at low flux rates if there is no barrier to the motion of atoms from higher to lower levels); on the other hand if the origin of the mobility is non-thermal (i.e. the motion is carried out at a smaller time scale than the time between atom deposition determined by the equilibration time of the incoming atom) then no flux dependence is expected.

The growth of Ag/Si(111) was studied as a function of flux rate F=1/125-1/4800 ML/sec at the low growth temperature T=150 K [1]. If the oscillations are plotted as a function of the rescaled time t/τ (where τ is the period of the first oscillation) then the intensity plots collapse on top of each other (within 10% accuracy) for the different flux rates (especially) for the first two oscillations. This flux independence is remarkable and demonstrates that the diffusion time for the atoms to move to the correct positions (to complete most of the lower layer before the next one starts to be populated) is much shorter than the time between the arrival of atoms, which is varied over a factor of 50 as a result of varying the flux rate.

It is interesting to ask why we have observed only weak, short-lived oscillations at the low temperature. Because of the low energy of the electron beam we use E=5 keV (relative to other RHEED studies) we were limited to a minimum ϑ=0.80 (while other RHEED studies at higher beam energies can be performed at ϑ less than 0.4) in varying the angle of incidence of the incoming beam. We have observed a strong dependence of the quality of the oscillations on ϑ, with decreasing oscillation quality as ϑ increases. For experiments at higher angle of incidence the oscillations gradually disappear until eventually we observe monotonic decrease of the intensity for ϑ>4.5. Such dependence has been documented in other experiments as well [14]. A possible explanation of this dependence has been proposed as an interference phenomenon between electron waves scattered at the top and bottom layers in the coverage-dependent potential experienced by the incident electron [15].

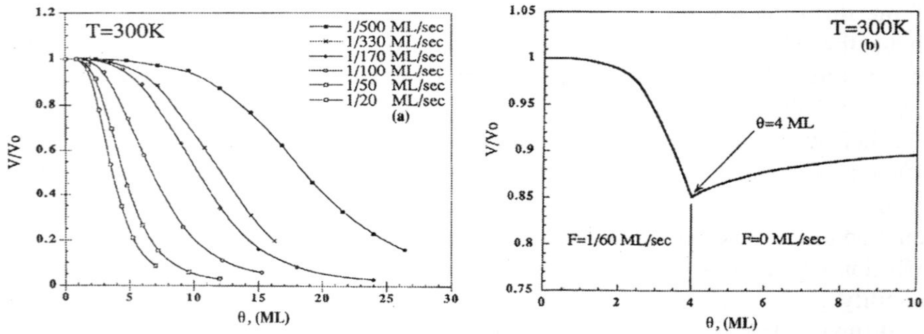

Fig 3: (a) Voltage traces vs. deposited amount for the growth of Ag/Si(111) at room temperature. The plateau coverage increases with decreasing flux rate. (b) Transient experiments showing that the voltage trace increases with time after the shutter of the Ag source is closed.

Fig 3(a) shows the room temperature results for the in situ transport studies of the Ag/Si(111) as a function of flux rate [6]. A 4-probe linear method was used with 10 mil W mechanically pressed on the substrate to measure the voltage drop while a 10 μA constant current is passed through the crystal. Since it is important to keep the contact resistance of the probes on the crystal unchanged and to avoid inhomogeneities in the film (especially around the contact region to the probes) it is easier to obtain information about the electron transport mechanism from flux dependent experiments that do not disturb the mass distribution in the film. Temperature dependent experiments can generate diffusion currents on the surface, substrate reconstructions or change in the contact resistance to the wires.

The measured voltage is plotted as a function of the deposited coverage for flux rates ranging from $F = \frac{1}{20} - \frac{1}{500}$ ML/sec. After an initial plateau where the voltage remains constant, it starts to decrease rapidly and quickly drops to 95% its initial value after a few (less than 10) monolayers have been deposited. The flux calibration is based on a quartz crystal monitor to measure the flux rate by varying the temperature in the Knudsen cell and is highly accurate. All the plateau values observed exceed a monolayer and range from 2-3 ML depending on the flux rate. It is surprising that larger plateau values are observed at the lower flux rates . If we define θ_c as the plateau value and plot the results at different flux rates the relation $\theta_c \sim F^{-0.4}$ holds. Both observations, i.e. why the plateau values exceed a monolayer and why they decrease with increasing flux rates provide important hints about the transport mechanism. Is the conductivity of the film simply controlled by the film connectivity and geometry (with a sharp onset once a percolating cluster spanning the surface is formed) or is the electronic structure of the substrate surface (whether metallic or semiconducting and changes in the carrier concentration close to the surface) responsible for the voltage drop?

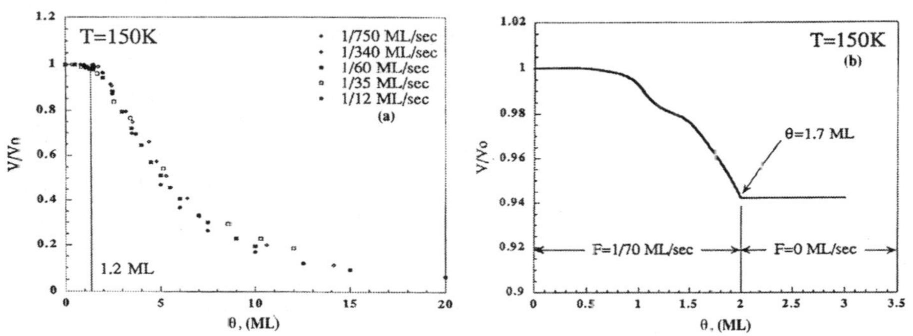

Fig 4: (a) Voltage trace vs. deposited amount for the growth of Ag/Si(111) at T=150 K. Smaller values for the coverage plateau are observed and the traces are flux rate independent. (b) Transient experiments showing that the voltage trace remains constant after the shutter of the Ag source is closed; this indicates the absence of thermal diffusion and the improved quality of the grown film.

Fig. 4(a) shows the corresponding traces at the lower growth temperature T=150 K carried over the same flux rate range as the room temperature experiments.

We observe similar traces with an initial flat dependence on Ag coverage followed by a fast decrease . However the length of the plateau is θ_c=1.2 ML, larger than a monolayer, but smaller than θ_c at room temperature and more importantly θ_c (and the full traces over the whole amount of deposited Ag) are independent of the flux rate.

The large values of the Ag coverage in the initial plateau regime, in excess of a monolayer, indicates that the transport mechanism is most likely related to the electronic structure of the substrate and not the percolating characteristics of the 2-dimensional film. If θ_c corresponds to the minimum coverage needed to form a percolating cluster (so a macroscopic path is established for electrons to move across the film), it is expected to have a value less than 1 ML (i.e. a square lattice has θ_c=0.5 for the case of classical percolation). Additionally for 2-dimensional transport and percolation transition a dependence of the shape of the traces on the flux rate should be expected, since the film morphology will be different for different flux rates [6].

Comparison between the room and low temperature voltage traces agrees with the corresponding comparison of the diffraction results supporting the change in growth mode at lower temperatures. This is further confirmed by the transient experiments shown in Fig.3(b) and Fig. 4(b), which monitor the voltage vs. coverage after the shutter is closed . At room temperature (Fig. 3(b)), the voltage gradually increases towards the initial value indicating an increase in the film resistivity while at T=150 K (Fig. 4(b)) the trace remains unchanged after the shutter is closed. Clearly atoms at room temperature have enough mobility to further roughen the film by joining the 3-d clusters present, while at T=150 K there is no change in the trace since thermal diffusion is low.

The low temperature quasi layer-by-layer growth has been also observed directly with STM studies [16] at low temperature T=90 K discussed in the first chapter of this section. The initial Ag clusters occupy both halves (faulted and unfaulted) of the 7×7 unit cell. As the Ag is increased up to 4-6 ML measurements of the interface width (defined by the mean square deviation of the interface from its average position) show that it is approximately half the Ag step height d=2.36 A (i.e. only two levels are exposed); the average terrace size increases to 30-40 A after 4-6 ML have been deposited. This amount of order in the film after several layers have been deposited is larger than what is expected for statistical growth. These results confirm the previous conclusions we have reached based on spot profile analysis of the diffraction spots (Figs. 1-2).

RHEED diffraction oscillations sustained over longer times have also been observed at low temperatures on Ag/Si(111) in other studies [2, 7]. (Part of the reason of the longer lived oscillations discussed before, is the smaller angle of incidence (ϑ=0.4-0.6) used because of the higher electron energy of the incident beam). All these studies confirm the improved quality in the grown film at low vs. room temperatures. In addition RHEED studies [4] at larger flux rates than the ones we have used ($F = 1/25 - 1$ ML/sec) show that oscillations can gradually appear with increasing flux rate and quasi later-by-layer growth is possible even at room temperature if a sufficiently high flux rate is used.

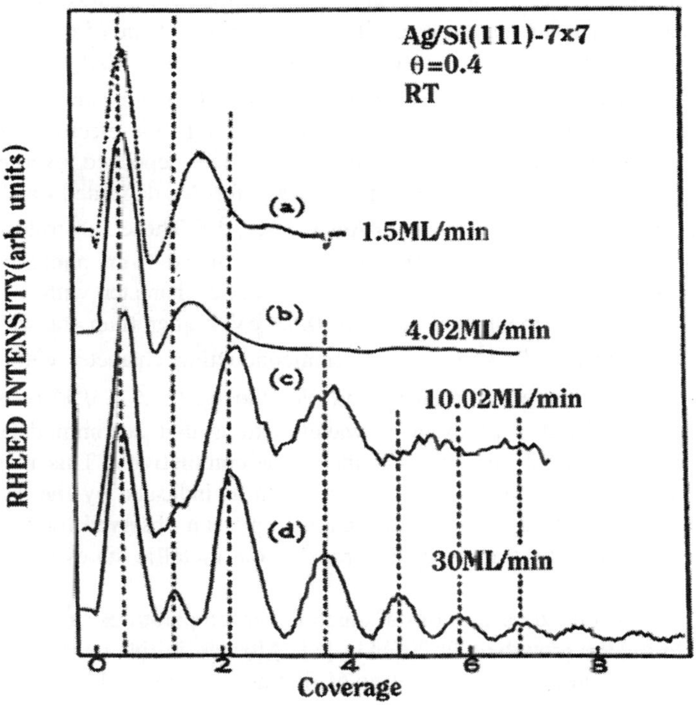

Fig 5: (a) RHEED intensity of the specular beam vs. Ag deposited amount for growth at room temperature indicating that oscillations and the quality of the film improve at high flux rates, in analogous fashion as the enhancement of growth at low temperatures.

Fig. 5 shows the RHEED specular intensity vs. coverage at room temperature with better oscillations present at the high flux rates in similar fashion as with experiments, performed by lowering the temperature at fixed flux rate, F. This at first is not surprising since one expects a duality between growth at high flux rates and growth at low temperatures; in both cases the time necessary for an atom to diffuse to a given site is longer than the time between successive depositions of incoming atoms . This implies that the degree of lateral order in the film is reduced since a higher island density (of smaller average island size at fixed coverage) is formed.

In situ conductivity studies at room temperature have also been carried out on Ag/Si(111) [8] with similar mechanical contacts of the probe wires to the film as in our experiments. The agreement is excellent between the shape of the observed trace voltage vs. deposited θ, the size of the plateau θ_c (2-3 ML) and the transient effects observed after the shutter is closed (Fig. 3(b)) with our experimental results. Stronger support that the resistivity changes are related to the electronic structure of the substrate (which depends on the structure of the deposited overlayer) [8] is obtained by comparing the 4-probe voltage change for Ag deposited on the Si(111)-(7×7) with Ag deposited on Si(111)-Ag($\sqrt{3}\times\sqrt{3}$). These substrates are known to have very different electronic structures as determined from photoemission and STM spectroscopic studies. The (7×7) has metallic character with surface states partially occupied and the Fermi level is 0.63 eV higher than the valence band, while the Ag($\sqrt{3}\times\sqrt{3}$) structure has semiconducting character with the Fermi level only 0.15 eV above the valence band. For the Ag-($\sqrt{3}\times\sqrt{3}$) overlayer there is a surface-state band S of extended wavefunctions that accommodates electrons donated by the deposited Ag atoms, enhancing the conductivity. This in turn affects the measured voltage at the start of the deposition as indicated by the dramatic drop of the voltage, in contrast, to the characteristic plateau observed for deposition on the (7×7) that indicates transfer of electrons to the metallic states and Fermi level depinning.

Strong sustained oscillations at much lower temperatures T=18 K have been observed during the growth of Pb/Si(111) [3]. After an initial irregular variation of the intensity that indicates the formation of first an amorphous Pb film (of 2-7 ML depending on growth temperature) strong RHEED intensity oscillations, shown in Fig. 6, develop that improved in strength and duration as the temperature is decreased, with the best quality oscillations observed at T=18 K. At the same time, spot profile analysis was used to extract an average domain size laterally in the range 15-45 A with larger domains (but worse oscillations) observed with increased temperature. The improved quality of the film at lower temperatures was also confirmed with in situ conductivity measurements (by measuring the voltage drop during Pb deposition across mechanical contact probes) that were used to extract the electron mean free path (after the temperature-dependent phonon scattering contribution is subtracted out) as a function of film thickness. The observed variations of the mean free path were explained in terms of quantum size effects (as a result of electron confinement in the film with stationary levels that depend on film thickness) and the dependence of the scattering time on film roughness. Higher mean free paths, for a given Ag deposited amount, were observed at lower than higher growth temperatures which agrees with the complementary observations of improved diffraction oscillations and smoother films at lower temperatures.

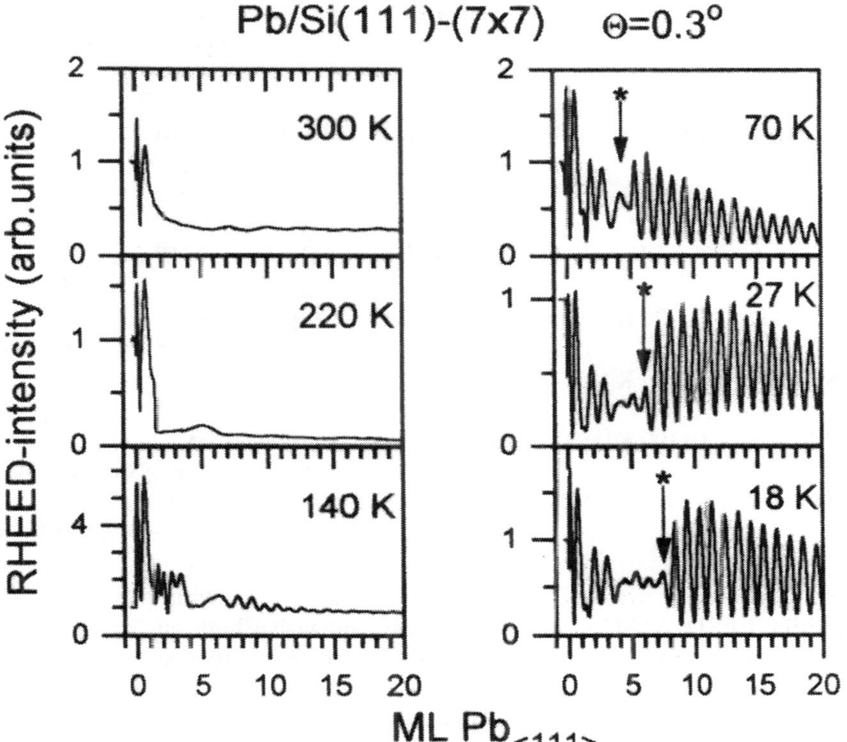

Fig 6: (a) RHEED intensity oscillations for the growth of Pb/Si(111). Stronger oscillations are observed at lower temperatures. The lowest temperature used T=18 K rules out thermal diffusion as the operating mechanism. In situ conductivity and the extracted electron mean free paths are larger at lower temperatures further supporting the improved quality (i.e. smaller interface width) of the film.

TABLE 1

SYSTEM	TEMPERATURE	METHOD	REFERENCE
Fe/Ag(100) Cu/Ag(100)	T=77 K	RHEED oscillations	**Egelhoff/Jacob** *Phys. Rev. Lett. 62, 921 (1989)*
Ni/W(110)	T=90 K	Sharp LEED spots	**Whitten/Gomer** *Surf. Sci. Lett. 275, L682 (1992)*
Cu/Cu(100)	T=100 K	Helium scattering oscillations	**Ernst *et al.*** *Surf. Sci. Lett. 275, L682 (1992)*
Ni/Mo(011)	T=5 K	Sharp LEED spots	**Fedorus *et al.*** *Surf. Sci. 317, 170 (1994)*
Ag/Ag(111)	T=150 K	RHEED oscillations	**Roos/Tringides** *Surf. Sci. 302, 37 (1994)*
Ag/Si(100)	T=300 K	STM islands	**Samsaver *et al.*** *Phys. Rev. Lett. 63, 2830 (1989)*
Pb/Si(111)	T=16 K	RHEED oscillations	**Jalochowski/Bauer** *Phys. Rev. B51, 7231 (1995)*
Ag/Si(111)	T=100 K	STM smooth multilayer	**Meyer/Rieder** *Appl. Phys. Lett. 64, 3560 (1994)*
Si/Si(100)	T=160 K	STM dimer formation	**Wolkow** *Phys. Rev. Lett. 74, 4448 (1995)*
O_2/Al(111)	T=300 K	STM islands	**Brune *et al.*** *Phys. Rev. Lett. 68, 624 (1992)*
Xe/Pt(111)	T=4 K	STM islands at steps	**Weiss/Eigler** *Phys. Rev. Lett. 69, 2240 (1992)*

Table 1: Different systems studied at low temperatures with different methods (RHEED, LEED, STM, in situ conductivity) show that high quality films can be grown at low temperatures despite the absence of thermal diffusion

Low temperature growth has also been studied in several other systems with similar conclusions about the improved quality of the film despite the absence of thermal diffusion. We list such studies in Table 1 by including both semiconducting and metallic substrates since the observation (and the lack of well documented operating mechanism) are equally puzzling. In some cases thermal diffusion barriers are known from independent studies (i.e. 0.47 eV for Ni/W(110), 0.36 eV for Cu/Cu(100), 0.67 eV for Si/Si(100), etc.) to imply at most 1 hop/year(!) at the corresponding growth temperatures clearly ruling out thermal diffusion as the explanation. The table is very concise listing only the growth temperature , method used to deduce the film morphology and the literature reference. A short clarification about the methods used: in addition to diffraction intensity oscillations and STM described previously, simply the quality of the LEED pattern can be used to deduce the smoothness of the film if the diffraction spots are sufficiently sharp (i.e. less than 20% of the Brillouin zone) and no diffuse intensity develops in the background.

4 Low Temperature Growth Mechanisms

It has been suggested [18] that one of the key parameters that controls the growth mode (whether layer-by-layer or 3-dimensional) is interlayer diffusion i.e. diffusion from higher to lower levels. In systems where interlayer diffusion is high it is expected that incoming atoms arriving at the top layer have a higher probability to move to the lower layer (and occupy vacant sites to complete the layer) before they can nucleate new islands on top and initiate the nucleation of a new layer. Several of the factors controlling interlayer diffusion have been identified: the size of an additional barrier at steps that partially reflects atoms approaching steps to remain on the top layer, suitable foreign low level impurities specific to a particular system (i.e. surfactants) that can modify both the barriers on the terrace and at steps or simply the size (and shape) of the initially formed islands (with atoms having a higher probability to reach the edges of small or irregular islands than the edges of large and compact islands). It is clear that this third possibility is enhanced during growth at low temperatures (or high flux rates) i.e. diffusion lengths are reduced so there is a higher probability to nucleate many islands which results in high island density of small and ramified islands. This increase in island density is further enhanced if a step edge barrier is present in the system. It is tempting to use this change in the nucleated island density as a generic explanation of all low temperature growth phenomena. However there are certain implications that are not fully consistent with all measured quantities. A high island density mechanism can produce films with excellent vertical order (i.e. sharp interfaces) but low lateral order at any stage of the growth (since it is exactly the "swiss cheese" geometry of the film that promotes interlayer diffusion). Poor lateral order in the film should imply broad diffraction spots of large FWHM (larger than 30% of the Brillouin zone corresponding approximately to an average island size of 3 lattice spacings) and low intensity oscillations (since the top level remains continuously ramified). Both these conclusions are inconsistent with the sharper spots observed (in most of the systems listed in Table 1 studied with diffration the FWHM is less than 20% of the Brillouin zone) and the strong amplitude of the oscillating intensity. Additionally within one oscillation certain coverage dependence should be present since one expects the high island density (and smaller islands size) to be especially present at the onset of the level but as the coverage of the level increases (for coverage higher than 0.5) more of the "holes" in the "swiss cheese" should be filled (i.e. the average island size is larger) so interlayer diffusion should be suppressed. No such coverage dependence has been observed.

For heteroepitaxial systems (and more specifically for Ag/Si(111)) the change in the lattice structure and energetics with Ag thickness should be considered. After the completion of a smooth Ag layer the growth of additional Ag is more analogous to the growth of Ag/Ag(111) than the growth of Ag/Si(111). It is known that for Ag/Ag(111) growth is 3-dimensional because interlayer diffusion is dramatically reduced, the diffusion barrier for Ag/Ag(111) is very low (E=0.1 eV) and a step edge barrier of approximately the same size favors large size nucleated islands and limited mass transport to the lower level. However most likely the transition from Ag/Si(111) to the Ag/Ag(111) should be a gradual one, over several layers. In such intermediate layer a considerable amount of strain is present that can be relieved through the formation of lattice imperfections. The low temperature STM study [18] has shown the presence of a pair of partial screw dislocations. It is possible that in such non-uniform layer the terrace diffusion barrier is not as low and the step edge barrier not as high as for the case of perfect Ag/Ag(111). Under these conditions a high nucleated island density and quasi layer-by-layer growth are still possible as observed experimentally. It is important to show the existence of this strained layer and to determine the critical thickness necessary before the Ag film behaves indistinguishably from a Ag(111) single crystal. These imperfections are most likely formed because of kinetic limitations and if they are annealed by heating the film to higher temperatures (T<480 K before the $\sqrt{3}\times\sqrt{3}$ is formed) the same energetic barriers as measured on Ag/Ag(111) are deduced [19].

A different explanation has been proposed to account for the presence of diffraction intensity oscillations and non-random growth at low temperatures despite the absence of thermal diffusion [12]. An incoming atom has considerable excess energy over an atom already adsorbed on the surface despite its modest thermal energy. The excess energy is determined by the desorption energy and for Ag/Si(111) is close to 3 eV. Eventually this energy is transferred to the substrate lattice in the form of phonons, electron-hole pairs and other low energy excitations. The transfer occurs over a very fast time scale (several picoseconds) so it is difficult to observe it directly in time resolved experiments. If the energy transfer is not very efficient and the incoming atom retains for some time a fraction of the excess energy, the atom can perform several lateral hops until finally adsorbs at a site further away from its original landing site. Depending on the jump length such motion can be sufficient to carry the atom at defect sites (i.e. edges of islands, vacancies, etc.) to eliminate film imperfections and complete the growing layer. Other mechanisms close variants to the concept of transient motion have been proposed: incoming atoms landing at step edges, island perimeter and other low coordination sites can "knock-off" loosely bound substrate atoms which in turn can move laterally over large distances to eliminate defects. The transient motion can be further assisted by attractive fields around steps that can guide the "flying" atoms towards islands to promote the completion of the layer. Since no direct experimental test of this transient mechanism has been documented with time resolving techniques, its validity has been mainly tested with molecular dynamics simulations that are best suited for this ultrafast picosecond time regime.

We have carried out such simulations to understand the growth of Ag/Si(111) at low temperatures [20]. Since no potential describing the interaction between Ag and Si atoms is available, we have used Si-Si potential with the mass of the incoming atom increased to the size corresponding to the heavier Ag atom. The trajectory of an incoming atom was followed from its landing point to the final adsorption site over several hundred picoseconds. In a fraction of the runs (25%) we observe that the distance traveled by the atom can be a few lattice constants (2-3) comparable to the lattice size used. We identify several factors that can affect the efficiency of energy transfer and the length of transient jumps: (i) The type of interatomic potential used (runs with the classical Stellinger-Weber potential show less transient motion than runs with a potential that includes three-body interactions). (ii) Landing site and the overall topology of the substrate unit cell. (If the atom lands at a point where the potential energy has large lateral variation then the incoming atom can move larger distances; while transient motion is suppressed if the atom lands at points of energy extrema, maximum or minimum. Landing sites with large lateral variation of the energy resemble steps, which implies that transient motion should be more pronounced at finite coverage where steps are more abundant). (iii) The implementation of the equilibration scheme. (If the number of substrate layers held at temperature of the heat bath (by velocity renormalization) increases then faster energy transfer and less transient motion is observed). The role of the 2-dimensionally of the substrate unit cell has also been emphasized in molecular dynamics simulations of Si deposition on Si(100) [21]. It is well-known that this surface dimerizes to lower its overall free energy and the atom diffusion barrier is E=0.67 eV along the channels between dimer rows. It was found however that diffusion during deposition is possible on top of the dimer rows (with a lower barrier, E=0.29 eV) because the incoming atoms can be trapped at metastable sites different from the equilibrium adsorption sites. The metastable sites are more populated because they are easily kinetically accessible to the deposited atom. This "population inversion" provides another diffusion path with a lower barrier than thermal diffusion and can explain why mass transport is possible at low temperatures.

The importance of the excess energy of an incoming atom has also been studied in another molecular dynamics simulation of epitaxial Si growth, with the use of the Stellinger-Weber potential [13]. A large degree of crystalline order was observed at growth temperatures as low as 500 K (depending on the grown surface orientation) which implies that atom movement and bond rearrangement is possible despite the low temperatures. This was attributed to a "local heating" effect where the substrate temperature around the landing site is higher than the temperature of the rest of the lattice (as measured by the time necessary for the energy of the incoming atom to dissipate to the lattice) and provides the thermal energy necessary to overcome the rearrangement barriers. Such processes are completed over the same equilibration time scale as transient motion(several picoseconds) and they should result in the same flux rate independence of the growth (as observed for the Ag/Si(111) system [1]).

5 Conclusions

Low temperature growth has been discussed mainly in connection to the Ag/Si(111) system at low temperatures (T=150 K). Surprisingly several experiments (RHEED intensity oscillations, in situ conductivity measurements, STM interface roughness analysis and molecular dynamics) indicate that quasi layer-by-layer growth can operate despite the low temperature. This suggests that better films can be grown at low than room temperature. Similar effects have also been observed in other systems at temperatures as low as 18 K.

The explanation for such phenomena is not fully developed. One possibility is the enhancement of interlayer diffusion because of the higher island nucleation density but not all observations (large lateral lengths) can be accounted for. Mechanisms related to the efficiency of the energy transfer of the atom to the substrate can provide non-thermal paths for mass transport but a more direct confirmation is necessary. Additional work especially on the theoretical side, to provide alternative mechanisms is needed to account for the universality of the unexpected phenomena in so many different systems.

Acknowledgements: Ames Laboratory is operated for the U. S. Department of Energy by Iowa State University under Contract No. W-7405-Eng-82. This work was supported by the Director for Energy Research Office of Basic Energy Science.

References

1a. K.R. Roos and M.C. Tringides, *Phys. Rev.* **B47** (1993) 12705.
1b. K.R. Roos and M.C. Tringides, *Surf. Sci.* **302** (1994) 37.
2. Z.H. Zhang, S. Hasegawa and S. Ino, *Phys. Rev.* **B55** (1997) 9983.
3. M. Jalochowski, M. Hoffman and E. Bauer, *Phys. Rev.* **B51** (1995) 7231.
4. M.C. Tringides, in *Handbook of Surface Science* vol. 1, "Physical Structure" ed. W.N. Unerlt (Elsevier,1996).
5. C.S. Lent and P.I. Cohen, *Surf. Sci.* **139** (1984) 121.
6. K.R. Kimberlin and M.C.Tringides, *Jour. Vac. Sci. Tech.* **A3(2)** (1995) 462.
7a. M. Jalochowski and E. Bauer, *Phys. Rev.* **B37** (1988) 8622.
7b. M. Jalochowski and E. Bauer, *Phys. Rev.* **B38** (1988) 52729.
8. S. Hasegawa and S. Ino, *Phys. Rev. Lett.* **68** (1992) 1192.
9. R. Schad, R. Heun, T. Heidenblut and M. Henzler, *Phys. Rev.* **B45** (1992) 1143.
10. E.P.Th. Suurmeijer, R. Benedictus, A. van der Stadt and T.M. Klapwijk, *Appl. Surf. Sci.* **70/71** (1993) 452.
11. S. Tosch and H. Neddemeyer, *Phys. Rev. Lett.* **61** (1988) 349.
12. W.F. Egelhoff and I. Jacob, *Phys. Rev. Lett.* **62** (1989) 921.
13. G.H. Gilmer and C. Roland, *Appl. Phys. Lett.* **65(7)** (1994) 824.
14. H. Xu, Y. Yang and T.Engel, *Surf. Sci.* **255** (1991) 73.
15. Y. Horio, A. Ichimiya, *Surf. Sci.* **298** (1993) 261.
16. D. Stauffer and A. Aharony, *Introduction to Percolation Theory, 2nd edition* (Taylor and Francis, London, 1992).
17a. G. Meyer and K.H. Rieder, *Appl. Phys. Lett.* **64** (1994) 3560.
17b. G. Meyer and K.H. Rieder, *Surf. Sci.* **331** (1995) 600.
18. R. Kunkel, B. Poelsema, L.K. Verheij and G. Comsa, *Phys. Rev. Lett.* **65** (1990) 733.
19. K.R. Roos and M.C. Tringides, *Surf. Sci.* (1997) in press.
20. R. Biswas, K. Roos and M.C. Tringides, *Phys. Rev.* **50** (1994) 10932.
21. Z. Zhang, Y. Lu and H. Metiu, *Surf. Sci. Lett.* **248** (1991) 2250.

Growing atomically flat metal films on semiconductor substrates

Chih-Kang Shih
Department of Physics, University of Texas, Austin, TX 78712

In thin film epitaxy, the growth of metallic film on semiconductor is particularly challenging because such a system is usually non-wetting. By using a two-step approach, low temperature deposition followed by room temperature annealing, we have discovered that atomically smooth Ag films can be grown on a *non-wetting* substrate of GaAs (110). Growth at low temperature (120 K) yields a rough film with small 3D clusters. Upon annealing to room temperature, however, a very striking phenomenon is observed: Instead of forming a film with larger clusters, an atomically smooth film is formed. If the deposition is below a critical thickness (~ 7 ML), all atoms in the overlayer can self-organize to form a film at this critical height and any deficiency from this critical amount is accommodated by creating some completely empty pits. If the deposition is above this critical thickness, then an flat film is formed and the excess atoms form one monolayer high terraces. This striking behavior could not be explained by traditional growth mechanisms and has stimulated the proposal of a novel "electronic growth" mechanism. Experimentally, Ag on many other III-V compounds exhibit a similar behavior: the existence of a thickness threshold above which atomically smooth metal films can be generated. The flat Ag-film on GaAs exhibits another very intriguing feature: The underlying Ag (111) network is modulated by a 1-D quasi-periodic superstructure. This represents a new type of quasicrystal that is fundamentally different from the commonly known quasicrystals which possess at least two-dimensional rotational order.

Introduction

Understanding the kinetic processes leading to flat layer-by-layer growth of epitaxial thin films [1,2] is important to areas such as quantum device engineering, where precise control of each atomic layer is desired. In terms of semiconductor on semiconductor, there has been significant advancements as a result of lasting accumulation of knowledge. On the other hand, formation of smooth films of metal on semiconductor has proven to be much more difficult because these systems tend to be non-wetting. Nevertheless, strong technological driving forces have kept alive vigorous efforts to find ways to achieve smooth film growth.

The growth mode which characterizes a non-wetting system is the Volmer-Webber growth where the overlayer forms 3-D islands. Ag on GaAs (110) has been considered as a prototypical non-wetting system as all the previous studies revealed 3-D island morphology for growth at room temperature or above [3]. For many years, the use of "surfactant" to suppress 3-D island growth has been at a focal point. Here we describe a method which produces smooth metal film growth on traditionally "non-wetting" semiconductor substrates without the use of surfactant. The method entails an intriguing two-step process: a certain amount of metal is first deposited onto a semiconductor substrate at low temperatures, followed by annealing at some higher temperature [4]. Most strikingly, there exists a thickness threshold above which "atomically flat" metal films can be generated. This novel growth mode was first discovered in the system of Ag on GaAs(110) surfaces [5], and was then noted with Ag on other III-V compounds [6]. This type of growth does not belong to any of the three known growth modes in heteroepitaxy. Furthermore, conventional theoretical approaches which emphasize atomic-level understanding of various kinetic processes and related stress effects can not explain this novel phenomena. This discovery has stimulated many theoretical efforts trying to explain this novel growth phenomenon [7]. Without digging into these new theoretical efforts, this paper focuses mostly on the experimental description of this novel growth phenomenon.

Another interesting phenomenon related to atomically flat Ag film on GaAs is the existence of a novel quasiperiodic structure. On this (111)-oriented two-dimensional Ag-film, the underlying lattice is modulated by a *one-dimensional* quasiperiodic superstructure. Until now all the known quasicrystals possessed at least two-dimensional rotational order. This new 1-D quasi-periodic arrangement of superstructure therefore can be considered as a new type of quasicrystal.

Experimental

The Ag growth on the cleaved III-V samples is described below. Atomically flat III-V (110) substrates were created by cleaving samples in ultra-high vacuum (UHV) with a base pressure $< 4 \times 10^{-11}$ torr. The Ag source, being the Ag bead on a filament type, is fully outgased before cleaving the sample. During the deposition, the chamber pressure is maintained below 1 x 10^{-10} torr. The deposition rate ranging from 0.3 Å/min to 1.5 Å/min. The thickness of the deposition is determined using a crystal thickness monitor. Accounting for the uncertainty in the geometrical factor, we estimate the absolute deposition thickness is accurate to within 15%. The relative accuracy of the thickness, however, is more accurate. Ag epitaxy was studied using scanning tunneling microscopy (STM) in UHV in two ways. First, Ag was deposited onto a GaAs substrate held at low temperature (120 K) but imaged only after it has been annealed up to room temperature. In this case, the typical warming up time to room temperature is about 6-10 hours. Second, Ag was deposited onto a GaAs substrate held at low temperature and its temperature dependent evolution is monitored by variable temperature STM [8] while warming up to room temperature. The warming up time in this later case is much longer, usually more than 24 hours. Low energy electron diffraction (LEED) measurements were also performed to obtain reciprocal space information.

Novel growth mode

For a nominal deposition of 15 Å of Ag on GaAs (110) at room temperature, we observe the characteristic 3-D island growth morphology which had already been reported [3]. The oblong-shaped clusters (Fig. 1(a)) are typically several hundred Å in length and 20-30 Å in height. For a similar thickness (15 Å) of Ag deposited at 135 K, the as-deposited film is composed of Ag nanoclusters with a fairly narrow size (about 2-3 nm diameter) distribution which completely cover the substrate (Fig. 1(b)). The formation of the nanocluster film is a direct consequence of a much reduced diffusion coefficient of Ag atoms at low temperature. Upon subsequent annealing to room temperature, a perfectly flat, epitaxial, single crystal film with a uniform thickness is formed with a few rectangular pits (Fig. 1(c)). These small pits extend down to the GaAs substrate. From their depth, the thickness of the film is measured to be about 15 Å as determined using STM. Inside these pits (Fig. 1(d)), atomic rows with spacings exactly matching the lattice constant of GaAs, 5.65 Å were observed. Additionally, there are a few Ag clusters scattered around inside the pit. Outside the pit on the surface of the Ag film, one sees stripes with two characteristic spacings of about 13 Å and 17 Å. Most interestingly, there exists a well-defined sequential arrangement of these two characteristic spacings, forming a quasi-periodic superstructure [9].

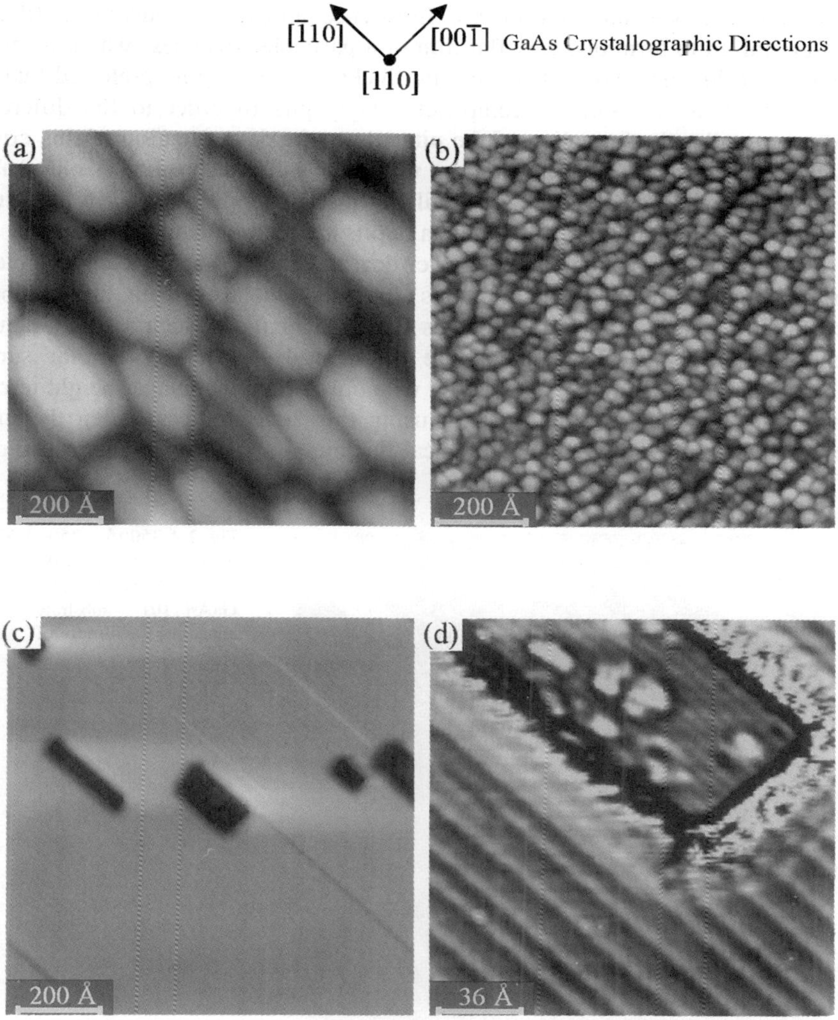

Fig. 1 (a) Silver 3-D islands created by depositing 15 Å of Ag at room temperature (RT) on a clean GaAs(110) substrate. (b) Low temperature (145 K) STM image showing the nanocluster morphology. (c) Silver 2-D film created by depositing 15 Å of Ag at 135 K on a clean GaAs(110) substrate and annealing the film to RT. (d) Curvature image of one of the pits in the film. The GaAs atomic rows are seen at the bottom of the pit; the quasi-periodic superstructure is revealed at the top.

The as-deposited Ag nanocluster film varies in thickness arising from the statistical nature of the deposition process, yet the final film after annealing is atomically flat over thousands of Å's. In essence, all the Ag atoms in the film are self-organized to form this very flat film of a particular thickness which is termed the critical thickness [10]. This specific thickness is so highly preferred that any deficiency from it results in completely empty pits in order to the difference. Experimentally, the formation of the pits allows the determination of this critical thickness to be 15 Å. Even in the case where the initial deposition is only half of this critical thickness, i.e. 7.5 Å, the film still has a tendency to form flat film at this particular thickness with larger area of pits to accommodate the deficiency.

If the initial deposition exceeds the critical thickness, the surface is completely covered by the Ag film, and 2-D islands of single monolayer height are distributed across the surface. Shown in Fig. 2 is the STM image of such a film when the deposition is about 17 Å, just above the critical thickness. Similar surface morphology (complete covered Ag-film with only single monolayer height terraces on the top) is found for films grown using this 2-step process when the initial coverage is much higher than the critical thickness (say, for example, a thickness of 50 Å).

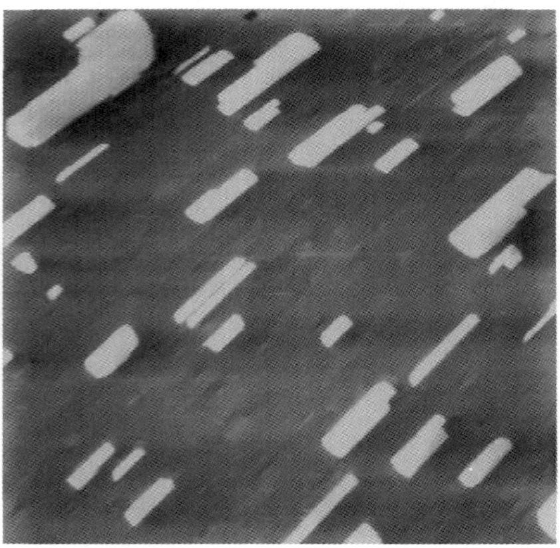

Fig. 2 A 3500Å x 3500Å STM image of Ag film by depositing 17 Å of Ag at 135 K on a clean GaAs(110) substrate and annealing the film to RT.

This novel growth behavior of atomically flat film growth **above** a thickness threshold does not fall into any of the existing three growth modes for heteroepitaxy. In fact the "critical thickness" described here has an opposite sense to the "critical thickness" in the Stranski-Krastonov growth mode. In S-K growth, flat film growth exists **below** the critical thickness [10].

The coverage dependent studies allow us to establish a qualitative behavior of the free energy of the flat Ag film versus the film thickness as shown in Figure 3. If the initial deposition is lower than the critical thickness, upon annealing to room temperature, the film will phase-separate into regions of flat film at the critical thickness and regions of zero thickness (pits).

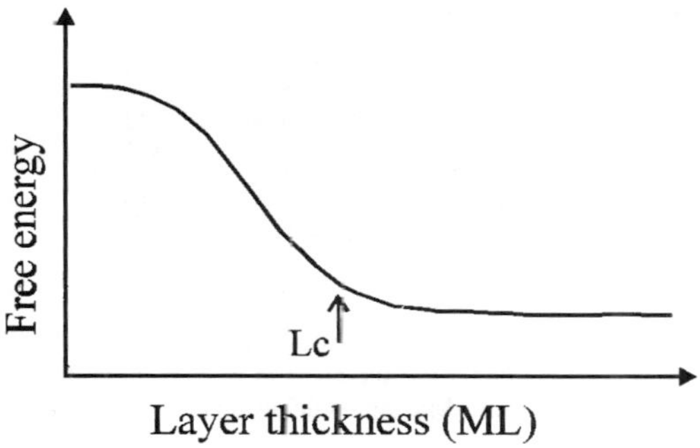

Fig. 3 Schematic diagram showing a qualitative behavior of the free energy as a function of the layer thickness, as suggested by the experimental data.

This novel growth phenomenon has stimulated the proposal of a novel "electronic growth" concept by Zhang, Niu and Shih [7]. In this theoretical model, by considering the energetic of the quantum confinement of the itinerant electrons and the charge transfer between the electronic system and the semiconductors, indeed a free energy curve similar to the one shown in Fig. 3 has been found for Ag/GaAs. The theoretical model further predicts other "electronic driven" novel growth phenomena. Such a novel "electronic growth" concept if proven to be true, will be the first compelling example in film growth in which the preference of the electrons to occupy certain quantum-mechanical states dominates the overall rearrangements of the atoms to prepare such states. Subsequent theoretical calculation based on the first principle density function theory appears to support this novel concept [11].

Experimentally, we have studied Ag on other III-V compounds. Many of them show a similar behavior: A thickness threshold above which atomically flat film can be generated by using the two-step process. Here we only show the example of Ag/GaP(110). Shown in Fig. 5 are two STM images of Ag film grown on GaP surfaces, one of them with a 10 Å deposition (Fig. 5a) and the other one with a 20Å deposition (Fig. 5(b)). In Fig. 5(a), one observes the film at 17 Å height with many voids exposing the GaP substrate and the area of the voids roughly account for the deficiency of the deposition. Apparently 17 Å is the critical thickness for this system. Furthermore, 20 Å Ag deposition results in a complete film with terraces of a single monolayer in height on the top as shown in Fig. 5(b).

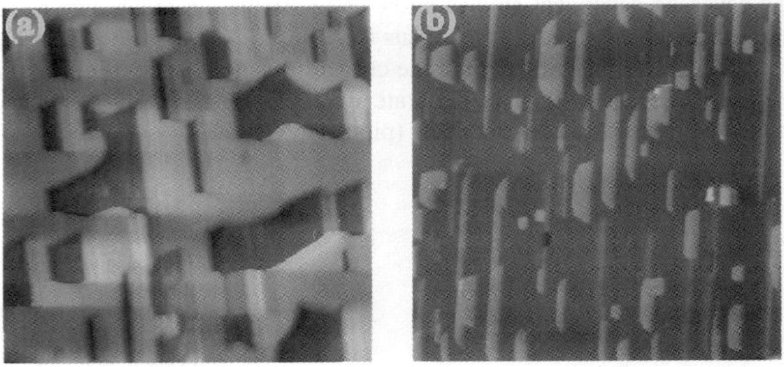

Fig. 4 (a) STM image of Ag/GaP(110) grown using the two-step process with a deposition thickness of 10 Å. The height difference between the bottom of the pits and the flat film surface is about 17 Å. (b) STM image of Ag/GaP when the deposition thickness is 20 Å.

Even though GaP and GaAs differ in band gap and lattice constants, the qualitative behavior of Ag growth is very similar: two-step process leads to the growth of atomically smooth film defined by a thickness threshold. In the case of Ag/GaAs (110), this thickness threshold is about 15 Å (~ 7 ML) and in the case of Ag/GaP (110) about 17 Å (~ 8 ML).

LEED analysis of the sample covered with flat silver film suggests a (111)-oriented surface and its lattice spacing is determined from the primary diffraction spots to agree closely with the value for bulk Ag(111) [12]. This is true for both Ag/GaAs and Ag/GaP. In addition to the main hexagonal diffraction spots, the LEED pattern contains many fractional spots between the main spots along the k-direction in parallel to the [001] direction of the substrate. Shown in Fig. 4 is LEED pattern for Ag/GaAs. Consistent with the superstructure modulation described in Fig. 1(d), these fractional spots are not regularly spaced. However, the positions of these fractional spots show a well-defined hierarchy which can be expanded as linear combinations of two basic units, a and b, with integer coefficients, where $a \approx 0.42$ and $b \approx 0.58$ (Fig. 4). This well-defined hierarchy of the diffraction spots signifies that there exists a well-defined real-space sequence of the quasi-periodic superstructure. For Ag/GaP, fractional spots are also observed, however, without a well-defined hierarchical structure. The discussion below will focus only on the quasiperiodic structure of Ag/GaAs.

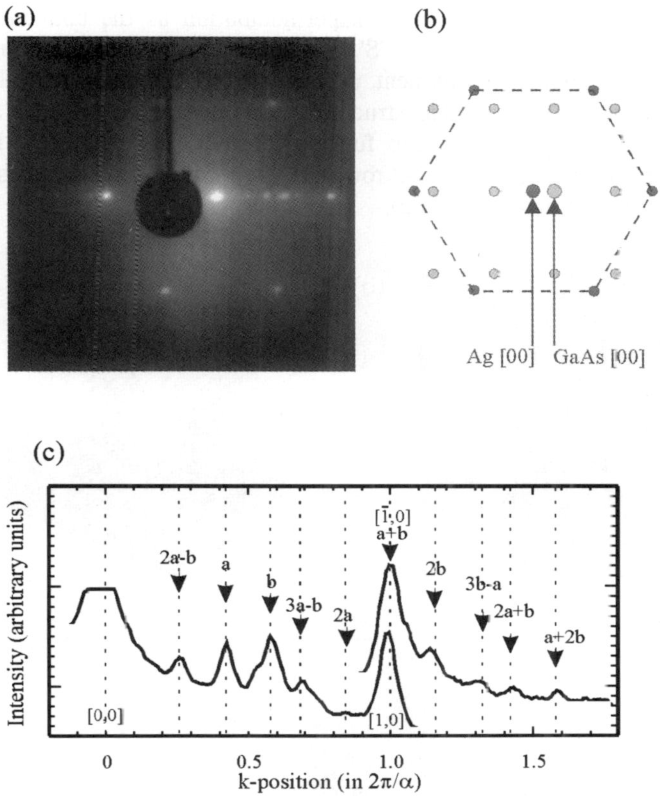

Fig. 5 (a) The LEED pattern acquired on the atomically flat silver film on GaAs(110). (b) The corresponding positions of the primary LEED spots for the Ag-film and the GaAs(110) substrate. Using the GaAs pattern as the reference, we determine that the primary LEED spots of the Ag film is the same as those of the Ag(111) surface. (c) The corresponding intensity profile along the direction of uniaxial modulation which is parallel to the substrate GaAs [001] direction. The spots between $[\bar{1},0]$ and $[\bar{2},0]$ are folded to be between [1,0] and [2,0]

Quasi-periodic superstructure

In the high resolution STM image shown in Fig. 6(a), the underlying hexagonal lattice is evident from the atomic row directions. Atomic rows appear to group together forming a chain-like structure running in parallel to the $[\bar{1}\,10]$ direction of the GaAs substrate. The apparent heights of these atomic chains are modulated by a superstructure with two characteristic spacings, one about 13 Å and the other about 17 Å, corresponding to 5 and 7 atomic row spacings in ideal bulk silver crystals. By using the valley of the supermodulation as the boundary the two characteristic spacings are labeled as 'S' for short segment and 'L' for long segment (Fig. 6(b)). Using such an assignment, one can also observe that the peak of the L-segment contains a double ridge structure whereas the peak of the S-segment contains a single ridge. One can further consider that the 7-row L-segment consists of smaller segments of 2-3-2 rows and the 5-row S-segment consists of 2-3 segments. This is shown in Fig. 6(c).

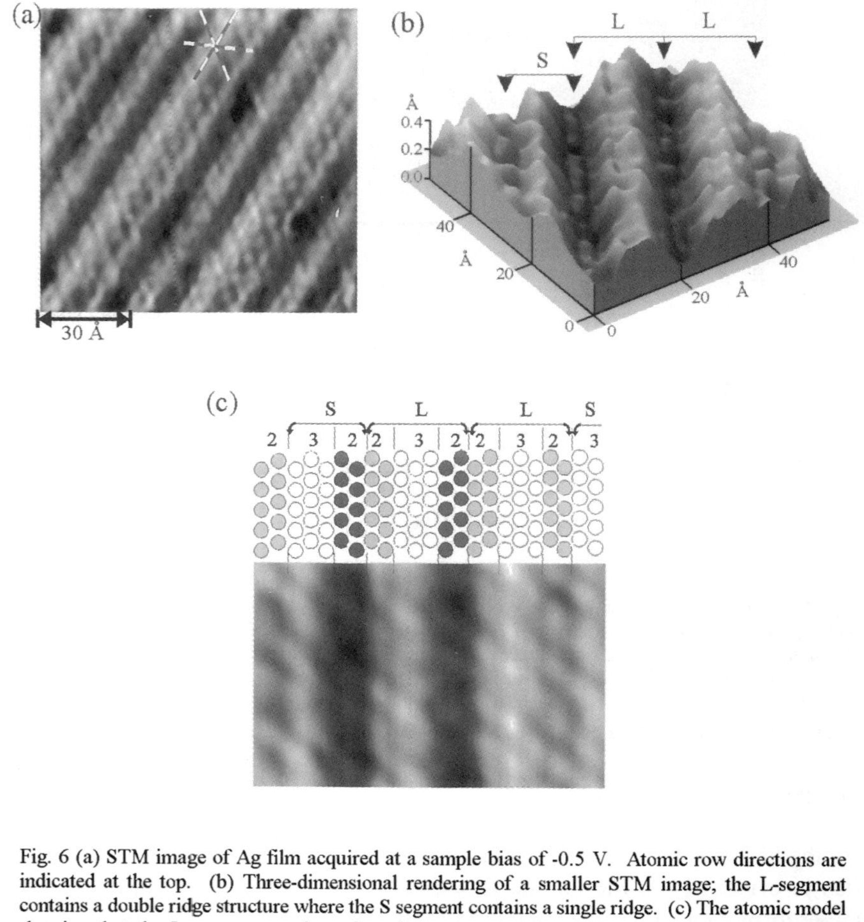

Fig. 6 (a) STM image of Ag film acquired at a sample bias of -0.5 V. Atomic row directions are indicated at the top. (b) Three-dimensional rendering of a smaller STM image; the L-segment contains a double ridge structure where the S segment contains a single ridge. (c) The atomic model showing that the L-segment consists of smaller segments of 2-3-2 rows and that the S-segment consists of 2-3 segments

Consistent with the LEED result, the arrangement of "S" and "L" segments are not periodic, rather they formed a very well defined real-space sequence. Shown in Fig. 7(a) is a 500Å x 500Å STM image showing a large area of quasi-periodic arrangement. Its corresponding reciprocal space image is shown in Fig. 7 (b).

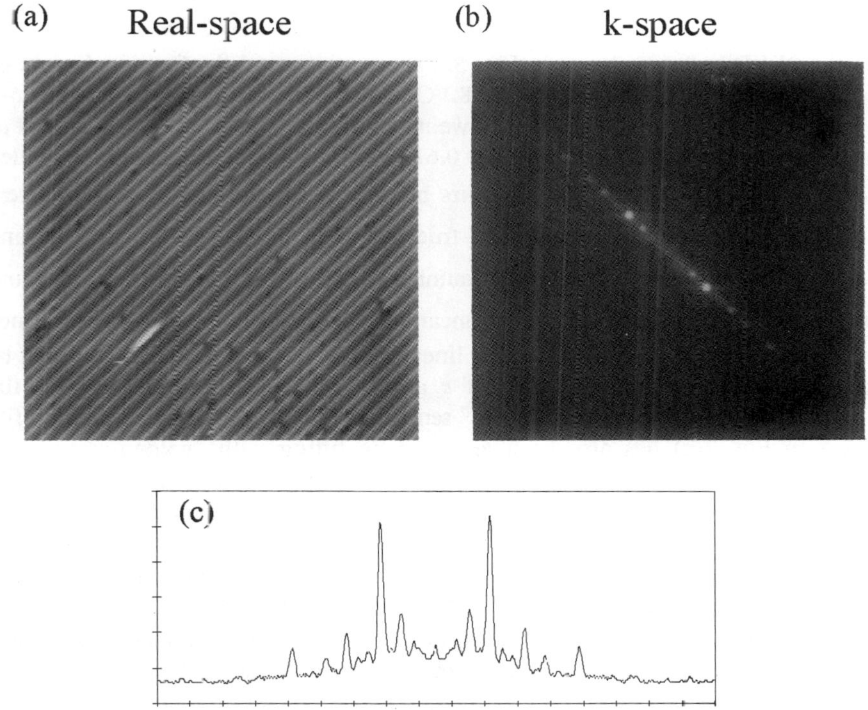

(d)

LSLLSLLSLLSLLLSLLSLLSLSLLSLLSLSLSLLSLSLLSLSLLSLLS

Fig. 7 (a) Real-space STM image (500Å x 500Å). (b) The corresponding Fourier-Transform k-space image. (c) The k-space line profile along the quasi-periodic direction. (d) The corresponding real-space sequence arrangement of "S" and "L" segments.

In the earlier publication [4], we characterized the quasi-periodicity as a "silver mean" sequence primarily based on the LEED pattern because the fractional spots can be expanded as linear combinations of the two numbers 0.58 and 0.42 which are close to two irrational numbers of $1/(\sqrt{2}+1)$ and $\sqrt{2}/(\sqrt{2}+1)$. Furthermore, the sequential arrangements of "S" and "L" segments also matches the "silver-mean" sequence if one also allows the existence of defects within the sequence.

However, subsequent analysis based on a more detailed real space sequence such as the one shown in Fig. 7(a), we have found that the sequence is actually closer to the "golden mean" sequence [13]. Most importantly, the property of "self-similarity" is also satisfied in the assignment of "golden mean."

In the "golden mean" sequence, the fractional spots should be expanded by 1 and $(\sqrt{5}+1)/2 \approx 1.618$ (or alternatively, by 1 and $(\sqrt{5}-1)/2 \approx 0.618$). Let us use the LEED patterns shown in Fig. 5. There we expand the fractional spots by the two numbers $a = 0.42$ and $b = 0.58$. One also find that a spot defined by $2a$-b $= 0.26$. Let us mark this spot as c. If we now use this "c" spot and the "a" spot as the basis, one can easily find that c/a = 0.619 which is actually closer to the golden mean ratio of $(\sqrt{5}-1)/2$. This occurs because the two numbers that we used, 0.42 and 0.58, still deviate from the true irrational numbers of $1/(\sqrt{2}+1)$ and $\sqrt{2}/(\sqrt{2}+1)$. If these two irrational numbers are used for a and b then c/a would still be $\sqrt{2}/(\sqrt{2}+1)$ and the "silver-mean" sequence would be retained. Since every spot that can be expanded as a linear combination of a and b can also be expanded as a linear combination of c and a, one therefore can say that the sequence is closer to the "golden mean" sequence. The reciprocal space spot profile shown in Fig. 7(c) has also been analyzed by fitting with "golden-mean" and "silver-mean" sequence separately. We have found that it can be fitted with "golden mean" sequence with much higher precision but deviates from "silver-mean" slightly.

From the sequence shown in Fig. 7(d), we find that the real-space sequence of "L" and "S" segments also matches with "golden mean" better. Furthermore, by further expanding the "L" segment with a "2-3-2" sequence of smaller segments and the "S" segment with a "3-2" sequence, we find that the "2" and "3" segments now also follow a very well-defined "golden mean" sequence. This indicates that the property of self-similarity is also satisfied in the assignment of the golden mean sequence. A more detailed analysis of the "quasiperiodic" structure will be published in the future [13].

Summary

In summary, by using a two-step process, we have discovered a novel growth mode which leads to formation of atomically flat silver film on various III-V semiconductor substrates. This type of growth does not belong to any of the three known growth modes in heteroepitaxy. Furthermore, the conventional theoretical approach which emphasizes atomic-level understanding of various kinetic processes and related stress effects can not explain this novel phenomena. However, new theoretical efforts which invoke a novel "electronic growth" concept do appear to explain this new phenomenon [7]. In addition, the flat Ag-film on

GaAs exhibits a very intriguing feature of a 1-D quasi-periodic superstructure. This represents a new type of quasicrystal that is fundamentally different from the commonly known quasicrystals which possess at least two-dimensional rotational order. In the future, it will be very interesting to explore if such a system also exhibit novel transport phenomena.

Acknowledgment

This work was supported by the National Science Foundation Grant # DMR-9402938, the NSF-Science and Technology Center, Grant # CHE-8920120. The experimental work was done in collaboration with A.R. Smith and K.-J. Chao. The theoretical work was done in collaboration with Z. Zhang and Q. Niu.

References

1. M. G. Lagally, *Physics Today* **46**, No. 11, 24 (1993); and references therein.
2. E.D. Williams and N.C. Bartelt, *Science* **251**, 393 (1991).
3. B. M. Trafas, Y.-N. Yang, R. L. Siefert, and J. H. Weaver, *Phys. Rev. B* **43**, 14107 (1991).
4. The growth process described has been used by D. A. Evans, M. Alonso, R. Cimino, K. Horn, *Phys. Rev. Lett.* **70**, 3483 (1993). However, it was interpreted that such a process resulted in a uniform size of Ag islands (i.e. quantum dots) which lead to a quantum size effects in photoemission.
5. A.R. Smith, K.-J. Chao, Q. Niu, and C.K. Shih, *Science* **273**, 226 (1996).
6. K.-J. Chao, Q. Niu, Z. Zhang, and C.K. Shih (unpublished).
7. Z. Zhang, Q. Niu, and C.K. Shih (unpublished).
8. A.R. Smith and C.K. Shih, *Rev. Sci. Instr.* **66**, 2499 (1995).
9. G. Gumbs and M.K. Ali, *Phys. Rev. Lett.* **60**, 1081 (1988); M. Holzer, *Phys. Rev. B* **38**, 1709 (1988).
10. As used here, critical thickness has an opposite meaning to the same term in the context of pseudomorphic growth where the epitaxial layer can only be grown up to a certain maximum critical thickness before strain relaxation occurs. For example, see J.J. Fritz, P.L. Gourley, and L.R. Dawson, *Appl. Phys. Lett.* **51**, 1004 (1987).
11. J.-H. Cho et. al. (unpublished).
12. We do, however, find a small shift in the position of the [0,0] beam of the Ag overlayer with respect to the [0,0] beam of the GaAs substrate. The offset angle is ~ 4°.
13. K.-J. Chao, Q. Niu, and C.K. Shih (unpublished).

REMOVAL

SPONTANEOUS HALOGEN ETCHING OF Si

J.H. Weaver

Department of Materials Science and Chemical Engineering
University of Minnesota, Minneapolis, MN 55455

C.M. Aldao

Institute of Materials Science and Technology (INTEMA)
Universidad Nacional de Mar del Plata-CONICET
Juan B. Justo 4302, 7600 Mar del Plata, Argentina

Abstract

This chapter focuses on thermally-activated halogen etching of Si(100)–2×1 at 300-1100 K. The emphasis is on visualization of atomic-scale surface morphologies and understanding the etching mechanisms. Using scanning tunneling microscopy, we examine the etching of steps and terraces and the formation of Si regrowth structures. Changes in these structures are controlled by site-specific desorption energies of silicon dihalides and the halogen concentration on the surface. Analysis of the STM data provides information about the energy differences of the various etching pathways. The protocol developed to analyze etching for Si should be applicable to other systems.

1. Introductory remarks

The dictionary defines etching as the art, process, or act of producing drawings or designs on plates of metal, glass, etc. by the action of an acid. Artists have long used such wet etching in their creative endeavors. Today, this definition of etching should be expanded to encompass dry etching, where the etchants are gaseous or air-borne, and it should take into account both physical and chemical processes. Sputtering by Ar^+ ion impact is an example of dry physical etching while layer-by-layer removal of Si atoms by halogens is an example of dry chemical etching. Combinations of physical and chemical processes are found in semiconductor device fabrication where, for example, patterns with high aspect ratios and submicron resolution are replicated in great quantities by reactive ion (or plasma assisted) etching [1,2]. To model surface modifications associated with such plasma processing, we must consider physical processes when energetic ions bombard the surface, chemical processes when reaction products are formed, and electronic processes involving electrons and holes created by photoabsorption. To date, the details of real-world applications of (industrial) dry etching are only partially understood. In large measure, this is because the processing parameters are coupled in unknown ways.

In general, there are two very different ways to develop an understanding of a complex process. One is to accumulate a large body of empirical information about the overall process so that descriptive models can be developed. The other is to reduce the problem into manageable components, to study each, and then to integrate the knowledge gained. Both approaches seek predictive capabilities, while the latter focuses on basic principles. In this chapter, we take the reductionist approach. By focusing on dry chemical etching of silicon under controlled conditions, we develop a detailed understanding of several key parameters. We consider etching of a particular surface, Si(100), by Cl, Br, and I in the range 300-1100 K as a function of fluence in the low flux regime, expecting that these model studies will serve as a reference point for investigations that consider the effects of photon irradiation and the impact of energetic species.

In this chapter, we review results that show the modification of step structures, the formation and growth of single-layer-deep etch pits on the terraces, and the appearance of Si regrowth islands [3-11]. The emphasis is on the surface morphology and its dependencies on temperature and the halogen concentration on the surface. These surface changes are associated with thermally-activated processes, and the relative importance of each can be explored by varying the temperature in the range 300-1100 K. From the etch patterns for Si(100), we determine reaction pathways and deduce etching energetics and kinetics. A discussion of the protocol involved in this analysis is an important part of this review.

2. Experimental considerations

The experiments reviewed here were conducted in a vacuum chamber equipped with a Park Scientific Instruments scanning tunneling microscope and related surface analytic capabilities (base pressure 7×10^{-11} Torr). The etching configuration depicted at the top of Fig. 1 shows a Si wafer positioned ~3 cm from a solid state source of molecular halogens. The temperature during exposure, 300-1100 K, was monitored with an optical pyrometer and checked periodically with a thermocouple (accuracy ±20 K, reproducibility ±10 K). The sources of molecular chlorine, bromine, and iodine were electrochemical cells formed by fusing mixtures of Ag- and Cd-halides. Embedded silver and platinum leads provide electrical contact. The

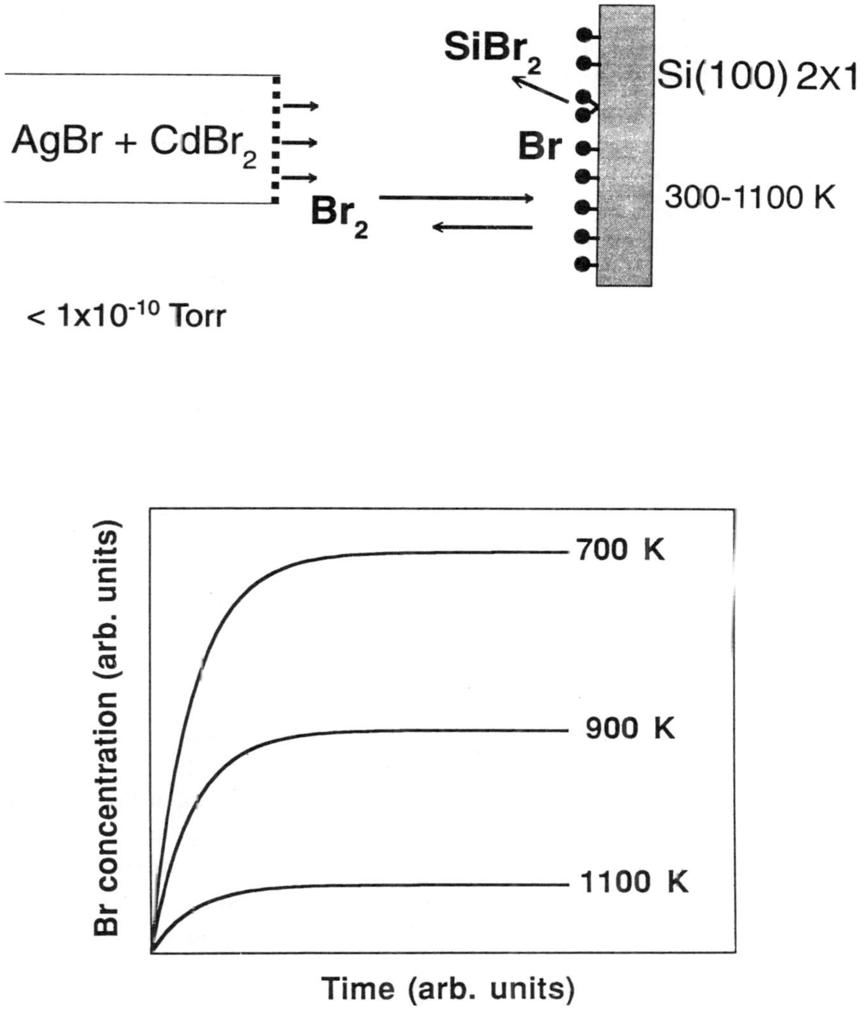

Figure 1. Schematic of the etching configuration emphasized in this review. Molecular bromine (or chlorine or iodine) desorbs from an electrochemical cell and impinges on a Si surface that is held at a temperature between 300 and 1100 K. During exposure, the surface concentration increases with time, ultimately reaching steady state. At 700 K, the etch rate is low and the concentration saturates. With increased temperature, the concentration drops as desorption of Si dihalides occurs. The flux will be important in the regime where there is active etching.

cell current represents the diffusion of halogen ions to the Pt mesh electrode where the ions combine and desorb as neutral molecules. We used cell currents of 2 to 120 μA, corresponding to the release of 6.2×10^{12} to 3.8×10^{14} molecules per second. Sample exposures were done for selected times so that exposures were given in units of mA-s. After exposure, the sample heater and the source were switched off simultaneously. The 1/e cooling rate was approximately 13 s. Once cool, the sample was transferred to the STM stage for imaging. The chamber pressure remained below 1×10^{-10} Torr, even when the sample was at 1100 K.

The Si wafers were oriented within 0.2–0.5° of (100) with a miscut toward [011]. Clean surfaces were prepared by degassing at 875 K for six hours and heating to 1475 K for 1–2 min. Images of clean surfaces showed defect densities of $\sim1.3\times10^{13}$ cm^{-2}, primarily in the form of missing dimers. STM images were obtained with tungsten tips that were produced by electrochemical etching and were cleaned by electron bombardment prior to use. Scan dimensions were calibrated with the structures of the surface and the heights of steps. All imaging was done at room temperature. The images reproduced here were not corrected for thermal drift. The sample bias was typically ±1-3 V. The tunneling currents were 0.3-2 nA.

3. Thermally-activated halogen etching

3.1 General considerations

Most of the measurements discussed here represent etching under constant halogen flux at a fixed temperature. They differ from what is considered in temperature programmed desorption where the surface is exposed at low temperature and subjected to thermal ramp to induce reaction. Procedurally, our samples are first cleaned to obtain well characterized starting surfaces. Their temperatures are then stabilized and they are exposed to halogen beams from source operating at constant current. The concentration of Cl, Br, or I on the surface increases during exposure until it reaches steady state, as depicted for Br in Fig. 1. To characterize the uptake of bromine, we exposed a surface at room temperature to a cell operating at 5 μA. The surface concentration increased to 0.08 ML after 10 s, to 0.23 ML after 30 s, to 0.56 after 90 s, and it had saturated by 600 s. Fitting this curve with a function of the form $dc/dt = f\ (1-\theta)$ gives f, defined as the incident flux times the sticking coefficient, of 0.009 ML/s.

The arrows in the upper part of Fig. 1 represent the flux of incident Br (or Cl or I) molecules, the flux of reflected molecules, and the flux of desorbed $SiBr_2$ molecules. The black dots represent adatoms on the surface. The three fluxes reach a balance at steady state. At 700 K, the etch rate is small and the surface concentration approaches 100%, based on STM imaging. An increase in flux in this slow-etching regime would have negligible effect. At 800 K, the steady state concentration is reduced to 95%. At 900 K, the surface concentration is ~65% (a lower limit because it is hard to image the adsorbates on this complex surface). At 1000 and 1100 K, the Br residence time is short, consistent with temperature programmed desorption measurements [12], and the etch rate is limited by the incident flux. We estimate the desorption energy of $SiBr_2$ to be 2.7 ± 0.1 eV, using the surface coverage measured after etching at 900 K, assuming second order kinetics, a prefactor of 10^{13} s^{-1}, and the same uptake as for the clean surface at 300 K. For Cl (I), the desorption energy should be lower (higher) but post-etch determinations of the halogen concentration have not been done. One of the important contributions of atomic scale imaging is to emphasize the site-specific character of desorption and the related morphology changes.

If measurements are done at room temperature on samples etched at high temperature, one should ask whether morphology changes are possible as the sample cools. In our studies, the rate of cooling is "fast." Since the etch rate is "slow" and the steady state Br concentration is at least 95%, we expect little change in morphology during the quench from temperatures below about 800 K. Changes in morphology are possible for etching above 1000 K, though a sample will cool from 1000 K to 750 K in ~3.5 s. Nevertheless, etching during this cooling time should follow the same temperature and concentration-dependent rules. Step profiles can change (but more slowly than for clean Si because of the presence of some Br), and vacancies can order (though ordering is minimal at 1000 K because of residual Br). The conclusions drawn in this review take these effects into account.

The steady-state etching regime is emphasized in many of our experiments. As far as the surface morphology is concerned, steady state denotes a surface structure that statistically does not change with exposure. Thus, the appearance of a particular feature does change but the statistical populations and ensemble averages do not. Steady state surface conditions were determined at a given temperature by exposures that were long enough to assure that the global morphology did not change.

The dimer character of Si(100) is maintained when exposed to Cl_2, Br_2, or I_2. Saturation coverage corresponds to one halogen per surface Si atom. For I/Si(100), there is a c(4×2) structure that is realized for a (local) coverage of 0.5 ML because of steric repulsion [8]. Saturation for I/Si(100) again corresponds to one adatom per surface atom. Surface concentrations could be higher if two halogens could bond to a single surface atom, as in the 3×1 structure of Si-H [13]. Evidence for the formation of the 3×1 structure has been found for Si-Br, but it occurs only at high temperature and it leaves behind a tell-tale etch pattern, as discussed in Section 3.3. Evidence for a 3×1 structure has not been detected for Cl or I.

Very different surface conditions can be achieved if halogen atoms are able to insert into the lattice. Fluorine, the smallest and most electronegative of the halogens, penetrates the surface to form a disordered reaction layer [2,14-16]. This introduces channels for material removal that allow etching via desorption of SiF_4 at room temperature. Such processes are limited by the kinetics of F-uptake and the reactions that lead to SiF_4 surface moieties. The interested reader is referred to the Si-F literature for details [2,14-17]. To our knowledge, there have been no STM studies for the F-Si(100) system that have extended to the etching regime.

Thermally-activated etching of Si by Cl, Br, and I occurs above ~700 K. One should keep in mind that there is exchange of Si atoms between step and terrace sites at these temperatures for clean Si(100)-2×1. As discussed by Zandvliet *et al.* [18], the equilibrium concentration of terrace adatoms reflects the energetics for the units that detach and attach at the step. For Si(100)-2×1, the relevant unit is made up of a double dimer of four atoms. Figure 2 depicts the attachment of one such unit at a corner site on a kinked step and the detachment of another from a

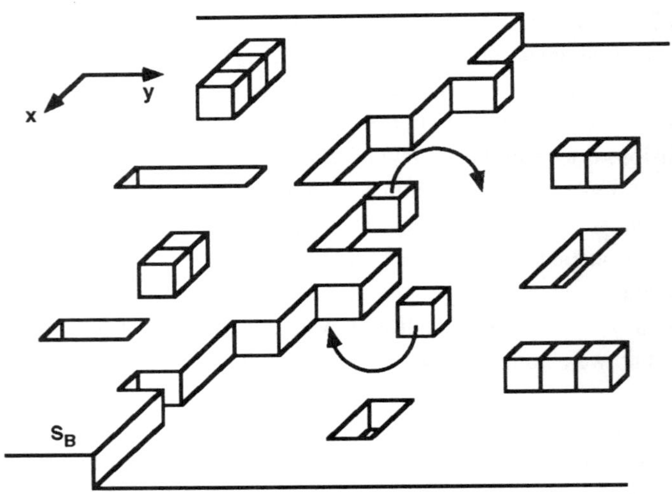

Figure 2. Schematic showing attachment and detachment of blocks from a step. The energies associated with removal are dependent on the number of neighbors and changes in morphology represent those energetics.

protruding position of the step. Block exchange is negligible at 300 K. Exchange at steps has been detected and measured for clean Si(100) above 550 K [19], and there will be Si adatoms on the terraces at 700 K. Bromine uptake during the accumulation stage at 700 K, Fig. 1, reduces the equilibrium terrace concentration of Si adatoms and hampers their diffusion to steps where they can be accommodated. When envisioning etching at 1000-1100 K, the exchange dynamics must always be taken into account because the halogen concentration is low.

Terrace vacancies are mobile at high temperature on clean surfaces. Studies of vacancies created by ion impact have established [20,21] that they diffuse along the dimer row direction and that there is an attractive interaction between vacancies in adjacent dimer rows [22,23]. The result is that vacancy line defects form perpendicular to the dimer row direction. While the impact of this phenomenon is negligible for a Br-covered surface at 700 K, it becomes more important with increasing temperature and correspondingly reduced Br concentration. Such vacancy-vacancy interactions produce extended vacancy lines at 1000-1100 K when the halogen concentration is low enough to allow vacancy diffusion (Section 3.3).

3.2 Bromine chemisorption on Si(100)-2×1

Atoms on the unreconstructed {100} surfaces of Si have two dangling bonds. Dimerization produces a 2×1 / 1×2 reconstruction, lowers the surface energy, and leads to dimer rows oriented along <011>. Dimer rows are perpendicular to each other on terraces that are separated by an odd number of single height steps. Adjacent steps on (100) are inequivalent if the surface is miscut with a surface normal tilted toward [011]. These surfaces have rows that run parallel or perpendicular to the step, defining S_A and S_B steps, respectively [24]. S_A steps are parallel to the rows of the upper terrace. They are smoother than those that are perpendicular to the rows, S_B steps, because S_A steps have lower step energy and a kink in an S_A step introduce a step of S_B character. Kink creation always increases the total step energy, but the energy associated with an S_B step is about 2.5 times greater than that of an S_A step. Thermal excitation produces kinks, and the density of kinks on S_B steps is much greater than that on S_A steps [25,26].

The STM image of Fig. 3 represents clean Si(100) in an area where the miscut was not strictly along <011>. Hence, the mean position of the imaged S_B step is not perpendicular to the row direction of the upper terrace (brighter portion of figure). The dimers and terrace defects can be identified in Fig. 3. Also evident are rows where the dynamic bond buckling has been stabilized and one atom of each dimer is "up" relative to the other. The up atoms are bright under our imaging

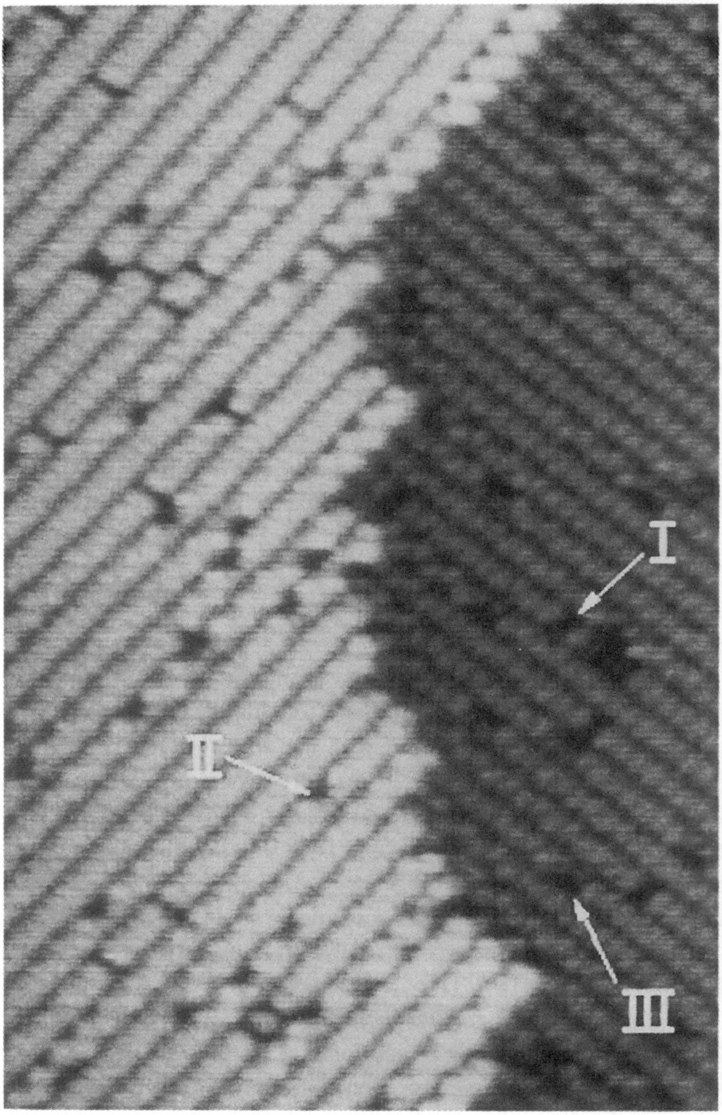

Figure 3. Clean Si(100)–2×1 showing a step that has mixed S_A and S_B character showing the dimer rows of the upper and lower terrace. Dimer buckling is fast at room temperature but defects or steps stabilize the buckling and introduce the pattern seen at the step at the top of the image. Arrow I points to a missing dimer near a complex of three missing dimers. Arrows II and III point to c-type defects.

conditions. Arrow I points to a dimer vacancy near a structure that appears to be derived from three missing dimers. Arrows II and III point to c-type defects [27].

Exposure of Si(100)-2×1 to molecular halogens results in dissociative chemisorption. While it is beyond the scope of this review to consider the dynamics of dissociation processes, we note that Kummel and coworkers [28] have used STM to determine the surface site distribution of adatoms from molecular beams of F_2, Cl_2, Br_2, and ICl. They have related chemisorption structures observed at room temperature to the uptake dynamics and the reactivities of different sites of Si(111)-7×7.

The energy released during dissociative chemisorption can cause transient short-range diffusion. Subsequent larger scale diffusion will be thermally activated. For Si(100)-2×1, there are inequivalent two directions for diffusion. These directions are evident in the top view of Fig. 4 where open circles represent Si atoms and solid circles represent halogens (Br). For Si on Si(100), diffusion is

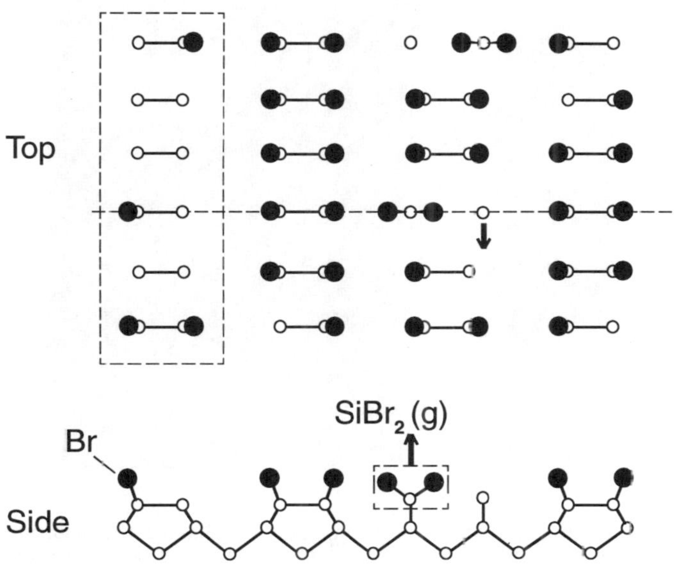

Figure 4. Schematic indicating Br (or Cl or I) adatoms on a Si(100)–2×1 surface in top and side view. The boxed area at the left represents a low Br concentration with Br bonded both singly and pairwise to Si dimers. Pairwise occupancy is favored. A halogen atom can be transferred to form a $SiBr_2$ (a) unit, depicted at the right. The desorption of this unit constitutes etching. Associated with etching is the release onto the surface of the unpaired Si atom (arrow at right). No attempt is made in this figure to show all possible bonding configurations that are involved at high temperature.

considerably more facile along the dimer row direction, with a difference in activation energy parallel and perpendicular to the row direction of 0.2-0.3 eV [29]. Halogen diffusivities on Si(100) have not been measured but calculations by Selloni and coworkers [30] indicate that Cl diffusion along the dimer rows has a barrier of 1 eV. For an isolated halogen as in the left column of Fig. 4, the jump rate along the row would then be $5\times10^{6}\ s^{-1}$ at 800 K for a barrier of 1 eV and an attempt frequency of $10^{13}\ s^{-1}$.

The boxed area of Fig. 4 represents a Br coverage of 0.33. Two isolated adatoms are shown, as are two that are bonded to a single dimer in monobromide configurations. High resolution unoccupied-state images such as that of Fig. 5 reveal the room temperature chemisorption structures. As shown , for a surface

Figure 5. STM image under conditions that emphasize the adatoms and show pairwise occupancy of dimers by Br (or Cl or I). The surface concentration is 8%, as determined by direct counting. The sample bias was 2.3 V.

with a concentration of 0.08, there is a strong tendency for two Br adatoms to bond to the same dimer as SiBr (a). Deviations from pairwise occupation occur during Br accumulation or at elevated temperatures. Pairwise occupancy produces a configuration that is favored by approximately the Si π-bond strength, less the repulsive interaction of the two adatoms on the same dimer [31]. With increasing halogen concentration, there are additional interactions among dimers that must be considered [32], as in the right-hand portion of Fig. 4 where the coverage is 0.85. This configuration represents the steady state concentration during etching at ~800 K. At 800 K, Br hoping between empty sites is possible, there can be bridge bonding, and unstable $SiBr_2$ (a) units can be formed[32]. The latter units are precursors to desorption.

Figure 6 represents a very schematic overall energy picture for halogen interaction with Si, using numbers for Si-Cl. For Cl-Si, initial state corresponds to Cl_2 (g) and Si(100)-2×1. Dissociative chemisorption releases approximately 5.2 eV

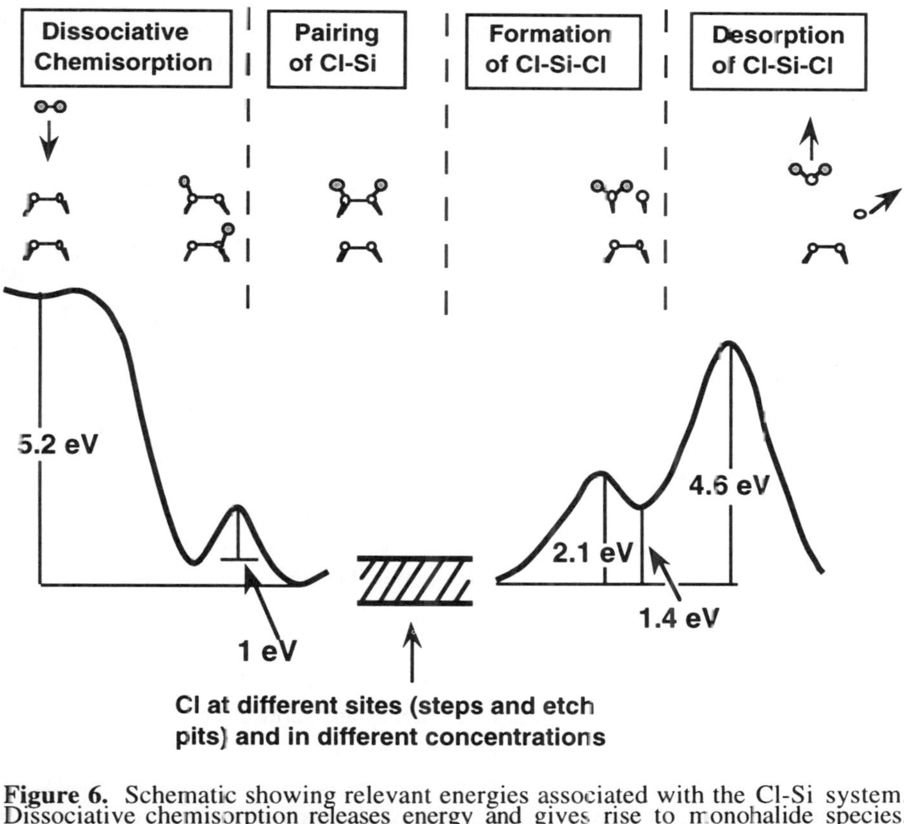

Figure 6. Schematic showing relevant energies associated with the Cl-Si system. Dissociative chemisorption releases energy and gives rise to monohalide species. Pairing occurs when the halogens can diffuse on the surface. The shaded area would include bonding states at steps and defects and would take into account the effects of concentration. The formation of a precursor to etching, $SiCl_2$ (a), and its desorption are depicted using energies calculated in Ref. 30.

per Cl_2 molecule [33]. Adatom diffusion facilitates pairwise bonding. The ratio between singly- and pairwise-occupied dimers in dynamic equilibrium would be $\exp(-\Delta H/kT)$ where ΔH is the enthalpy difference for the two states. Increasing the temperature would increase the number density of single adatoms and also their diffusivity. Adsorbate interactions associated with increased halogen concentration will give rise to energy states within the shaded range of Fig. 6. Also within that range will be configurations associated with steps and etch pits. The right hand side of Fig. 6 is accessed during etching, where $SiCl_2$ moeities form and desorb, and unpaired Si atoms from etched dimers are released onto the terrace. The energy values for Cl-Si(100) are those from Ref. 33 for chemisorption and Ref. 30 for diffusion and etching. Different values for Br and I will account for different rates of reaction and steady state surface concentrations. For I, the interaction energy of pairwise-occupied dimers introduces a c(4×2) structure that is not observed for Cl or Br [8].

3.3 Si(100)-2×1 steady-state etching by bromine

The morphologies for Br etched Si(100)-2×1 can be visualized through the STM images of Fig. 7(a)-(c). Figure 7(a) was obtained after exposure at 700 K and (b) and (c) were obtained after reaching steady state etching conditions at 800 and 900 K [5]. As anticipated in Fig. 2, the surface is modified most easily at steps and defects. This has been demonstrated clearly by Chander *et al.* [3] through studies leading up to and including steady state at 900 K. They reported roughening of both types of steps and an increasing areal ratio of S_A to S_B terraces. Preferential etching at S_B steps can be rationalized through the step formation energies, as for clean surface steps. Similar arguments explain why etching is easier at kinks than on straight steps (in the former, there is no net increase in step length). Dimers that protrude from the step edge have an even lower barrier because their removal reduces the step length, Fig. 2. Terrace pit formation has the highest barrier since it produces both S_A- and S_B-like step units. The importance of these processes can be tuned by varying the temperature.

Figure 7(a) was obtained after exposure to Br at 700 K. Removal at this temperature is dominated by processes at steps, and terrace structures are largely those of the chemisorbed surface. Noteworthy are bright elongated features that are not seen on clean Si(100)-2×1. They represent Si regrowth islands surrounded by 2×1 Si dimers saturated with Br. These chains are short with no features wider than two rows. They are evenly distributed when produced by exposure to halogens at 700 K. Their appearance is more complex when they are produced at higher temperature, an effect that almost certainly reflects their origin (Section 3.4). As

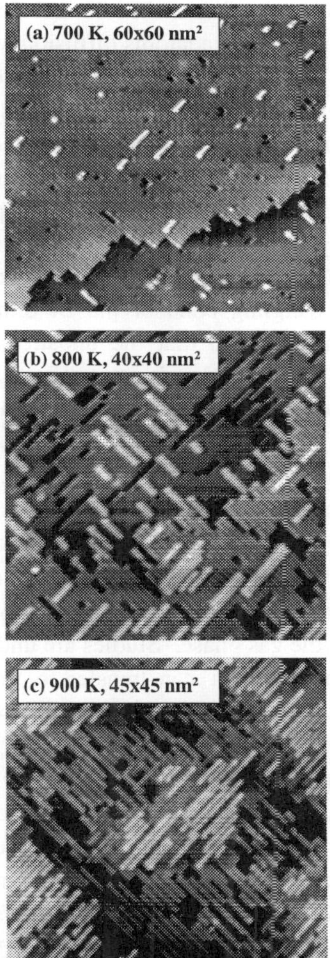

Figure 7. STM images of Si(100)–2×1 exposed to Br_2. The etch pits and regrowth chains increase in number and size with increased exposure temperature. An S_B step crosses image (a) and another is evident in (b). Etching at 700 K, and especially at 800 K, roughens the steps. The terraces remain relatively defect free at 700 K but enhanced terrace etching is evident at 800 K where remnants of the terrace resemble regrowth structures. The morphology achieved by etching at 900 K is quite different because it reflects the formation of a 3×1 phase derived from rows of $SiBr_2$ (a) separated by SiBr dimer rows. $SiBr_2$ desorption results in missing atom rows.

noted in Section 3.1, a clean surface at 700 K includes Si adatoms on the terraces. Halogen accumulation impedes Si adatoms diffusion to the steps and it pushes them into supersaturation. This is probably the only source for regrowth far from the steps at 700 K. No obvious regrowth was observed when the surface was saturated with Br at room temperature and then heated to 700 K.

Etching via desorption of $SiBr_2$ is depicted in Figs. 4 and 6. We have described desorption in terms of the formation of a precursor $SiBr_2$ (a) structure, Fig. 4, accomplished by the transfer of Br from the surface or from the Si atom partner of the dimer. This $SiBr_2$ (a) unit usually decays into Si-Br (a) as the Br atom is returned to the dangling bond of the Si dimer or to the surface. When desorption occurs, we have argued that the unpaired, destabilized Si atom moves onto the terrace. These processes are shown in Fig. 6 for Si-Cl.

Selloni and coworkers [30] have calculated the energies of different Si-Cl configurations and have suggested a scenario in which two of the processes occur in a different order. They have proposed that Cl transfer to form $SiCl_2$ (a) destabilizes the dimer Si atom. This atom has only two backbonds and can escape to the terrace (arrow in center of Fig. 4). The de-excitation pathway of $SiCl_2$ (a) is thermally changed in a way that increases its lifetime. Hence, it has a higher desorption probability. Both pictures result in a dimer vacancy with a Si atom on the terrace and a $SiBr_2$ molecule in the gas phase. Studies are underway to examine the effects of halogen concentration (how Si can transfer to the terrace when the terrace is halogen-saturated) and the implications vis-à-vis the 3×1 patterning of Si(100).

From Figs. 7(a) and (b), it is clear that the morphology changes substantially when the exposure temperature is increased from 700 K to 800 K. In a general sense, the differences can be understood in terms of an increase in all of the thermally-activated processes, with changes in relative rates that favor the higher-energy channels. Terrace pitting and pit growth become much more likely, and step etching is enhanced. From the STM images, surfaces etched at 800 K show significant local height variations because pits and islands constitute ~30% of the terrace area. In steady state, the steps appear rougher because retreating steps encounter already-pitted terrace areas. In steady state etching at 800 K, the islands represent remnants of the higher layer as well as regrowth from Si atoms released by step and pit etching since the ratio of island area to pit area on a given terrace is ~1.5. On a surface like Fig. 7(b) the Br concentration is ~95%. Three-dimensional roughening is inhibited by the energy difference between pit creation and pit growth and the fact that the dimer rows in each exposed layer alternate in direction. No more than three layers are exposed on a given terrace unless a pit is very large.

Figure 7(c) shows a terrace produced by steady state etching at 900 K. Higher resolution images show that it is made up of mixed domains of 2×1 and 3×1 character. The 3×1 domains are composed of Si dimer rows separated by missing rows of Si atoms [4]. This remarkable structure is achieved by continuous exposure in the range ~850–900 K. It can also be produced by saturating at room temperature and then heating to ~850 K [34]. Under these conditions, limited areas develop in which $SiBr_2$ (a) rows alternate with Br-saturated dimer rows. The four Br atoms bonded to three Si atoms in the side view of Fig. 4 are the primary unit (unpaired Si atom in Fig. 4 would form a dimer with the atom to its right). The transition to a 3×1 surface phase requires coordinated bond breaking and redimerization but its formation assures its demise. Desorption of $SiBr_2$ leaves rows of Br-terminated dimers alternating with missing-atom rows. The fact that patterned areas are observed indicates that the desorption of $SiBr_2$ units is slow enough that Br can be redistributed to form 3×1 domains but fast enough that domains are not degraded as the missing rows develop. Once patterned, these areas undergo conventional etching until there are large enough new areas where the halogen concentration can build up and 3×1 conversion can again occur. Of the halogens, only Br has been observed to produce this 3×1 pattern. For Cl, the etch rate is higher at a given temperature and the surface concentration is not sufficient for 3×1 formation, at least under our conditions. For I, the adatom is so large that steric repulsion precludes 3×1 formation. Surfaces etched with F have not been investigated with STM.

A different steady state surface morphology is achieved during Br exposure above ~900 K, and this can be related to the fact that the mean surface concentration decreases, Fig. 1. At 1000 K, the weakened Si-Si bonds facilitate pitting of the terraces. As shown in Fig. 8(a), the surface is characterized by local structures of scattered small pits and patches of clean Si among Br-terminated areas. These single-layer deep defects can expand rapidly, ultimately coalescing to produce large pits. Here, the ratio between pitting and step etching is much greater than it was at 700 K or even 800 K. These processes can expose several layers, depending on the terrace width, because steps continue to retreat across pitted terraces. In this regime, the lower surface concentration limits the overall rate, and etching would increase with increased halogen flux. This is in contrast to the behavior at lower temperature where the surface is saturated and desorption is the rate limiting factor.

Etching at 1100 K again results in a distinctive morphology. At 1100 K, dimer vacancy creation is even more readily accomplished and these vacancies have a reasonably high mobility. As noted in Section 3.1, dimer vacancies in adjacent rows experience an attractive interaction [20-23] and line defects (LD) form perpendicular to the dimer rows, Fig. 8(b). The nucleation of a line defect requires a

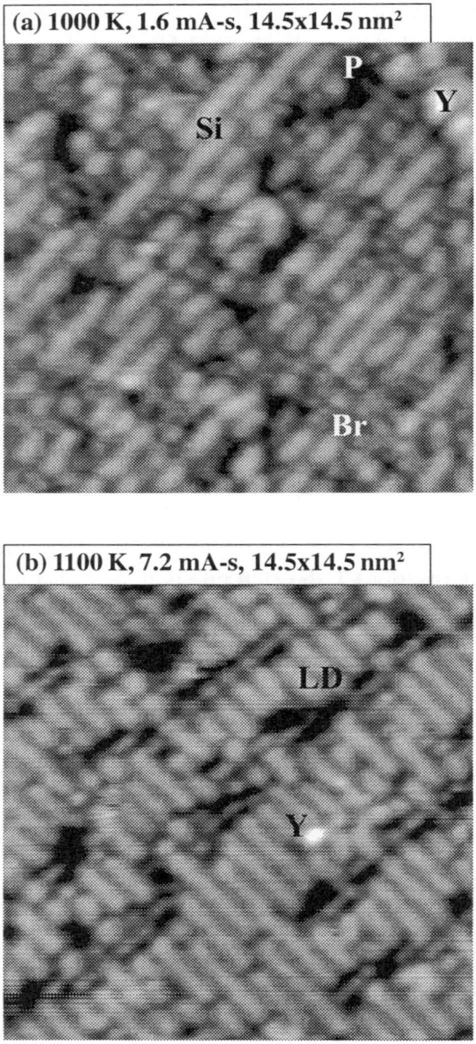

Figure 8. Small-scale images taken after steady state etching at 1000 and 1100 K. Evident in (a) are Br-free Si dimers (Si) and Br adsorption sites (Br) that are mixed with irregularly shaped pits (P). The bright feature labeled Y represents a disordered regrowth structure. (b) Dimer vacancies interact to form line defects (LD) that elongated perpendicular to the dimer row direction. Very little Br remains after etching at 1100 K.

minimum vacancy density and sufficient time and temperature for them to organize into lines. The average length of the line defects increases with exposure at 1100 K and the average spacing decreases. As a consequence, the surface evolves into a steady state 2×n network where $6 \leq n \leq 12$. These line defects are 25 to 50 Å apart and they can reach lengths of ~30 dimer rows. Flashing a sample that has reached the configuration shown in Fig. 8(b) will reduce the defect density. Line defects similar to those of Fig. 8(b) have been reported on Si(100)–2×1 surfaces that have been ion sputtered and annealed at 875-1125 K [20, 21].

While etching at 1000 K produces a large number of vacancies, these vacancies do not assemble into line defects. This suggests that vacancy diffusion is reduced by the presence of Br adatoms. Support for this kinetic constraint comes from images obtained after annealing surfaces that had been etched at 1000 K but without an incident flux to replenish the Br loss in etching. Line defects did form in this Br-depletion/anneal experiment.

3.3 Si(100) steps

We noted above that S_A and S_B steps are distinguished according to their orientation relative to the dimer row direction. In fact, there are two different B-type steps, depending on whether the dimer row of the upper terrace ends on a dimer row or between two dimers rows of the lower terrace. The latter rebonded S_B step is energetically favored because non-rebonded S_B steps are characterized by extra dangling bonds [24].

Many recent studies [18, 19, 25, 26, 35-41] have focused on steps on Si(100), emphasizing homoepitaxial growth via step flow and equilibrium exchange of atoms between step sites and terraces. These studies have demonstrated that fluctuations in step structure occur through attachment and detachment of blocks of two dimers because exchange of these blocks maintains the rebonding character at step sites and minimizes the dangling bonds. Recently, Pearson *et al.* [19] used high temperature STM to determine attachment and detachment rates at specific sites of S_B steps. Subsequently, Aldao *et al.* [42,43] examined the atomic dynamics associated with equilibrium step profiles, using Monte Carlo simulations to reproduce the observed step patterns of the clean surface.

One would expect that the dynamics associated with attachment and detachment should be altered when a step is exposed to an etchant at elevated temperature [44]. In particular, the overall detachment rate should be enhanced when an etch channel supplements the original thermal channel. Figures 9(a) and (b) show occupied-state images before and after exposure to 1.2 mA-s of Br_2 at 700 K. Comparison confirms that the S_B steps are rougher after etching and that the surface

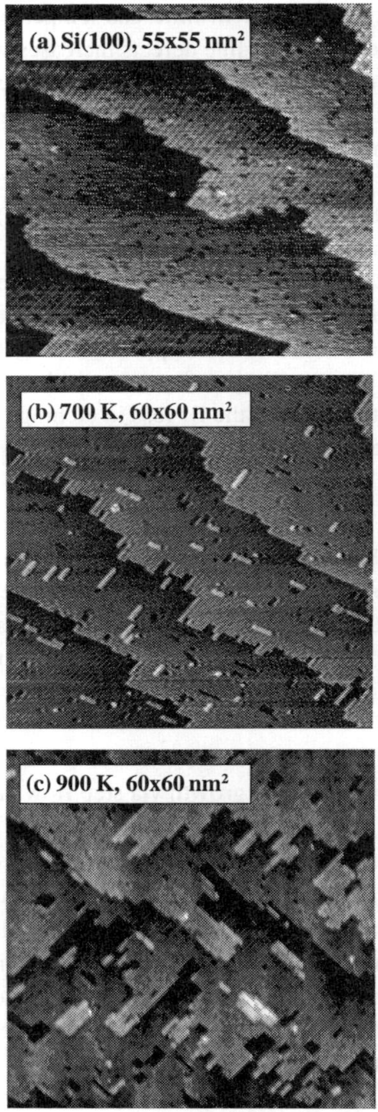

Figure 9. Occupied-state images of Si(100)-2×1 showing changes in step roughness. (a) Clean surface after flashing at 1475 K and cooling to room temperature in ~10 min. (b) Image taken after etching at 700 K with exposure to the equivalent of 1.2 Br atoms per Si atom. (c) Image taken after etching at 900 K where the bromine fluence was below the threshold for 3×1 etching

is saturated. Figure 9(c) shows the etch patterns associated with early stages of etching that were produced at 900 K by exposure to 0.4 mA-s (a fluence well below the steady state regime that produces the 3×1 structure).

The step roughness can be described quantitatively in terms of the separation between kinks and the kink lengths perpendicular to a step, as depicted in Fig. 10(a). The kink separations and lengths are measured in dimer-row units or blocks, denoted by the integers *s* and *n*. The values for $<s>$ and $<n>$ are 2.8 and 1.8 for a surface at 300 K that had been cooled over a period of 10 min from 1475 K [6]. These values of $<s>$ and $<n>$ correspond to steps in equilibrium with terraces at ~700 K. This sample cooling procedure was similar to that of Lagally and coworkers [45]. It produces a slightly smaller step roughness than that reported by Zandvliet and coworkers [26]. The latter authors held their samples at 775 K for several hours before cooling rapidly to room temperature. They estimated that the step configuration was frozen at 775 K. Williams *et al.* [44] have reported an even greater roughness after etching at 700 K for 18 min, with a decrease in $<s>$ to 2.3 and an increase in $<n>$ to 2.33.

The S_B step patterns can be modeled with a Monte Carlo method that describes attachment and detachment of units of two Si dimers (four atoms) at a step [43,44]. The probabilities for such processes are dictated by rules based on first and second neighbor interaction energies. Block 1 in Fig. 10(a) has one such nearest neighbor along x, block 2 has 2, and block 3 has none. All blocks have a filled block beneath them so the interaction along y is ignored. Second nearest neighbors are those reached by diagonal or z-displacement (block 1 has 3 corner or second nearest neighbors, block 2 has 2, and block 3 has none). There are no adjustable parameters since the rules (parameters) are derived from experiment.

In the simulation, every occupied block site must be directly above another occupied site [the y-direction in Fig. 10(a)]. This is the solid-on-solid approximation because it excludes vacancies and overhangs. It should be valid as long as the landscape is reasonably smooth. The model can be described as an array of interacting columns of varying integer heights. The step profile is represented by a 1D array that specifies the number of blocks in every column relative to a flat reference. Attachments and detachments can occur only at the tops of the columns.

Scanning tunneling microscopy makes it possible to assess the dynamic processes responsible for step fluctuations. Pearson *et al.* [19] demonstrated that attachment events are preferred at those sites that have a higher number of first neighbors. Using their results and considering the numbers of appropriate neighbors, Williams *et al.* [43] deduced that the detaching probabilities decreased by a factor of exp (-0.12 eV / kT) for every first neighbor and increased by a factor of

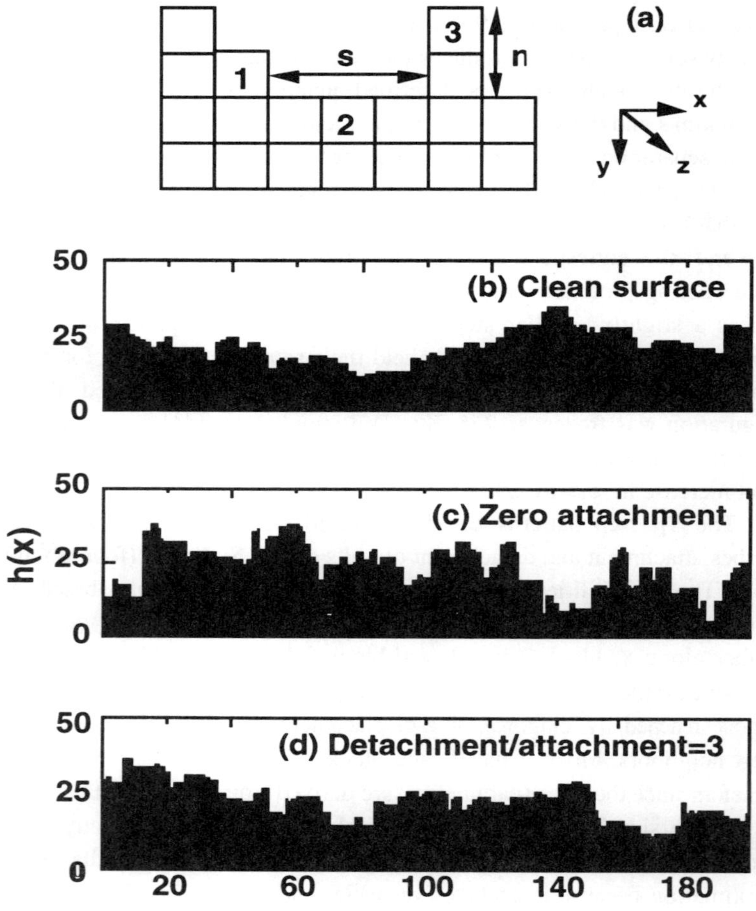

Figure 10. (a) Schematic of an S_B step represented by blocks of two dimers. The step sites all have one neighbor in the y-direction, but the number of nearest neighbors ranges from 0 to 2 and the number of next-nearest neighbors ranges from 0 to 4. The kink separation is defined by s and the kink length is defined by n. (b) S_B step profile from the model in which the number of detachments equals the number of attachments (d/a=1). This corresponds to equilibrium between the steps and the adatoms on the terrace. (c) Step profile in the limit of zero attachments, d/a=∞. (d) Step profile obtained when the d/a=3. This ratio corresponds to steady state etching at 700 K.

exp [0.037 eV / kT] for every second neighbor. Conversely, attaching probabilities increased by a factor of exp (0.12 eV / kT) for every first neighbor and decreased by a factor of exp (-0.037 eV / kT for every second neighbor.

The simulations start with a kinkless step of 2000 sites. The exchange events are then performed until steady-state is reached. By making one attachment for every detachment, the net number of particles leaving the step will be the same as those that attach. The pattern of Fig. 10(b) reproduces the experimental step structure very well. Etching breaks this balance between detachment and attachment because it introduces step retreat. It therefore increases the rate of detachment from steps (via $SiBr_2$ desorption) and, because there is Br on the terrace, it impedes Si diffusion to the steps. Nonetheless, the step patterns are controlled by the same processes that control equilibrium patterns, with account taken of the different ratio between attachments and detachments. For high enough etch rates, the only relevant processes are detachment events.

Figure 10(c) shows the outcome of modeling for the limiting case of zero attachments (high etch rate). Comparison indicates that this simulated roughness exceeds the experimental roughness. However, the trend toward enhanced local roughness is apparent in the experiment, Figs. 9(b) and (c), and we found that the step roughness after etching at 700 K can be reproduced with a detachment/attachment ratio, d/a, of 3 [Fig. 10(c)]. For d/a=3, the average kink length is 2.19, compared to 2.3 for etching at 700 K. Experiment gives an average kink separation of 2.33 after etching at 700 K, in agreement with 2.38 from the model for d/a=3.

Rioux *et al.* [6] analyzed the step roughness as a function of flux at 900 K. They showed that roughening increased with flux from 5 to 40 μA. Step roughening was taken to reflect the ability of the surface to heal damage produced by etching, i.e. the ability of Si adatoms to reattach. Those experiments were done at fixed fluence so that the amount of time at elevated temperature decreased as the flux increased. Accordingly, the ratio between detachment and attachment events was skewed to favor detachment. Rapid etching then creates a configuration that is farther from equilibrium.

This model cannot describe everything in the patterns at 900 K because terrace etching is important. Hence, step roughness is influenced by break-through of an advancing step with pits on the terrace. Likewise, overhangs develop via kink formation at long peninsulas, as is evident in Fig. 9(c). These phenomena account for the greater roughness, the larger values for kink lengths, and the smaller values for kink separations relative to what is predicted by the model.

3.4. Dimer chains

The chains and islands of Fig. 7 are similar in appearance to the growth structures observed in Si homoepitaxy [46-49]. For etching, however, they can have formed from Si atoms released by desorption events or they can be the remnants of etched terraces. In either case, the terminations of single width chains reflect the bonding energies of Si on Si(100), influenced by terrace irregularities and the presence of the halogens.

Figure 11(a) shows dimer chains and etch pits after etching at 800 K [see also Fig. 7(b)]. The chains are perpendicular to the terrace dimer rows. Within a chain, the bright elliptical features are individual dimers. They appear brighter at the ends of chains, most likely due to a slight relaxation. Examples of chains with even (E) and odd (O) numbers of dimers are labeled in Fig. 11(a) and the implications with regard to bonding are evident from Fig. 11(b). Chains with an odd number of dimers greatly outnumber those with an even number for etching between 700 and 900 K. Most even-length chains appear in pairs (EP) separated by a single missing dimer. This suggests that they are segments of odd-length chains that were split.

Removal of a dimer from an odd-length chain could yield two even-length segments or two odd-length segments. The STM results reveal nearly equal numbers of EP and OP segments, demonstrating that neither the trough nor the row sites, labeled T and R in Fig. 11(b), were vacated preferentially. Odd length chains with end dimers that were both atop rows of the terrace were rare. Only ~5% of the chains were of the isolated even-number type after etching 700 K (where regrowth probably accounts for their creation) and 800 K (where there is regrowth and extensive terrace removal). For the 3×1 surface produced at 900 K, the fraction of even-length chains is ~10%. This increase can be attributed to the fact that the chains develop through terrace fragmentation. The dominance of odd-length chains is explained by step energetics since rebonded S_B steps are favored over non-bonded or S_B' steps [24].

Statistical distributions for odd length chains indicate a shift to greater length with increasing temperature. At 700 K, the chain lengths are restricted by the amount of material available for regrowth. At 900 K, chain formation relies on the processes that create 3×1 patterns. At 800 K, the chain length is limited by the distance between next-layer etch pits. They are longer at 800 K than at 700 K because a larger amount of Si is liberated by etching.

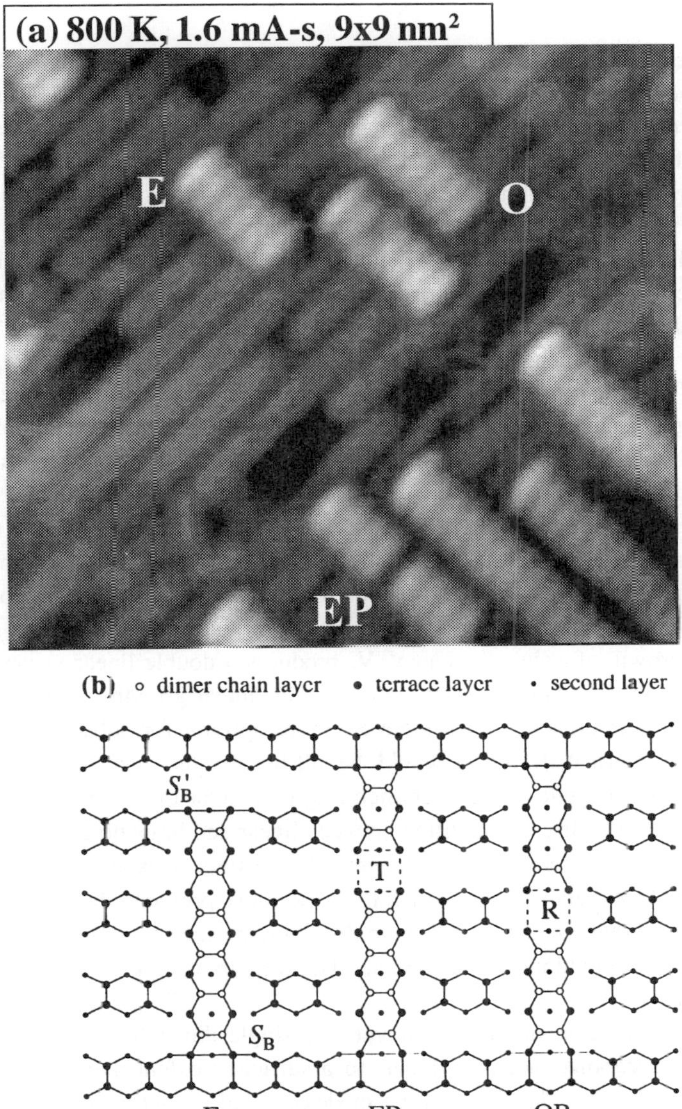

Figure 11. (a) Occupied states image showing even- and odd-numbered dimer chains, E and O respectively. Chain termination is favored in the troughs between dimer rows of the underlying terrace, resulting in odd-length (O) chains. Other structures such as even-length paired segments (EP) separated by a single missing dimer are thought to originate as odd-length chains. (b) Schematic showing rebonded S_B and nonbonded S_B' steps. Configurations for even- and odd-length paired segments (EP and OP) are shown to result from removal of dimers that were atop troughs (T) or dimer rows (R) of the underlying terrace.

3.5 Pit growth analysis

The reaction pathways responsible for the morphology changes discussed above involve atomic level interactions, and we can analyze the STM images to extract information about etching dynamics. When the analyses were first undertaken, there was no guarantee that they would lead to quantitative conclusions. The etching phenomena might have been too complicated for simple analysis, or the energies for the different processes might have been too similar, giving rise to random pit distributions. Fortunately, it turned out that the experimental results could be described quantitatively in terms of atomic interactions, and these successes suggest that the protocols could be extended to other systems.

From Fig. 6, the rate limiting steps for etching are the formation and desorption of volatile molecules. The relevant energy differences reflect interactions between neighboring dimers. Accordingly, the single-layer-deep etch pits will exhibit distributions in size and shape that arise from energy anisotropies for dimer removal at different sites around a pit. These effects give rise to linear pits that are primarily one row wide, at least until the pits start to overlap. The fact that they are linear indicates that they are produced by dimer removal along the dimer row direction. In fact, two possibilities for pit growth exist, namely linear growth via removal along the row direction and branch formation via removal from an adjacent row. Linear growth of a single vacancy, V, produces a double linear vacancy, VV, as depicted in Fig. 12. This unit can continue with linear growth to form VVV or it can branch. If it branches, then there are more options for subsequent growth. The simplest pattern distribution would be all-linear pits, and one would then conclude that the energetics were overwhelmingly in favor of dimer removal at the ends of the pits. In reality, competition between linear and branching growth would give rise to the observed areal density for pits with distinct sizes and shapes. While it is reasonable to expect that pits will follow the most probable distribution, there are complications that could randomize the distribution, such as limited availability of halogens and interactions between pits. The test, then, lies in analysis of the STM results [7].

Etch pit initiation involves removal of a single dimer from a terrace. This dimer vacancy destabilizes adjacent dimers to a sufficient extent that halogen attack will lead to pit growth. Etching can be modeled based on activation energies that depend on the configuration of a given site. In particular, the areal distributions of linear pits on Si(100) can be described in terms of interaction energies defined in Fig. 13. The second-neighbor interaction corresponds to the so-called corner energy, E_C [25,26].

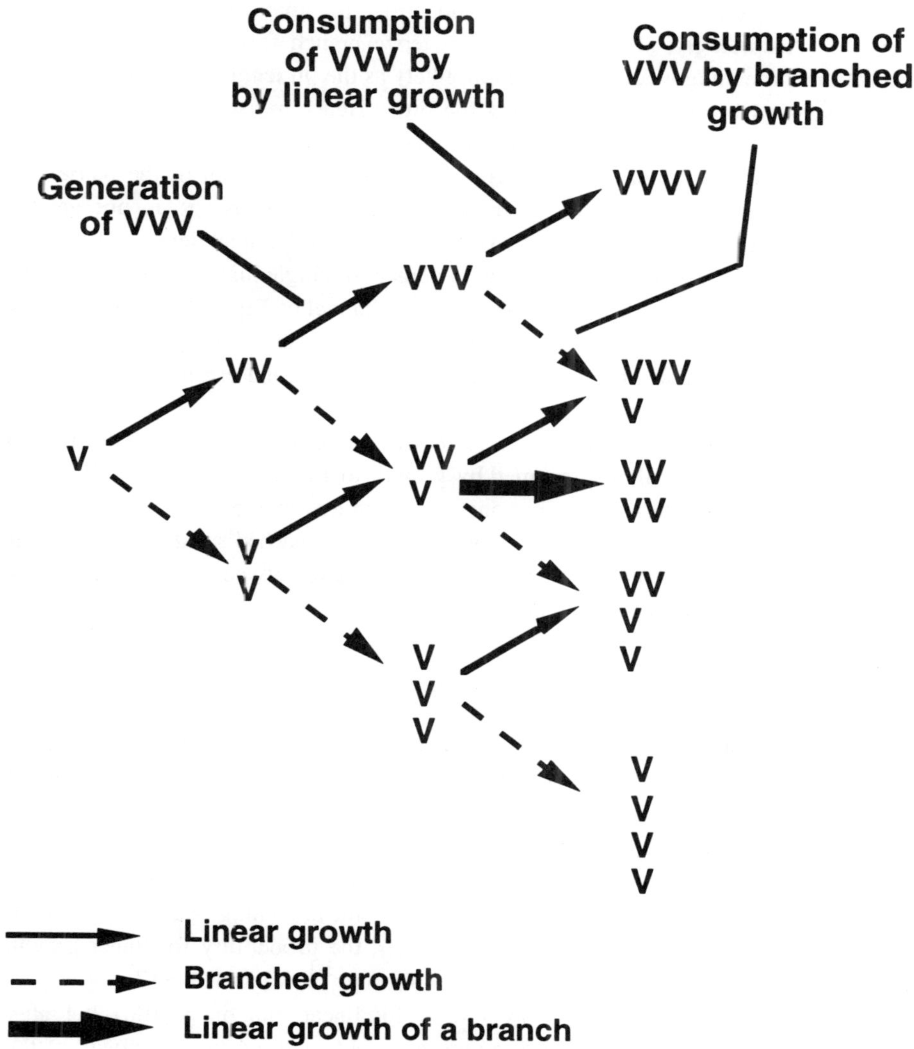

Figure 12. A schematic depicting growth of etch pits. Competition between pathways that lead to the generation and consumption of pits.

For a given linear pit, there is a probability, p_L, that will describe its linear growth. Accordingly, $(1 - p_L)$ will characterize branch formation. One would expect that branching would become more likely as the pit length increases (as more branching sites become available along the pit). Hence, p_L would decrease due to this increased competition. It has been observed, however, that branching occurs primarily at the ends, implying that there is a significant reduction in the barrier for branching at the pit ends relative to the mid-points [7]. Thus, p_L is not greatly affected by chain length. It is then possible to write that the areal density of a linear chain with i units $[V_i]$ is related to the areal density of single dimer vacancies [V] by $[V_i] = p_L^{i-1}[V]$. This retraces the linear growth channel of Fig. 12 backward to its origin, with a constant p_L. While correct, additional insight into the energetics of branching can be obtained by considering the general case and examining deviations of p_L from a constant value.

On a terrace, pits of a specific size will be generated by growth from a smaller unit and they will be consumed by growth to the next larger unit, Fig. 12. In quasi-steady-state, these rates are constant, regardless of the size. (In this case, quasi-steady-state means that most configurations have reached steady state but that there may be some which are slow to mature.) The ratio between the areal densities of chains with lengths i+1 and i can be written

$$[V_{i+1}] / [V_i] = 1 / [1 + 2k_{EB}/k_L + (i-2)\, k_{MB}/k_L] \qquad (1)$$

where the generation rate k_L accounts for linear growth at the end of the chain, k_{EB} accounts for branching at the end of the chain, and k_{MB} accounts for branching at middle sites along the chain (see arrows defining these growth sites in Fig. 13). For a linear pit, there are two sites for linear growth, and a statistical factor of two is inherent in k_L. There are two possible sites for branching at every dimer vacancy. Accordingly, the statistical factor is four for end point branching and 2(i-2) for mid-chain branching. If k_{MB} is much less than k_{MB}, the probability for linear growth is independent of i and Eq. (1) is reduced to $[V_i] = p_L^{i-1}[V]$ as above [7, 50].

Figure 14 shows $\ln([V_i] / [V])$ vs. i for linear pits on Si(100)-2×1 after etching at 850 K with Cl. The data points were obtained by direct analysis of the STM images. The solid line is a fit to the data. The straight dashed line would describe the situation where all branching occurred at chain ends, $k_{MB} = 0$. Hence, the probability p_L is constant and given by the slope of the line. The concave-downward dashed line would describe the case where branching is equally probable at all chain sites, $k_{MB} = k_{EB}$. In this case, p_L must decrease with chain length. The experimental results indicate that reality is closer to the $k_{MB} = 0$ case with a ≈

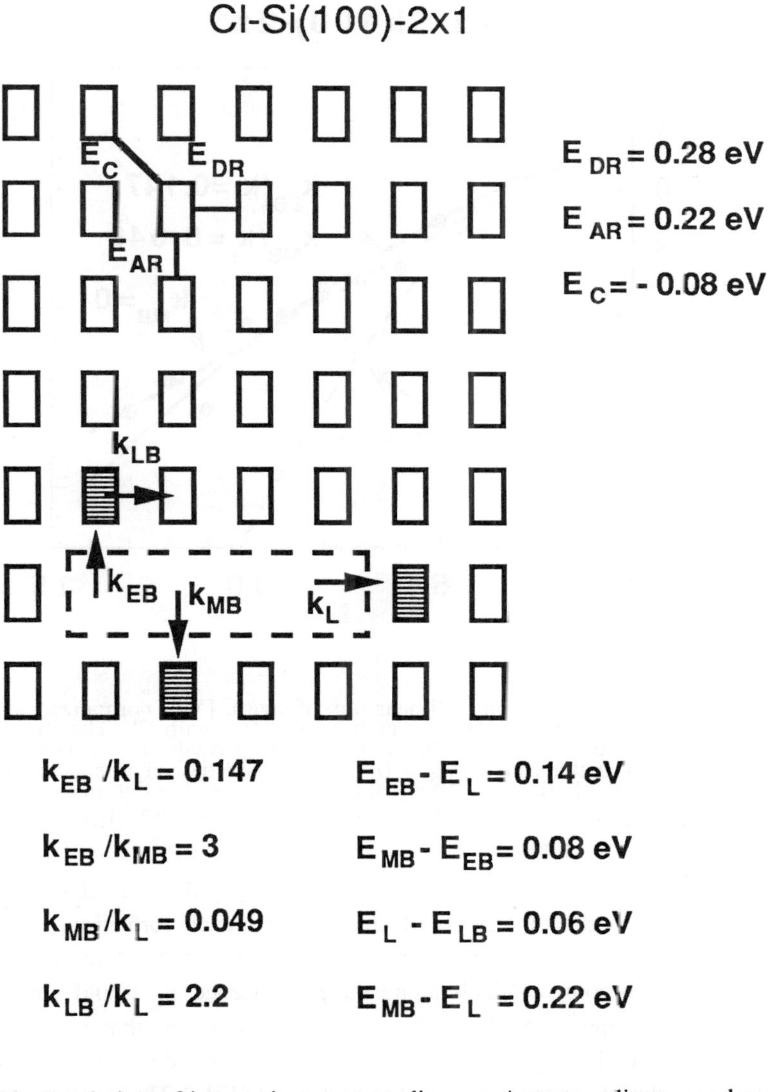

Figure 13. Depiction of interactions among dimers. A terrace dimer can have up to two dimers in the same row, two in adjacent rows, and four in second nearest neighbors positions with interaction energies E_{DR}, E_{AR}, and E_C. The values for the interaction energies were deduced from the STM data. Those fits gave ratios in rate constants for events depicted with arrows for linear branching (k_L), end branching (k_{EB}), mid-chain branching (k_{MB}), and linear growth from a branch (k_{LB}). Rate constants and energy differences dictate the atomic scale patterning.

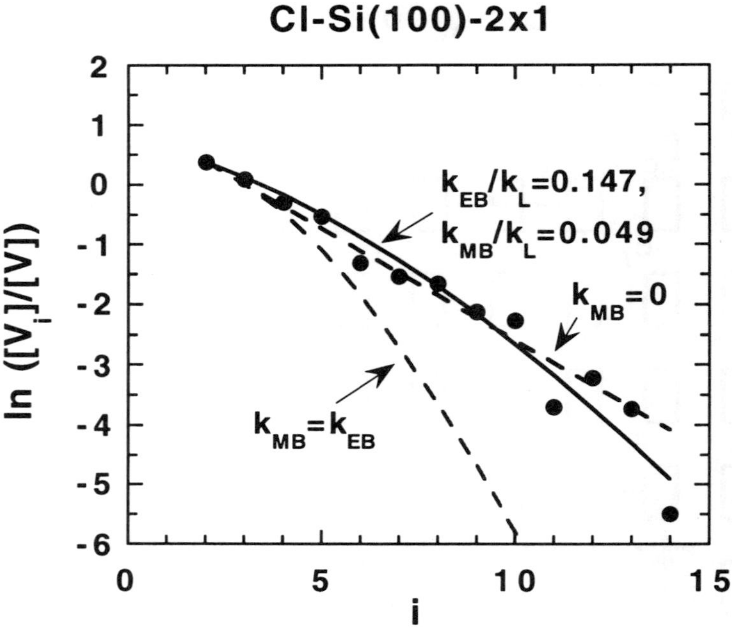

Figure 14. Areal distribution of linear pits of size i, $[V_i]$, normalized to the areal density of single vacancies, [V], after etching at 850 K with Cl. The straight dashed line represents a constant probability for linear growth, $p_L = 0.69$, with no branching except at the end ($k_{MB} = 0$). The curved dashed line accounts for p_L varying with chain length, but with equal chance of branching at mid-points and end points, $k_{MB} = k_{EB}$. The solid line is a fit using Eq. 1 with $k_{EB}/k_L = 0.147$ and $k_{MB}/k_L = 0.049$.

For a given linear pit, there is a probability, p_L, that will describe its linear growth. Accordingly, $(1 - p_L)$ will characterize branch formation. One would expect that branching would become more likely as the pit length increases (as more branching sites become available along the pit). Hence, p_L would decrease due to this increased competition. It has been observed, however, that branching occurs primarily at the ends, implying that there is a significant reduction in the barrier for branching at the pit ends relative to the mid-points [7]. Thus, p_L is not greatly removal to form a branch at a mid-chain site vs. an end site. Branching occurs preferentially at chain ends with such events being three times more likely than mid-chain branching, $k_{EB}/k_{MB} = 3$. These energies and rates are summarized in Fig. 13.

From these results, we conclude that pit growth on terraces can be quantified, the size distribution and shape anisotropy of the pits can be deduced, and activation energies for Si removal along and perpendicular to the dimer row direction can be determined. This removal of Si dimers has now been simulated through a Monte Carlo scheme in which events take place according to probabilities that are dictated by the dimer interaction energies. The fact that the experimental observations are reproduced very well by the simulation indicates that the essentials of etching are included [51].

It is important to point out that terraces are in quasi-steady-state conditions in the above analysis for etching at 850 K. Results obtained under similar flux/fluence at 700 K demonstrate that this distribution is not reached. The difference reflects the rate of step retreat relative to the rate of pit growth. At 700 K, pits that develop on terraces are erased before their length distributions reach quasi-steady-state values [52].

4. Concluding Remarks

The morphologies for Br-, Cl-, and I-etched Si(100)–2×1 reflect temperature-dependent diffusion of the halogens, silicon, and the vacancies produced by etching. With changes in temperature, we have related the evolution of the surface morphology to the interplay between etching at different sites of the surface. We have found that etching with Br_2, Cl_2, and I_2 gives rise to the same basic structures (single-layer etch pits and regrowth chains and islands) but that these appear at different temperatures. This should be expected, given the dependence of etch rates on the controlling energies and the way these energies influence halogen surface concentrations. For Si(100), the step patterns associated with etching can be described using the same atomic level interactions that control the step patterns for the clean surface. From the terrace etch patterns on Si(100)–2×1, we have developed a protocol to quantify the energies associated with pit growth. Such knowledge of kinetic parameters can be used to develop a comprehensive kinetic model that would allow the prediction of atomic details of surfaces exposed to various conditions of etchant flux, fluence, and temperature, and hence for the production of nanopatterned surfaces.

Recently, we have reported on the etching of Si(100) with double height steps. The interested reader should see Ref. 53 for similarities and differences relative to Si(100)–2×1. For a detailed discussion of etching of the Si(111) surface, the reader is directed to Ref. 54 and citations therein.

5. Acknowledgments

This work was supported by the U.S. Office of Naval Research and the Army Research Office. We gratefully acknowledge the contributions from the Electronic Materials Group at the University of Minnesota, past and present.

6. References

1. J. Schultz, in VLSI Technology, edited by S.M. Sze (McGraw-Hill, New York, 1988).
2. H.F. Winters and J.W. Coburn, Surf. Sci. Rep. **14**, 161 (1992), and references therein.
3. M. Chander, Y.Z. Li, J.C. Patrin, and J.H. Weaver, Phys. Rev. B **47**, 13035 (1993).
4. M. Chander, Y.Z. Li, D. Rioux, and J.H. Weaver, Phys. Rev. Lett. **71**, 4154 (1993).
5. D. Rioux, R.J. Pechman, M. Chander, and J.H. Weaver, Phys. Rev. B **50**, 4430 (1994).
6. D. Rioux, M. Chander, Y.Z. Li, and J.H. Weaver, Phys. Rev. B **49**, 11071 (1994).
7. M. Chander, D.A. Goetsch, C.M. Aldao, and J.H. Weaver, Phys. Rev. Lett. **74**, 2014 (1995).
8. D. Rioux, F. Stepniak, R.J. Pechman, and J.H. Weaver, Phys. Rev. B **51**, 10981 (1995).
9. J.V. Seiple and J.P. Pelz, Phys. Rev. Lett. **73**, 999 (1994).
10. K.E. Johnson and T. Engel, Phys. Rev. Lett. **69**, 339 (1992)
11. K. Wurm, R. Kliese, Y. Hong, B. Röttger, Y. Wei, H. Neddermeyer, and I.S.T. Tong, Phys. Rev. B **50**, 1567 (1994).
12. R.B. Jackman, R.J. Price, and J.S. Foord, Appl. Surf. Sci. **36**, 296 (1989).
13. J.J. Boland, Phys. Rev. Lett. **65**, 3325 (1990) and Surf. Sci. **261**, 17 (1992). See also J.J. Boland, Phys. Rev. Lett. **67**, 1539 (1991) and J. Vac. Sci. Technol. A 10, 2458 (1992).
14. H.F. Winters and D. Haarer, Phys. Rev. B **36**, 6613 (1987).
15. J.A. Yarmoff and F.R. McFeely, Phys. Rev. B **38**, 2057 (1988); C.W. Lo, D.K. Shuh, and J.A. Yarmoff, J. Vac. Sci. Technol. A **11**, 2054 (1993); C.W. Lo, D.K. Shuh, V. Chakarian, T.D. Durbin, P.R. Varekamp, and J.A. Yarmoff, Phys. Rev. B. **47**, 15648 (1993).
16. M.J.M. Vugts, G.J.P. Joosten, A. van Oosterum, H.A.J. Senhorst, and H.C.W. Beijerinck, J. Vac. Sci. Technol. A **12**, 2999 (1994).
17. J.R. Engstrom, M.M. Nelson, and T. Engel, Surf. Sci. **215**, 437 (1989).

18. H.J.W. Zandvliet, B. Poelsema, and H.B. Elswijk, Phys. Rev. B **51**, 5465 (1995).
19. C. Pearson, B. Borovsky, M. Krueger, R. Curtis, and E. Ganz, Phys. Rev. Lett. **74**, 2710 (1995).
20. P. Bedrossian and T. Klitsner, Phys. Rev. Lett. 68, 646 (1992).
21. H.J.W. Zandvliet, H.B. Elswijk, E.J. van Loenen, and I.S.T. Tsong, Phys. Rev. B 46, 7581 (1992).
22. P.C. Weakliem, Z. Zhang, and H. Metiu, Surf. Sci. **336**, 303 (1995).
23. H.J.W. Zandvliet, H.K. Louwsma, P.E. Hegeman, and B. Poelsma, Phys. Rev. Lett. **75**, 3890 (1995).
24. D.J. Chadi, Phys. Rev. Lett. **59**, 1691 (1987).
25. B.S. Swartzentruber, Y.-W. Mo, R. Kariotis, M.G. Lagally, and M.B. Webb, Phys. Rev. Lett. **65**, 1913 (1990)
26. H.J.W. Zandvliet, H.B. Elswijk, E.J. van Loenen, and D. Dijkkamp, Phys. Rev. B **45**, 5965 (1992).
27. The origin of the C-type defect has attracted considerable attention. See, M. Chander, Y.Z. Li, J.C. Patrin, and J.H. Weaver, Phys. Rev. B **48**, 2493 (1993).
28. J.A. Jensen, C. Yan, and A.C. Kummel, Science **267**, 493 (1995); J. Chem. Phys. **102**, 3381 (1995); Y. Liu, D.P. Masson, and A.C. Kummel, Science **276**, 1681 (1997).
29. J.L. Iguain, H.O. Mártin, and C.M. Aldao, Phys. Rev. B **54**, 8751 (1996).
30. G.A. de Wijs, A. De Vita, and A. Selloni, Phys. Rev. Lett. **78**, 4877 (1997).
31. U. Höfer, L. Li, and T.F. Heinz, Phys. Rev. B **45**, 9485 (1992); M.P. D'Evelyn, Y.L. Yang, and L.F. Sutcu, J. Chem. Phys. **96**, 852 (1991).
32. A. Vittadini, A. Selloni, and M. Casarin, Phys. Rev. B **49**, 11191 (1994); G. Thornton, P.L. Wincott, R. McGrath, I.T. McGovern, F.M. Quinn, D. Norman, and D.D. Vvedensky, Surf. Sci. **211/212**, 959 (1989); S.L. Bennett, C.L. Greenwood, and E.M. Williams, Surf. Sci. **290**, 267 (1993); Q. Gao, C.C. Cheng, P.J. Chen, W.J. Choyke, and J. Yates J.T., J. Chem. Phys. **98**,8308 (1993); D. Purdie, N.S. Prakash, K.G. Purcell, P.L. Wincott, and G. Thornton, Phys. Rev. B **48**, 2275 (1993); J.J. Boland, Science **262**, 1703 (1993).
33. D.J.D. Sullivan, H.C. Flaum, and A.C. Kummel, J. Phys. Chem. **97**, 12051 (1993).
34. S.J. Chey, Y. Gong, and J.H. Weaver, to be published.
35. B.S. Swartzentruber, N. Kitamura, M.G. Lagally, and M.B. Webb, Phys. Rev. B **47**, 13432 (1993).

36. N. Kitamura, B.S. Swartzentruber, M.G. Lagally and M.B. Webb, Phys. Rev. B **48**, 5704 (1993).
37. H.J.W. Zandvliet, H.B. Elswijk, and E.J. van Loenen, Surf. Sci. **272**, 264 (1992).
38. H.J.W. Zandvliet, H. Wormeester, D.J. Wentink, A. van Silfhout, and H.B. Elswijk, Phys. Rev. Lett. **70**, 2122 (1993).
39. E.D. Williams and N.C. Bartelt, Science **251**, 393 (1991).
40. N.C. Bartelt, J.L. Goldberg, T.L. Einstein, and E.D. Williams, Surf. Sci. **273**, 252 (1992).
41. N.C. Bartelt, R.M. Tromp, and E.D. Williams, Phys. Rev. Lett. **73**, 1656 (1994).
42. J.R. Sánchez and C.M. Aldao, Phys. Rev. B **54**, R11058 (1996).
43. F.J. Williams, J.R. Sánchez, and C.M. Aldao, Surf. Sci. (in press).
44. F.J. Williams, C.M. Aldao, Y. Gong, and J.H. Weaver, Phys. Rev. B **55**, 13829 (1997).
45. B.S. Swartzentruber, Y.-W. Mo, M.B. Webb, and M.G. Lagally, J. Vac. Sci. Technol. **A7**, 2801 (1989).
46. R.J. Hamers, U.K. Köhler, and J.E. Demuth, Ultramicroscopy **31**, 10 (1989).
47. J.Y. Tsao, E. Chason, U. Koehler, and R. Hamers, Phys. Rev. B **40**, 11951 (1989).
48. Y.-W. Mo, B.S. Swartzentruber, R. Kariotis, M.B. Webb, and M.G. Lagally, Phys. Rev. Lett. **63**, 2393 (1989).
49. Y.-W. Mo, R. Kariotis, B.S. Swartzentruber, M.B. Webb, and M.G. Lagally, J. Vac. Sci. Technol. **A 8**, 201 (1990).
50. M. Chander, D.A. Goetsch, C.M. Aldao, and J.H. Weaver, Phys. Rev. B **52**, 8288 (1995).
51. J.R. Sánchez, C.M. Aldao, and J.H. Weaver, J. Vac. Sci. Technol. B **13**, 2230 (1995).
52. F.J. Williams, C.M. Aldao, and J.H. Weaver, J. Vac. Sci. Technol. B **14**, 2519 (1996).
53. Y. Gong, D.W. Owens, and J.H. Weaver, Phys. Rev. B **53**, R16144 (1996).
54. R.J. Pechman, X.-S. Wang, and J.H. Weaver, Phys. Rev. B **52**, 11412 (1995) and R.J. Pechman, T. Moriwaki, J.H. Weaver, and G.S. Khoo, Surf. Sci. Lett. **341**, L1085 (1995).

SURFACE MORPHOLOGY OF ION BOMBARDED Si(001) AND Ge(001) SURFACES

H.J.W. ZANDVLIET
Faculty of Applied Physics and Centre of Materials Research (CMO), University of Twente, P.O. Box 217, NL-7500 AE Enschede, The Netherlands

I.S.T. TSONG
Department of Physics and Astronomy, Arizona State University, Tempe, Arizona 85287, USA

1 Introduction

The study of nonequilibrium processes at surfaces, such as epitaxial growth and etching, has received a lot of attention. Both growth and etching are technologically important processes because they often play a key role in the fabrication of microelectronic devices. Recently, the effect of low energetic ion bombardment of metal and semiconductor surfaces has been studied by several techniques. High dose experiments to study the transition from the formation of vacancy islands to layer-by-layer sputtering have been conducted by workers using averaging techniques such as Thermal Energy Atom Scattering (TEAS), [1,2] Reflection High Energy Electron Diffraction (RHEED) [3] and Reflection Electron Microscopy [4]. With the advent of the Scanning Tunneling Microscope (STM) it became possible to study the surface morphology on an atomic scale [5,6,7,8,9,10,11]. Although it seems reasonable to assume that etching with low energy ions at elevated temperatures can be considered as inverse growth there are some remarkable differences. First, during ion bombardment both adatoms as well as vacancies are created, whereas during epitaxial growth adatoms are in most cases deposited individually. At sufficient high temperatures these ion sputtering induced adatoms could either incorporate at step edges or hop between levels, annihilating a single mobile vacancy or filling a vacancy island. Second, the vacancy and adatom clusters, which develop immediately after the ion impact, might even have a size larger than the critical nucleus size. Third, the created adatom and vacancy clusters can, due to the ion impact, have an energy with exceeds the thermal energy. This increase in energy of the clusters enhances the intra- and interlayer mass transport and hence results, in general, into smoother surfaces. Finally, during growth as well as etching of Si(001) and Ge(001) there is an additional complication related to the tendency of these surfaces to dimerize. Because (vacancy) islands nucleate independently the dimer rows of the neighbouring (vacancy) islands have a probability of 50 % to be in the correct registry. The coalescence of (vacancy) islands with an out-of-phase dimer row registry gives rise to an antiphase boundary, whereas an in-phase registry does not. Due to the symmetry of the diamond lattice one expects in principle two different types of antiphase boundaries, i.e. along or perpendicular to the substrate dimer row direction. The formation of antiphase boundaries, which often act as nucleation centres of new islands, gives rise to an additional roughening of the surface. In this chapter the surface morphology of group IV semiconductor (001) surfaces which develops after low-dose (up to a removal of several monolayers) low energy ion bombardment is briefly reviewed. Special attention is paid to the characteristic surface morphological features, which develop after ion bombardment at elevated temperatures, e.g. the ordering of vacancies into vacancy islands and the formation and stability of antiphase boundaries.

2 Experimental

The 5×20 mm^2 Si(001) and Ge(001) substrates were cut from commercially available wafers and ultrasonically rinsed in ethanol before loading into the ultrahigh- vacuum (UHV) system. The misorientation of the surface with respect to (001) was 0.5° for the Si(001) surface and 0.8° for the Ge(001) surface, respectively. The samples were heated resistively inside the Ultra High Vacuum (UHV) chamber, which has a base pressure in the mid 10^{-11} Torr range. Temperatures were measured with an infrared pyrometer (calibrated against a NiCr-NiAl thermocouple) with an accuracy of about 25 K. After outgassing the Si(001) sample and holder for several hours the sample was thermally cleaned at 1500 K during several seconds. This procedure results in a (2×1) reconstructed atomically clean Si(001) surface with an average terrace length of about 150 Å. The Ge(001) surface was cleaned by using cycles of Argon ion bombardment at 800 eV followed by annealing at 1100 K. The (2×1) reconstructed Ge(001) was, after cleaning, nearly defect free, in marked contrast to the cleaned Si(001) surface which typically contained 0.5–1 % missing dimer defects. After the annealing treatment, the samples were transferred to the Scanning Tunneling Microscope (STM) where the usual (2×1) reconstruction was observed. The sample was then transferred to another part of the chamber were in situ bombardment took place. The ion current was measured by a Faraday cup and it was adjusted to give various doses ranging from 7×10^{11} ions cm^{-2} to 1×10^{16} ions cm^{-2}. After bombardment,

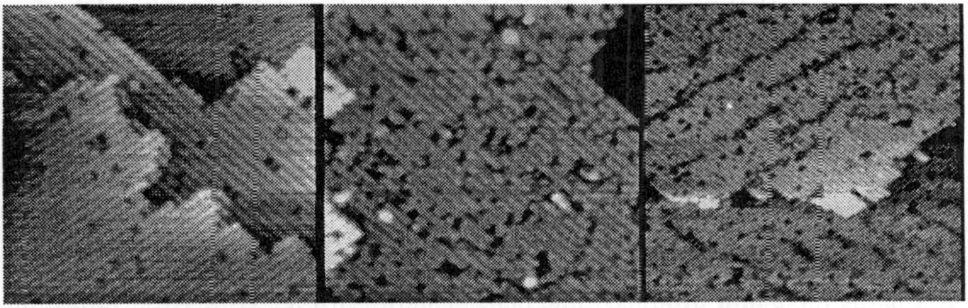

Figure 1: Filled-state STM images of Si(001)-(2 × 1) surfaces taken at sample bias of -2 V and tunneling current 0.5 nA. (a) Clean surface prior to ion bombardment. Scan area is 280 Å × 280 Å. (b) After bombardment by 3-keV Ar^+ ions at a dose of 1.5×10^{12} ions cm^{-2}. Scan area is 260 Å × 260 Å. (c) Bombarded surface after annealing at 1025 K for 2 min. Scan area is 400 Å × 400 Å.

the sample was transferred back to the STM for imaging. The Si(001)- (2 × 1) and Ge(001)-(2 × 1) surfaces are easily contaminated by e.g. oxygen and the contamination produced surface features, which could be mistaken as ion-bombardment damage. To ensure that the STM images obtained were not the result of oxygen contamination, we performed two tests. The first test was to leak the 99.9999 % pure argon gas via the ion gun into the UHV chamber, and then image the Si/Ge surface with the STM to confirm that the reconstruction was unaffected. The second test was to place the sample in the proximity of, but not actually in, the path of the argon ion beam for periods longer than the typical dose period. The unbombarded surface was then examined by the scanning tunneling microscope to verify that the reconstruction did not show signs of being contaminated.

3 Results and Discussion

3.1 Low-dose ion bombardment of Si(001)

In Fig. 1 a sequence of STM images of Si(001) showing (a) the clean reconstructed surface with a low defect density prior to ion bombardment; (b) the surface after room temperature ion bombardment with 3-keV Ar^+ ions at a dose of 1.5×10^{12} ions cm^{-2}; (c) a different scanned area of the surface as in (b), after annealing to 1025 K for 2 min. The starting Si(001) surface is relatively defect-free. Most of the defects present in Fig. 1(a) appear to be type-C defects as defined by Hamers and Köhler [12] with some type-A single dimer vacancy defects. Very few type-B double dimer vacancy defects were found on this surface and there were no Ni-contaminated line defects. After low-dose bombardment by 3-KeV Argon ions at a dose of 1.5×10^{12} ions cm^{-2}, which translated into six ions per 20 nm × 20 nm imaged area, the density of defects (including adatom clusters) increased dramatically. A survey of many images reveals that there appears to be a greater number of defects than the expected ion impacts. This contrasts sharply with the approximate 1:1 craters to the ion-impact ratio found in ion-bombarded Si(111) 7×7 surface [7]. It is, however, conceivable that some of the defects shown in Fig. 1(b) are electronic rather than topographic in nature. Annealing the bombarded surface as in image 1(b) at 1025 K for 2 min produced line defects shown in Fig. 1(c). After annealing at temperatures above 1200 K for 2 min., however, the line defects completely vanished (ruling out Ni contamination) and the (2×1) reconstruction was restored. Line scans across these line defects yield an average depth of 1.4 Å, which is close to the single-step height on the (2×1) surface. The formation of a step edge along the dimer row direction is energetically very favourable, whereas the formation of a step edge perpendicular to the dimer row direction requires substantially more energy. [13] One should therefore expect that the vacancy island are, at least in thermodynamic equilibrium, elongated in a direction along the dimer row direction. At low vacancy concentration (< 0.2-0.3 ML), our annealing experiments, however, reveal that an elongation of the vacancy islands perpendicular to the dimer row direction is favoured. Interestingly, at vacancy concentrations above about 0.3 ML the vacancy island shape rotates by 90° and the vacancy islands are elongated along the dimer row direction. [14,15] A more detailed analysis reveals that for a sufficient diluted vacancy island pattern the sum of the strain relaxation energy along the dimer row direction, which falls off logarithmically with the island spacing, and the energy reduction due to the specific rebonding of the narrow vacancy lines is large enough to compensate for the difference in island edge formation energy. [8,15,16] As we will discuss below ordering of the defects into defect lines oriented perpendicular to the substrate dimer rows is driven by a short-range attractive interaction between defects located in adjacent dimer rows and a long-range repulsive interaction, which is related to the strain relaxation and decays as L^{-2} (L refers to the vacancy line-vacancy line spacing), along the dimer row direction. Let us introduce a simple one-dimensional model, which is based on the energy balance between the creation of domain walls and the repulsive interaction, which arises due to the strain relaxation, between the walls. [17] The energy E per unit length can be written as:

$$E = \frac{l\rho_l}{l+L} + \frac{L\rho_L}{l+L} + \frac{2\gamma}{l+L} - \frac{2\sigma}{l+L}\ln\left(\frac{l+L}{2\pi a_0}\sin\left(\frac{\pi l}{l+L}\right)\right) \quad , \qquad (1)$$

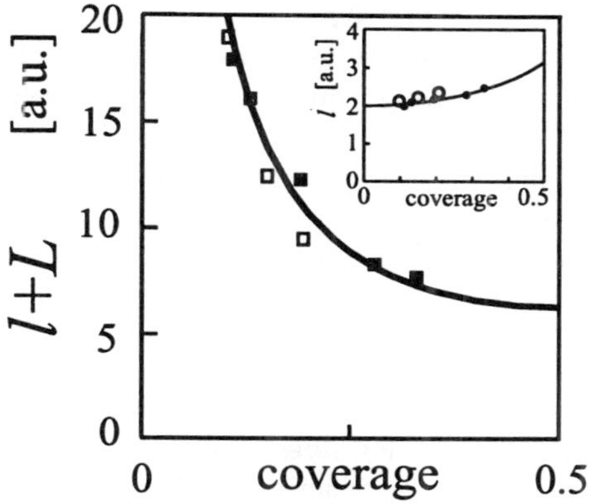

Figure 2: Periodicity, $l + L$, in units of a (=3.84 Å) of the vacancy line pattern versus coverage ($\theta = l/(l + L)$). Solid line: $l_0\theta/\sin(\theta)$. Inset: width, l, in units of a versus the vacancy line pattern versus coverage. Solid line: $l_0\theta/\sin(\theta)$. Filled squares and circles refer to the Ni induced vacancy line pattern. Open squares and circles refer to the Ar^+ induced vacancy line pattern.

where γ is the domain creation energy, ρ_l and ρ_L the energy per unit along the vacancy line for the minority and majority domain respectively, l the averaged width of a vacancy line and L the averaged spacing between neighboring vacany lines. The fourth term refers to the elastic strain relaxation: a_0 is a microscopic cutoff (e.g. the lattice constant) and σ is a function of the elastic constants and the difference of the surface stress within the two types of domains. Expansion of the logarithmic term in Eq. (1) around the minimum energy configuration results into a leading term which is proportional to L^{-2}. From the minimalization of the energy, Zeppenfeld et al.[17] found that, independent of the actual values of γ and σ, the following relation between the width of the vacancy line, l, and the periodicity, $l + L$, at a certain vacancy coverage $\theta = l/(l + L)$ and the minimal size of the width of the vacancy at a coverage approaching zero, i.e., l_0, must hold: $l = l_0\pi\theta/\sin(\pi\theta)$ and $l + L = l_0\theta/\sin(\pi\theta)$. In Fig. 2 plots of the periodicity, $l + L$, and the width of the vacancy line, l, versus coverage θ are shown. Interestingly, etching with oxygen [18,19], bromine [20,21] and iodine[22] or deposition of 1 ML Ge on Si(001)[23], 1 ML Bi on Si(001) [24,25] and 1 ML on Ge(001)[26] gives rise to a similar vacancy line pattern[27]. But also small amounts of Ni[28,29], Cu or Ag[30] contamination give rise to a vacancy line pattern. In Fig. 2 we have, for comparison, included the data of a Ni induced vacancy line pattern. Although the data points of the Ar^- induced vacancy line pattern don't collapse perfectly on the theoretical curve the Ni induced vacancy line pattern points do. We ascribe this difference to the fact that the Ni induced vacancy line pattern is fully equilibrated, but the Ar^+ induced vacancy line pattern is not fully equilibrated (an increase in the annealing temperature eventually results in a disappearance of the Ar^+ induced vacancy line pattern, whereas the Ni induced vacancy line pattern cannot be annealed away). Because this simple one-dimensional model is obeyed quite well, we conclude that the vacancy line structure is indeed in (local) thermodynamical equilibrium and is stabilised by an effective long-range repulsive interaction decaying as L^{-2} (a more detailed analysis of the thermally induced wandering of the vacancy lines also clearly points to an effective long-range repulsive interaction, decaying as L^{-2}, between neighbouring vacancy lines [16,31]). The ordering of vacancies into vacancy lines is further driven by a (short-range) attractive interaction between vacancies in adjacent dimer rows. The exact strength of this attractive interaction can be extracted from the kink and kink length distribution of the vacancy lines (see for more details ref. 15). With increasing vacancy concentration the spacing, L, between the vacancy lines decreases until eventually the strain relaxation term (proportional to $\ln(L)$) becomes so small that this term including the energy term related to the specific rebonding at the island edge is no longer large enough to compensate for the difference in step edge formation energy. Bedrossian [14] studied in a particular detailed work the development of the surface morphology of Si(001) upon low energy Xe^+ bombardment at elevated temperatures in the removal regime from 1/4 to 1/2 monolayers. The shape of the vacancy islands in this removal regime turns out to be indeed elongated in a direction along the dimer row direction.

3.2 Low-dose ion bombardment of Ge(001)

Recently several ion bombardment STM studies [32,33,35] of Ge(001) have been performed. Chey, Van Nostrand and Cahill [32] have studied the surface morphology of Ge(001) after 240 eV Xe^+ bombardment. The authors investigated a wide range of removed layers ranging from 0.2 ML up to 130 ML in the temperature range from 440 K to 570 K. A maximum surface roughness is reached for etching at 525 K. A transition in the surface morphology from a disordered mound like pattern at temperatures below 545 K to a quite regular pattern of pits above 545 K was found. As has been pointed out by these authors the asymmetric kinetics for the attachment of dimer vacancies at ascending versus descending steps may play a key role. In a latter study [33] Chey, Van Nostrand and Cahill also studied the smoothening kinetics of ion bombarded Ge(001). They showed that the dependence of the relaxation rate on the in-plane length

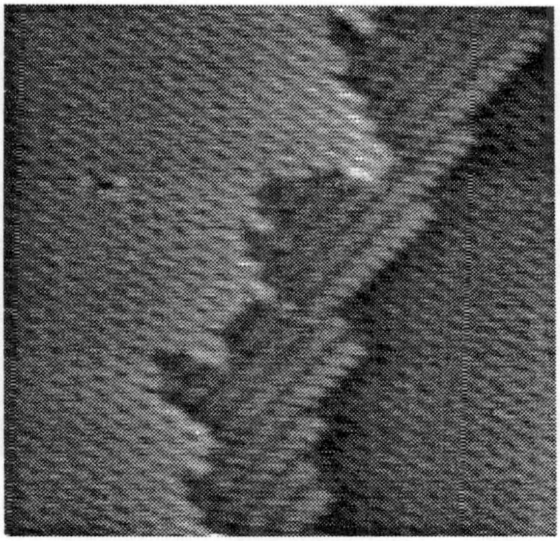

Figure 3: Clean Ge(001) prior to the ion bombardment experiment. Scan size 200 Å x 300 Å.

scale is not in agreement with the well-known continuum model of Mullins [34]. More recently, Zandvliet and de Groot [35] have analyzed the surface morphology of Ge(001) in the low-dose regime.

In Fig. 3 an STM image of a Ge(001) surface prior to ion bombardment is shown. In marked contrast to the clean Si(001) surface the Ge(001) surface is nearly defect free. After room temperature 800 eV Ar^+ bombardment at a flux of 7×10^{11} ions cm^{-2} s^{-1} for 200 s, resulting in the removal of about 0.2 monolayers both vacancy and adatom islands are resolved in Fig. 4(a). The bombarded surface in Fig. 4(a) looks a little disordered but subsequent annealing results in a decrease of the disorder. The vacancy islands are again only one monolayer deep and high-resolution images reveal that the atoms within the vacancy islands are dimerized [32,35].

The elongation of the vacancy island in the direction parallel to the dimer rows of the upper terrace is consistent with previous observations on Si(001) and has been shown to result from preferential annihilation of mobile surface vacancies at the ends, rather than the sides, of the dimer rows [5]. There is, however, one remarkable difference with the Si(001) surface: low-dose (vacancy concentration <0.2 monolayers) sputtering of Si(001) by 3 keV Argon ions and subsequent annealing at sufficient high temperatures (>800 K) was found to introduce line defects which are oriented perpendicular to the dimer rows of the upper terrace [6,7]. We have never found such line defects on the Ge(001) after low-dose sputtering and annealing. Small differences of the anisotropic

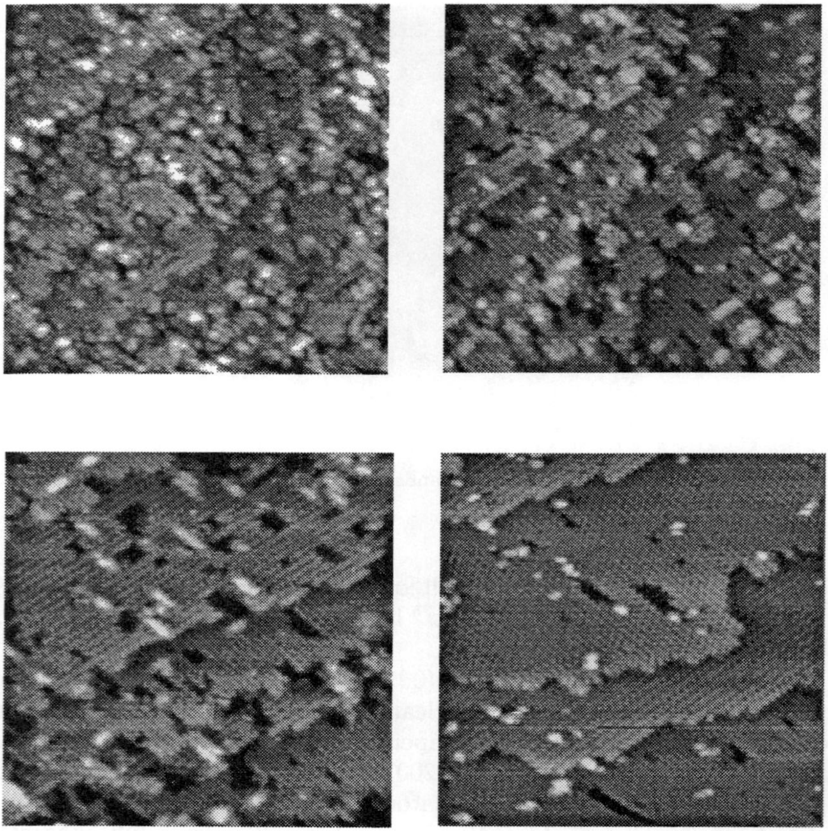

Figure 4: (a) Ge(001) surface after room temperature 800 eV Ar^+ bombardment at a flux of 7×10^{11} ions cm^{-2} s^{-1} for 200 sec. Scan size 500 Å × 500 Å. (b) As in (a) after 2 min. anneal at 380 K. Scan size 500 Å × 500 Å. (c) As in (b) after 2 min. anneal at 430 K. Scan size 500 Å × 500 Å. (d) As in (c) after 2 min. anneal at 500 K. Scan size 500 Å × 500 Å.

vacancy island edge formation energy and surface stress tensor might explain this different behaviour. The number density of adatom and vacancy islands decreases, whereas the averaged vacancy island size and spacing increases with increasing anneal temperature (see Figs. 4(b)- (d)). These observations imply that intralayer transport of vacancies and adatom clusters must occur at tem-

Figure 5: (a) Ge(001) surface after room temperature 800 eV Ar^+ bombardment at a flux of 7×10^{11} ions cm^{-2} s^{-1} for 2000 sec. and 2 min. anneal at 500 K. Scan size 1000 Å × 1000 Å. (b) Ge(001) surface after room temperature 800 eV Ar^+ bombardment at a flux of 7×10^{11} ions cm^{-2} s^{-1} for 5000 sec. and 2 min. anneal at 500 K. Scan size 1000 Å × 1000 Å. (c) Ge(001) surface after room temperature 800 eV Ar^+ bombardment at a flux of 7×10^{11} ions cm^{-2} s^{-1} for 5000 sec. and 2 min. anneal at 450 K. Scan size 1000 Å × 1000 Å.

peratures as low as 380 K. Annealing at temperatures above 600 K is sufficient to restore the original surface morphology.

In Figs. 5(a)-(c) the total removed dose is increased from 2 monlayers (Fig. 5(a)) to 5 monolayers (Figs. 5(b)- (c)). The characteristic network of valley's with depths up to 10 monolayers can clearly be resolved in Fig. 5(c). Annealing of this rough surface for 2 minutes at 500 K already results in a significant amount of smoothening. Chey, Van Nostrand and Cahill [32] proposed that the valley-like pattern might be due to a preference for vacancy annihilation at descending steps versus ascending steps. We agree with these authors on this point but we believe that also the formation of antiphase boundaries (discussed in detail below) may play a role in this roughening process.

In Fig. 6 a Ge(001) surface bombarded at 400 K by 800 eV Ar^+ is shown, the total removed dose is about 0.5 monolayers and the flux is again 7×10^{11} ions cm^{-2} s^{-1}. The number density of adatom clusters after the 400 K bombardment experiment is much lower as compared to the a surface bombarded at room temperature followed by annealing at 400K. This low number density of adatom clusters suggest that interlayer transport and adatom-vacancy annihilation does occur at sputtering temperatures above 400 K. Probably, the adatom and vacancy clusters created at elevated temperatures have an additional energy component due to the ion impact which enhances the mass transport process at the surface. Anyway, we must conclude that ion bombardment at elevated temperatures is not equivalent to ion bombardment at

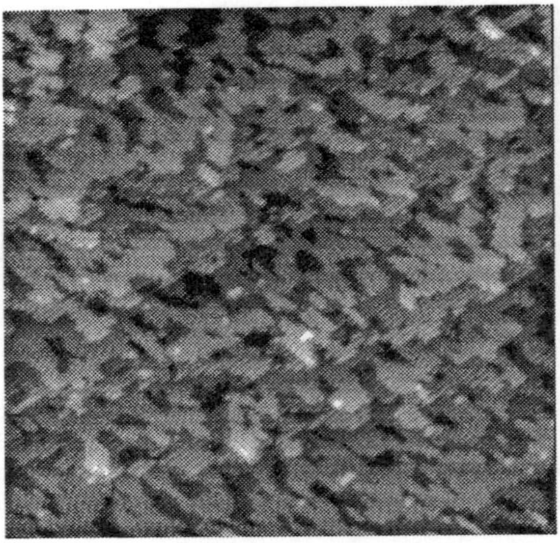

Figure 6: Ge(001) surface after 800 eV Ar^+ bombardment at a flux of 7×10^{11} ions cm^{-2} s^{-1} for 500 sec. at a temperature of 400 K. Scan size 1000 Å × 1000 Å.

room temperature followed by annealing at this temperature. Further annealing of the bombarded surface leads finally to the coalescence of neighbouring vacancy islands. The vacancy islands nucleate separately and independently and the dimerization within the vacancy islands has a probability of 50 % to be in the right (or wrong) registry. If the dimer rows in the two neighbouring vacancy islands do not align an antiphase boundary will develop. In Figs. 7-10 three different types of antiphase boundaries are displayed. In Figs. 7-8 there is an undecorated antiphase boundary parallel and perpendicular to the dimer row direction, whereas in Fig. 9 an decorated antiphase boundary positioned in a direction perpendicular to the dimer rows is present. The buckling registry of the dimers in the two dimer rows which form the antiphase boundary in Fig. 7 are out-of-phase along the antiphase boundary direction. In Fig. 10 a decorated antiphase boundary, which is similar to the antiphase boundary on Si(001), is displayed. On Si(001) only one type has been reported by Bedrossian and Kaxiras[36].

In the case of Si(001) these decorated antiphase boundaries are as narrow as one dimer row and as long as the substrate's terrace width. On the Ge(001) two of the three types of antiphase boundaries are not decorated with a dimer row. Annealing above temperatures of 600 K removes these antiphase boundaries completely. Especially the "zipper" type of antiphase boundary (Fig. 10) is very stable. The high stability of this type of antiphase boundary as well as

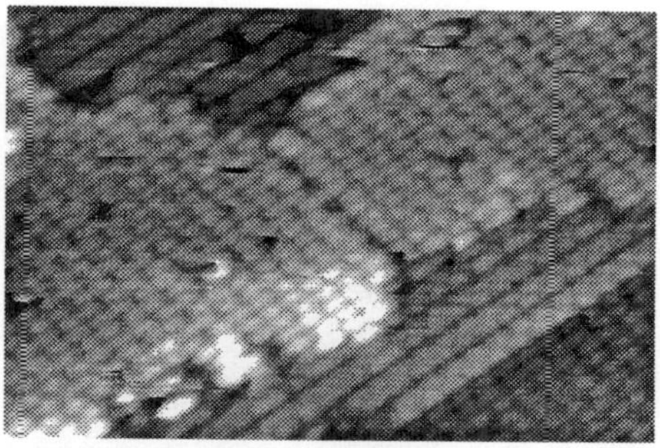

Figure 7: Undecorated antiphase boundaries aligned along the substrate dimer row direction on the Ge(001) surface after 800 eV Argon ion bombardment and subsequent annealing. Sample bias -2 V and tunneling current 1 nA.

Figure 8: Undecorated antiphase boundaries aligned perpendicular to the substrate dimer row direction on the Ge(001) surface after 800 eV Argon ion bombardment and subsequent annealing. Sample bias -2 V and tunneling current 1 nA.

the preference of vacancies to annihilate at descending steps [32] might explain the development of the valley-like surface morphology as shown in Fig. 5(c). Also during Si homoepitaxial growth the Si islands nucleate independently and when growing islands meet the dimer row registry can be in phase or out of phase [37,38]. Hamers et al. [37,38] showed that these antiphase boundaries act as preferential nucleation sites in perfect agreement with Bedrossian and Kaxiras experimental and theoretical findings [36] on Si(001) as well as our experimental data of Ge(001). Interestingly, the antiphase boundary depicted in Fig. 7 cannot compared with the AP2-type antiphase boundary (notation of Hamers et al.) found during homoepitaxial growth by Hamers et al. [37], whereas the antiphase boundary in Fig. 9 is the same as the AP1-type antiphase boundary.

Figure 9: Decorated antiphase boundaries aligned perpendicular to the substrate dimer row direction on the Ge(001) surface after 800 eV Argon ion bombardment and subsequent annealing. Sample bias -2 V and tunneling current 1 nA.

Figure 10: Decorated antiphase boundaries (Zipper) on the Ge(001) surface after 800 eV Argon ion bombardment and subsequent annealing. Sample bias -2 V and tunneling current 1 nA.

4 Conclusions

In summary, we have briefly reviewed the influence of low-dose ion bombardment on the surface morphology of the group IV semiconductor (001) surfaces. STM images present the atomistic details of the creation, ripening and coalescence of vacancy islands. In the case of Si(001) we have paid special attention to the influence of surface strain on the shape and ordering of the vacancy islands, whereas in the case of Ge(001) the formation of antiphase boundaries due to coalescence of out-of-phase vacancy islands is discussed. The preference for mobile vacancies to annihilate at descending steps versus ascending steps as well as the formation of antiphase boundaries drives the roughening of the semiconductor group IV (001) surfaces upon low-dose ion bombardment.

References

1. B. Poelsema, L.K. Verheij and G. Comsa, Phys. Rev. Lett. 53, 2500 (1983).
2. B. Poelsema, R. Kunkel, L.K. Verheij and G. Comsa, Phys. Rev. B 41, 11609 (1990).
3. P. Bedrossian, J.E. Houston, J.Y. Tsao, E. Chason and S.T. Picraux, Phys. Rev. Lett. 67, 124 (1991).
4. H. Kahata and K. Yagi, Jpnn. J. Appl. Phys. 28, L1042 (1989).
5. P. Bedrossian and T. Klitsner, Phys. Rev. B44, 13783 (1991).
6. P. Bedrossian and T. Klitsner, Phys. Rev. Lett. 68, 646 (1992).
7. H.J.W. Zandvliet, H.B. Elswijk, E.J. van Loenen and I.S.T. Tsong, Phys. Rev. B 46, 7581 (1992).
8. H. Feil, H.J.W. Zandvliet, M.-H. Tsai, J.J. Dow and I.S.T. Tsong, Phys. Rev. Lett. 69, 3076 (1992).
9. T. Michely, K.H. Besocke and G. Comsa, Surf. Sci. Lett. 230, L135 (1990).
10. T. Michely and G. Comsa, Surf. Sci. 256, 217 (1991).
11. B.S. Swartzentruber, C.M. Matzke, D.L. Kendall and J.E. Houston, Surf. Sci. 329, 83 (1995).
12. R.J. Hamers and U.K. Köhler, J. Vac. Sci. Technol. A 7, 2854 (1989).
13. H.J.W. Zandvliet, H.B. Elswijk, E.J. van Loenen and D. Dijkkamp, Phys. Rev. B 45, 5965 (1992).
14. P. Bedrossian, Surf. Sci. 301, 223 (1994).
15. H.J.W. Zandvliet, Surf. Sci. 377-379, 1 (1997).
16. H.J.W. Zandvliet, H.K. Louwsma, P.E. Hegeman and B. Poelsema, Phys. Rev. Lett. 75, 3890 (1995).
17. P. Zeppenfeld, M. Krzyzowski, C. Romainczyk, G. Comsa and M.G. Lagally, Phys. Rev. Lett. 72, 2737 (1994).
18. K. Wurm, R. Kliese, Y. Hong, B. Rttger, H. Neddermeyer and I.S.T. Tsong, Phys. Rev. B 50, 1567 (1994).
19. Y. Wei, L. Li and I.S.T. Tsong, Appl. Phys. Lett. 66, 1818 (1995).
20. D. Rioux, M. Chander, Y.Z. Li and J.H. Weaver, Phys. Rev. B 49, 11071

(1994).
21. D. Rioux, R.J. Pechman, M. Chander and J.H. Weaver, Phys. Rev. B 50, 4430 (1994).
22. D. Rioux, F. Stepniak, R.J. Pechman and J.H. Weaver, Phys. Rev. B 51, 10891 (1995).
23. F. Wu, X. Chen, Z. Zhang and M.G. Lagally, Phys. Rev. Lett. 74, 574 (1995).
24. Ch. Park, R.Z. Bakhtizin, T. Hashizume and T. Sakurai, J. Vac. Sci. Technol. B12, 2049 (1994).
25. Ch. Park, R.Z. Bakhtizin, T. Hashizume and T. Sakurai, Jpn. J. Appl. Phys. 32, 528 (1993).
26. H.K. Louwsma, H.J.W. Zandvliet, B.A.G. Kersten, J. Chesnau, A. van Silfhout and B. Poelsema, Surf. Sci. Lett. 381, L594 (1997).
27. Y. Wei and I.S.T. Tsong, Nucl. Instr. Meth. B115, 572 (1996).
28. H. Niehus, Köhler, M. Copel and J.E. Demuth, J. Microsc. 152, 735 (1988).
29. F.K. Men, A.R. Smith, K.J. Chao, Z. Zhang and C.K. Shih, Phys. Rev. B 52, R8650 (1995).
30. C.S. Chang, Y.M. Huang, C.C. Chen and T.T. Tsong, Surf. Sci. Lett. 367, L8 (1996).
31. P.C. Weakliem, Z. Zhang and H. Metiu, Surf. Sci. 336, 303 (1995).
32. S. Jay Chey, J.E. Van Nostrand and D.G. Cahill, Phys. Rev. B52, 16696 (1995).
33. S. Jay Chey, J.E. Van Nostrand and D.G. Cahill, Phys. Rev. Lett. 76, 3995 (1996).
34. W.W. Mullins, J. Appl. Phys. 30, 77 (1959).
35. H.J.W. Zandvliet and E. de Groot, Surf. Sci. 371, 79 (1996).
36. P. Bedrossian and E. Kaxiras, Phys. Rev. Lett. 70, 2589 (1993).
37. R. Hamers, U. Köhler and J. Demuth, J. Vac. Sci. Technol. A8, 195 (1990).
38. R. Hamers, U. Köhler and J. Demuth, Ultramicroscopy 31, 10 (1989).